Organic Synthesis
an interdisciplinary challenge

International Union of Pure and Applied Chemistry
Organic Chemistry Division
in conjunction with
Gesellschaft Deutscher Chemiker and
University of Freiburg i.Br.

Organizing Committee

CHAIRMAN H. Prinzbach *Freiburg*

SECRETARY GENERAL W. Eberbach *Freiburg*

C. A. Grob *Basel*   G. Schill *Freiburg*

H. Musso *Karlsruhe*   J. Streith *Mulhouse*

W. Reif *Ludwigshafen*   E. Vogel *Köln*

Ch. Rüchardt *Freiburg*

Acknowledgements

Thanks are due to the following institutions
for financial support for the Conference:

Deutsche Forschungsgemeinschaft, *FRG*
Ministerium für Wissenschaft und Kunst des
Landes, *Baden-Württemberg*
BASF AG, *Ludwigshafen*
Bayer AG, *Leverkusen*
Chemische Werke Hüls AG, *Marl*
Ciba-Geigy AG, *Basel*
Goedecke AG, *Freiburg*
Hoechst AG, *Frankfurt*
Hoffmann-La Roche AG, *Basel*
Lonza AG, *Basel*
Paul-Martini-Stiftung, *Mainz*
Sandoz AG, *Basel*
Schering AG, *Berlin*

Organic Synthesis
an interdisciplinary challenge

Proceedings of the 5th IUPAC Symposium on Organic Synthesis
Freiburg i.Br., FRG, 27–30 August 1984

edited by

JACQUES STREITH

University of Muhlhouse, Muhlhouse, France

HORST PRINZBACH and GOTTFRIED SCHILL

Both of the University of Freiburg, Freiburg i.Br., FRG

PUBLISHED FOR THE

INTERNATIONAL UNION OF
PURE AND APPLIED CHEMISTRY

BLACKWELL SCIENTIFIC PUBLICATIONS

OXFORD   LONDON   EDINBURGH

BOSTON   PALO ALTO   MELBOURNE

© 1985 International Union of Pure and Applied
Chemistry and published for them by
Blackwell Scientific Publications
Editorial offices:
Osney Mead, Oxford, OX2 0EL
8 John Street, London WC1N 2ES
23 Ainslie Place, Edinburgh, EH3 6AJ
52 Beacon Street, Boston, Massachusetts
  02108, USA
667 Lytton Avenue, Palo Alto, California
  94301, USA
107 Barry Street, Carlton, Victoria 3053, Australia

First published 1985

Printed and bound in Great Britain

DISTRIBUTORS

USA and Canada
  Blackwell Scientific Publications Inc
  PO Box 50009, Palo Alto, California 94303

Australia
  Blackwell Scientific Publications
  (Australia) Pty Ltd
  107 Barry Street, Carlton, Victoria 3053

British Library
Cataloguing in Publication Data

IUPAC Symposium on Organic Synthesis (5th :
  1984 : Freiburg)
  Organic synthesis: an interdisciplinary
  challenge: proceedings of the 5th IUPAC
  Symposium on Organic Synthesis, Freiburg,
  i. Br., FRG 27–30 August 1984.
  1. Chemistry, Organic—Synthesis
  I. Title   II. Streith, Jacques   III. Prinzbach,
  Horst   IV. Schill, Gottfried   V. International
  Union of Pure and Applied Chemistry, Organic
  Chemistry Division   VI. Gesellschaft Deutscher
  Chemiker
  547'.2   QD262

  ISBN 0-632-01441-5

Contents

Preface

Since 1974 the IUPAC 'International Conference on Organic Synthesis' has been a biennial event. Following meetings in Louvain-la-Neuve (Belgium), Haifa/Jerusalem (Israel), Madison (USA) and Tokyo (Japan), Freiburg i. Br. hosted this year's conference 27th–30th August). Chemistry has a long tradition in this city: the first local chemistry experimentalist was mostly likely the Franciscan Friar Bertholdus Schwarz, to whom a monument (see p. viii) in front of the townhall is dedicated. About the time that Archduke Albrecht of Austria founded (1457) the University of Freiburg, the monk carried out his fundamental experiments, producing gunpowder by mixing saltpetre, sulphur and charcoal. In more recent times, research in organic chemistry at the University of Freiburg has been intimately connected with the names of L. Gatterman (1900–1920). H. Wieland (1920–1926), H. Staudinger (1926–1951), A. Lüttringhaus (1951–1971) and K. Wallenfels (1953–1978).

Besides the 32 plenary lectures covering four main themes

MODERN METHODOLOGY IN ORGANIC SYNTHESIS
TRANSITION METALS IN ORGANIC SYNTHESIS
ENZYMATIC PROCESSES IN ORGANIC SYNTHESIS
CARBOHYDRATES IN ORGANIC SYNTHESIS

180 posters were presented. Posters were not subjected to any selection process, rather all contributions were displayed, in order to also give those researchers from less fortunate countries an opportunity to present their results.

The future of organic chemistry, in particular organic synthesis, has been the subject of frequent and, depending on the views and/or interests of those involved, differing commentary in recent years. The Organising Committee chose the above formulation of the conference themes in order to stress the worldwide attempts of synthetic chemists to open up new territories and to reclaim long neglected territories, to pursue increasingly interdisciplinary goals and activities.

The conference attracted almost 1,100 participants from 35 countries, for whom there was no doubt that the representatives of this discipline were able to successfully integrate the above aims, which lead beyond the boundaries of classical organic synthesis. The sixth conference in this series (Moscow, 10th–15th August, 1986) will have to show to what extent promises and expectations in this direction have been fulfilled.

H. Prinzbach   G. Schill   J. Streith
Freiburg/Muhlhouse, September 1984

Monument to Schwarz at Freiburg

The monument to Schwarz is erected in the Augustine-place. It consists of a fountain, which sends its crystalline water through four brass pipes. Above these are four niches; in one of which, on the front side, a bas-relief is to be seen showing the terrified monk at the moment when the explosion surprises him. In the three other niches inscriptions are to be read concerning the name, life, and condition of the inventor, and the time and manner of his invention. The inventor himself, chiseled in grey stone, stands upon the column of the fountain, in his monastic habit, absorbed in deep study, holding his right hand (in which he keeps a book) upon the mortar, which stands upon a pedestal by the side of him. The sculptor is M. Knittel.

Berthold Schwarz was born at Freiburg, and became a monk in the monastery there of the order of St. Francis.

Bas-relief from the monument to
Berthold Schwarz, Freiburg i. Br, Rathausplatz

Monument to Schwarz at Freiburg

In old books Bertholdus Schwarz is also called Constantine Angklitzer, and his common name seems to have been given him on account of his physical and chemical experiments, unusual for the age in which he lived; which gained for him the appellation of 'the Black Berthold' (der Schwarze Berthold), Bertholdus being his friar name. His favourite study was chemistry, in which he made great progress. Once, in the year 1340, he had stamped in a mortar a mixture of saltpetre, sulphur and charcoal, and had covered it by a stone. A spark of fire got into the mortar, the stone flew up into the ceiling of the room, and with such power that a part of the building fell into ruins. The monk searched after the cause, and repeated the experiment: the result was always the same; and thus gunpowder was invented; but Schwarz himself is believed to have lost his life in pursuing this study about 1354.

The Chinese are said, indeed, to have invented the composition of gunpowder, and the monks to have learnt it from them; for some men of that order had, already in the thirteenth century, visited China as missionaries, and returned to Europe in 1295.

(*London News*, November 24, 1853. By courtesy of Augustiner Museum, Freiburg)

Transition metal complexes as catalysts in organic synthesis

Günther Wilke

Max-Planck-Institut für Kohlenforschung
Kaiser-Wilhelm-Platz 1, D-4330 Mülheim a. d. Ruhr, Germany

Abstract - The aim of this review is to describe some historical
fundamental discoveries as well as developments achieved in the
Max-Planck-Institut für Kohlenforschung regarding transition metal
catalysed reactions. These opened up new pathways for organic
synthesis. The review includes examples based on the following
elements: Pt, Pd, Ni, Ti, Zr, Hf, Cr, Mo, Mn, Co, Rh.

For this introductory lecture entitled "Transition Metal Complexes as Catalysts in Organic Synthesis" it seemed appropriate to me to remind you of
historical fundamental discoveries as well as to describe developments
achieved in our institute.

1. For this purpose I will mention first the prototype of all π-complexes
which was discovered in 1827 by Zeise [1], when he reacted platinumchloride
with ethanol and obtained an ethene complex, which gave with KCl the so-
called "sal kalico platinicus inflammabilis". Today this complex is known as
"Zeise's-salt", and its correct structure was not determined precisely by
neutron diffraction until 1975 [2].

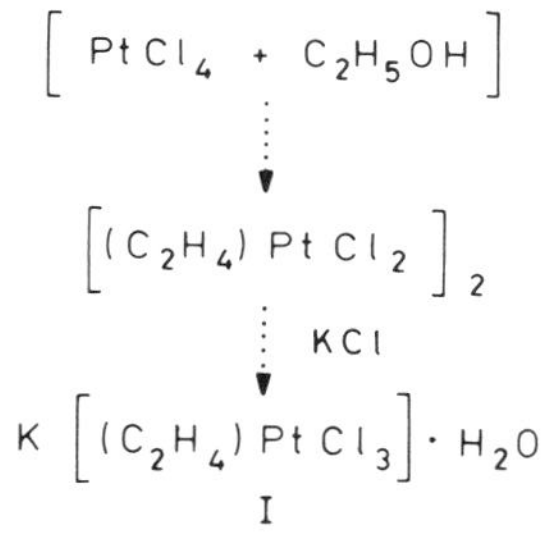

$$\left[PtCl_4 + C_2H_5OH \right]$$

$$\downarrow$$

$$\left[(C_2H_4) PtCl_2 \right]_2$$

$$\downarrow KCl$$

$$K \left[(C_2H_4) PtCl_3 \right] \cdot H_2O$$

$$I$$

„ sal kalico platinicus inflammabilis"

Fig. 1   Formation of Zeise's salt

A very recent development renewed the importance of the Zeise's-salt. The
problem arises in asymmetric catalytic synthesis yielding chiral olefins to
determine the enantiomeric excess (e.e.) of the products immediately after
finishing the experiment. J. Köhler and G. Schomburg [3] developed such a
method following a concept by Gil-Av [4]. Optically active Zeise's-salts are
reacted with the mixture of enantiomeric olefins to form diastereomeric complexes, which are separated by HPLC.

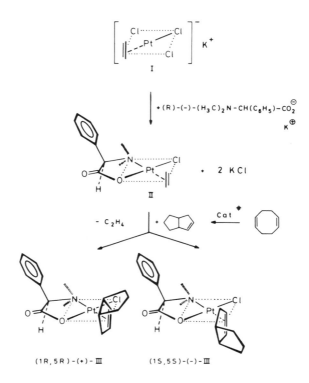

Fig. 2   Synthesis of diastereomeric platinum complexes

[3.3.0]-bicyclooctene was obtained by asymmetric catalytic isomerisation of
1,5-cyclooctadiene with an e.e. of 13 %. After separation of the diastereomeric complexes by HPLC one diastereomer could be crystallised and analysed
by X-ray diffraction. The absolute configuration was determined to be
(1R,5R)-(+)-[3.3.0]-bicyclooctene [5].

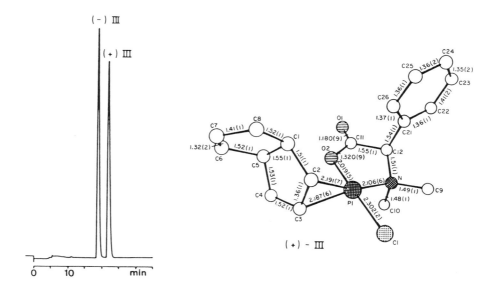

Fig. 3   HPLC separation and X-ray structure

This procedure seems to be applicable to the resolution of enantiomeric olefins in a very general sense, i.e. for the determination of e.e. as well as
for the preparative isolation of samples of one diastereomer or enantiomer
respectively when displaced by an excess of ethene.

2. Very early palladium played an important role in organo transition metal chemistry when in 1959 J. Smidt and W. Hafner [6] discovered an important technical process which was christened the "Wacker-process" and became the basis for the large scale production of acetaldehyde and acetic acid respectively from ethene.

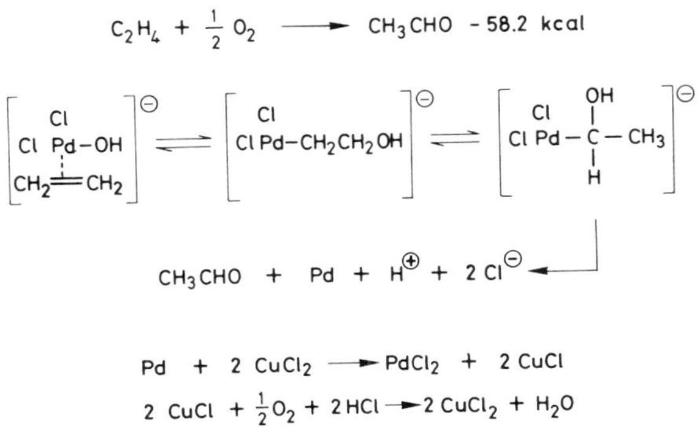

$$C_2H_4 + \tfrac{1}{2} O_2 \longrightarrow CH_3CHO \; - 58.2 \; kcal$$

$$
\begin{bmatrix} Cl \\ Cl \; Pd-OH \\ CH_2 = CH_2 \end{bmatrix}^{\ominus}
\rightleftharpoons
\begin{bmatrix} Cl \\ Cl \, Pd-CH_2CH_2OH \end{bmatrix}^{\ominus}
\rightleftharpoons
\begin{bmatrix} Cl \quad OH \\ Cl \, Pd - C - CH_3 \\ \quad\quad H \end{bmatrix}^{\ominus}
$$

$$CH_3CHO \; + \; Pd \; + \; H^{\oplus} \; + \; 2 \, Cl^{\ominus} \longleftarrow$$

$$Pd \; + \; 2 \, CuCl_2 \longrightarrow PdCl_2 \; + \; 2 \, CuCl$$

$$2 \, CuCl + \tfrac{1}{2} O_2 + 2 \, HCl \longrightarrow 2 \, CuCl_2 + H_2O$$

Fig. 4   The "Wacker-process"

In our days this process becomes substituted by the carbonylation of methanol catalysed by rhodium complexes. I will discuss this point later on.

At about the same time Smidt and Hafner [7] made another important contribution to organo transition metal chemistry, because they were able to isolate independently from a Russian group [8] the first π-allyl metal system, the dimer of allylpalladium chloride characterised by a delocalised allyl ligand.

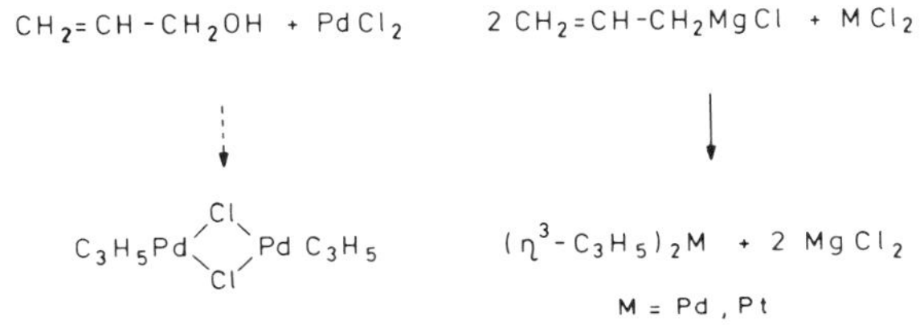

$$CH_2 = CH - CH_2OH + PdCl_2 \qquad\qquad 2 \, CH_2 = CH - CH_2MgCl + MCl_2$$

$$C_3H_5Pd \overset{Cl}{\underset{Cl}{\diagup\!\!\diagdown}} Pd \, C_3H_5 \qquad\qquad (\eta^3 - C_3H_5)_2M \; + \; 2 \, MgCl_2$$

$$M = Pd \, , \, Pt$$

Fig. 5   Synthesis of $[C_3H_5PdCl]_2$ and $(\eta^3 - C_3H_5)_2M$ (M = Pd, Pt)

The mechanism of the formation of this allyl system was not elucidated. Only two years later we synthesised the first so-called "pure" π-allyl metal system, bis-π-allyl-nickel [9] which is the prototype of homoleptic η^3-allyl transition metal complexes. Shortly afterwards the corresponding complexes of palladium and platinum were obtained [10]. As will be shown later, the η^3-allyl system is of fundamental importance for catalytic reactions of 1,3-dienes.

In the following scheme six complexes having structures, which are regarded to correspond to intermediates of the catalytic cyclooligomerization of butadiene are mentioned. The η^3,η^3-allyl systems as well as the η^1,η^3-system are well-known from organonickel chemistry [11]. The η^1,η^1-divinylplatinacyclopentane was described by F.G.A. Stone et al. [12]. Nevertheless two gaps still existed the η^1,η^1-vinylmetallacycloheptene and the η^1,η^1-metallacyclononadiene.

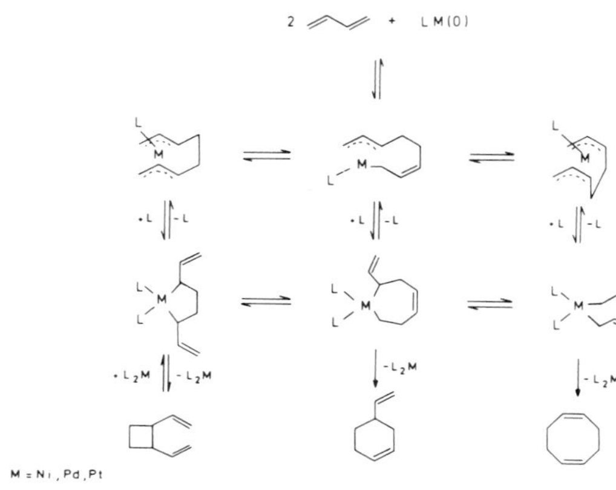

Fig. 6   Intermediates in the catalytic cyclooligomerizations
         of butadiene

Only recently P. Binger and H.M. Büch were able to isolate two palladium
complexes which belong to this series and fill the gaps [13].

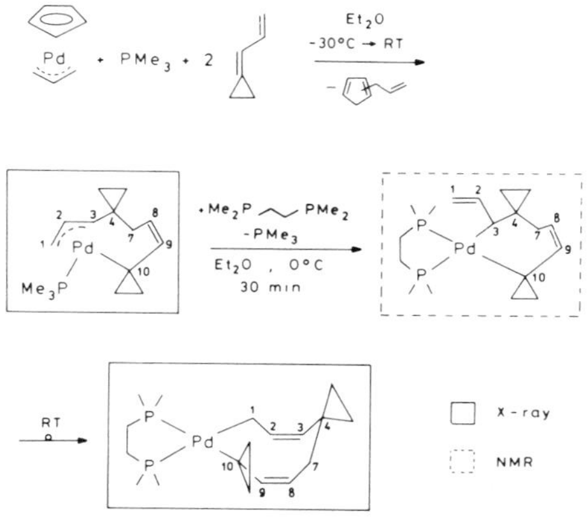

Fig. 7 η^1,η^1-pallada-cycles

These complexes were obtained from a very special butadiene and the complicated structures were elucidated by X-ray diffraction and/or NMR-spectroscopy.

Remarkable palladium catalysed reactions have been found by P. Binger. One
of the most fascinating examples is the cyclooligomerization of 1,1-disubsti-
tuted cyclopropenes yielding dimers, trimers and tetramers [14].

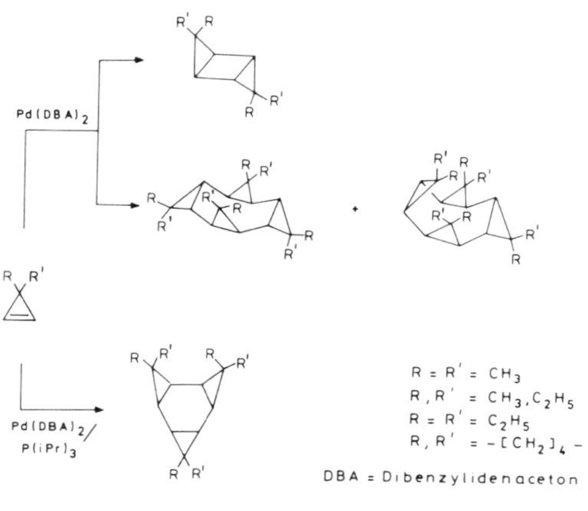

Fig. 8   Cyclooligomerization of cyclopropenes

3. Nickel is one of the most popular transition metals in organic synthesis and it already played an important role at the end of the last century. The discovery by Mond [15] that nickel metal reacts with CO giving Ni(CO)$_4$ prompted Sabatier and Senderens to try to synthesise an analogous nickel ethene complex [16]. They failed but discovered the catalytic hydrogenation by nickel metal [17]. More than 70 years later we were able to isolate such a complex, tris-ethene-nickel(O) from cyclododecatrienenickel(O) and ethene [18].

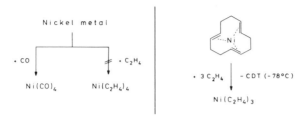

Fig. 9 - Synthesis of Ni(CO)$_4$ and Ni(C$_2$H$_4$)$_3$

All our developments in organonickel chemistry trace back to the so-called "nickel effect", which was discovered by K. Ziegler et al. and published 30 years ago [19].

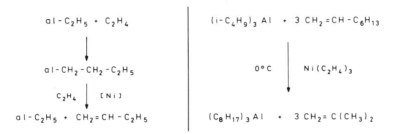

Fig. 10 The "nickel effect"

Originally the nickel effect was assumed to be caused by nickel metal probably in colloidal form. Later on we were able to show that nickel(O) complexes catalyse the homogeneous transalkylation of aluminiumtrialkyl with α-olefins and a special version of technical interest was developed just recently. Tri-n-octylaluminium can be obtained under extremely mild conditions in very pure form by transalkylation in the presence of traces of a nickel(O) complex [20].

Nickel(O) complexes in which the nickel atom is surrounded by ligands which can be completely displaced by substrates contain "naked nickel" according to our definition [21]. Naked nickel catalysts are extremely versatile catalysts which allow complicated organic molecules to be easily synthesised [22].

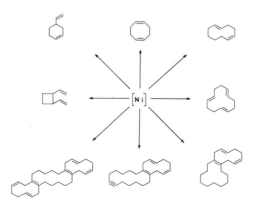

Fig. 11 Cyclooligomerization catalysed by nickel(O)

W. Reppe et al. already used nickel in form of e.g. nickelacetylacetonate as a catalyst for the most elegant cyclotetramerization of acetylene [23].

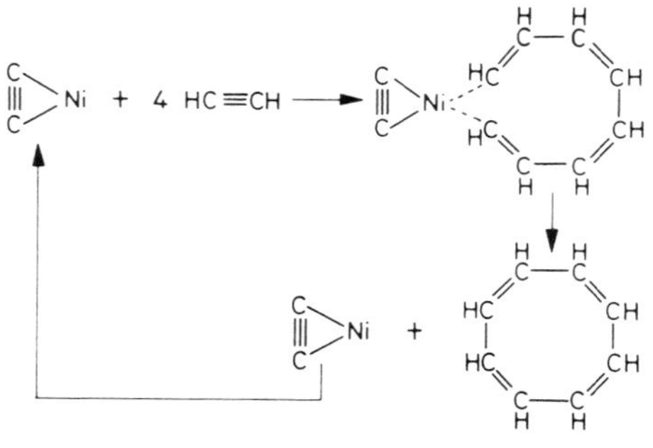

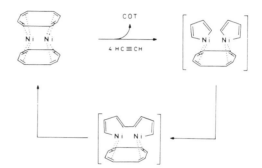

Fig. 12   Reppe's scheme of COT-synthesis

Today we formulate the catalytic synthesis of cyclooctatetraene on the basis of some experimental results in the following manner [24].

Fig. 13   Cyclotetramerization of acetylene on a binuclear catalyst

We assume a binuclear catalyst to be involved. In this sense $(COT)_2Ni_2$, a binuclear sandwich complex is indeed a very active catalyst, as shown by experiments carried out in our pressure laboratory as well as at the BASF.

	CATALYST		
	Ni(acac)$_2$	Ni$_2$(COT)$_2$	
	BASF	BASF	MPI
COT	0.56 kg	0.66 kg	1.24 kg
CUPRENE	8 kg	0.6 kg	} 0.2 kg
RESIDUE	1 kg	0.2 kg	
SELECTIVITY	6 %	45 %	66 %
CYCLES	21	49	93
(COT/Ni)			

Fig. 14   COT-synthesis catalysed by Ni(acac)$_2$ and $(COT)_2Ni_2$ respectively

The comparison shows that $(COT)_2Ni_2$ is more active than Ni(acac)$_2$, gives less cuprene and produces COT in much higher selectivity. Furthermore, no incubation period is observed with $(COT)_2Ni_2$ in contrast to Ni(acac)$_2$. These results strongly support the above-mentioned mechanism.

In the context with organonickel chemistry another type of catalysts should be described, which allows to dimerise or to codimerise unsaturated molecules. The catalysts are formed from η^3-allylnickel halides and Lewis-acids. The catalysis can be directed by the phosphanes coordinated to the nickel atom. A special catalyst of this kind dimerises propene at temperatures as low as -60°C giving 2,3-dimethylbutene with high selectivity.

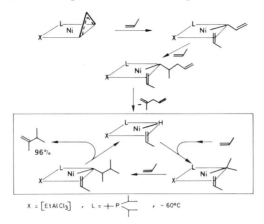

Fig. 15 Dimerization of propene

These catalysts display the highest activity of the synthetic homogeneous catalysts known today. The activity is comparable with that of the most active enzymes [25].

Dimerization of Propene

Catalyst : $\eta^3\text{-}C_3H_5NiBr$ / $P(C_6H_{11})_3$ / $EtAlCl_2$

Solvent : C_6H_5Cl

Temp. : $-55°C$

Rate : $35\,g\ C_6/\mu mol\ catalyst \cdot h$

$35\,t\ C_6/\ mol\ catalyst \cdot h$

$0{,}6\,t\ C_6/1g\ Ni \cdot h$

Turn-over number $\cdot sec^{-1}$: 230

Fig. 16 Activity of a catalyst for the dimerization of propene

If these catalysts are modified with optically active ligands, than it is possible to catalyse asymmetric reactions, such as the codimerization of bicycloheptene and ethene [26].

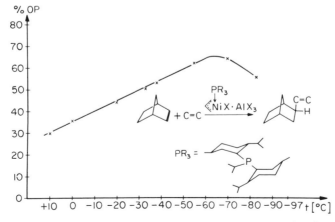

Fig. 17 Asymmetric catalytic synthesis of exo-vinylnorbornan

Very recent modifications of the catalyst now enable us to run this reaction with a high number of catalytic cycles and with high selectivity [27].

It was a problem to react functionalised olefins with these catalysts, because the functionalities deactivated the catalysts by complexing the Lewis-acid. But recently a catalyst was developed [28], with which this problem could be overcome. Instead of with an alumium organic Lewis-acid the catalyst was modified by a stable complex anion such as BF_4. Using this catalyst, acrylic acid ester can be dimerised under formation of monounsaturated adipic acid ester. This product seems to have technical interest as a new monomer for the synthesis of homo- or copolymeric polyesters or polyamides.

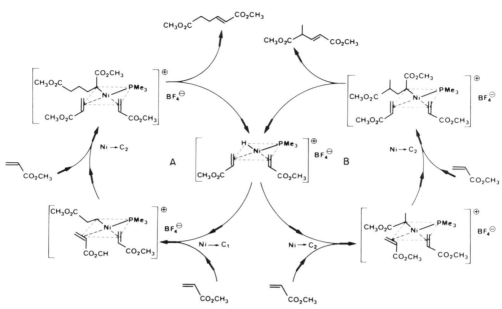

Fig. 18   Dimerization of acrylic acid ester

4. Coming to the group of titanium, zirconium and hafnium I would like to remind you of the Ziegler-catalysts, demonstrated by the patent situation of the years 1953-55. This discovery caused a revolution of the organic synthesis of polyolefins.

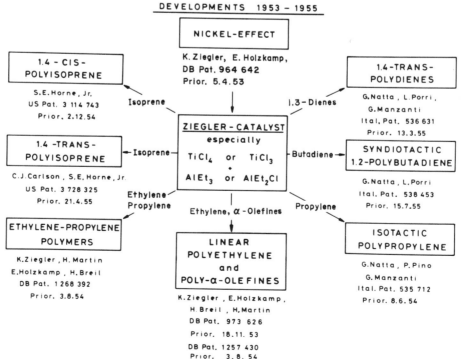

Fig. 19   Polyolefin patents based on Ziegler-catalysts

The Ziegler-catalysts are defined to be composed of a transition metal component and a main group organometallic.

Our investigations on Ti, Zr and Hf led to some well-defined complexes which exhibit catalytic activities alone [29].

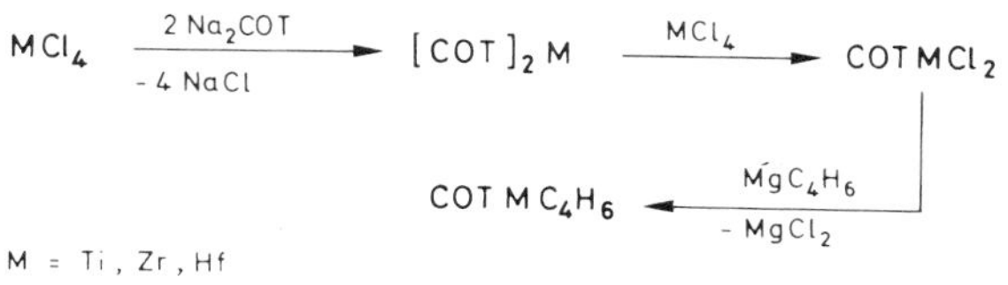

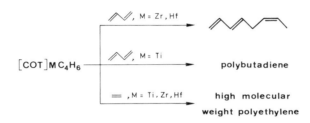

$$MCl_4 \xrightarrow[- 4\,NaCl]{2\,Na_2COT} [COT]_2M \xrightarrow{MCl_4} COTMCl_2$$

$$COT\,M\,C_4H_6 \xleftarrow[- MgCl_2]{\dot{M}g\,C_4H_6}$$

M = Ti, Zr, Hf

Fig. 20   Synthesis of $COT \cdot M \cdot C_4H_6$ complexes (M = Ti, Zr, Hf)

These complexes show remarkable catalytic activity in e.g. the dimerization or polymerization of butadiene or ethene [29].

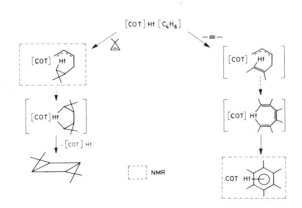

Fig. 21   Dimerization and polymerization of butadiene and ethene respectively by $COT \cdot M \cdot C_4H_6$ complexes

The hafnium complex $COTHfC_4H_6$ on the other hand catalyses the cyclooligomerization of cyclopropenes and alkynes. Intermediates could be isolated in these reactions [30].

Fig. 22   Cyclodimerization of dimethylcyclopropene and cyclotrimerization of butyne

5. After the discovery of ferrocene the synthesis of dibenzene chromium caused a great surprise, as neutral molecules were coordinated to a zero-valent metal atom [31]. In a very early state of our research on cyclooligomerizations we found that on the basis of chromium catalysts can be prepared, which are active for the cyclotrimerization of butadiene. In this context we tried to trap the active species in form of a bis-arene complex by reacting the catalyst with 2-butyne. Indeed we obtained our first π-complex [32] which we thought to be bis-hexamethylbenzenechromium(O). Only later did we become aware of the phenomenon which we christened "dichotomy". The complex isolated was pentamethylcyclopentadienylhexamethylbenzene-chromium(I) [33].

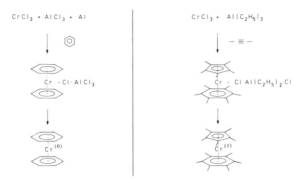

Fig. 23 Sandwich complexes of chromium

Of general interest is a recent finding in this context. We were interested in simulating the formation of the arene part of our complex. To our surprise an intermediate could be isolated indicating that two chromium atoms are involved in the formation of the six-membered ring [34].

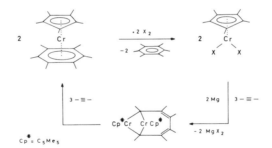

Fig. 24 A binuclear intermediate of a cyclotrimerization of butyne

The complex was characterised by X-ray diffraction [35].

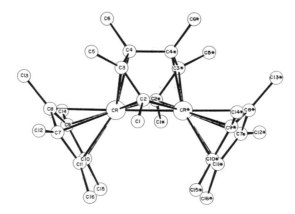

Fig. 25 X-ray structure of the intermediate

In the course of our studies on homoleptic η^3-allyl metal systems, tris-η^3-allyl chromium was most interesting because it was the first organo transition metal species not containing halogen atoms or any organo main group metal component, which catalysed the polymerization of ethene to high molecular weight polyethene [36].

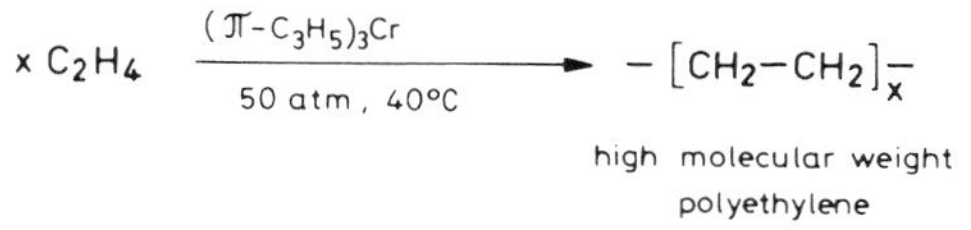

$$x\ C_2H_4 \quad \xrightarrow[\text{50 atm, 40°C}]{(\pi\text{-}C_3H_5)_3Cr} \quad -[CH_2\text{—}CH_2]^-_x$$

high molecular weight
polyethylene

Fig. 26 $(\eta^3$-$C_3H_5)_3Cr$ as catalyst for the polymerization of ethene

$(\eta^3$-$C_3H_5)_3Cr$ was the first complex of this series which we were able to synthesise followed by $(\eta^3$-$C_3H_5)_4Mo$ and $(\eta^3$-$C_3H_5)_4W$ [36]. Our investigations on the cyclotrimerization of butadiene gave rise to the knowledge of intermediate complexes having C_8- or C_{12}-chains coordinated to the central metal atom. But in all cases the butadiene molecules were linked together already. Molybdenum, however, enabled us to learn something about initial complexes formed by the metal and butadiene. $Mo(C_4H_6)_3$ was first synthesised by P.S. Skell et al. [37] by metal vaporization. The compound can also be synthesised by a conventional method [38]. Now it was most interesting to observe that the coordinated butadiene molecules link together under the influence of CO yielding a C_{12}-chain. Finally under more drastic conditions cyclododecatriene is formed preferentially as ttc and tcc-CDT [38]. Here again we have good proof for a stepwise reaction mechanism.

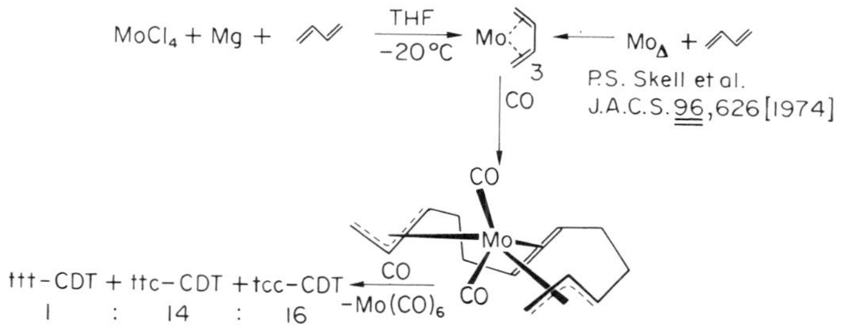

$$MoCl_4 + Mg + \mathbin{/\!\!\!\diagup} \xrightarrow[-20°C]{THF} Mo\left(\!\!\!\right)_3 \longleftarrow Mo_\Delta + \mathbin{/\!\!\!\diagup}$$

P.S. Skell et al.
J.A.C.S. <u>96</u>,626[1974]

$$ttt\text{-}CDT + ttc\text{-}CDT + tcc\text{-}CDT \xleftarrow[-Mo(CO)_6]{CO}$$
1 : 14 : 16

Fig. 27 Stepwise formation of CDT

6. At this stage I like to point to the remarkable phenomenon, that the catalytic matrix which enables us to cyclotrimerise butadiene was found at a number of fairly different metals: Ti, Cr, Mo, Ni and very recently Mn. In this context I should mention that manganese was the metal from which the first η^1-allyl complex was isolated in 1960 already [39]. This year K. Jonas et al. found that a manganese complex is also a highly effective catalyst for the synthesis of CDT [40].

$CH_2=CH-CH_2-Mn(CO)_5$ first η^1-allylmetalsystem

H.D. KAESZ , R.B. KING and F.G.A. STONE , Z. NATURFORSCH
<u>156</u> 682 [1960]

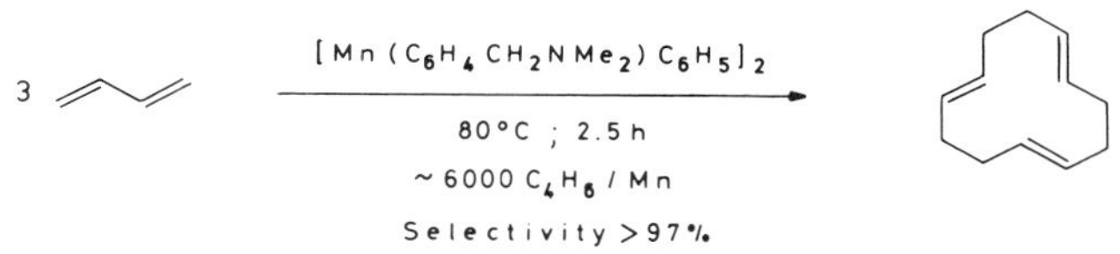

$$3 \mathbin{/\!\!\!\diagup} \xrightarrow[\substack{80°C\ ;\ 2.5\ h \\ \sim 6000\ C_4H_8\ /\ Mn \\ \text{Selectivity} > 97\%}]{[Mn(C_6H_4CH_2NMe_2)C_6H_5]_2}$$

Fig. 28 η^1-allyl Mn(CO)$_5$ and the cyclotrimerization of butadiene catalysed by a Mn-complex

It has to be recognized that all the cyclooligomerizations of 1,3-dienes can
generally be described by oxidative additions and formation of bis-allyl
systems followed by reductive eliminations under C-C bond formation yielding
the ring system. The metal is oxidised by about two valency states and afterwards reduced again to the initial valency state. This is also the case if
monoolefins are cyclooligomerised.

7. For catalytic organic synthesis the element cobalt is of greatest importance even in nature if we think of vitamin-B. In synthetic organic chemistry
cobalt was also recognised to be a very useful catalyst component at an early
stage of the art already. When I discussed the "Wacker-process" I pointed out
that it is beginning to be substituted by a new version of an old process.
The carbonylation of methanol was discovered by Reppe et al. in 1953 [41].
A cobalt catalyst was used, but the reaction conditions were rather severe.
In recent years the carbonylation of methanol was redeveloped on the basis
of rhodium at Monsanto [42]. This process now seems to be the optimal one
for the large scale production of acetic acid.

$$CH_3OH + CO \longrightarrow CH_3COOH$$

	Co (BASF)	Rh (Monsanto)
Temp.	230 °C	180 °C
Pressure	600 - 700 Atm	30 - 40 Atm
Selectivity	~ 90 %	> 99 %
M-conc.	~ 10^{-1} M	~ 10^{-3} M
H-effect	CH_4 , CH_3CHO C_2H_5OH , CH_3COOH	——

Fig. 29   Comparison of conditions of the Co/Rh-catalysed
          carbonylation of methanol

Not only the carbonylation, but also the oxosynthesis, today called hydroformylation, which was discovered by O. Roelen 1938 [43] a former coworker
of Franz Fischer becomes shifted to a certain extent from cobalt to rhodium
because of higher selectivity observed.

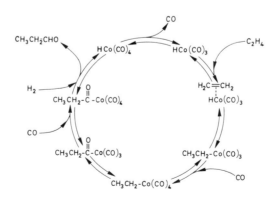

Fig. 30   Catalytic cycle of hydroformylation

Rhodium catalysts led to a higher ratio of n/iso-products, which is commercially of great value.

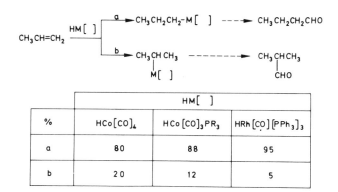

%	HM[ ]		
	$HCo[CO]_4$	$HCo[CO]_3PR_3$	$HRh[CO][PPh_3]_3$
a	80	88	95
b	20	12	5

Fig. 31 Selectivity of hydroformylation

The catalytic cycles of the carbonylation as well as of the hydroformylation are both known to proceed through metal hydride species as intermediates. In this context must be mentioned a most remarkable finding by K. Jonas and his coworkers [44]. In the course of his investigations on certain complexes of cobalt it was found that a Co-H-bond can be added to benzene in the sense of a hydrocobaltation. The addition product was isolated and characterised. It reacts with hydrogen, releases cyclohexane and reforms the hydride moiety ready for the next cycle. Benzene is hydrogenated to cyclohexane homogeneously under mild conditions in this way. It must be emphasised that this is the first case of a successful isolation of an intermediate in the hydrogenation of benzene.

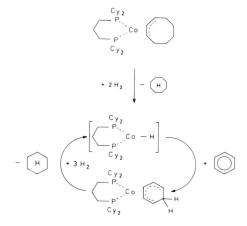

Fig. 32 Catalytic cycle of benzene hydrogenation

The last example shall demonstrate that cobalt does not necessarily form Co-H groups in catalytic reactions. In recent years H. Bönnemann and his group developed a versatile catalytic process by which heterocyclic compounds can be synthesised [45]. The process is based on cyclopentadienylcobalt complexes and is here demonstrated by the cocyclooligomerization of acrylonitrile and acetylene yielding vinylpyridine. It is a high pressure acetylene process. Here again the catalytic cycles are based on oxidative additions and reductive eliminations but without the appearance of Co-H systems as intermediates.

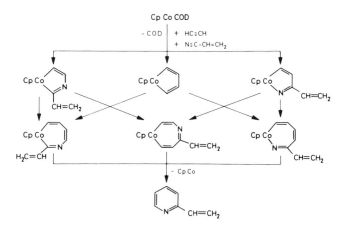

Fig. 33   Catalytic synthesis of vinylpyridine

In my lecture I presented examples of historical interest as well as developments originating from our institute well knowing that I could have mentioned numerous results from other laboratories. The aim of this review was to show the organic chemists participating in this meeting, who are not so familiar with organo transition metal chemistry, that homogeneous catalysis by transition metal complexes can be extremely useful for organic synthesis.

REFERENCES

1. W.C. Zeise, Pogg. Ann. Phys. 9, 632 (1827); ibid. 21, 497 (1831)
2. R.A. Love, T.F. Koetzle, G.J.B. Williams, L.C. Andrews, and R. Bau, Inorg. Chem. 14, 2653 (1975)
3. J. Köhler and G. Schomburg, Chromatographia 14, 559 (1981)
4. M. Goldman, Z. Kustanovich, S. Weinstein, A. Tishbee, and E. Gil-Av, Poster, Vth International Symposium on Column Liquid Chromatography, Avignon (1981)
5. J. Gajowski, Dissertation, Universität Bochum (1981), and C. Krüger and R. Goddard, (1983) unpublished
6. J. Smidt, W. Hafner et al., Angew. Chem. 71, 176 (1959)
7. J. Smidt and W. Hafner, Angew. Chem. 71, 284 (1959)
8. I.I. Moiseev, E.A. Fedorowskaya, and Ya.K. Syrkin, Zhur. Neorg. Khim. 4, 2641 (1959)
9. G. Wilke and B. Bogdanović, Angew. Chem. 73, 756 (1961)
10. a) G. Wilke, B. Bogdanović, P. Hardt, P. Heimbach, W. Keim, M. Kröner, W. Oberkirch, K. Tanaka, E. Steinrücke, D. Walter, and H. Zimmermann, Angew. Chem. 78, 157 (1966); Angew. Chem. Int. Ed. Engl. 5, 151 (1966)
    b) W. Keim, Dissertation, Techn. Hochschule Aachen (1963)
11. a) P.W. Jolly and G. Wilke, "The Organic Chemistry of Nickel", Academic Press, New York, London, Vol. 1 and 2 (1974/75)
    b) A. Döhring, P.W. Jolly, R. Mynott, K.-P. Schick, and G. Wilke, Z. Naturforsch. 36b, 1198 (1981)
12. G.K. Barker, M. Green, J.A.K. Howard, J.L. Spencer, and F.G.A. Stone, J. Chem. Soc. Dalton Trans. 1978, 1839
13. H.M. Büch, P. Binger, R. Benn, C. Krüger, and A. Rufinska, Angew. Chem. 95, 814 (1983); Angew. Chem. Int. Ed. Engl. 22, 774 (1983)
14. P. Binger and U. Schuchardt, Chem. Ber. 114, 1649 (1981)
15. L. Mond, DP 57 320 (1890)

16. P. Sabatier, Bull. Soc. Chim. France 1261 (1939)
17. P. Sabatier and J.-B. Senderens, Compt. rend. 124, 616, 1358 (1897)
18. K. Fischer, K. Jonas, and G. Wilke, Angew. Chem. 85, 620 (1973); Angew. Chem. Int. Ed. Engl. 12, 565 (1973)
19. K. Ziegler, H.G. Gellert, E. Holzkamp, and G. Wilke, Brennstoffchem. 35, 321 (1954)
20. K. Fischer, K. Jonas, A. Mollbach, and G. Wilke, Z. Naturforsch. 39b, 1011 (1984)
21. G. Wilke et al., Angew. Chem. 75, 10 (1963); Angew. Chem. Int. Ed. Engl. 2, 105 (1963)
22. G. Wilke, Jahrbuch MPG (1967) 156
23. W. Reppe, O. Schlichting, K. Klager, and T. Toepel, Liebigs Ann. Chem. 560, 1 (1948)
24. G. Wilke, Pure and Appl. Chem. 50, 677 (1978)
25. B. Bogdanović, B. Spliethoff, and G. Wilke, Angew. Chem. 92, 633 (1980); Angew. Chem. Int. Ed. Engl. 19, 622 (1980)
26. B. Bogdanović, B. Henc, A. Lösler, B. Meister, H. Pauling, and G. Wilke, Angew. Chem. 85, 1013 (1973); Angew. Chem. Int. Ed. Engl. 12, 954 (1973)
27. H. Kuhn and G. Wilke, (1984) unpublished
28. K. Sperling, Dissertation, Universität Bochum (1983)
29. G. Wilke, Fundamental Research in Homogeneous Catalysis, M. Tsutsui, Plenum Press, New York, London, Vol. 3, 1 (1979)
30. L. Stehling and G. Wilke, (1983) unpublished
31. E.O. Fischer and W. Hafner, Z. Naturforsch. 10b, 665 (1955)
32. G. Wilke and M. Kröner, Angew. Chem. 71, 574 (1959)
33. H. Benn, G. Wilke, and D. Henneberg, Angew. Chem. 85, 1052 (1973); Angew. Chem. Int. Ed. Engl. 12, 1001 (1973)
34. H. Benn and G. Wilke, (1984) unpublished
35. C. Krüger and R. Goddard, (1984) unpublished
36. a) W. Oberkirch, Dissertation, Techn. Hochschule Aachen (1963)
    b) G. Wilke, J. Organomet. Chem. 200, 349 (1980)
37. P.S. Skell, E.M. Van Dam, and M.P. Silvon, J. Am. Chem. Soc. 96, 626 (1974)
38. W. Gausing and G. Wilke, Angew. Chem. 93, 201 (1981); Angew. Chem. Int. Ed. Engl. 20, 186 (1981)
39. H.D. Kaesz, R.B. King, and F.G.A. Stone, Z. Naturforsch. 15b, 682 (1960)
40. K. Jonas, G. Burkhart, and U. Nienhaus, (1984) unpublished
41. W. Reppe, H. Kröper, N. v. Kutepow, and H.J. Pistor, Liebigs Ann. Chem. 582, 72 (1953)
42. D. Foster, Ann. N.Y. Acad. Sci. 295, 79 (1977)
43. O. Roelen, DBP 849 548 (1938)
44. a) K. Jonas, Pure and Appl. Chem. 56, 63 (1984)
    b) K. Jonas and H. Priemer, (1984) unpublished
45. H. Bönnemann, Angew. Chem. 90, 517 (1978); Angew. Chem. Int. Ed. Engl. 18, 505 (1978)

The synthesis of heterocycles via transition metal chemistry

Louis S. Hegedus

Department of Chemistry, Colorado State University, Fort Collins, Colorado 80523, USA

Abstract – An efficient synthesis of 4-bromoindole-1-tosamide based on the Pd(II)-catalyzed cyclization of an o-ethenylaniline p-toluenesulfonamide has been developed. A Pd(0) oxidative addition-olefin insertion-β-hydride elimination cycle produced a number of 4-substituted indole-1-tosamides. Selective electrophilic substitutions at the 3-position provided access to the 3-chloromercuri- and 3-iodoindole-1-tosamides. Transmetallation to palladium and allyl chloride insertion produced 3-allyl-4-bromoindole-1-tosamide which could be cyclized to the benz[c,d]indoline. A Pd(0) oxidative addition-olefin insertion-β-hydride elimination cycle converted the 3-iodo compound to a number of 4-bromo-3-substituted indole-1-tosamides including potential precursors to optically active tryptophans, ergot alkaloids, and to clavicipitic acid.

A variety of substituted β-lactams, including a cepham analog, were synthesized by the photochemical reaction of [(methoxy)(methyl)carbene]-chromium complexes with substituted imines. Oxazines and oxazolines were inert towards chromium carbene complexes. Oxazines were converted to bicyclic β-lactams by the photolytic reaction with molybdenum carbene complexes. Oxazolines were considerably less reactive and produced only low yields of β-lactam product and an equivalent amount of the corresponding oxazinone, incorporating two (MeO)(Me)C(CO) groups. Azobenzenes form imidates and 1,2- and 1,3-diazetidinones when irradiated with chromium carbene complexes.

INTRODUCTION

Synthetic methodology involving transition metals has been evolving over the last twenty years, but only recently has the area become sufficiently sophisticated to permit its application to the synthesis of highly functionalized complex organic molecules. This is particularly true of organopalladium chemistry which had its beginnings well over twenty years ago, but only recently has found application in the synthesis of complex molecules. In this presentation the application of well established palladium(II) and palladium(0) catalyzed processes to synthesize 3,4-disubstituted indoles related to the ergots will be described as will the use of some quite new chemistry of Group VI transition metal carbenes for the synthesis of cephams, oxacephams, and oxapenams.

3,4-DISUBSTITUTED INDOLES

Ergot alkaloids (1)[1] and the related clavicipitic acid (2)[2] pose several synthetic problems. Both are 3,4-disubstituted indoles, and while the 3 position of the indole nucleus is quite reactive toward electrophiles, the 4-position is the least reactive position on the indole ring toward conventional electrophilic aromatic substitution (Figure 1). In addition the ergots have a strained six-membered ring bridging the 3- and 4-positions. The approach outlined in equation 1 uses organopalladium chemistry to solve both of these problems.

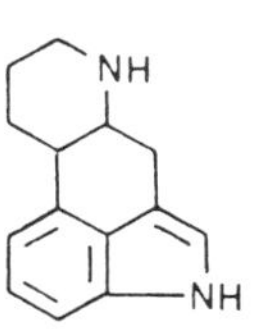

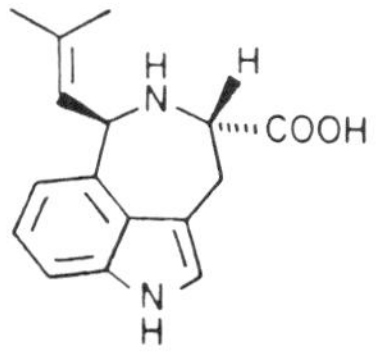

clavicipitic acid

1   2

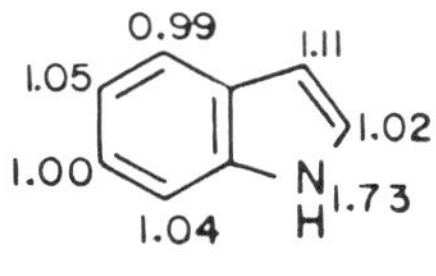

Figure I.   Electron Densities of the Indole Ring System

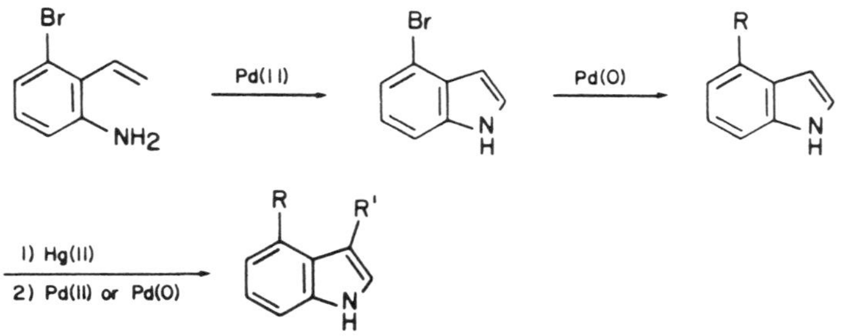

(1)

2-Amino-6-bromostyrene was synthesized in excellent overall yield using the conventional organic chemistry described in equation 2. Although the palladium(II) catalyzed cyclization

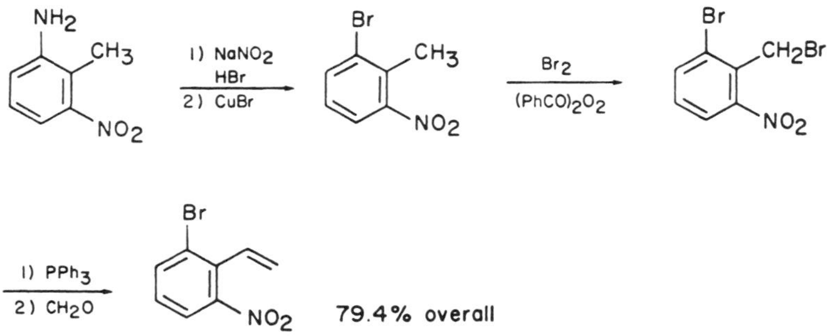

(2)

79.4% overall

of 2-aminostyrene to indole itself (a palladium(II) catalyzed intramolecular amination of an olefin) was an efficient process,[3] the 4-bromosubstituted indole apparently interferred with this process, as was evidenced by the low overall yield for cyclization (equation 3). To circumvent this problem the free amino group was converted to the tosamide[4] and this substrate cyclized in excellent yield (on >30 g scale) to give the highly crystalline 4-bromo-N-tosyl indole. The tosyl group was easily removed in high yield by basic hydrolysis to give pure 4-bromoindole. Although the ensuing reactions proceeded well with the free indole, the N-tosylindole was more convenient to handle and was used for the further transformations discussed below.

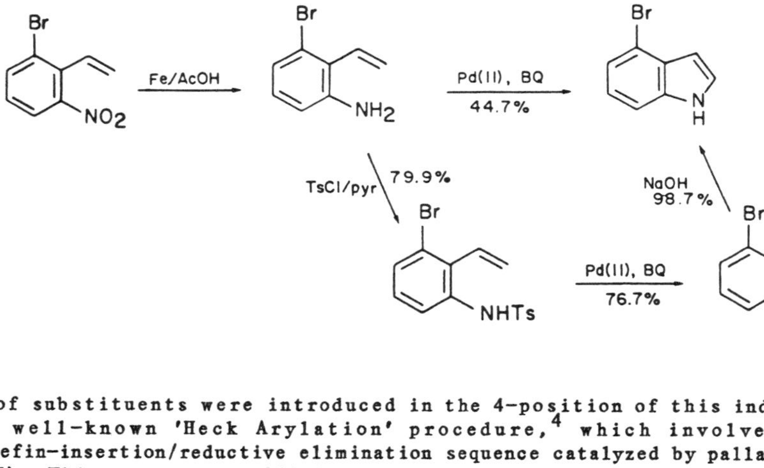

(3)

A variety of substituents were introduced in the 4-position of this indole (equation 4) using the well-known 'Heck Arylation' procedure,[4] which involves an oxidative addition/olefin-insertion/reductive elimination sequence catalyzed by palladium(0) complexes (equation 5). This was a very efficient procedure for electron rich, electron poor, and 'normal' olefins. Particularly important was the high yield achieved with 2-methyl-3-buten-2-ol, since this side chain provides the elements of the left portion of both ergots and clavicipitic acid.

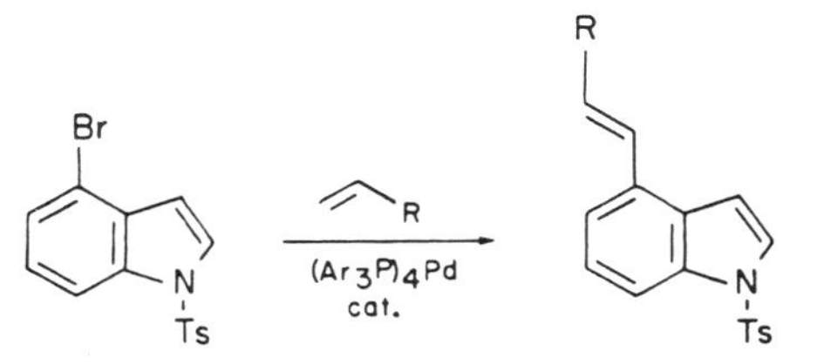

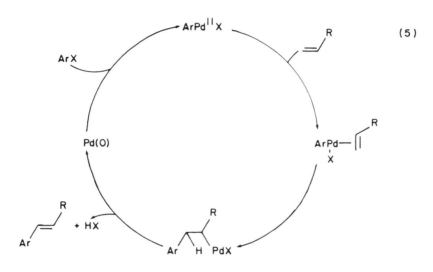

$$(4)$$

R = COOCH_3   85.7 %

R = -C(CH_3)_2OH   97.3 %

R = Ph   74.0 %

R = NPhth   74.2 %

$$(5)$$

THE "HECK REACTION"

Although indoles themselves are generally reactive toward electrophiles at the 3-position N-acylation or tosylation results in reduced reactivity. However, the 4-bromo-N-tosyl indole underwent clean 3-mercuration in quantitative yield, to produce an intermediate that was further functionalized in two ways. Cleavage of the mercurio compound with iodine produced the 3-iodo-4-bromo-1-tosylindole. This substrate underwent exclusive reaction at the iodide bearing position when subjected to palladium(0) catalyzed olefin insertions in the absence of phosphines (equation 6). Again both electron deficient and electron rich olefins inserted in high yield making a variety of 3-substituted-4-bromo indoles available. Particularly important was the successful introduction of the N-acetamidoacrylate unit in good yield, since this fragment is common to the right hand portion of both the ergots and clavicipitic acid.

Alternatively, treatment of the 3-mercurioindole with lithium tetrachloropalladate resulted in a facile transmetallation from mercury to palladium.[5] Methyl acrylate inserted efficiently, although stoichiometric amounts of palladium(II) were required. However, allyl halides regenerated the active palladium(II) catalyst by β-halide elimination,[6] resulting in an efficient catalytic allylation at the 3-position (equation 7). These substrates were perfectly constituted for an intramolecular oxidative addition-insertion, to form the strained C-ring of the ergot alkaloids. Again a palladium(0) catalyzed process proved efficient for this transformation. Note that rearrangement from the indole to the naphthalene occurred during this transformation. Studies are in progress to append the D-ring of the ergots by a heteroatom Diels-Alder process.[7]

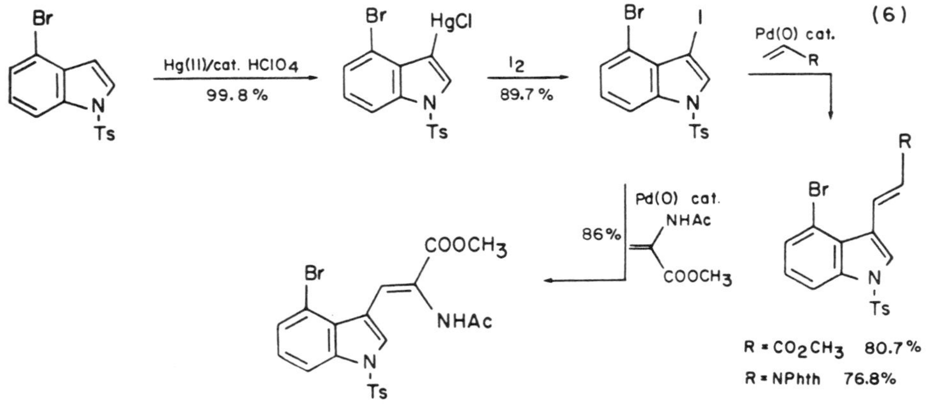

(6)

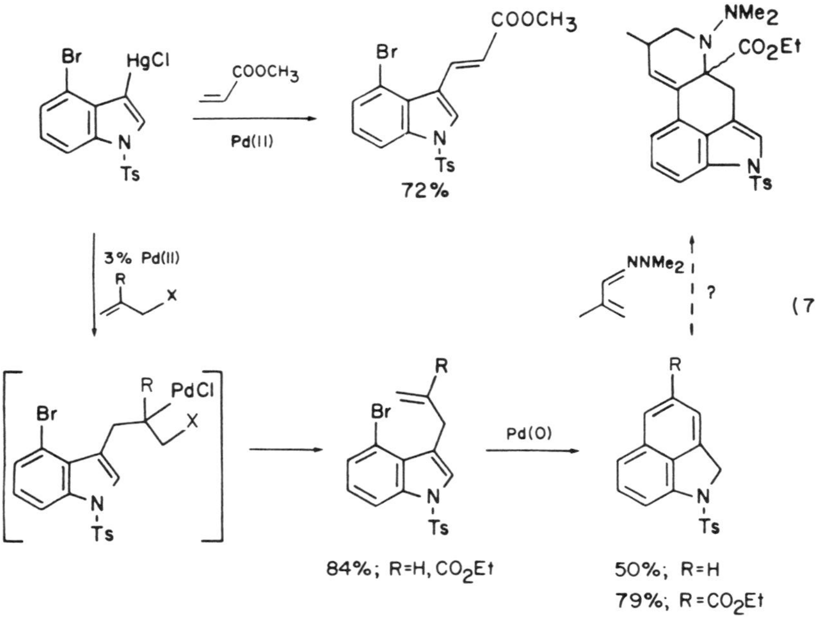

R = CO$_2$CH$_3$ 80.7%
R = NPhth   76.8%

(7)

84%; R=H, CO$_2$Et

50%; R=H
79%; R=CO$_2$Et

Having developed methods for the introduction of alkyl substituents at the 3- or 4-position of the indole nucleus, it remained to make the disubstituted system which is the precursor to clavicipitic acid. This was achieved in high yield by taking the acetamido acrylate product from above, reducing it using Wilkinsons catalyst, and then alkylating the 4-position with 2-methyl-3-buten-2-ol producing material having all the elements of clavicipitic acid (equation 8). Cyclization studies are in progress.

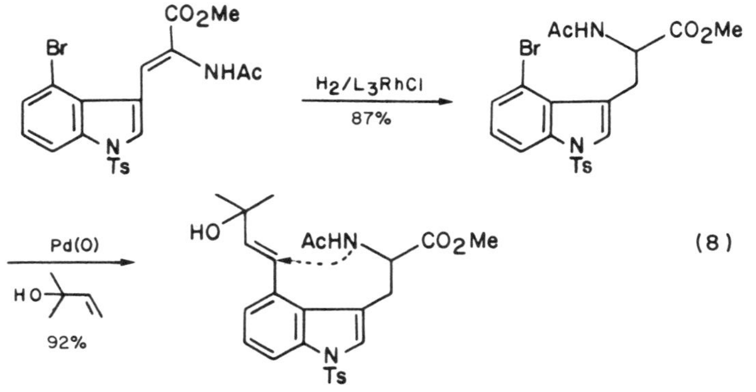

(8)

NEW SYNTHESIS OF β-LACTAMS AND DIAZETIDINONES

In contrast to the well established organopalladium chemistry discussed above, the uses of Group VI carbene complexes in organic synthesis is relatively new. Group VI carbene complexes are most readily prepared by the reaction of organolithium reagents with the metal hexacarbonyl, followed by O-alkylation of the 'ate' complex with active alkylating agents such as Meerwin's reagent (equation 9).[8] These carbene complexes are relatively stable

$$M(CO)_6 + RLi \longrightarrow (CO)_5 \overset{(-)}{M}-\overset{\overset{O}{\|}}{C}-R \xrightarrow{Me_3{}^+OBF_4{}^-} (CO)_5 M=\overset{\overset{OMe}{|}}{C}-R \quad (9)$$

$$M = Cr, Mo, W$$

crystalline solids which are easy to prepare and to handle. We recently discovered that irradiation of chromium carbene complexes in the presence of imines led to the production of β-lactams in excellent yield (equation 10).[9] The reaction was remarkable in several

$$(CO)_5 Cr=C \overset{OMe}{\underset{R^1}{<}} \;+\; \overset{R^2}{\underset{R^3}{>}}C=N-R^4 \xrightarrow[Et_2O]{h\nu} \quad \begin{matrix} MeO & R^3 \\ R^1 \cdots\!\!-\!\!\!-\!\! R^2 \\ O \!\!=\!\! \underset{N}{\big|} \\ R^4 \end{matrix} \quad (10)$$

respects. It was regio- and stereospecific. Only β-lactams were formed, and only one diastereoisomer was formed. When a chiral thiazoline was used, the reaction was enantiospecific giving a penicillin analog as a single enantiomer (equation 11). (Lower

$$(CO)_5 Cr=C \overset{OMe}{\underset{Me}{<}} \;+\; \begin{matrix} S \\ N \underset{CO_2Me}{\overset{\diagdown}{\diagup}} H \end{matrix} \xrightarrow[CH_2Cl_2]{h\nu} \quad \begin{matrix} Me & H & S \\ MeO \cdots & & \\ O\!\!=\!\! N & & H \\ & & CO_2Me \end{matrix} \quad (11)$$

$$(R, S, R)$$

$$[\alpha]_D = +218.1°$$

enantioselectivity was observed with nonrigid acyclic chiral imines). A variety of polyfunctional imines were also converted to β-lactams in fair to excellent yield (Table 1). These included a thiazine precursor to a cepham derivative, and seven N-functionalized imines. In these cases, the reaction was again regio- and stereospecific, giving only β-lactams, and only a single diastereoisomer at the two centers formed in the reaction. In contrast, α-ketoimines and α-diimines gave mixtures of diastereoisomers, the ratio of which depended on the substituents.

In an attempt to prepare oxapenam and oxacephem analogous, chromium carbene complexes were treated with oxazolines and oxazoles under the standard reaction conditions (visible light, 25°, ether solvent). No β-lactam formation was observed. Instead the nitrogen heterocycle was recovered unchanged, and the carbene complex had decomposed. The more stable tungsten carbene complexes similarly were unreactive. However, irradiation of the considerably less stable (thermally) molybdenum methoxymethyl carbene complex in the presence of oxazoles at 0° in THF produced oxacepham derivatives in modest yield. Again only a single diastereoisomer at the two centers formed in the reaction was obtained (equation 12).

Table I. β-Lactams from Imines and [(Methoxy)(methyl)carbene]chromium Complexes

IMINE	PRODUCT	YIELD %[a]
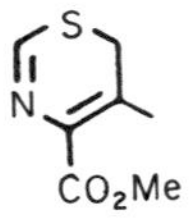	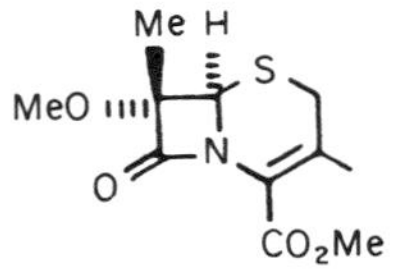	52
PhCH = NR R = CH$_2$P(O)(OEt)$_2$ R = CH(CO$_2$Me) P(O)(OEt)$_2$ R = CH = CH$_2$	 R = CH$_2$P(O)(OEt)$_2$ R = CH (CO$_2$Me) P(O)(OEt)$_2$ R = CH = CH$_2$	90 80 41[b]
 R = Ph R = Me	 R = Ph R = Me	52[b] (A/B = 2) 52[b] (A/B = 1.15)
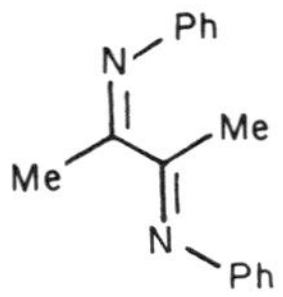	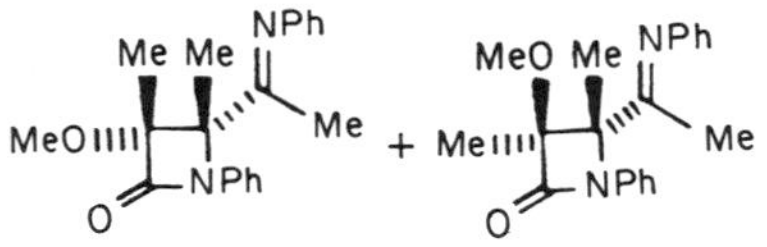	40[b] (A/B = 0.25)

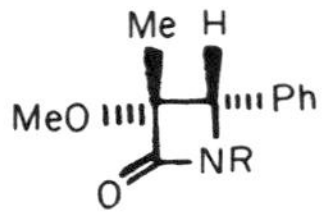

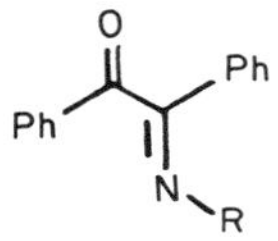

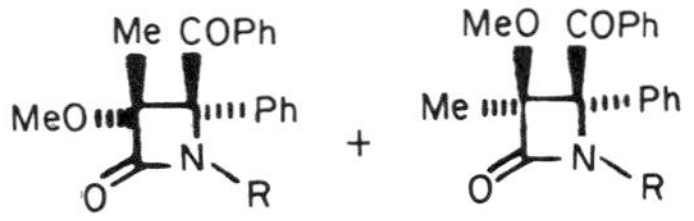

[a]Reported yields are for isolated, analytically pure product.

[b]Yields are not optimized.

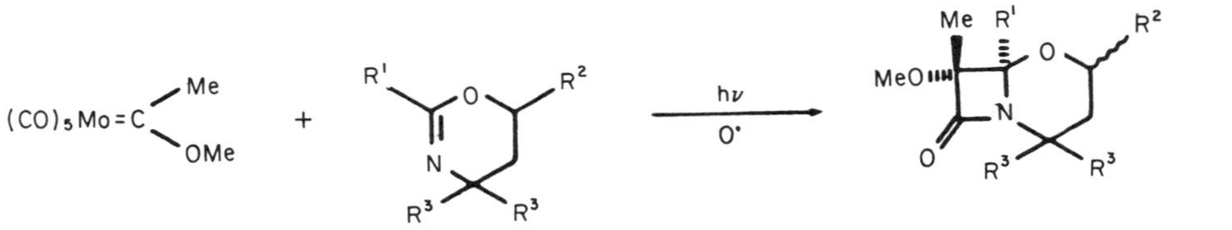

$$(CO)_5Mo = C\langle \overset{Me}{OMe} \quad + \quad \text{[oxazine]} \quad \xrightarrow[O^\bullet]{h\nu} \quad \text{[product]} \qquad (12)$$

R^1, R^2, R^3 = H   R^1, R^2, R^3 = H 41%

R^1 = Et, R^2, R^3 = Me   R^1 = Et, R^2, R^3 = Me 42%

R^1 = PhCH$_2$, R^2, R^3 = Me   R^1 = PhCH$_2$, R^2, R^3 = Me 46%

Oxazolines behaved in yet a different manner. They were considerably less reactive than the oxazines, and a substantial amount of unreacted starting material was recovered even when excess carbene complex was used. A low yield of the desired oxapenam analog was obtained, again as a single diastereoisomer. In addition, an oxazinone, incorporating two 'carbene' units was formed in equal amount (equation 13), as a single isomer whose geometry cannot be assigned. Although oxazinones are the major products from the reaction of free ketenes with oxazolines, it is unlikely that free ketene is involved in the process described in equation 13 since oxazinones have not been observed in other reactions wherein large quantities of molybdenum carbene complexes have photolytically decomposed.

(13)

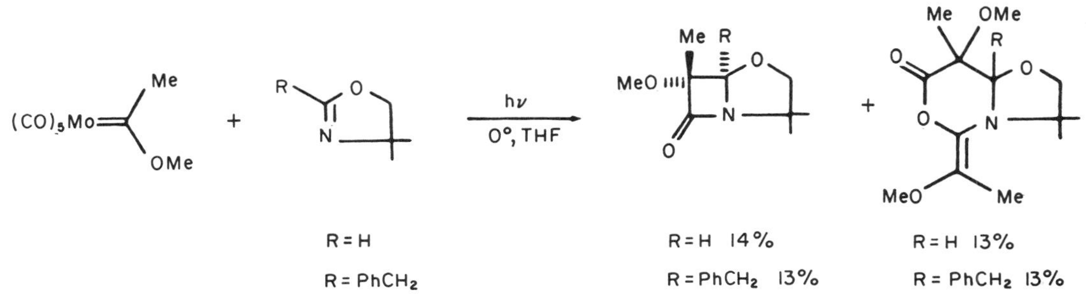

R = H

R = PhCH₂

R = H 14%

R = PhCH₂ 13%

R = H 13%

R = PhCH₂ 13%

Azobenzenes underwent a remarkable transformation when irradiated (sunlight, 30°C) in the presence of methoxymethyl chromium carbene complex (equation 14). The major products were

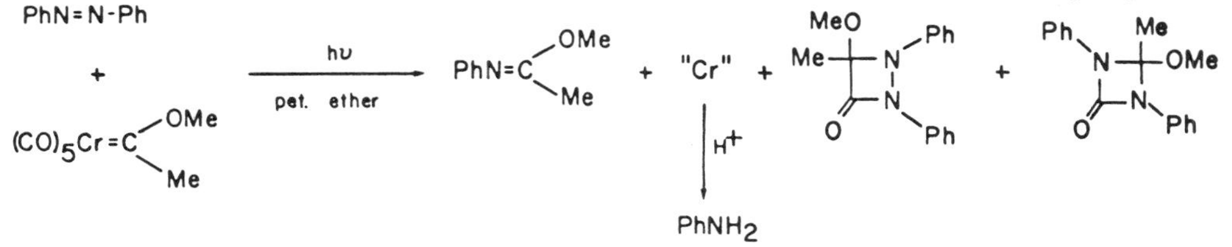

an imidate containing one half of the azobenzene and the carbene carbon, and a chromium containing residue which liberated aniline upon acidification. A small amount of diazetidinone product (~10%) was obtained, and this consisted of a 1:1 mixture of the 1,2- and the 1,3-diazetidinone. When carried out under milder conditions, up to 35% diazetidinone material could be obtained, with the 1,3-diazetidinone predominating (6:1). p-Methoxy azobenzene gave exclusively the 1,3- product (equation 15), as did p-dimethylaminoazobenzene (equation 16) while p-nitroazobenzene was virtually unreactive. Although no mechanistic studies have been carried out, the reaction is thought to proceed as in Scheme I, involving photocycloaddition, metallacyclobutane fragmentation, recombination, CO insertion and reductive elimination.

(15)

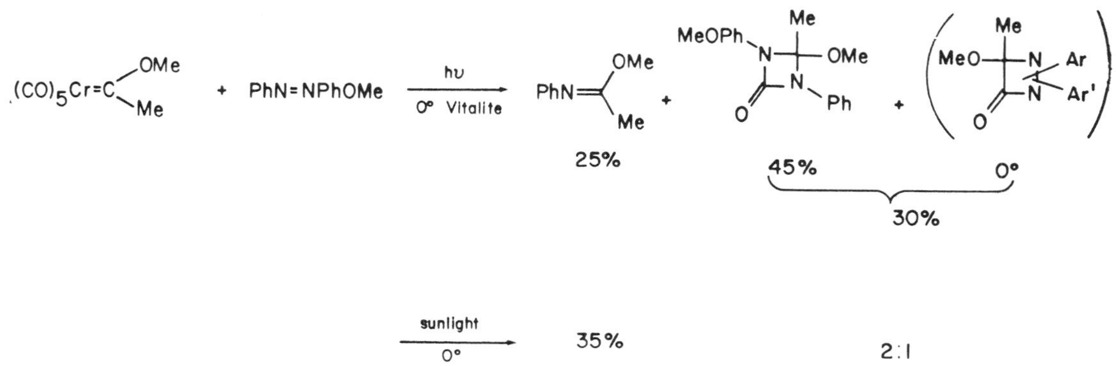

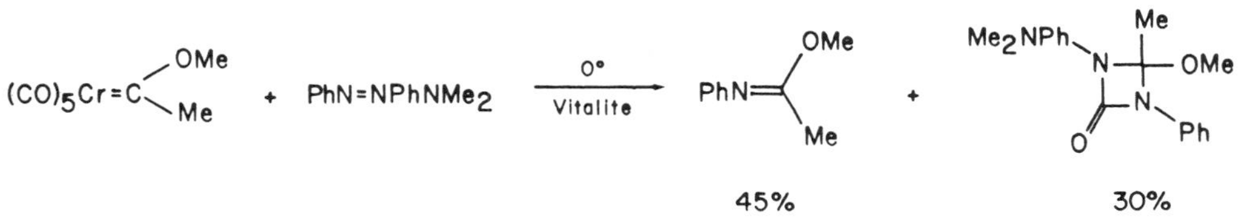

SCHEME I

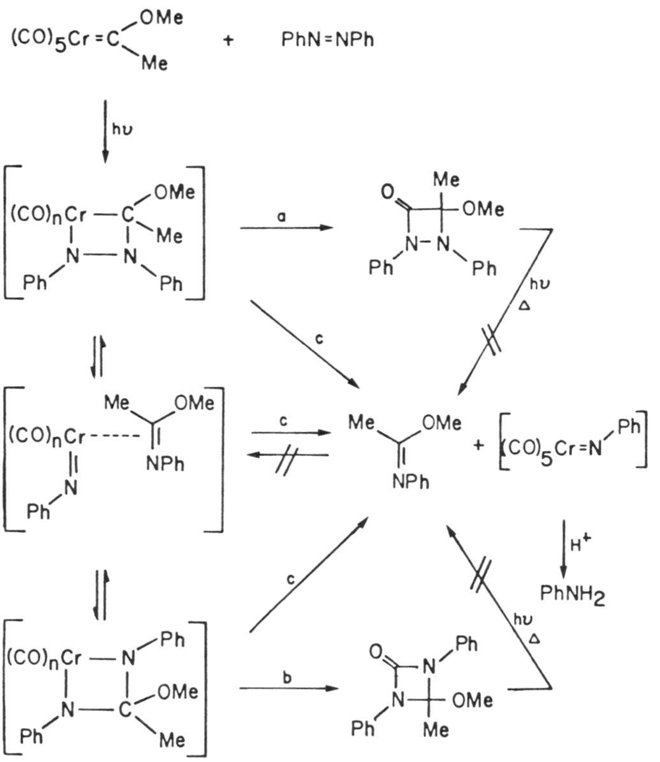

CONCLUSIONS

The above discussion demonstrates that organotransition metal chemistry can offer useful and unusual tools to the synthetic organic chemist. Unconventional products can be made from both conventional and unconventional starting materials. The reactions often proceed under mild conditions, and are very specific, obviating the need for protacted protection, deprotection, and reactivity adjustment sequences. These organometallic reactions can be made to work with quite complex substrates, and many should be incorporated into the arsenal of synthetic organic chemists.

Acknowledgement – This work was supported by grants from the National Science Foundation and the Public Health Service, National Institute of General Medical Sciences. This research was most ably carried out by Dr. Peter Harrington, Mr. Eric Michalson, Ms. Lisa Schultze, Mr. Jose Toro, Mr. Chen Yijun, Dr. Michael McGuire, and Dr. Andreas Kramer.

REFERENCES

1. Kozikowski, A. P. Heterocycles, 1981, 16, 267.
2. Kozikowski, A. P.; Greco, M. N. J. Org. Chem., 1984, 49, 0000.
3. Hegedus, L. S.; Allen, G. F.; Bozell, J. J.; Waterman, E. L. J. Am. Chem. Soc., 1978, 100, 5800.
4. Heck, R. F. Accts. Chem. Res., 1979, 12, 146.
5. Larock, R. C. Tetrahedron, 1982, 38, 1713.
6. Heck, R. F. J. Am. Chem. Soc., 1968, 90, 5531; Heck, R. F. J. Organometal. Chem., 1971, 33, 399; Bergstrom, D. E.; Ruth, J. L. J. Am. Chem. Soc., 1976, 98, 1587; Ruth, J. L.; Bergstrom, D. E. J. Org. Chem., 1978, 43, 2870.
7. Bozer, D. L. Tetrahedron, 1983, 39, 2869.
8. Aumann, R.; Fischer, E. O. Chem. Ber., 1968, 101, 9541.
9. Hegedus, L. S.; McGuire, M. A.; Schultze, L. M.; Yijun, C.; Anderson, O. P. J. Am. Chem. Soc., 1984, 106, 2680.

The activation and selective functionalisation of alkanes by some soluble transition metal systems

Denise Baudry, Michel Ephritikhine, Hugh Felkin[*], Tauqir Fillebeen-Khan, Yvonne Gault, Rupert Holmes-Smith, Lin Yingrui and Janusz Zakrzewski

Institut de Chimie des Substances Naturelles, C.N.R.S.,
91190 Gif-sur-Yvette, France

A variety of soluble transition metal polyhydrides (L_2ReH_7, L_3RuH_4, L_2IrH_5; L = tertiary phosphine), in the presence of an olefin (e.g., 3,3-dimethylbutene) as a hydrogen acceptor, have recently been found to bring about the activation of C-H bonds in saturated hydrocarbons under mild conditions, in some cases with selective attack at methyl (rather than methine and methylene) groups. The following homogeneous reactions are described: (a) the one-pot formation of 1-alkenes from the corresponding n-alkanes $C_nH_{2n + 2}$ (n = 5-8) via the bis(phosphine) diene rhenium trihydrides $(Ph_3P)_2(\eta\text{-diene})ReH_3$ (60-80% overall yield, 95% selectivity); (b) the selective, catalytic, formation of methylenecyclohexane from methylcyclohexane; (c) the catalytic conversion of cyclo-octane into cyclo-octene (up to 70 catalytic turnovers). The mechanism of these reactions is briefly discussed: the key step is thought to involve the insertion of coordinatively unsaturated (14e) intermediates (e.g., L_2ReH_3, L_2IrH) into a C-H bond of the hydrocarbon. Such systems may ultimately be developed into useful methods for functionalising isolated, unactivated, methyl groups. At the moment, however, they do not constitute practical propositions for synthetic purposes; although the yields with respect to the metal are generally quite high, the conversions are still very low (< 1%), since the hydrocarbon substrates also serve as solvents and are therefore present in large excess.

INTRODUCTION

In the laboratory, saturated hydrocarbons are at present more useful as solvents, and as substrates for nomenclature, than as starting materials for synthesis. The problem with the "paraffins", however, is not so much, as their name suggests, their lack of reactivity, as the lack of selectivity of the reactions they undergo.

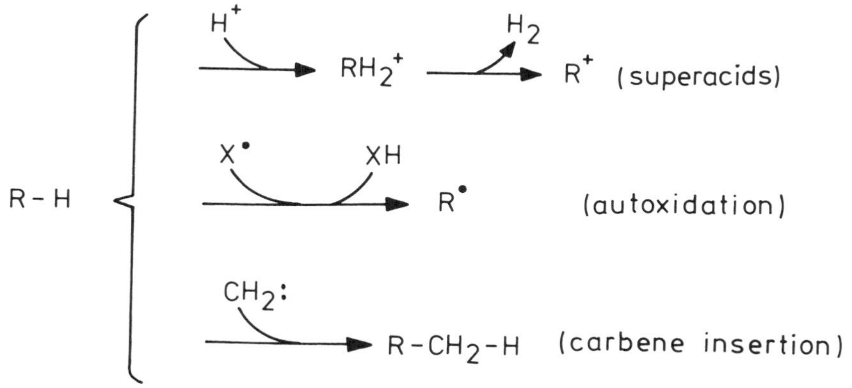

Scheme 1. Examples of Reagents which Effect the Activation of C-H Bonds in Saturated Hydrocarbons.

There are essentially three types of reagents which quite readily attack the unactivated C-H bonds of saturated hydrocarbons (Scheme 1): (a) strong electrophiles, such as superacids and carbocations; (b) radical reagents, which lead for example to autoxidation and chlorination (a number of transition metal systems are known which catalyse such radical reactions); and (c) carbenes, which insert into C-H bonds. Insofar as these reactions show any selectivity, they generally lead to preferential attack upon the weakest C-H bonds (i.e., at methine and methylene, rather than methyl, groups), of which there are often many in the same molecule. With a few exceptions, they are therefore not very useful.

In this talk, I shall discuss some homogeneous reactions which appear to involve the insertion, into unactivated C-H bonds, of ligand-deficient transition metal complexes, i.e., insertions by co-ordinatively unsaturated transition metal atoms rather than by co-ordinatively unsaturated, ligand-deficient, carbon atoms (carbenes). These reactions can lead to the functionalisation of saturated hydrocarbons and show a selectivity which is different from that generally found in the reactions shown in Scheme 1.

Heterogeneous transition metal catalysts are used industrially for the treatment, on a very large scale (megatons/day), of oil feedstocks (Scheme 2).

Reforming: Al_2O_3 - Pt - Re, $\sim 500^{\circ}C$

$\longrightarrow$ isomerisation, cyclisation, + H_2, - H_2

Hydrocracking: SiO_2 - Al_2O_3 - Ni or Pt, $\sim 300^{\circ}C$

$\longrightarrow$ fragmentation, isomerisation, cyclisation, + H_2

Scheme 2.   Activation of C-H Bonds in Alkanes: Industrial
Heterogeneous Transition Metal Systems.

The reactions which take place on the catalyst surface (which can be regarded as being made up of ligand-deficient transition metal atoms) are complex and not perfectly understood, but some of them are thought to be triggered by the insertion of a metal atom into a C-H bond of the alkane substrate. These important industrial processes are carried out at high temperatures, but other heterogeneous systems react with saturated hydrocarbons under far milder conditions; thus certain clean transition metal films catalyse H/D exchange in alkanes below room temperature, and totally ligand-deficient ("naked") transition metal atoms have been shown to insert into the C-H bonds of saturated hydrocarbons in argon matrices at 15K. These heterogeneous, carbene-like, insertions are, however, all rather unselective.

Homogeneous transition metal catalysts are often more selective than their heterogeneous counterparts, and their selectivity can in many cases be modulated by varying the ligands around the metal; moreover, more efficient use is

made of the metal in homogeneous reactions. There is thus some current interest in the "activation" of C-H bonds in saturated hydrocarbons by soluble
transition metal compounds, and this is sustained by the prospect of finding
a system which would allow the efficient, selective, and if possible catalytic,
functionalisation of these rather unreactive substrates.

Despite a fair amount of effort, until a few years ago the outlook for
finding such a system did not look very encouraging. Intermolecular insertion
of ligand-deficient transition metals into alkane C-H bonds (Scheme 3, top
line) was known to occur readily with certain heterogeneous systems (see above)
and, in 1969, A.E. Shilov and his group discovered a soluble platinum system
comprising only inorganic ligands which catalyses H/D exchange in alkanes. Co-
ordinatively unsaturated transition metal complexes with organic ligands, however (Scheme 3, bottom line), were all found to undergo intramolecular insertion into a C-H bond of one of their own ligands (cyclometallation, left-hand
arrow), rather than intermolecular insertion into a C-H bond of an alkane substrate (right-hand arrow). Intermolecular C-H insertion only took place with
compounds possessing activated C-H bonds (acetone, acetonitrile, toluene, etc.).
Such a preference for intra- versus intermolecular reactions is, of course, a
well known phenomenon in organic chemistry.

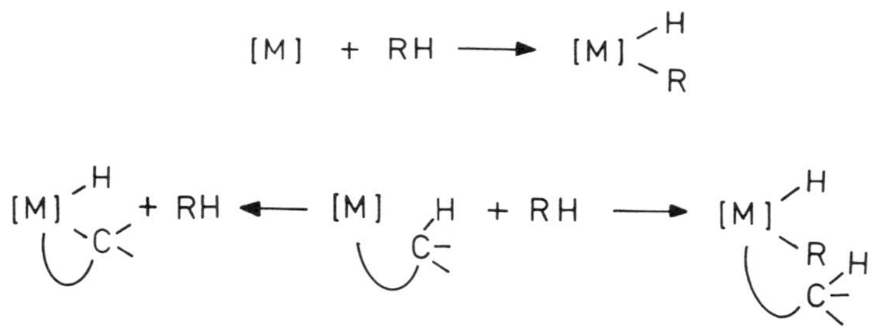

$$[M] \ + \ RH \longrightarrow [M] \overset{\diagup H}{\diagdown R}$$

$$[M] \overset{\diagup H}{\diagdown C} \ + \ RH \longleftarrow [M] \diagup C \ + \ RH \longrightarrow [M] \overset{\diagup H}{\diagdown R \diagdown H}$$

Scheme 3.   Intermolecular versus Intramolecular C-H Activation.

Very recently, however, a few soluble transition metal complexes (containing Ir, Re, Ru and
Rh) have been discovered which bring about the intermolecular insertion of the transition metal into C-H bonds of saturated hydrocarbons, despite the presence of organic ligands (Scheme
3, bottom line, right-hand arrow). The discovery of the first such system in 1979 by R.H.
Crabtree and his group, and the isolation and characterisation, in 1982, by R.G. Bergman and
W.A.G. Graham and their respective groups, of the first-formed C-H insertion products, represented major advances in the field of alkane C-H bond activation by soluble transition metal
systems. The reasons why certain systems lead to intermolecular reactions, whereas in the
majority of cases cyclometallation occurs, are not yet understood.

Some of these new soluble transition metal systems not only activate C-H bonds but also allow
the functionalisation (sometimes selective, sometimes catalytic, sometimes both) of n-alkanes
and cycloalkanes, and it is these systems that I shall discuss in this talk.

BIS(PHOSPHINE)RHENIUM HEPTAHYDRIDES

About four years ago, we observed the activation of the C-H bonds of cyclopentane under mild

conditions by means of bis(triarylphosphine)rhenium heptahydrides in the presence of an olefin (typically 3,3-dimethylbutene = neohexene) as a hydrogen acceptor (Scheme 4). When a solution of bis(triphenylphosphine)rhenium heptahydride and neohexene in cyclopentane was

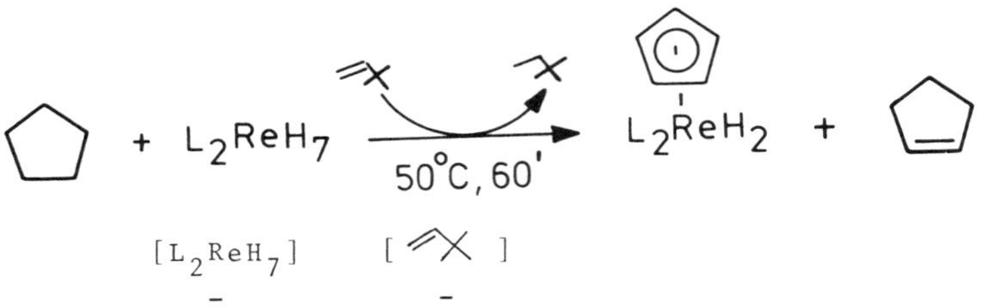

[L$_2$ReH$_7$]	[⤳✗]		
–	–		
30 mM	300 mM	0.45	–
20 mM	200 mM	0.45	0.15 – 0.45
2 mM	20 mM	0.7	1.0 – 1.1

L = (p-X-C$_6$H$_4$)$_3$P, X = Me, H, F.

Scheme 4. Activation of Cyclopentane by Bis(phosphine)rhenium Heptahydrides.

heated under the conditions shown in Scheme 4, cyclopentadienylbis(triphenylphosphine)rhenium
dihydride was formed in 45% yield (i.e., 0.45 mole per mole of heptahydride, penultimate
column in Scheme 4). Further work showed that cyclopentene is also formed [0.15 (X = Me) –
0.45 (X = F) mole per mole of heptahydride, last column in Scheme 4], and that the yields
(with respect to rhenium) are increased when the solutions are made more dilute; in fact,
using 2 mM heptahydride (X = F), the formation of cyclopentene verges on the catalytic (1.1
mole formed per mole of heptahydride, bottom line in Scheme 4). The "activation" of cyclopentane by this system thus affords a mixture of two products: an organometallic hydride, and an
olefin.

A number of other saturated hydrocarbons are also "activated" under the same conditions,
and it turns out that cyclopentane represents an intermediate case between the n-alkanes on
the one hand, which give only organometallic products with at most a trace of olefin, and the
larger cycloalkanes on the other, which are transformed catalytically into the corresponding
cycloalkenes.

The organometallic product obtained from n-pentane is shown in Scheme 5. Bis(phosphine)
(η-diene)rhenium trihydrides of this type are air-stable compounds; they can also be obtained

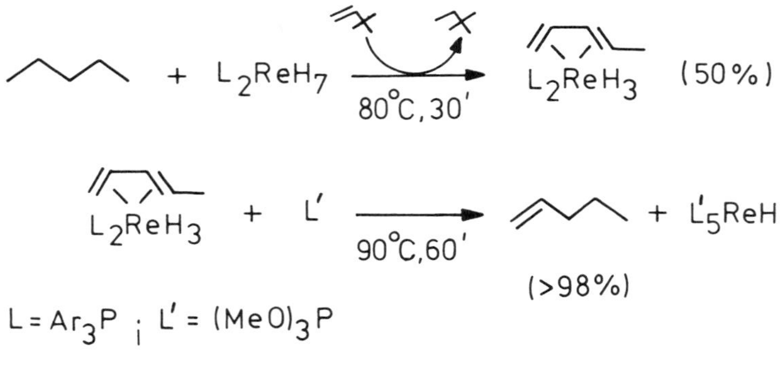

L = Ar$_3$P ; L′ = (MeO)$_3$P

Scheme 5. Selective Functionalisation of n-Pentane.

by reaction of bis(phosphine)rhenium heptahydrides with the corresponding dienes or olefins. We have found that, as shown in Scheme 5, the (η-pentadiene)rhenium trihydride formed from n-pentane can be converted into 1-pentene in essentially quantitative (> 95%) yield and with high (> 98%) selectivity by treatment with trimethyl phosphite. This selective conversion of n-pentane into 1-pentene can be carried out as a "one-pot" reaction, without isolating the intermediate (η-diene)rhenium trihydride. We do not understand the reasons for the remarkably high regioselectivity.

Whereas the organometallic compound shown in Scheme 5 is the only (η-conjugated-<u>trans</u>-diene)rhenium trihydride which can be formed from n-pentane, two isomers could be formed from n-hexane and n-heptane, and three from n-octane (Scheme 6, R = H, Me, Et, respectively). In the event, perhaps not unexpectedly, this is exactly what occurs; these three n-alkanes are converted, in 60-80% yield, into unseparable mixtures of all the possible (η-diene)rhenium

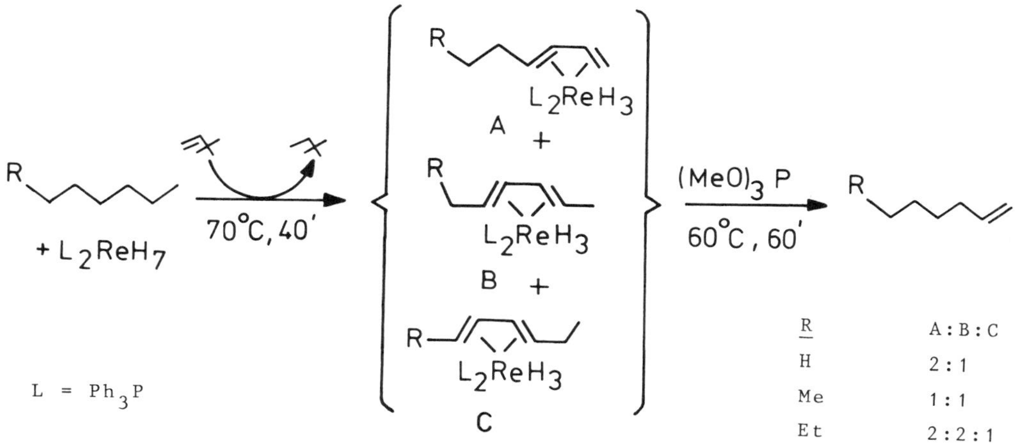

Scheme 6. Selective Functionalisation of n-Alkanes.

trihydride isomers in approximately the statistical proportions (Scheme 6, bottom right-hand corner). Unfortunately, these proportions tell us nothing about where in the n-alkane the initial activation of a C-H bond takes place, since we have found (by n.m.r. spin saturation transfer) that the various isomers shown in Scheme 6 are interconverting in solution and that the mixtures we obtain are therefore simply equilibrium mixtures. The pentadiene trihydride (Scheme 5) undergoes a similar (in this case degenerate) rearrangement.

We were, however, agreeably surprised to find that these diene trihydride mixtures (Scheme 6) are converted, by treatment with trimethyl phosphite, into the corresponding 1-alkenes in the same high yield (> 95%), and with the same high regioselectivity (> 98%), as the pentadiene trihydride shown in Scheme 5. This selective functionalisation of a terminal carbon atom in an n-alkane is thus a general reaction.

With the same system, cycloalkanes are transformed catalytically (and selectively; it is noteworthy, for example, that only cyclohexene, and no benzene, is formed from cyclohexane) into the corresponding cycloalkenes. The number of catalytic turnovers is low (the last column in Scheme 7 shows the number of moles of cycloalkene formed per mole of heptahydride). No organometallic product is formed from the larger cycloalkanes (cyclohexane, cycloheptane and cyclo-octane). The presence of a hydrogen acceptor is necessary in these catalytic reactions, which would otherwise be thermodynamically impossible at 80°C, but it need not be neohexene;

using cyclohexene, for example cyclo-octene is formed from cyclo-octane with 6 catalytic

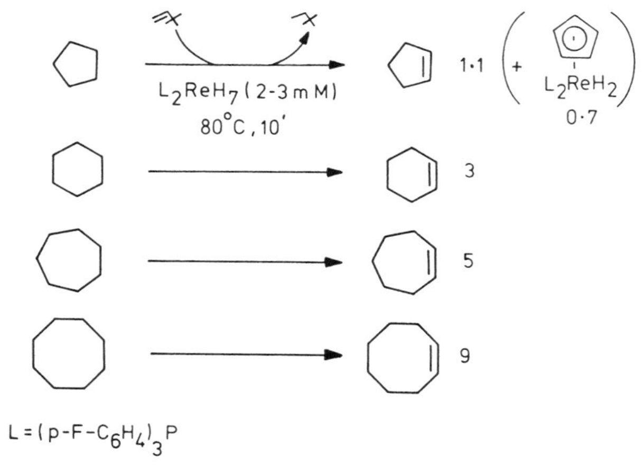

$L_2ReH_7 (2-3\,mM)$
$80°C, 10'$

1·1   (+ L_2ReH_2)
                    0·7

3

5

9

$L = (p-F-C_6H_4)_3 P$

Scheme 7. Catalytic Dehydrogenation of Cycloalkanes.

turnovers (instead of 9 using neohexene, Scheme 7). This soluble rhenium system is thus able to effect the catalytic dehydrogenation of certain saturated hydrocarbons.

MECHANISTIC SPECULATIONS

The reactions of bis(phosphine)rhenium heptahydrides with saturated hydrocarbons which I have just described (Schemes 4-7) do not appear to involve electrophilic or radical processes of the type shown in Scheme 1 (first two lines). The presence of a strong electrophile would cause at least some of the neohexene to rearrange to tetramethylethylene, and this is not observed. Radical intermediates would be expected to dimerise, at least to a small extent, but no trace of dicyclohexyl or dicyclo-octyl could be detected in the reaction products from cyclohexane and cyclo-octane; moreover, the selectivities observed with methylcyclohexane as substrate (preferential attack at methyl rather than at methine C-H bonds), to which I shall return later (see Scheme 11), are not those expected for radical reagents. We therefore assume that the key step in these reactions involves the carbene-like insertion of a ligand-deficient rhenium intermediate into a C-H bond of the hydrocarbon substrate (Scheme 3, bottom line, right-hand arrow).

Based upon this assumption, a possible, speculative, reaction mechanism is shown in Scheme 8 (L = tertiary phosphine). This scheme is compatible with the facts as we know them, the various individual steps have precedent in other systems, and it involves no unnecessary complications ("entia non sunt multiplicanda praeter necessitatem"; William of Occam, ca 1300). Further work may, of course, show that this scheme is too simple; for example, rhenium clusters, or cyclometallated intermediates, may well be involved, but at the moment we have no data which suggest that this is so. We imagine that the initial bis(phosphine)rhenium heptahydride loses two hydrogens to form the co-ordinatively unsaturated 16e pentahydride shown on the left-hand side of Scheme 8. This pentahydride is apparently not reactive enough to insert into unactivated C-H bonds, and we think it undergoes further dehydrogenation (by neohexene) to form the bis(phosphine)rhenium trihydride shown on the right-hand side of the Scheme. This hypothetical intermediate is a 14e, highly ligand-deficient, compound (a sort of "super-carbene"), and we assume that it is capable of inserting into a C-H bond of a saturated hydrocarbon to form a 16e alkylrhenium tetrahydride which then undergoes further reaction (intramolecular β-elimination) leading to an 18e, co-ordinatively saturated, alkenerhenium pentahydride. Finally, dissociation of the alkene regenerates the initial 16e bis(phosphine)rhenium pentahydride, thus completing the catalytic cycle. This dissociation of alkene apparently

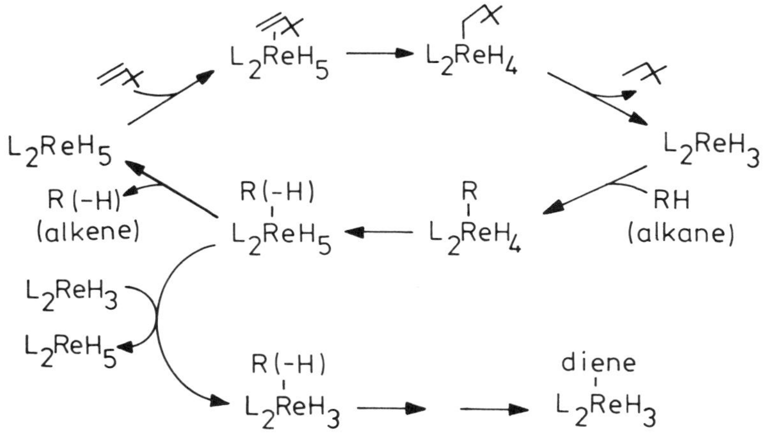

Scheme 8.   Postulated Catalytic Cycle (Speculation).

takes place very readily in the case of relatively bulky alkenes [e.g., R(-H) = cyclohexene, cyclo-octene], and the overall reaction is catalytic (Scheme 7). With less bulky olefins [e. g., R(-H) = 1-alkene], however, we presume that the less sterically crowded alkenerhenium pentahydride intermediates survive long enough in solution to undergo further dehydrogenation (possibly by the route indicated in Scheme 8), and the reaction then leads ultimately to the (non-catalytic) formation of alkadienerhenium trihydrides (Schemes 5 and 6). The stability of the postulated cyclopentenerhenium pentahydride (formed from cyclopentane) appears to be intermediate, since cyclopentane leads to a mixture of cyclopentene and the cyclopentadienyl-rhenium dihydride shown in Scheme 4.

All the steps in our postulated catalytic cycle (Scheme 8) must be reversible, and we imagine that the system is driven clockwise (as shown) by the large concentration gradient (> 100) between the alkane RH and neohexane. The relatively large free energy of hydrogenation of neohexene (21.9 kcal/mol) is not a determining factor, since the conversion of cyclo-octane into cyclo-octene is still catalytic (6 turnovers instead of 9) with cyclohexene (free energy of hydrogenation 18.7 kcal/mol) as the hydrogen acceptor instead of neohexene. The maximum number of catalytic turnovers which can be achieved with this rhenium system (about 10 with cyclo-octane, Scheme 9) is presumably limited by competition between the key C-H insertion step and various deactivation processes (e.g., further dehydrogenation, cyclo-metallation, and oligomerisation of the various co-ordinatively unsaturated intermediates).

I should emphasise again that this mechanism is purely speculative; we have neither isolated nor identified any of the intermediates shown in Scheme 8. Speculative as it is, however, this scheme suggested to us that other 18e transition metal polyhydride systems, which should be capable of affording reactive, ligand-deficient, 14e intermediates in the same way, might also catalyse the conversion of alkanes into alkenes. We have therefore examined a variety of such systems in the hope of finding one which would deactivate less readily and thus lead to a greater number of catalytic turnovers, and we have found that some iridium and ruthenium polyhydrides are indeed more effective (and in some cases more selective) catalysts for the dehydrogenation of saturated hydrocarbons.

IRIDIUM AND RUTHENIUM POLYHYDRIDES

The three best systems we have tried so far are shown in Scheme 9 and compared with the
rhenium system. We have mainly used cyclo-octane as the substrate, since it gave the best results with the rhenium system, it has a convenient boiling point, and cyclo-octene is particularly easy to determine by g.c. The last column in Scheme 9 shows the number of catalytic

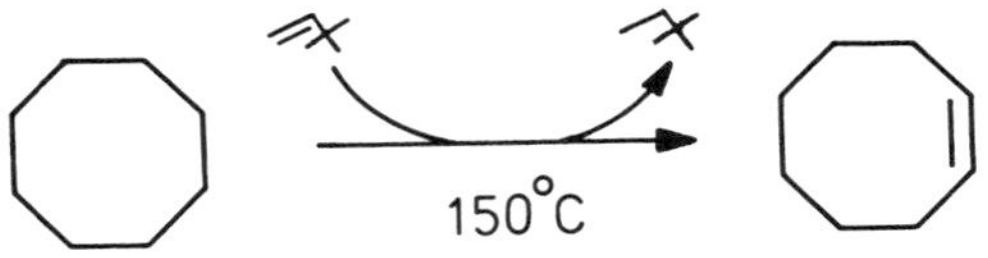

$(Ar_3P)_2ReH_7$	$<$ 10 min	$\sim$ 10
$(Ar_3P)_3RuH_4$	10 days	45 - 55
$(Ar_3P)_2IrH_5$	5 days	30 - 45
$(iPr_3P)_2IrH_5$	3 hours	23
	5 days	70

Ar = $p-F-C_6H_4$

Scheme 9.   Catalytic Dehydrogenation of Cyclo-octane by Various
Transition Metal Polyhydride Systems.

turnovers achieved (moles of cyclo-octene formed per mole of transition metal polyhydride).
The rhenium system is more reactive (the reaction is essentially instantaneous at $150^{\circ}C$), but
the iridium and ruthenium systems are more efficient in that they achieve a greater number of
catalytic turnovers before becoming deactivated. We imagine that these catalytic dehydrogena-
tions all take place by the same type of mechanism as is shown in Scheme 8 for the rhenium
system.

Unlike the rhenium system, which converts n-alkanes non-catalytically into dienerhenium trihydrides (Schemes 5 and 6), bis(tri-isopropylphosphine)iridium pentahydride catalyses the dehydrogenation of n-alkanes into alkenes. Thus (Scheme 10), n-hexane gives a mixture of n-he-
xenes (with about 9 catalytic turnovers), the major product (about 75%) being trans-2-hexene.
Unfortunately, under the same conditions, this system catalyses the isomerisation of 1-hexene

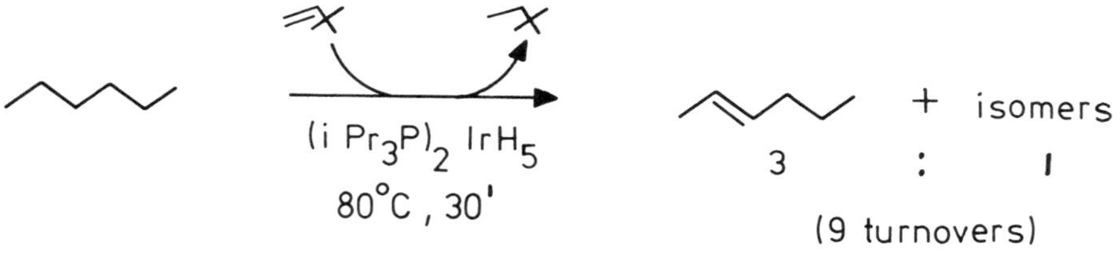

Scheme 10.   Catalytic Dehydrogenation of n-Hexane.

into trans-2-hexene, so that here again, as in the rhenium system, we cannot tell whether the
initial insertion into a C-H bond of the n-alkane takes place preferentially at a methyl or
at a methylene group.

SELECTIVITY

With methylcyclohexane as the substrate, isomerisation of the various olefins formed does not occur at all with the rhenium system, and is relatively slow with the iridium system; it is therefore possible to deduce the site of initial attack from the structure of the olefins. The results are shown in Scheme 11.

In the case of the rhenium system, the proportions of olefins formed suggest that the "intrinsic" reactivity sequence is methyl>methylene>methine, and that this is counteracted by a strong steric effect which directs the reagents, especially those containing bulky (triaryl) phosphines, away from the more hindered sites (i.e., the methyl and 2- and 6-methylene groups in methylcyclohexane).

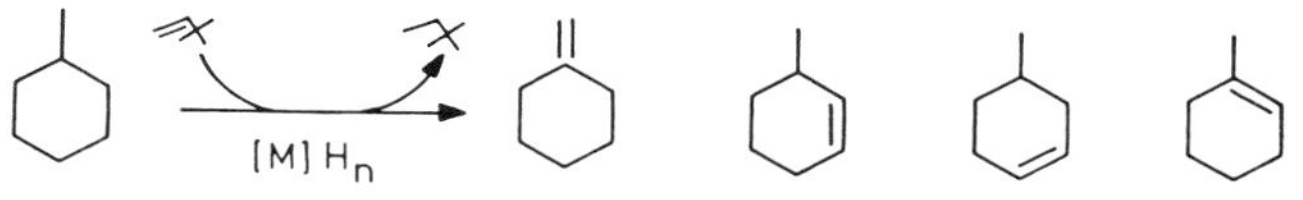

$[M]H_n$	conditions	turn-overs		proportions			
$(Ar_3P)_2ReH_7$	80°C, 10'	3	6	29	65	<1	
$(Et_3P)_2ReH_7$	80°C, 2h	0.2	28	27	45	<1	
$(iPr_3P)_2IrH_5$	100°C, 15'	3	64	10	16	9	
	100°C, 1h	3.5	55	12	18	15	
	150°C, 65h	10	1	13	24	62	

Scheme 11. Dehydrogenation of Methylcyclohexane by Transition Metal Polyhydride Systems: Selectivities.

In the case of the iridium system, the reactivity sequence methyl>methylene>methine is clearcut. Extrapolation to zero time shows that the kinetic mixture of olefins consists of >70% of methylenecyclohexane and <5% of 1-methylcyclohexene.

CONCLUSIONS AND OUTLOOK

The work I have outlined establishes that the selective, non-radical, "activation" and functionalisation of unstrained saturated hydrocarbons by means of soluble transition metal catalysts under mild conditions is a feasible process. Our approach to the conversion of alkanes into alkenes by means of transition metal polyhydrides appears to be quite general. We regard catalysis in these reactions as a desirable feature (especially in view of the cost of some transition metals), and we believe that systems such as the ones I have described, which are capable of forming 14e intermediates (Scheme 8), are on the whole more likely to lead to efficient catalysis than systems involving only 16e intermediates. Not only are 14e complexes expected to be more reactive than related 16e complexes in the key C-H insertion step, but catalysis requires the regeneration of a reactive intermediate, and this can take place when the product of the C-H insertion step is itself still co-ordinatively unsaturated (16e) and therefore able to undergo further reaction (e.g., β-elimination, Scheme 8).

Although the selectivities achieved (preferential functionalisation of methyl groups) are in some cases remarkably high, the systems I have described do not at the moment constitute practical propositions for synthetic purposes; the number of catalytic turnovers is not large, and the conversions are still abysmally low (< 1%), since the hydrocarbon substrates serve at the same time as solvents for the reaction and are therefore necessarily present in very large excess. What we would like to find is a system which does not readily undergo deactivation, which does not isomerise olefins, and which will, if possible, tolerate at least some functional groups. We obviously also need to find a solvent which will not itself react with a reagent capable of activating C-H bonds in saturated hydrocarbons (preliminary experiments leading to the functionalisation of cholestane suggest that hexamethyldisiloxane may be such a solvent).

Soluble transition metal systems may ultimately be developed into useful reagents for functionalising isolated, unactivated, methyl groups, but we clearly still have a long way to go before reaching this goal, and I am afraid that my presence today at a conference on organic synthesis is slightly premature.

Asymmetric synthesis catalyzed by chiral ferrocenylphosphine-transition metal complexes

Tamio Hayashi

Department of Synthetic Chemistry, Faculty of Engineering
Kyoto University, Kyoto 606, Japan

Abstract – Optically active allylsilanes (up to 95% ee) which contain an asymmetric carbon atom directly bonded to the silicon atom have been prepared by asymmetric cross-coupling of α-(trimethylsilyl)benzyl or 1-(trialkylsilyl)ethyl Grignard reagent with alkenyl bromides in the presence of a chiral ferrocenylphosphine-palladium complex ((R)-(S)-PPFA-Pd) as a catalyst. New chiral ferrocenylphosphine ligands containing a functional group remote from the phosphino groups have been found to be effective for the palladium-catalyzed allylation of the enolate anion of prochiral active methine compounds such as 2-acetylcyclohexanone to give the allylated products of up to 82% ee.

INTRODUCTION

Asymmetric carbon-carbon bond-forming reactions are of great significance for the synthesis of optically active compounds, and the use of chiral transition-metal catalysts for such reactions has attracted considerable attention owing to a number of advantages of catalytic asymmetric synthesis (Ref. 1). One of the most crucial points in obtaining high stereoselectivity in the catalytic asymmetric synthesis is the choice of the chiral ligand which will enable the catalyst for a given reaction to be as efficient in stereoselectivity as possible. We have designed and prepared several kinds of optically active phosphine ligands (Ref. 2 & 3) and shown that some of the phosphine ligands are effective for asymmetric hydrogenation of olefins (Ref. 4), ketones (Ref. 5), and imines (Ref. 6), asymmetric hydrosilylation of ketones (Ref. 7) and olefins (Ref. 8), asymmetric cross-coupling (Ref. 3, 9, 10, & 11), and asymmetric allylation (Ref. 12). Here we describe some recent results of the asymmetric cross-coupling forming optically active allylsilanes and asymmetric allylation catalyzed by chiral ferrocenylphosphine-palladium complexes with emphasis on the design of chiral phosphine ligands.

CHIRAL FERROCENYLPHOSPHINES

Chiral ferrocenylphosphines (Ref. 2 & 13) were readily prepared by way of lithiation of optically resolved N,N-dimethyl-1-ferrocenylethylamine (FA). The lithiation was previously reported by Ugi and coworkers to proceed with high stereoselectivity (Ref. 14). (R)-N,N-Dimethyl-1-[(S)-2-(diphenylphosphino)ferrocenyl]ethylamine ((R)-(S)-PPFA) was obtained by

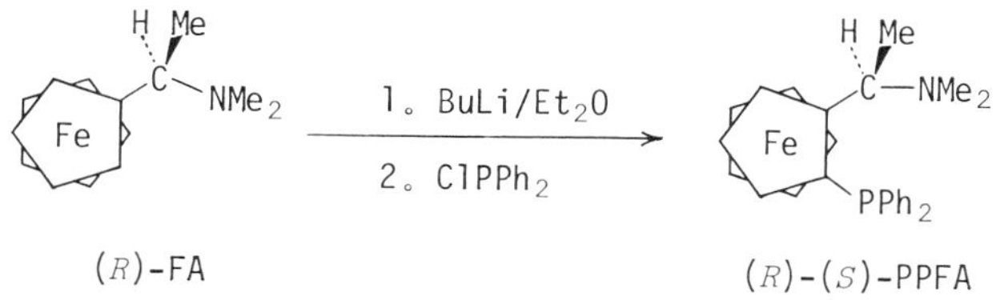

$$(1)$$

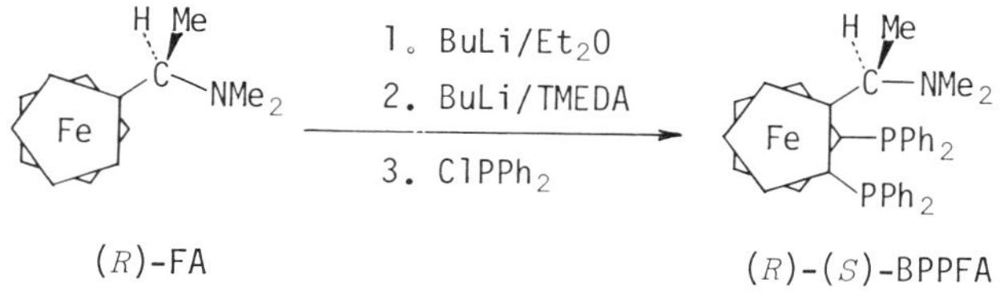

$$(2)$$

35

diphenylphosphination of the lithiated ferrocene generated by the reaction of (R)-FA with butyllithium (eq. 1). The stepwise lithiation of (R)-FA with butyllithium in ether and with butyllithium in TMEDA followed by treatment with chlorodiphenylphosphine led to the introduction of two diphenylphosphino groups, one onto each of the cyclopentadienyl rings to give (R)-(S)-BPPFA (eq. 2).

In designing chiral phosphine ligands we have attached importance to introduction of a functional group into the ligand that, where possible, would be expected to interact attractively with a functional group in the substrate. The chiral ferrocenylphosphines are quite suitable for the modification (Ref. 13). Thus, the dimethylamino group in PPFA or BPPFA can be readily substituted by other amino groups via the acetate ($\underline{1}$) or ($\underline{2}$), respectively, as shown in eq. 3. Hydroxy and alkoxy groups can be also introduced into the side chain of the ferrocene. The nucleophilic substitution reactions proceeded with complete retention of configuration, as have been usually observed in nucleophilic substitution at the α position of ferrocenylethane derivatives.

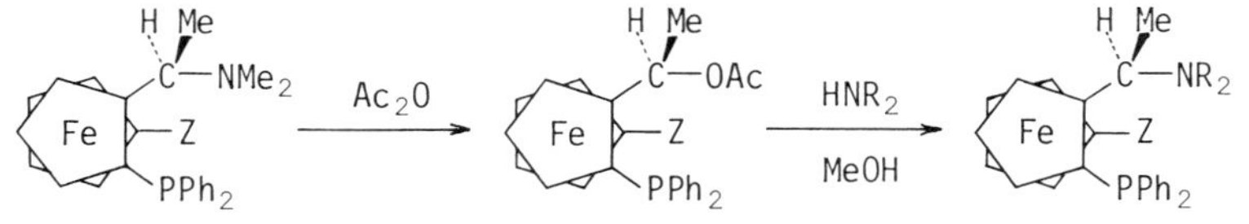

$$Z = H, \quad PPFA \qquad (\underline{1})$$
$$Z = PPh_2, BPPFA \qquad (\underline{2})$$

In addition to the easy modification of the functional groups on the ferrocene side chain, the chiral ferrocenylphosphines have the following unique and significant features: (1) They all contain a planar element of chirality that does not racemize or epimerize under usual reaction conditions. (2) Both mono- and bisphosphines can be prepared from the same chiral source, simply by changing the lithiation procedure. (3) They can be isolated and purified very readily because of their stability in air, their good crystallizability, and orange color, making chromatographic separation easy. (4) They may be regarded as triarylphosphines, which are often more favorable ligands for transition metal-catalyzed reactions than alkylphosphines.

ASYMMETRIC CROSS-COUPLING

Asymmetric cross-coupling of secondary alkyl Grignard reagents with alkenyl halides which has been effected by chiral phosphine-nickel or -palladium catalyst is now recognized to provide an efficient route to the synthesis of optically active olefins, which could hardly be obtained by other methods (Ref. 3, 9, 10, 11, & 15). The asymmetric cross-coupling may be classified as kinetic resolution of the racemic reagent, but the optical purity of the coupling product is not affected appreciably by the degree of conversion of the Grignard reagent since the inversion of the Grignard reagent is relatively fast compared with the coupling reaction (Scheme 1).

Scheme 1

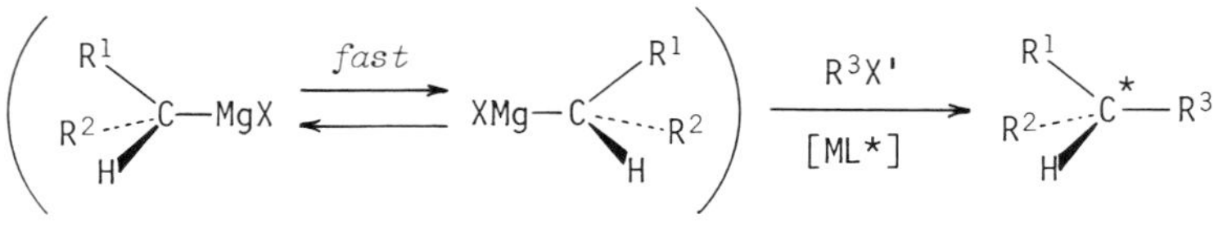

racemic   optically active

$$PhMeCHMgCl \quad + \quad CH_2=CHBr \xrightarrow{\;L*/NiCl_2\;} PhMe\overset{*}{C}HCH=CH_2 \qquad (4)$$

$$L* = (R)-(S)-PPFA: 68\% \text{ ee } (S)$$
$$(S)-Valphos: \quad 81\% \text{ ee } (S)$$

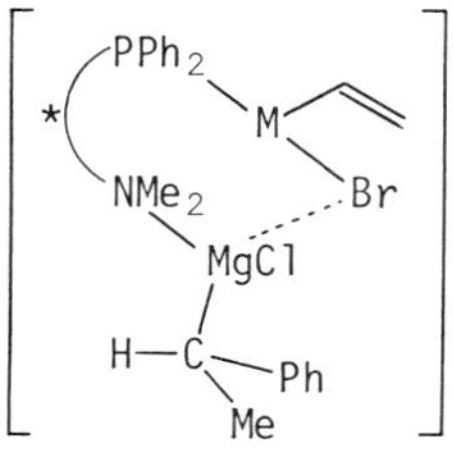

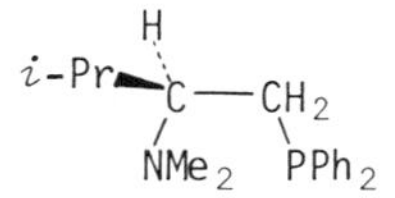

(S)-Valphos

We have examined various types of chiral phosphine ligands for the reaction of secondary alkyl Grignard reagents represented by (1-phenylethyl)magnesium chloride and found that chiral ferrocenylphosphines with secondary amino groups such as dimethylamino and some of (β-aminoalkyl)phosphines are very effective giving rise to the coupling products of over 60% ee (Ref. 3 & 10) (eq. 4).

The important role of the amino group on the ligands may be visualized by its strong ability to coordinate with the magnesium atom in the Grignard reagent. The coordination in a dia-stereomeric transition state, as exemplified by (3), must increase the stereoselectivity via the enhanced steric interaction. It is most probable that the optical purity and configuration of the coupling product are mainly determined by the transmetallation of alkyl group from the Grignard reagent to the transition-metal catalyst.

By using the asymmetric cross-coupling, we have succeeded in an efficient synthesis of optically active allylsilanes with high optical purity, which contain an asymmetric carbon atom directly bonded to the silicon atom and could hardly be obtained by other methods (Ref. 16). For the cross-coupling of α-trimethylsilylbenzylmagnesium bromide (4) with vinyl bromide (5a), we have examined several nickel and palladium catalysts with chiral phosphine ligands. Dichloro[(R)-N,N-dimethyl-1-[(S)-2-(diphenylphosphino)ferrocenyl]ethylamine]-palladium(II) (PdCl$_2$[(R)-(S)-PPFA]) (Ref. 10) was found to be the most effective catalyst giving in a high yield the coupling product (-)-3-phenyl-3-trimethylsilylpropene (6a) with the highest optical rotation. Dichloro[(S)-2-dimethylamino-3-methyl-1-diphenylphosphino-butane]palladium(II) (PdCl$_2$[(S)-Valphos]) (Ref. 3) was also an active catalyst, but the stereoselectivity was much lower. Nickel complexes NiCl$_2$[(S)-prophos] (Ref. 17) and NiCl$_2$[(-)-DIOP] (Ref. 18) showed lower catalytic activity and stereoselectivity. Nickel complexes with (S)-Valphos and (R)-(S)-PPFA, both of which have been successfully used for the cross-coupling of 1-phenylethylmagnesium chloride (Ref. 3 & 9), were almost inactive for the present reaction. The stereoselectivity of the cross-coupling catalyzed by PdCl$_2$[(R)-(S)-PPFA] was strongly dependent on the reaction temperature, the lower temperature giving the higher selectivity. Thus, the reaction at 50°C, 30°C, 19°C, and 0°C gave (6a) of 56%, 60%, 85%, and 95% ee, respectively. The palladium-(R)-(S)-PPFA catalyst was also effective for the reaction of (4) with (E)- and (Z)-1-bromopropene (5b) and (E)- and (Z)-β-bromostyrene (5c) to afford the corresponding coupling products (5b) and (5c) in an optically active form

$$Me_3Si\text{-}CH(Ph)\text{-}MgBr \quad + \quad \underset{R^2}{\overset{R^1}{>}}\!\!=\!\!\text{Br} \quad \xrightarrow{PdCl_2[(R)\text{-}(S)\text{-}PPFA]} \quad (5)$$

(4)

(5a): R^1 = R^2 = H

(E)-(5b): R^1 = Me, R^2 = H

(Z)-(5b): R^1 = H, R^2 = Me

(E)-(5c): R^1 = Ph, R^2 = H

(Z)-(5c): R^1 = H, R^2 = Ph

(6a): R^1 = R^2 = H

(E)-(6b): R^1 = Me, R^2 = H

(Z)-(6b): R^1 = H, R^2 = Me

(E)-(6c): R^1 = Ph, R^2 = H

(Z)-(6c): R^1 = H, R^2 = Ph

(7a): R = H

(7b): R = Me

(7c): R = Ph

PdCl$_2$[(R)-(S)-PPFA]

TABLE 1. Asymmetric cross-coupling of the Grignard reagent (4) with bromides (5) catalyzed by PdCl$_2$[(R)-(S)-PPFA]

alkenyl bromide (5)	temp (°C)	time (days)	product (yield %)	$[\alpha]_D^{20}$ of (6) (c in C$_6$H$_6$)	% ee (config'n)	$[\alpha]_D^{20}$ of (7) (c in C$_6$H$_6$)
(5a)	30	4	(6a) (79)	-42.9° (5.0)	66 (R)	
(5a)	19	4	(6a) (63)	-55.2 (5.0)	85 (R)	
(5a)	0	4	(6a) (42)	-61.8 (5.0)	95 (R)	-1.35° (2.6)
(E)-(5b)	0	5	(E)-(6b) (77)	-36.3 (5.1)	85 (R)	+8.10 (3.8)
(Z)-(5b)	0	5	(Z)-(6b) (38)	-21.3 (4.5)	24 (R)	+3.30 (6.4)
(E)-(5c)	0	2	(E)-(6c) (93)	-43.9 (1.0)	95 (R)	-2.3 (5.6)
(Z)-(5c)	15	2	(Z)-(6c) (95)	-44.3 (1.0)	13 (R)	-0.3 (1.5)

without $\underline{E}$-$\underline{Z}$ isomerization of the olefinic double bond (eq. 5). The results are summarized in Table 1. The $\underline{R}$ isomer was preferentially produced in every case. Bromides of $\underline{E}$ configuration, $(\underline{E})$-$(\underline{5b})$ and -$(\underline{5c})$ led to the $(\underline{E})$-allylsilanes $(\underline{E})$-$(\underline{6b})$ and -$(\underline{6c})$ with high optical purity (85% and 95%, respectively) while $(\underline{Z})$-bromides to $(\underline{Z})$-allylsilanes with lower optical purity. The selectivity attained here in the reaction of $(\underline{5a})$ and $(\underline{E})$-$(\underline{5c})$, 97% $\underline{R}$ selectively, is among the highest of asymmetric reactions by means of chiral catalysis, especially for carbon-carbon bond forming reactions (Ref. 1).

The configuration $\underline{R}$ and enantiomeric purity of the allylsilanes $(\underline{6})$ were determined by comparing the optical rotation data of alkyltrimethylsilanes $(\underline{7})$ obtained by hydrogenation of $(\underline{6})$ with those of $(\underline{7})$ obtained by palladium-catalyzed asymmetric hydrosilylation of styrene derivatives $(\underline{8})$ (eq. 6). Since the oxidative cleavage of the carbon-silicon bond in trifluorosilylalkanes with m-chloroperoxybenzoic acid (MCPBA) forming the carbon-oxygen bond has been known to proceed with complete retention of configuration (Ref. 19), the trimethylalkylsilanes $(\underline{7a\text{-}c})$ derived from trichloroalkylsilanes $(\underline{9a\text{-}c})$ should have the same configuration and enantiomeric purity as the alcohols $(\underline{8a\text{-}c})$ have.

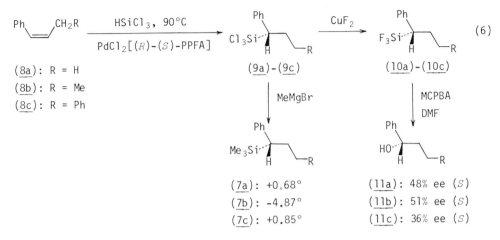

$(\underline{8a})$: R = H
$(\underline{8b})$: R = Me
$(\underline{8c})$: R = Ph

$(\underline{9a})$-$(\underline{9c})$

$(\underline{10a})$-$(\underline{10c})$

(6)

$(\underline{7a})$: +0.68°
$(\underline{7b})$: -4.87°
$(\underline{7c})$: +0.85°

$(\underline{11a})$: 48% ee $(\underline{S})$
$(\underline{11b})$: 51% ee $(\underline{S})$
$(\underline{11c})$: 36% ee $(\underline{S})$

As an example, the case of $(-)$-$(\underline{3a})$ is described. Firstly, $(\underline{Z})$-1-phenylpropene $(\underline{8a})$ was hydrosilylated with trichlorosilane in the presence of $PdCl_2[(\underline{R})$-$(\underline{S})$-PPFA] as catalyst at 90°C. The resultant 1-phenyl-1-trichlorosilylpropane $(\underline{9a})$ was devided in two parts. One part was converted by treatment with CuF_2 to the corresponding trifluorosilyl derivative $(\underline{10a})$, which in turn was transformed into 1-phenyl-1-propanol $(\underline{11a})$ by oxidation with MCPBA in DMF. From this reaction, $(\underline{S})$-$(-)$-$(\underline{11a})$ with 48% ee was obtained. The other half of the trichlorosilyl derivative $(\underline{9a})$ was treated with MeMgBr to give $(\underline{7a})$ with specific rotation $[\alpha]_D^{20}$ +0.68°, having the same configuration $\underline{S}$ as the alcohol. From those two series of reactions it follows that the maximum rotation value $[\alpha]_D^{20}$ of $(\underline{S})$-$(+)$-$(\underline{7a})$ should be +1.42°. Next, the cross-coupling product $(-)$-$(\underline{3a})$ with specific rotation $[\alpha]_D^{20}$ -60.1° was hydrogenated over the Pd/C catalyst to give $(\underline{7a})$ with $[\alpha]_D^{20}$ -1.35°. Consequently, the $(-)$-$(\underline{6a})$ has $\underline{R}$ configuration and 95% ee. In the same way as above, the maximum rotations of the alkylsilanes $(\underline{S})$-$(\underline{7b})$ and $(\underline{S})$-$(\underline{7c})$ are calculated to be $[\alpha]_D^{20}$ -9.55°, and +2.36° ($\underline{c}$ 3-6, benzene), respectively.

Optically active allylsilanes $(\underline{13})$ with methyl group at the chiral carbon atom have been also prepared by the asymmetric cross-coupling of 1-(trialkylsilyl)ethylmagnesium chloride $(\underline{12})$ in the presence of the palladium-$(\underline{R})$-$(\underline{S})$-PPFA catalyst (Ref. 20) (eq. 7). The stereoselectivity

(12)

(13)

(7)

R^1 = Ph, R^2 = H, R$_3$Si = Me$_3$Si: 71% ee $(\underline{S})$
                              PhMe$_2$Si: 68% ee $(\underline{S})$
                              Ph$_3$Si: 36% ee $(\underline{S})$
                              Et$_3$Si: 93% ee $(\underline{S})$

R^1 = H, R^2 = Ph, R$_3$Si = Me$_3$Si: 59% ee $(\underline{S})$
R^1 = Me, R^2 = H, R$_3$Si = Et$_3$Si: 85% ee $(\underline{S})$

of the reaction was affected by substituents on the silyl group of the Grignard reagent ($\underline{12}$), triethylsilyl group being generally the best to give the allylsilane ($\underline{13}$) of 93% ee.

Allylsilanes are well-recognized to be useful intermediates in organic synthesis, reacting with a wide range of electrophiles in a regiospecific manner (Ref. 21). Use of the optically active allylsilanes obtained above for the electrophilic substitution reaction (S_E') produced various kinds of optically active compounds by an asymmetric induction and demonstrated that the S_E' reaction proceeds with anti stereochemistry (Ref. 16 & 22).

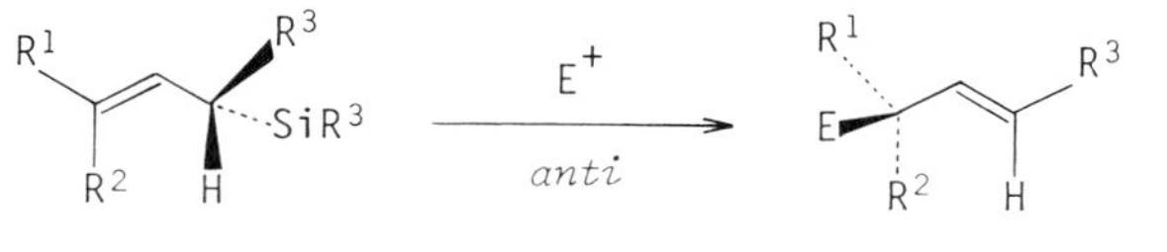

$$E = t\text{-Bu } (t\text{-BuCl/TiCl}_4)$$
$$MeCO \ (MeCOCl/AlCl_3)$$
$$HOCH_2 \ (trioxane/TiCl_4)$$
$$D \ (CF_3COOD), \ HO \ (MCPBA)$$

ASYMMETRIC ALLYLATION

There has been considerable synthetic and mechanistic interest in palladium-catalyzed allylation of nucleophiles with allylic compounds represented by allyl acetates (Ref. 23). The catalytic cycle of the allylation is generally accepted to involve a π-allylpalladium(II) complex as a key intermediate, which is formed by oxidative addition of an allylic substrate to palladium(0) and undergoes nucleophilic attack to yield allylation product and to regenerate palladium(0) (Scheme 2). Studies on stereochemistry of the catalytic allylation and stoichiometric reaction of π-allylpalladium complexes with nucleophiles have demonstrated that soft nucleophiles such as dimethyl sodiomalonate attack a carbon atom of the π-ally from the side opposite to the palladium (Ref. 24).

Scheme 2

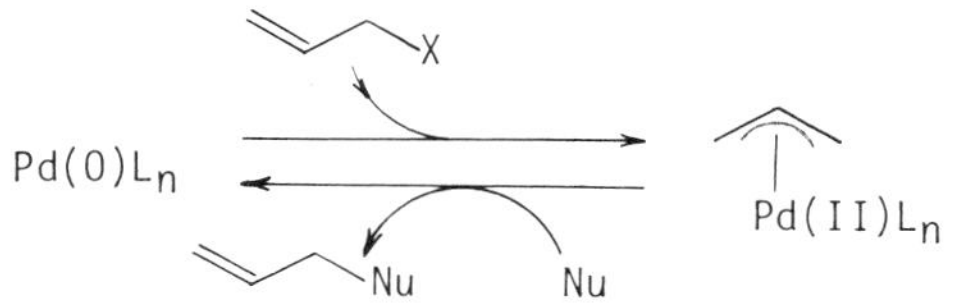

In 1978, Kagan and coworkers have reported that reaction of prochiral active methine compounds with allylic ethers in the presence of a DIOP-palladium catalyst gave optically active allylated products of up to 10% ee (Ref. 25). The low stereoselectivity may be ascribed to the use of DIOP as a chiral source. DIOP is the ligand that possibly controls the stereochemistry by orientation of the phenyl rings on the phosphorus and the phenyl rings are too distant from the developing asymmetric center since the prochiral substrate is the soft nucleophile attacking the face of π-allyl opposite to the palladium (($\underline{A}$) in Scheme 3).

Scheme 3

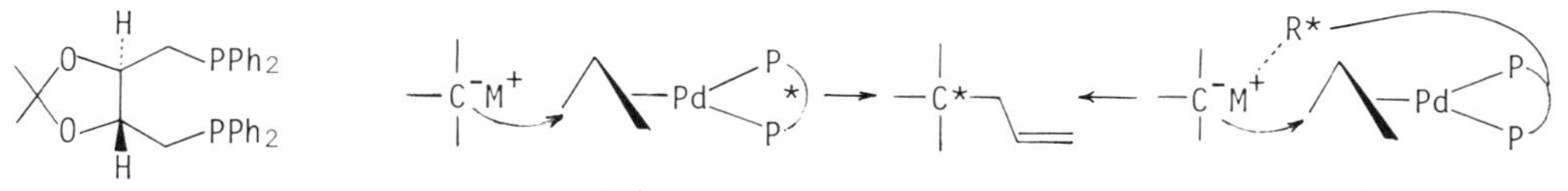

(-)-DIOP   ($\underline{A}$)   ($\underline{B}$)

We have designed and prepared, for the asymmetric allylation, new phosphine ligands ($\underline{13}$) and ($\underline{14}$) which possess a chiral functional group at an appropriate distance from the coordinated phosphino groups (Ref. 12). The chiral functional group is expected to interact with the nucleophile to bring about high stereoselectivity (($\underline{B}$) in Scheme 3). The phosphine ligands ($\underline{13a}$) and ($\underline{13b}$), which contain valine methyl ester and ephedrine, respectively, were used for the palladium-catalyzed reaction (at $-30\sim-50°C$) of sodium enolate of 2-acetylcyclohexane ($\underline{15}$) with allyl acetate (eq. 8), to give ($\underline{16}$) with 52% ee ($\underline{R}$) and 31% ee ($\underline{R}$), respectively. The importance of the distance between the chiral functional group and the phosphino groups coordinated to the palladium is indicated by lower selectivity (5% ee) of ($\underline{14}$) which is analogous to ($\underline{13a}$) but with glutaramide instead of succinamide.

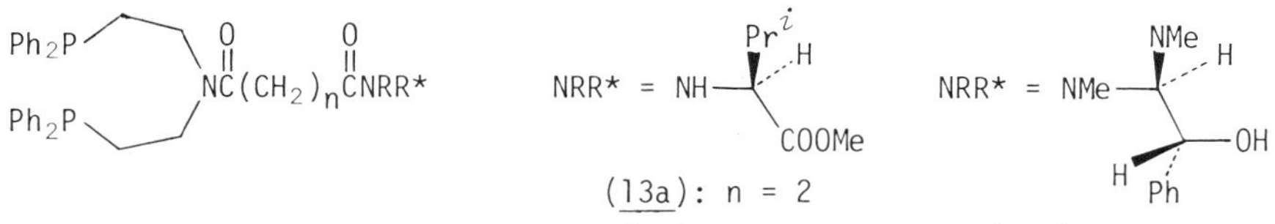

($\underline{13a}$): n = 2
($\underline{14}$): n = 3

($\underline{13b}$): n = 2

$$(8)$$

To obtain higher stereoselectivity, we have prepared chiral ferrocenylphosphine ligands with various kinds of substituents on the side chain, and examined them for their stereoselectivity in the palladium-catalyzed asymmetric allylation of (15) (eq. 8). The results are summarized in Table 2. The ligand (17a) which contains (2-hydroxyethyl)amino group was found most active and stereoselective to produce (16) of up to 81% ee (S). Table 2 contains the following significant features. (1) The phosphine ligands (17a)-(17d) gave (S)-(16) of over 50% ee. They all have 2-hydroxyethyl group on nitrogen or oxygen atom on the ferrocene side chain. (2) Lower stereoselectivity was observed in the reaction with (17g) which is analogous to (17b) but has hydroxy group on the tertiary carbon atom. (3) The longer distance between hydroxy group and the ferrocene moiety lowered the stereoselectivity. Thus, the ligand (17f) with 3-hydroxypropylamino group gave (S)-(16) of 46% ee, and the ligand (17j) with 2-(2-hydroxyethyloxy)ethylamino group gave racemic (16). The shorter distance also lowered the stereoselectivity, as shown in the reaction with (17h) where the hydroxy group is located in the ferrocenylmethyl position. (4) Replacement of hydroxy group on (17a) or (17b) by methoxy or amino group resulted in the formation of (16) with opposite configuration R in low optical yield. The similar selectivity was observed with the ligands (2), (BPPFA), and (17l), the last lacking functional groups on the side chain.

(R)-(S)-BPPF-X (17)

X = NR$_2$, OR, etc.

TABLE 2. Asymmetric allylation of 2-acetylcyclohexanone (15) with allyl acetate catalyzed by Pd/(R)-(S)-BPPF-X.

-X in (R)-(S)-BPPF-X (17)		temp (°C)	time (h)	conversion (%)	$[\alpha]_D^{20}$(deg) of (16) (chloroform)	% ee (config'n)
-NMeCH$_2$CH$_2$OH	(17a)	15	21	100	+40	16 (S)
		0	38	100	+69	27 (S)
		-10	44	100	+107	42 (S)
		-30	18	100	+134	53 (S)
		-50	20	100	+185	73 (S)
		-60	44	100	+205	81 (S)
-NHCH$_2$CH$_2$OH	(17b)	-50	17	13	+156	62 (S)
-N(CH$_2$CH$_2$OH)$_2$	(17c)	-50	17	61	+156	62 (S)
-OCH$_2$CH$_2$OH	(17d)	-50	20	81	+135	53 (S)
-NMeCH(CH$_2$OH)$_2$	(17e)	-50	18	55	+124	49 (S)
-NH(CH$_2$)$_3$OH	(17f)	-50	17	11	+117	46 (S)
-NHCH$_2$CMe$_2$OH	(17g)	-30	42	93	+79	31 (S)
-OH	(17h)	-30	17	89	+75	30 (S)
-NH$_2$	(17i)	-30	17	64	+39	15 (S)
-NHCH$_2$CH$_2$OCH$_2$CH$_2$OH	(17j)	-30	17	36	+1	0
-NHCH$_2$CH$_2$OMe	(17k)	-30	16	100	-15	6 (R)
-OCOMe	(2)	-30	17	44	-41	16 (R)
-Me	(17l)	-30	50	100	-47	19 (R)
-NMeCH$_2$CH$_2$NHMe	(17m)	-30	18	100	-50	20 (R)
-NMe$_2$	(BPPFA)	-30	16	100	-57	22 (R)

The features described above indicate that the hydroxy group located in a position appropriately apart from the ferrocenylbisphosphine moiety is playing a key role for high stereoselectivity in the asymmetric allylation. The high efficiency of the ligand (17a) may be illustrated by Fig. 1 where the enantioface of the enolate of (15) is differentiated effectively by the coordination of the hydroxy group to the sodium ion.

Fig. 1

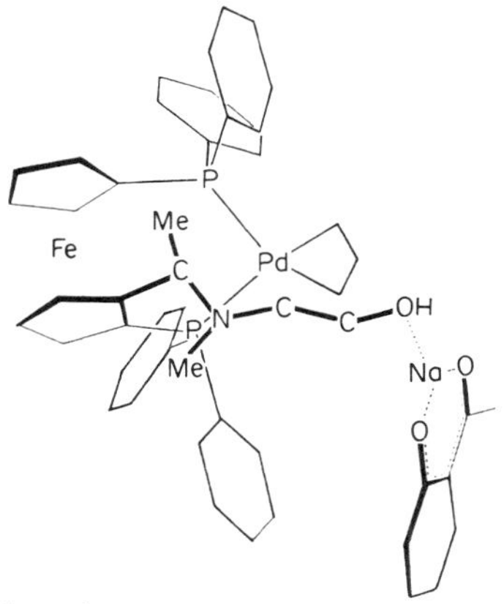

The palladium catalyst complexed with the ferrocenylphosphine (17a) was also effective for the asymmetric allylation of several other methine compounds. The products and their enantiomeric purities are shown below.

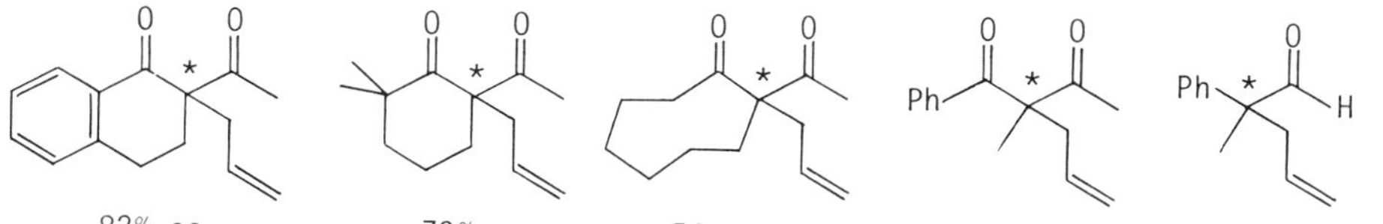

82% ee   70% ee   58% ee   60% ee   53% ee

REFERENCES

1. Reviews: H.B. Kagan and J.C. Fiaud, Top. Stereochem. 10, 175-285 (1978). B. Bosnich and M.D. Fryzuk, Top. Stereochem. 12, 119-154 (1981).
2. T. Hayashi, T. Mise, M. Fukushima, M. Kagotani, N. Nagashima, Y. Hamada, A. Matsumoto, S. Kawakami, M. Konishi, K. Yamamoto and M. Kumada, Bull. Chem. Soc. Jpn. 53, 1138-1151 (1980).
3. T. Hayashi, M. Konishi, M. Fukushima, K. Kanehira, T. Hioki and M. Kumada, J. Org. Chem. 48, 2195-2202 (1983).
4. T. Hayashi, T. Mise, S. Mitachi, K. Yamamoto and M. Kumada, Tetrahedron Lett. 1133-1134 (1976).
5. T. Hayashi, T. Mise and M. Kumada, Tetrahedron Lett. 4351-4354 (1976). T. Hayashi, A. Katsumura, M. Konishi and M. Kumada, Tetrahedron Lett. 425-428 (1979).
6. T. Hayashi and M. Kumada, in Fundamental Research in Homogeneous Catalysis, Vol. 2, Y. Ishii and M. Tsutsui, Ed., Plenum, New York, 159-180 (1978).
7. T. Hayashi, K. Yamamoto and M. Kumada, Tetrahedron Lett. 4405-4408 (1974).
8. T. Hayashi, K. Tamao, Y. Katsuro, I. Nakae and M. Kumada, Tetrahedron Lett. 21, 1871-1874 (1980). T. Hayashi, K. Kabeta, T. Yamamoto, K. Tamao and M. Kumada, Tetrahedron Lett. 24, 5661-5664 (1983).
9. T. Hayashi, M. Tajika, K. Tamao and M. Kumada, J. Am. Chem. Soc. 98, 3718-3719 (1976). T. Hayashi, M. Konishi, M. Fukushima, T. Mise, M. Kagotani, M. Tajika and M. Kumada, J. Am. Chem. Soc. 104, 180-186 (1982). T. Hayashi, M. Konishi, T. Hioki, M. Kumada, A. Ratajczak and H. Niedbała, Bull. Chem. Soc. Jpn. 54, 3615-3616 (1981). T. Hayashi, in Asymmetric Reactions and Processes in Chemistry, E.L. Eliel and S. Otsuka, Ed., American Chemical Society, Washington, D.C., ACS Symp. Ser. No. 185, Chap. 12. (1982).
10. T. Hayashi, T. Hagihara, Y. Katsuro and M. Kumada, Bull. Chem. Soc. Jpn. 56, 363-364 (1983).
11. T. Hayashi, N. Nagashima and M. Kumada, Tetrahedron Lett. 21, 4623-4626 (1980). T. Hayashi, K. Kanehira, T. Hioki and M. Kumada, Tetrahedron Lett. 22, 137-140 (1981).
12. T. Hayashi, K. Kanehira, H. Tsuchiya and M. Kumada, J. Chem. Soc., Chem. Commun. 1162-1164 (1982).
13. T. Hayashi and M. Kumada, Acc. Chem. Res. 15, 395-401 (1982).
14. D. Marquarding, H. Klusacek, G. Gokel, P. Hoffmann and I. Ugi, J. Am. Chem. Soc. 92, 5389-5393 (1970).

15. G. Consiglio, F. Morandini and O. Piccolo, Tetrahedron 39, 2699-2707 (1983).
16. T. Hayashi, M. Konishi, H. Ito and M. Kumada, J. Am. Chem. Soc. 104, 4962-4963 (1982).
17. M.D. Fryzuk and B. Bosnich, J. Am. Chem. Soc. 100, 5491-5494 (1978).
18. H.B. Kagan and T.P. Dang, J. Am. Chem. Soc. 94, 6429-6433 (1972).
19. K. Tamao, T. Kakui, M. Akita, T. Iwahara, R. Kanatani, J. Yoshida and M. Kumada, Tetrahedron 39, 983 (1983).
20. T. Hayashi, Y. Okamoto and M. Kumada, unpublished.
21. Reviews: T.H. Chan and I. Fleming, Synthesis 761-786 (1979). H. Sakurai, Pure Appl. Chem. 54, 1-22 (1982).
22. T. Hayashi, M. Konishi and M. Kumada, J. Am. Chem. Soc. 104, 4963-4965 (1982). T. Hayashi, H. Ito and M. Kumada, Tetrahedron Lett. 23, 4605-4606 (1982). T. Hayashi, Y. Okamoto and M. Kumada, Tetrahedron Lett. 24, 807-808 (1983). T. Hayashi, M. Konishi and M. Kumada, J. Org. Chem. 48, 281-282 (1983). T. Hayashi, M. Konishi and M. Kumada, J. Chem. Soc., Chem. Commun. 736-737 (1983).
23. Reviews: B.M. Trost, Acc. Chem. Res. 13, 385-393 (1980). J. Tsuji, Organic Synthesis with Palladium Compounds, Springer-Verlag, New York (1980).
24. T. Hayashi, T. Hagihara, M. Konishi and M. Kumada, J. Am. Chem. Soc. 105, 7767-7768 (1983). T. Hayashi, M. Konishi and M. Kumada, J. Chem. Soc., Chem. Commun. 107-108 (1984), and references cited therein.
25. J.C. Fiaud, A. Hibon de Gournay, M. Larcheveque and H.B. Kagan, J. Organomet. Chem. 154, 175-185 (1978).
26. T. Hayashi, T. Hagihara, K. Kanehira and M. Kumada, unpublished.

Palladium-catalyzed reactions of organometallic compounds

I.P. Beletskaya

Department of Chemistry, Lomonosov Moscow State University,
Moscow, SU-119899 GSP, U.S.S.R.

Abstract - Various reactions of organotin, organomercury and organocopper compounds with organic halides resulting in C-C bond formation have been carried out using Pd-complexes as catalysts. It has
been found that reactions of organotin compounds with aryl iodides,
allyl bromide, acyl halides and also with carbon monoxide and aryl
iodide as well as the reaction of hexamethyldistannanes with aryl
iodide proceed with high yield under very mild conditions when Pd-
complexes without phosphine ligands are used as catalysts. The reactions of arylmercury halides (or diarylmercury) with aryliodide,
acyl chloride and carbon monoxide require phosphine palladium complexes and the presence of nucleophilic catalyst such as iodide-ion.
Both aryl groups of diarylmercury are transformed in these reactions.
The comparison of reactivity of organometallic compounds of magnesium, zinc, aluminium in reactions with allyl bromide, p-iodoanisole,
benzoyl chloride and its p-nitro derivative has been performed.

In 1972 two research groups in France and in Japan discovered a cross-coupling reaction of Grignard reagents with organic halides in the presence of
nickel complexes (1,2), three years later similar reaction had been carried
out with zero-valent palladium (3). It turned out to be highly stereospec-

$$RM + R'X \xrightarrow[\text{Et}_2O]{\text{NiX}_2L_2} R-R' + MX \qquad (1)$$

$$M = MgX, \text{Li}; \quad R' = \text{alkenyl, aryl}$$

ific. These reactions, which were inspired by Kharash reaction, considerably
enlarged the scope of reactions, leading to the C-C bond formation. Further
step was aimed at involvement of various organometallic compounds in these
reactions (4-6). Introduction into these reactions of organometallics toler-
ating various functional groups is of special interest. Such reactions and
possibilities of their execution under mild conditions are the subject of
the present report.

RM (M = Hg, Sn, Cu) REACTIONS WITH R'X

Reactions of organotin compounds

It has been shown earlier, that organotin compounds react with aryl and
benzyl bromides (7,8) with catalysis by palladium phosphine complexes, however these reactions require rather rigid conditions (eq.2,3).

$$ArBr + Bu_3Sn\diagup\diagdown \xrightarrow[\text{PhH, }100°]{\text{Pd(PPh}_3)_4} Ar\diagup\diagdown \qquad (2)$$
$$72 - 100\%$$

$$RBr + R'_4Sn \xrightarrow[\text{HMPA, }65°]{\text{PhCH}_2\text{PdCl(PPh}_3)_2} R-R' \qquad (3)$$
$$62 - 100\%$$

We have carried out such a reaction for a large number of organic halides
and organotins (9,10). The most active of them (vinyl- and phenylethynyltrimethyltin) react in HMPA at room temperature. The reaction of trifluoro-
vinyltrimethyltin with aryl iodides yielded substituted trifluorostyren-
es (11).

$$RX + R'SnMe_3 \xrightarrow[\text{HMPA, } 70°]{PhPdI(PPh_3)_2} \begin{array}{l} R\text{-}R' \\ 75 - 100\% \end{array} \tag{4}$$

$R = p\text{-}NO_2C_6H_4$ $X = I$ $R' = Me, CH_2=CH\ (20°), PhC\equiv C\ (20°),$
                                                  $p\text{-}YC_6H_4\ (Y = MeO, Me, H, Cl)$

$R = p\text{-}MeCOC_6H_4,$ $X = I$ $R' = Ph$
    $p\text{-}MeO_2CC_6H_4,$
    $PhC\equiv C,$
    $2\text{-}C_4H_3S$

$R = trans\text{-}PhCH=CH$ $X = Br$ $R' = Ph$
$R = p\text{-}YC_6H_4$ $X = I$ $R' = CF_2=CF$
    $(Y = MeO, Me, H, Cl)$

It turned out that this reaction could be carried out under extremely mild conditions not only in HMPA, but also in DMF and even in acetone, using more active catalyst without electron-donating phosphine ligands (eq.5) (12). We named such catalyst a "ligandless" one, though, of course, it contained ligands - solvent molecules. Such a complex can be obtained in situ from $LiPdCl_3$.

$$ArI + RSnMe_3 \xrightarrow[\text{solv., } 20°]{"Pd"} \begin{array}{l} Ar\text{-}R + Me_3SnI \\ 77 - 100\% \end{array} \tag{5}$$

$Ar = p\text{-}NO_2C_6H_4$ $R = CH_2=CH, Ph, PhC\equiv C$
$Ar = Ph, p\text{-}CNC_6H_4,$ $R = Ph$
    $2,4\text{-}(NO_2)_2C_6H_3$
$Ar = Ph$ $R = p\text{-}MeC_6H_4$
$"Pd" = ArPdIL_2$ $(from\ LiPdCl_3)$
$L = solvent$

In some cases the reaction in acetone proceeds faster, than in HMPA, probably due to facilitation of transmetallation stage in the absence of strong electron-donating ligands, which is in accord with an assumption, that for organotin compounds this stage is the rate-determining one.

A few words about the reaction mechanism. We compared the results of catalytic and stoichiometric reactions (eq.6) and found that in the latter case the cross-coupling product yield decreased sharply along with formation of some unusual products - nitrobenzene and p-methyldiphenyl. We attribute this to participation in the reaction of unsaturated complex of zero-valent palladium (eq.7). In the presence of triphenylphosphine or p-nitroiodobenzene, capable of bonding with this complex, the reaction proceeds smoothly yielding only the cross-coupling product, i.e. it doesn't differ from the catalytic version (13).

$$p\text{-}MeC_6H_4SnMe_3 + p\text{-}NO_2C_6H_4PdIL_2 \xrightarrow{-Me_3SnI} p\text{-}NO_2C_6H_4Pd(p\text{-}MeC_6H_4)L_2 \longrightarrow$$

$$\longrightarrow \underset{43\%}{p\text{-}MeC_6H_4C_6H_4NO_2\text{-}p} + PdL_2 \qquad L = PPh_3 \tag{6}$$

$$(7)$$

Another aspect is the mechanism of the reductive elimination. It is assumed that it doesn't proceed spontaneously, but through the addition of organic halides to Pd(II) with formation of hexacoordinated palladium(IV) (14-16). We have shown that in our reaction this was not the case. We compared the results of two stoichiometric reactions between organotin and organopalladium compounds in the presence of p-nitroiodobenzene. The organometallic compounds contained similar organic moieties - phenyl and m-tolyl; in one of the reactions phenyl radical was bonded with tin and m-tolyl - with palladium (eq.8), and in the other reaction - vice versa (eq.9). It turned out that the product composition was different for these two reactions, which contradicted the assumption that Pd(IV) was formed as an intermediate.

$$PhSnMe_3 + m\text{-}MeC_6H_4PdIL_2 \xrightarrow[-Me_3SnI]{p\text{-}NO_2C_6H_4I} PhC_6H_4Me\text{-}m + PhC_6H_4NO_2\text{-}p$$
$$\qquad\qquad\qquad\qquad\qquad\qquad\qquad\qquad\qquad 33\% \qquad\qquad 67\% \qquad (8)$$

$$m\text{-}MeC_6H_4SnMe_3 + PhPdIL_2 \xrightarrow[-Me_3SnI]{p\text{-}NO_2C_6H_4I} PhC_6H_4Me\text{-}m + m\text{-}MeC_6H_4C_6H_4NO_2\text{-}p$$
$$\qquad\qquad\qquad\qquad\qquad\qquad\qquad\qquad\qquad 32\% \qquad\qquad 62\% \qquad (9)$$

The product composition and ratio, and the fact that the cross-coupling product of p-nitroiodobenzene contains only the organic moiety bonded with tin can be explained by reactions (10) and (11), their overall rate being higher than that of reactions, leading to m-methyldiphenyl.

$$p\text{-}NO_2C_6H_4I + PdL_2 \longrightarrow p\text{-}NO_2C_6H_4PdIL_2 \qquad\qquad (10)$$

$$p\text{-}NO_2C_6H_4PdIL_2 + ArSnMe_3 \longrightarrow p\text{-}NO_2C_6H_4PdArL_2 \xrightarrow{-PdL_2} \qquad (11)$$
$$\longrightarrow p\text{-}NO_2C_6H_4Ar$$

So one can consider the following catalytic cycle comprised of oxidative addition to Pd(0), transmetallation and reductive elimination.

$$(12)$$

$$L = PPh_3$$

Another important group of cross-coupling reactions with carbon-carbon bond formation includes allyllation of carbanions with participation of π-allyl palladium complexes. The reactions were discovered by Tsuji (17) and developed by Trost (18). It is known that allyllation is easily performed for soft carbanions such as malonic ester anion. However it turned out that various organotins equally and easily undergo Pd-catalyzed reactions with allyl halides and allyl acetates (19-23). The use of "ligandless" palladium with allyl halides in DMF or HMPA enables to carry out the reaction under mild conditions with high product yield. For allyl acetates, which are less active in the reaction of oxidative addition (this can result in Pd-black precipitation) phosphine complexes or palladium should be used again.

$$\diagup\!\!\diagdown\!\!\diagup Br + RSnMe_3 \xrightarrow[HMPA, 20°]{"Pd"} \diagup\!\!\diagdown\!\!\diagup R \qquad (13)$$
$$\qquad\qquad\qquad\qquad\qquad\qquad 77 - 95\%$$

$$"Pd" = (\pi - C_3H_5PdCl)_2; \quad R = \diagup\!\!\diagdown, \quad p\text{-}YC_6H_4 \quad (Y = MeO, Me, H, Cl),$$
$$9\text{-}C_{13}H_9$$

$$\diagup\!\!\diagdown\!\!\diagup OAc + RSnMe_3 \xrightarrow[HMPA, 20°]{Pd(PPh_3)_4} \diagup\!\!\diagdown\!\!\diagup R \qquad (14)$$
$$\qquad\qquad\qquad\qquad\qquad\qquad 68 - 100\%$$

$$R = \text{\Large\char"2040} \quad , \quad Ph\text{\Large\char"2040}, \quad p\text{-}YC_6H_4 \quad (Y = Me, H, Cl), \; 9\text{-}C_{13}H_9, \; C_9H_7,$$

$$c\text{-}C_5H_5, \quad MeCOCH_2, \quad PhCH(COOEt),$$

Various organotin compounds were introduced into the reaction with allyl acetates, among them vinyl- and allyl-derivatives, leading to 1,4- and 1,5-dienes, aromatic compounds, fluorenyl, indenyl, cyclopentadienyl derivatives, stannilated ketones and esters, which gave products only of mono-allylsubstitution (note, that use of enolstannanes has an advantage in this respect against alkali metal enolates, giving usually a mixture of mono- and disubstituted allylation products).

The reaction with allylhalides is regioselective - one isomer is formed predominantly (23) with attack at the least hindered site. If acetone or THF are used as solvents, the reaction requires somewhat higher temperatures (55° and 68°), but proceeds without by-product formation.

$$\text{allyl-Cl} \; \text{or} \; \text{allyl-Cl} \; + \; PhSnMe_3 \; \xrightarrow{\text{"Pd"}} \; \text{$\sim$77\%} \; + \; \text{$\sim$21\%} \; + \; Ph_2 \qquad (15)$$

It is known, that participation of π-allylpalladium intermediates is assumed for Pd-catalyzed allylation. These intermediates in solution can exist in various forms (1 - 4) (eq.16). We found (24), that in the reactions of allyldemetallation of organotin compounds (the initial catalyst - tetrakis(triphenylphosphine)palladium) common ion additions (in the form of tetrabutylammonium salts) caused sharp deceleration of the reaction (eq.17). This prompts to suggest, that π-allyl intermediate takes part in the reaction in the dissociated cationic form (2). When complex (4) with one phosphine ligand was used as initial catalyst the common ion effect was very small. A conclusion can be derived from here, that complex (4) reacts in non-dissociated form and its contribution to the reaction with participation of tetrakis(triphenylphosphine)palladium is insignificant. Earlier we discovered an unusual behavior of thienyl and phenylethynyltrimethyltin in the reactions with allyl acetates and allyl halides. These compounds, usually highly reactive in electrophilic substitution (25), turned out to be quite inactive in the reactions with allyl halides and completely inert towards allyl acetate. We did not find the reason for this. However only in case of these compounds we found the positive effect of the common ion addition - the cross-coupling product yield in the reaction with allyl iodide increased more than two-fold. These results can be explained easily assuming σ-allyl complex (3) to be the active species.

$$\begin{array}{c}
Pd(PPh_3)_4 \\
\updownarrow \; -2PPh_3 \\
Pd(PPh_3)_2
\end{array}$$

$$R\text{\char"2040} + Me_3SnX \qquad \qquad \text{allyl-X} \qquad (16)$$

$$RSnMe_3 \qquad \left[\text{allyl-}Pd(PPh_3)_2\right]^+ X^-$$

$$\updownarrow \; (1)$$

$$\text{allyl-}Pd(PPh_3)_2X \qquad \left[\text{allyl-}Pd(PPh_3)_2\right]^+ + X^- \qquad \text{allyl-}Pd(PPh_3)X + PPh_3$$

$$(\underline{3}) \qquad\qquad (\underline{2}) \qquad\qquad (\underline{4})$$

RSnMe$_3$ + allyl-X		$\xrightarrow[\text{HMPA, 20°}]{\text{"Pd", Bu}_4\text{NX}}$	R$\text{\char"2040}$ + Me$_3$SnX		(17)
R	X	"Pd"	time(h)	R$\text{\char"2040}$	(%)
p-MeC$_6$H$_4$	OAc	Pd(PPh$_3$)$_4$	6	27	(100)*
p-MeC$_6$H$_4$	I	Pd(PPh$_3$)$_4$	6	0	(56)
Ph	OAC	π-C$_3$H$_5$PdCl(PPh$_3$)	25	81	(88)
Ph	I	π-C$_3$H$_5$PdCl(PPh$_3$)	30	40	(55)
2-C$_4$H$_3$S	I	Pd(PPh$_3$)$_4$	6	46	(18)

*-number in parentheses - yield without Bu$_4$NX

Organotin compounds themselves can easily be obtained by the reaction of organic halides with hexalkyldistannanes (26-29). Usage of "ligandless" palladium enables to carry out these reactions with arylhalides containing various (electron-donating and -withdrawing) substituents under mild conditions. This is a very convenient method for synthesis of nitroaryl derivatives of tin, which are difficult to obtain otherwise. It should be noted, that distannanes themselves are also easily formed by Pd-catalyzed reaction of trialkylstannanes (30) and can be used without isolation in the synthesis of organotins.

$$R_3SnCl \xrightarrow[\text{2."Pd"}]{\text{1.LiAlH}_4, \text{ Et}_2O} R_6Sn_2 \xrightarrow[\text{HMPA, }20°]{R'I, \text{ "Pd"}} R'SnR_3 \qquad (18)$$
$$68 - 100\%$$

"Pd" = $PdCl_2(MeCN)_2$

R = Me $\quad$ R' = YC_6H_4 (Y = p-MeO, H, o-, m-, p-NO$_2$, p-MeCO,

$\qquad\qquad$ o-, p-MeO$_2$C, p-CN, 2,4-(NO$_2$)$_2$)

R = Et $\quad$ R' = C_6H_5, p-NO$_2$C$_6$H$_4$

R = Bu $\quad$ R' = Me, Et, p-YC$_6$H$_4$ (Y = H, Cl, MeO$_2$C, NO$_2$), 2-C$_4$H$_3$S

Allyl derivatives of tin can be obtained by reaction of hexalkyldistannanes with allyl halides or acetates, hence there are two pathways leading to 1,5-dienes, as it is shown by reaction (19), where allyl moieties in organotin compound and in allylation agent change places (29).

(19)

Reactions of organomercury compounds

Cross-coupling with participation of organomercury compounds, catalyzed by palladium complexes, proceeds with much difficulty and results in extensive homocoupling (catalytic demercuration of organomercurials). The role of this pathway is greater when more active catalyst is used (31).

$$p\text{-NO}_2C_6H_4I + Ph_2Hg \xrightarrow[\text{HMPA,}20°,3\text{ h}]{PhPdI(PPh_3)_2} p\text{-NO}_2C_6H_4Ph + Ph_2 \qquad (20)$$
$$35\% \qquad\qquad 10\%$$

We have found that in the presence of nucleophilic catalyst (iodide ion is the best for organomercurials) the cross-coupling reaction with aryl iodides (eq.21) proceeds under mild conditions with high yield of the cross-coupling product. It is important, that both organic groups R in R_2Hg take part in the reaction.

$$2YC_6H_4I + R_2Hg \text{ (2RHgX)} \xrightarrow[\text{solv., }20\text{-}70°]{PhPdI(PPh_3)_2,I^-} 2Ar\text{-}R + R_2 \qquad (21)$$
$$70\text{-}100\% \quad 0\text{-}30\%$$

solv. = HMPA, DMSO, DMF, Me$_2$CO

Y = p-NO$_2$ $\qquad\qquad$ R = Me, 2,4,6-Me$_3$C$_6$H$_2$, Y'C$_6$H$_4$

$\qquad\qquad\qquad\qquad$ (Y' = o-, p-MeO, p-Me, H, p-Cl)

$\qquad\qquad\qquad\qquad$ R = p-NH$_2$C$_6$H$_4$, p-MeCONHC$_6$H$_4$, X = OAc;

$\qquad\qquad\qquad\qquad$ o-HOC$_6$H$_4$, X = Cl

Y = p-CN, p-MeCO, $\qquad$ R = 2,4,6-Me$_3$C$_6$H$_2$

$\quad$ p-MeO$_2$C

Y = o-, m-, p-NO$_2$, $\qquad$ R = Ph

$\quad$ 2,4-(NO$_2$)$_2$,

$\quad$ 2,4,6-(NO$_2$)$_3$

In contrast to organotin compounds, the reaction of organomercurials should be performed in oxygen-free atmosphere (both selectivity and reaction rate are increased in this case).

Cross-coupling reactions of organotin and organomercury compounds are useful for synthesis of arylated heterocyclic compounds (32).

$$\text{ArI} + \underset{X}{\boxed{\quad}}\text{—HgY} \quad \xrightarrow[\text{solv.,}\ 20°]{\text{"Pd",}\ I^-} \quad \underset{X}{\boxed{\quad}}\text{—Ar} \qquad (22)$$

$$70 - 98\%$$

X = O, S;   Y = Cl, Br, 2-thienyl;
Ar = p-Y'C$_6$H$_4$ (Y' = p-NO$_2$, p-CN, p-MeCO, p-MeO$_2$C),

        2-, 3-, 4-pyridyl;
"Pd" = PhPdI(PPh$_3$)$_2$;   solv. = HMPA, THF, acetone

$$\underset{N}{\boxed{\quad}}\text{—I} + \underset{S}{\boxed{\quad}}\text{—SnMe}_3 \quad \xrightarrow[\text{HMPA,}\ 20\text{-}70°]{\text{"Pd"}} \quad \underset{N}{\boxed{\quad}}\text{—}\underset{S}{\boxed{\quad}} \qquad (23)$$

$$80 - 96\%$$

"Pd" = (MeCN)$_2$PdCl$_2$, PdCl$_2$

Heteroaromatic moiety can be either in aryl halide or in organometallic compound or in both as in reactions of 2-thienyl- or 2-furyl-mercury derivatives with 2-, 3-, 4-iodopyridines. The same reaction with organotins does not require inert atmosphere and proceeds without by-product formation. However organomercurials are more readily available, and besides, in this case both organic moieties are utilized in synthesis.

The presence of iodide ion facilitates considerably palladium- and rhodium-catalyzed demercuration of aromatic mercury derivatives with formation of symmetrical biaryls (33) (the reaction time is decreased ca. 5-fold, and instead of 80° (34) the reaction requires only 25°; the reaction proceeds smoothly in various solvents. Rhodium complex is a better catalyst for ArHgX, and palladium complex - for Ar$_2$Hg.

$$\text{Ar}_2\text{Hg (Ar'HgX)} \quad \xrightarrow[\text{solv., Hal}]{\text{"Pd" ("Rh")}} \quad \text{Ar}_2 \qquad (24)$$

$$81 - 99\%$$

solv. = HMPA, DMF, THF, acetone, MeCN, PhH
"Pd" = PhPdI(PPh$_3$)$_2$, "Rh" = [Rh(CO)$_2$Cl]$_2$
Hal$^-$ = Cl$^-$, Br$^-$, I$^-$
Ar = p-YC$_6$H$_4$ (Y = MeO, H, Cl), 2-C$_4$H$_3$S
Ar' = p-Y'C$_6$H$_4$ (Y' = Me$_2$N, NH$_2$, MeO, Me, H)
X = Cl, OAc

Reactions of organocopper compounds

Application of palladium catalysis facilitates cross-coupling reactions of organocopper compounds. Here the presence of halide ion is also required, otherwise higher temperature is needed for the reaction, but the main point is that in this case extensive homocoupling is observed (35).

$$\text{RX} + \text{PhC}\equiv\text{CCu} \quad \xrightarrow[\text{solv.,}\ 20°]{\text{"Pd", X}^-} \quad \text{PhC}\equiv\text{CR} \qquad (25)$$

$$84 - 98\%$$

solv. = HMPA, DMSO, DMF, THF, Et$_2$O, EtOH, Me$_2$CO, PhH
"Pd" = PhPdI(PPh$_3$)$_2$, (Ph$_3$As)$_2$PdCl$_2$, (Ph$_3$Sb)$_2$PdCl$_2$, (MeCN)$_2$PdCl$_2$,
        PdCl$_2$, LiPdCl$_3$, (π-C$_3$H$_5$PdCl)$_2$
R = p-YC$_6$H$_4$ (Y = H, MeO$_2$C, MeCO, CN, NO$_2$), 2-C$_4$H$_3$S, 2-C$_5$H$_4$N; X = I
R = 2,4-(NO$_2$)$_2$C$_6$H$_3$, trans-PhCH=CH, CH$_2$=CHCH$_2$; X = Br

High yields of styryl- and allylphenylacetylenes were obtained (94 - 96%). In the reaction with methyl iodide even in the presence of tetrabutylammonium iodide homocoupling is not suppressed and 54% of (PhC≡C)$_2$ and 46% phenylmethylacetylene are formed. In the presence of iodide ion the reaction only slightly depends on the nature of the catalyst used. That the most readily available and cheap palladium(II) chloride can be used in the reaction. HMPA, DMF, THF, acetone, ether, MeCN and benzene serve as solvents. In acetone sodium iodide may be used instead of tetrabutylammonium salts. Nickel catalyst proved to be less active.

$$p-NO_2C_6H_4I + PhC\equiv CCu \xrightarrow[NaI]{acetone} \boxed{\begin{array}{c} \xrightarrow[\text{20 min}]{PhPdI(PPh_3)_2} \\ \xrightarrow[\text{17 h}]{NiCl_2(PPh_3)_2} \end{array}} \cdot \begin{array}{c} 97\% \\ p-NO_2C_6H_4C\equiv CPh \\ 92\% \end{array} \quad (26)$$

The role of tetrabutylammonium iodide (MX) lies not only in solubility increase for CuX salt, which is important for a heterogeneous reaction of PhC≡CCu (it leads to surface cleaning). We assume, that halide ion plays a role of nucleophilic catalyst and activates organometallic compounds at a transmetallation stage. This view is supported by the fact, that in the homogeneous reaction of mesitylcopper a sharp acceleration in the presence of tetrabutylammonium iodide is also observed. This acceleration is most vivid in THF (36).

$$p-NO_2C_6H_4I + MesCu \xrightarrow[\text{solv., } 25°]{"Pd"} p-NO_2C_6H_4-Mes \quad (27)$$

in the absence of I⁻: 60% in 17 h in HMPA-THF,
  13% in 42 h in THF,
with I⁻: 95% in 20 min in HMPA-THF,
  90% in 10 min in THF

The reaction occurs smoothly with a variety of aryl iodides.

$$ArI + MesCu \xrightarrow[\text{solv., } 25°]{"Pd", I^-} \begin{array}{c} Ar-Mes \\ 70 - 90\% \end{array} \quad (28)$$

solv. = HMPA-THF
"Pd" = PhPdI(PPh₃)₂
Ar = p-YC₆H₄ (Y = CN, MeCO, MeO₂C, Cl, MeO)

Propargyl alcohol turned out to be a convenient substrate for synthesis of terminal arylacetylenes. It reacts easily with ArI in benzene in the presence of CuX and triethylamine. The obtained arylpropargyl alcohol without isolation and purification can be oxidized in benzene-MnO₂-KOH system. Arylacetylenes are obtained with high yields.

$$ArI + CH\equiv CCH_2OH \xrightarrow[PhH]{"Pd", Et_3N, CuI} \begin{array}{c} ArC\equiv CCH_2OH \\ 81 - 97\% \end{array} \xrightarrow[PhH]{MnO_2, KOH} \begin{array}{c} ArC\equiv CH \\ 79 - 99\% \end{array}$$

Ar = p-YC₆H₄ (Y = n-C₃H₇O, Cl, MeCO, MeO₂C, CN, NO₂) (29)

REACTIONS OF ORGANOTIN AND ORGANOMERCURY COMPOUNDS WITH RCOX

A considerable attention has been paid in the recent years to transition metal complex catalyzed reaction of organotins with RCOX, leading to unsymmetrical ketones (37-39).

$$RCOCl + R'SnMe_3 \xrightarrow[\substack{Me_2CO \text{ or} \\ HMPA, 20°}]{"Pd"} \begin{array}{c} RCOR' \\ 70 - 100\% \end{array} \quad (30)$$

"Pd" = (MeCN)₂PdCl₂, (π-C₃H₅PdCl)₂
R = Ph   R' = Me, PhCH₂, CH₂=CH, 2-C₄H₃S,
                    p-YC₆H₄ (Y = H, Cl, CN, NO₂)
R = p-NO₂C₆H₄   R' = Me, CH₂=CH, p-YC₆H₄ (Y = MeO, Me, H),
                    C₆F₅, 2-C₄H₃S

$$R = Me, \ o\text{-}FC_6H_4, \ PhCH_2, \qquad R' = Ph$$
$$PhCH=CH, \ p\text{-}ClCOC_6H_4$$
$$R = p\text{-}ClCOC_6H_4 \qquad\qquad\qquad R' = Me$$

Use of palladium complexes without phosphine ligands enables to carry out this reaction at room temperature not only in HMPA, but also in acetone. Ketones are obtained in high yields. Acyl chlorides of aromatic acids and of cinnamic acid react considerably faster, than the corresponding derivatives of the aliphatic acids (39).

The reaction of aroyl chlorides with hexalkyldistannanes in the presence of triphenylphosphine palladium complexes leads to symmetrical biarylketones as the main product. Substitution of triethylphosphite ligand for triphenylphosphine and execution of the reaction under slight CO pressure allowed to suppress decarbonylation process and to obtain in high yields products of reductive coupling of acyl chlorides - symmetrical α-diketones.

$$RCOCl + Bu_6Sn_2 \xrightarrow[\substack{\text{"Pd"-P(OEt)}_3 \\ 8 \text{ atm CO}}]{\text{"Pd"-PPh}_3} \quad \begin{array}{ll} R_2CO & 47 - 68\% \\ (RCO)_2 & 63 - 78\% \end{array} \qquad (31)$$

$$R = p\text{-}YC_6H_4 \ (Y = MeO, Me, H, Cl), \ C_7H_{15}$$
$$\text{"Pd"-PPh}_3 = PhPdI(PPh_3)_2; \quad \text{"Pd"-P(OEt)}_3 = (\pi\text{-}C_3H_5PdCl)_2 + 4P(OEt)_3$$

In α-diketone synthesis organic halides can probably substitute for acyl chlorides. We have demonstrated such a possibility by synthesis of p-anisyl from p-iodoanisole.

$$p\text{-}MeOC_6H_4I + Bu_6Sn_2 \xrightarrow[\substack{8 \text{ atm CO}}]{\text{"Pd"-P(OEt)}_3} \quad \underset{71\%}{(p\text{-}MeOC_6H_4CO)_2} \qquad (32)$$

In the reactions of organomercurials iodide ion again has a significant influence. Thus, it is known, that acyldemercuration, catalyzed by tetrakis-(triphenylphosphine) palladium, proceeds under action of acyl bromides in HMPA at 60° only for symmetrical organomercury compounds. In the reaction of diphenylmercury with PhCOBr 66% of benzophenone and 12% of biphenyl (40) are formed in 30 minutes. In the presence of iodide ion acyldemercuration proceeds easily at room temperature, and it is important that both aryl groups of biarylmercury take part in the reaction (41). THF and acetone can be successfully used instead of HMPA (in THF the reaction occurs not only faster, than in other solvents, but also with the highest selectivity). $PdCl_2$ proved to be a good catalyst in acetone.

$$2RCOCl + R_2'Hg \ (2R'HgX) \xrightarrow[\text{solv.}, \ 20°]{\text{"Pd"}, \ I^-} \quad \underset{72 - 96\%}{2RCOR'} \qquad (33)$$

$$\text{solv.} = HMPA, \ THF, \ Me_2CO$$
$$\text{"Pd"} = PhPdI(PPh_3)_2, \ (MeCN)_2PdCl_2, \ PdCl_2$$
$$R = Me, \ n\text{-}C_3H_7, \ PhCH_2, \ PhCH=CH, \ YC_6H_4 \ (Y = H, \ p\text{-}F, \ p\text{-}Cl, \ p\text{-}Br, p\text{-}NO_2)$$
$$R' \text{ in } R_2'Hg = YC_6H_4 \ (Y = H, \ o\text{-}, \ p\text{-}MeO), \ PhC\equiv C \ (40\%), \ 2\text{-}C_4H_3S$$
$$R' \text{ in } RHgX = Me, \ n\text{-}C_3H_7 \ (60\%), \ p\text{-}YC_6H_4 \ (Y = H, \ MeO, \ Me, \ Cl)$$

CARBONYLATION

It is known, that Pd-catalyzed carbonylation in RX - organotin system results in unsymmetrical ketones with high yields, but requires rather rigid conditions - 30 atm CO and 120° (42).

$$RX + R_4'Sn + CO \xrightarrow[\substack{HMPA, \ 120°, \\ 30 \text{ atm CO}}]{PhPdI(PPh_3)_2} \quad \underset{62 - 85\%}{RCOR'} \qquad (34)$$

R = Ph, PhCH$_2$, PhCH=CH, EtO$_2$CCH$_2$

R' = Me, Bu, Ph

When our catalytic system is used, carbonylation can be carried out under extremely mild conditions - at room temperature, 1 atm CO (43).

$$RI + R'SnMe_3 + CO \xrightarrow[\substack{solv., \ 20° \\ 1 \ atm \ CO}]{(\pi-C_3H_5PdCl)_2}} RCOR' \qquad (35)$$
$$78 - 98\%$$

solv. = HMPA, DMF, Me$_2$CO

R = p-NO$_2$C$_6$H$_4$ R' = Me, CH$_2$=CH, 2-C$_4$H$_3$S, C$_6$F$_5$, PhC≡C,
                                        p-YC$_6$H$_4$ (Y = MeO, Me, H, Cl, NO$_2$)

R = Me, p-NO$_2$C$_6$H$_4$CH$_2$,   R' = Ph
    2-C$_4$H$_3$S, 2-C$_5$H$_4$N,
    p-YC$_6$H$_4$ (Cl, I, MeO$_2$C, CN)

The most active organotin compounds in HMPA give also cross-coupling products, because complexes of aryl($\underline{5}$) and acyl ($\underline{6}$) type participate in the reaction and its selectivity can be enhanced by CO pressure increase. But it turned out, that the same effect is produced when acetone is used as a solvent: the ketone yield becomes virtually quantitative. Obviously, the difference in the transmetallation reaction rate for these complexes grows in this case.

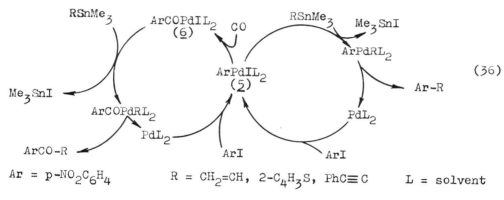

Ar = p-NO$_2$C$_6$H$_4$ R = CH$_2$=CH, 2-C$_4$H$_3$S, PhC≡C   L = solvent

In the presence of palladium phosphine complex carbonylation of ArI - Alk$_3$SnNu system is easy and leads to esters and amides of substituted benzoic acids (44).

$$p-XC_6H_4I + Alk_3SnNu + CO \xrightarrow[\substack{HMPA, \ 20°, \\ 1 \ atm \ CO}]{PhPdI(PPh_3)_2} p-YC_6H_4CONu \qquad (37)$$
$$78 - 100\%$$

Alk = Me, Et
Y = NO$_2$ Nu = MeO, Et$_2$N, PhS (6% and 90% of p-NO$_2$C$_6$H$_4$SPh)
Y = CN, MeO$_2$C   Nu = MeO
Y = H   Nu = Et$_2$N, PhS

When Nu = SPh the direction of the reaction is strongly dependent on the nature of ArI. Thus the reaction of p-nitroiodobenzene results mainly in the cross-coupling product (90%) and only 6% of thioester. In the reaction of PhI only carbonylation product is formed.

In a mixture of diarylmercury (or ArHgX) - RX only organomercurials are subjected to carbonylation, catalyzed by palladium or rhodium complexes. This reaction leads to symmetrical ketones. It also requires rigid conditions

(68-104 atm CO at 60° for reaction of ArHgCl with Rh-catalyst (45)). We have shown (46), that in the presence of iodide ion at 1 atm and 25° both biaryl- and monoarylmercury derivatives react readily, the rhodium catalyst being here better for biarylmercurials, and palladium one for arylmercury halides.

$$Ar_2Hg \ (2Ar'HgX) + CO \xrightarrow[\substack{HMPA, \ 20°, \\ 1 \text{ atm CO}}]{\text{"Rh" ("Pd"), } I^-} ArCOAr \qquad (38)$$

$$70 - 98\%$$

"Rh" = $[Rh(CO)_2Cl]_2$ for Ar_2Hg; "Pd" = $PhPdI(PPh_3)_2$ for Ar'HgX

Ar = $p\text{-}MeC_6H_4$, Ph, $2\text{-}C_4H_3S$

Ar' = YC_6H_4 (Y = p-Cl, H, p-MeO, Me, $p\text{-}NH_2$, $p\text{-}Me_2N$, o-OH), $2\text{-}C_4H_3S$

It is interesting to compare behavior of various organometallic compounds in Pd-catalyzed cross-coupling. We made such a comparison for a variety of organometallic compounds, obtained _in situ_ from Grignard reagents (mainly, phenyl derivatives). Four electrophilic agents were used: allyl bromide, p-iodoanisole, benzoyl chloride and p-nitrobenzoyl chloride. Besides, the ligands in palladium complexes were also varied. Tetrakis(triphenylphosphine)palladium catalyzed reactions of allyl bromide with organomagnesium and organozinc compounds, which proceeded with high rate and selectivity (eq.39,40, Table 1) (cf. non-catalyzed reactions).

$$RMgBr + \diagup\!\!\!\!\diagdown Br \xrightarrow{THF, \ 20°} \begin{array}{c} \boxed{\dfrac{Pd(PPh_3)_4}{15\text{-}30 \text{ min}}} \xrightarrow{} R\diagup\!\!\!\!\diagdown \quad 90\% \\ \\ \xrightarrow{} \quad 30\% \\ 0.5\text{-}3 \text{ hrs} \end{array} \qquad (39)$$

R = Ph, $p\text{-}MeC_6H_4$, $p\text{-}BrC_6H_4$, $PhCH_2$, $9\text{-}C_{13}H_9$

$$PhM + \diagup\!\!\!\!\diagdown Br \xrightarrow[THF, \ 20°]{\text{"Pd", 5 mol.\%}} Ph\diagup\!\!\!\!\diagdown + Ph_2 \qquad (40)$$

TABLE 1. Reactions of organometallic compounds with allyl bromide

M	"Pd"	t,(hrs)	Ph$\diagup\!\!\!\!\diagdown$ (%)	Ph_2 (%)
MgBr	$Pd(PPh_3)_4$	0.25	96	0
	$PdCl_2(PPh_3)_2$	0.25	63	35
	$(\pi\text{-}C_3H_5PdCl)_2$	0.25	20	74
ZnCl	$Pd(PPh_3)_4$	3.5	100	0
	$(\pi\text{-}C_3H_5PdCl)_2$	3.0	55	40
CdBr	$Pd(PPh_3)_4$	12.0	71	24
$AlCl_2$[a]	$Pd(PPh_3)_4$	20.0	trace	trace
	$(\pi\text{-}C_3H_5PdCl)_2$	8.0	94	trace

Note a: in ether

Reactions of p-iodoanisole (which is less active than allyl bromide) with various organometallic compounds are of low selectivity (eq.41) and an exchange of organic moieties takes place giving homocoupling products along with products of cross-coupling (the only exception is phenylzinc chloride).

$$PhM + p\text{-}MeOC_6H_4I \xrightarrow[THF, \ 20°, \ 1 \text{ h}]{\text{"Pd", 1 mol.\%}} p\text{-}MeOC_6H_4Ph + Ph_2 + (p\text{-}MeOC_6H_4)_2 \qquad (41)$$

M = Li, Cu, MgCl, ZnCl, CdCl, $AlPh_2$ (95% for M = ZnCl)

In the acyldemetallation reactions the highest ketone yields were achieved for zinc and mercury derivatives (when p-nitrobenzoyl chloride was used)(see eq.42, Table 2).

$$RM + ArCOCl \xrightarrow[THF, \ 0\text{-}20°]{\text{"Pd", 1 mol.\%}} ArCOR + R_2 \qquad (42)$$

TABLE 2. Reactions of organometallic compounds with ArCOCl (Ar = phenyl and p-nitrophenyl; numbers in parentheses for the latter)

M	R	"Pd"	t (min)		PhCOR (%)		R_2 (%)	
Cu	p-MeC$_6$H$_4$	A[a]	20		68		27	
		B[a]	30		60		22	
MgBr	p-MeC$_6$H$_4$	–	4	(1)	67	(22)	3	(5)
		A	4	(1)	71[b]	(21)	7	(12)
		B	4	(1)	60	(21)	21	(20)
ZnCl	p-MeC$_6$H$_4$	–		(15)		(43)		(54)
		A	3	(10)	95	(69)	5	(24)
		B	3	(3)	72	(49)	21	(51)
	(from PhLi)	B	1		94		6	–
	n-Pr	A	5	(7)	77	(60)	–	–
		B	5	(3)	81	(66)	–	–
CdCl	p-MeC$_6$H$_4$	A	4	(20)	97	(56)	3	(20)
		B	15	(10)	12	(20)	66	(80)
	n-Pr	A	15		52[c]		–	–
		B	10		72		–	–
AlR$_2$	p-MeC$_6$H$_4$	A	240		46		3	–
		B	60		76		–	–
HgI$_2$[d]	p-MeC$_6$H$_4$	B		(10)		(72)		(18)
	Me	B		(5)		(74)		–
	n-Pr	B		(10)		(90)		–

Notes: a – A = PdCl$_2$(PPh$_3$)$_2$, B = PdCl$_2$(MeCN)$_2$; b – 20% of PhC(OH)R$_2$; c – 48% of PhCHO; d – Me$_2$CO, 20°, NaI (4 eqv.)

The fact that zinc derivatives showed the best results in allylation and arylation reactions prompted us to perform allylation of the Reformatsky reagent (47) and to carry out a synthesis of trifluorostyrene from trifluorovinylzinc chloride, obtained _in situ_ from organolithium compound (48).

Acknowledgement – I thank my coworkers, whose names appear in our papers, cited here; my special thanks to Dr. N.A. Bumagin.

REFERENCES

1. R.J.P. Corriu, J.P. Masse, J.C.S. Chem. Comm., 144 (1972).
2. K. Tamao, K. Sumitani, M. Kumada, J. Am. Chem. Soc., 94, 4374-4376 (1972).
3. M. Yamamura, I. Moritani, S.-I. Murahashi, J. Organometal. Chem., 91, C39-C42 (1975).
4. E. Negishi, A.O. King, N. Okukado, J. Org. Chem., 42, 1821-1822 (1977).
5. E. Negishi, S. Baba, J.C.S. Chem. Comm., 596-597 (1976).
6. E. Negishi, D.E. Van Horn, J. Am. Chem. Soc., 99, 3168-3170 (1977).
7. M. Kosugi, K. Kasazawa, I. Shimizu, T. Migita, Chem. Lett., 301-302 (1977).
8. D. Milstein, J.K. Stille, J. Am. Chem. Soc., 101, 4992-4998 (1979).
9. A.N. Kashin, I.G. Bumagina, N.A. Bumagin, I.P. Beletskaya, Zh. Org. Khim., 17, 21-28 (1981).
10. N.A. Bumagin, I.G. Bumagina, I.P. Beletskaya, Dokl. Akad. Nauk SSSR, 272, 1384-1388 (1983).
11. R.S. Sorokina, I.P. Beletskaya, I.O. Kalinovskii, L.F. Rybakova, II Vsesouznaya Konferentsia po metalloorganitcheskoi khimii, Tezisi dokl., SSSR, Gorkii, 1982, p. 100.
12. N.A. Bumagin, I.G. Bumagina, I.P. Beletskaya, Dokl. Akad. Nauk SSSR, 274, 818-822 (1984).
13. I.G. Bumagina, Dissertation, Moscow, 1981.
14. D. Milstein, J.K. Stille, J. Am. Chem. Soc., 101, 4981-4991 (1979).
15. A. Gille, J.K. Stille, J. Am. Chem. Soc., 102, 4933-4941 (1980).
16. A. Moravskiy, J.K. Stille, J. Am. Chem. Soc., 103, 4182-4186 (1981).
17. J. Tsuji, H. Takahasi, Tetrahedron Lett., 4387-4388 (1965).
18. B.M. Trost, Tetrahedron, 33, 2615-2649 (1977).
19. J. Godschalx, J.K. Stille, Tetrahedron Lett., 21, 2599-2602 (1980).

20. B.M. Trost, E. Keinan, Tetrahedron Lett., 21, 2591-2594 (1980); 21, 2595-2598 (1980).

21. N.A. Bumagin, I.G. Bumagina, A.N. Kashin, I.P. Beletskaya, Zh. Obshch. Khim., LII, 714 (1982).

22. N.A. Bumagin, A.N. Kasatkin, I.P. Beletskaya, Dokl. Akad. Nauk SSSR, 266, 862-866 (1982).

23. A.N. Kasatkin, Dissertation, Moscow, 1983.

24. N.A. Bumagin, A.N. Kasatkin, I.P. Beletskaya, Izv. Akad. Nauk SSSR, Ser. Khim., 912-918 (1983).

25. O.A. Reutov, I.P. Beletskaya, V.I. Sokolov, Mekhanizmi reaktsii metallo-organitcheskikh soedinenii (Reaction Mechanisms of Organometallic Compounds), Moscow, 1972.

26. A.N. Kashin, I.G. Bumagina, N.A. Bumagin, I.P. Beletskaya, O.A. Reutov, Izv. Akad. Nauk SSSR, Ser. Khim., 2185 (1981).

27. A.N. Kashin, I.G. Bumagina, N.A. Bumagin, V.M. Bakunin, I.P. Beletskaya, Zh. Org. Khim., 17, 905-912 (1981).

28. N.A. Bumagin, I.G. Bumagina, I.P. Beletskaya, Dokl. Akad. Nauk SSSR, 274, 1103-1106 (1984).

29. N.A. Bumagin, A.N. Kasatkin, I.P. Beletskaya, Izv. Akad. Nauk SSSR, Ser. Khim., 636-642 (1984).

30. N.A. Bumagin, Yu.V. Gulevich, I.P. Beletskaya, Izv. Akad. Nauk SSSR,Ser. Khim., 1137-1142 (1984).

31. N.A. Bumagin, I.O. Kalinovskii, I.P. Beletskaya, Izv. Akad. Nauk SSSR, Ser. Khim., 1619-1624 (1983).

32. N.A. Bumagin, I.O. Kalinovskii, I.P. Beletskaya, Zh. Geterotsikl. Soed., 1467-1470 (1983).

33. N.A. Bumagin, I.O. Kalinovskii, I.P. Beletskaya, Zh. Org. Khim., 18, 1324 (1982).

34. R.C. Larock, J.C. Bernhard, J. Org. Chem., 42, 1680-1684 (1977).

35. N.A. Bumagin, I.O. Kalinovskii, I.P. Beletskaya, Dokl. Akad. Nauk SSSR, 265, 1138-1143 (1982).

36. N.A. Bumagin, I.O. Kalinovskii, I.P. Beletskaya, J. Organometal. Chem., 267, C1-C3 (1984).

37. M. Kosugi, I. Shimizu, T. Migita, J. Organometal. Chem., 129, C36-C38 (1977).

38. D. Milstein, J.K. Stille, J. Org. Chem., 44, 1613-1618 (1979).

39. N.A. Bumagin, I.G. Bumagina, A.N. Kashin, I.P. Beletskaya, Zh. Org. Khim., 18, 1131-1137 (1982).

40. K. Tagaki, T. Okamoto, Y. Sakakibara, A. Ohno, S. Oka, N. Hayama, Chem. Lett., 951-954 (1975).

41. N.A. Bumagin, I.O. Kalinovskii, I.P. Beletskaya, Zh. Org. Khim., 18, 1325 (1982).

42. M. Tanaka, Tetrahedron Lett., 2601-2602 (1979).

43. N.A. Bumagin, I.G. Bumagina, A.N. Kashin, I.P. Beletskaya, Dokl. Akad. Nauk SSSR, 261, 1141-1144 (1981).

44. N.A. Bumagin, Yu.V. Gulevich, I.P. Beletskaya, Izv. Akad. Nauk SSSR, Ser. Khim., 953 (1984).

45. R.L. Larock, S.S. Hershberger, J. Org. Chem., 45, 3840-3846 (1980).

46. N.A. Bumagin, I.O. Kalinovskii, I.P. Beletskaya, Izv. Akad. Nauk SSSR, Ser. Khim., 221 (1982).

47. N.A. Bumagin, A.N. Kasatkin, I.P. Beletskaya, Izv. Akad. Nauk SSSR, Ser. Khim., 1858-1865 (1984).

48. R.S. Sorokina, L.F. Rybakova, I.O. Kalinovskii, I.P. Beletskaya, Izv. Akad. Nauk SSSR, Ser. Khim., in press (1984).

Metal-catalysed stereoselective additions of hydrogen cyanide and other small molecules to alkenes and alkynes

W. Roy Jackson, Patrick Perlmutter, Patricia S. Elmes, Craig G. Lovel,
Robin J. Thompson, David Haarburger, Michael K.S. Probert, Andrew J. Smallridge,
Eva M. Campi, Neil J. Fitzmaurice and Michael A. Kertesz

Department of Chemistry, Monash University, Clayton, Victoria, Australia 3168.

Abstract – The nickel(0) catalysed hydrocyanation of alkynes gives high
yields of α,β-unsaturated nitriles with stereospecific syn-addition of
hydrogen cyanide in most cases. Acetone cyanohydrin can be used as a
source of hydrogen cyanide in many of these reactions. The regioselectivity of addition can be controlled in many compounds, e.g., by
increasing the bulk of the groups substituted on silicon in silylalkynes or
by chelative effects in alkynes bearing substituents containing oxygen
functionality. Hydrocyanation of a wide range of dienes has been shown to
give products whose formation can be rationalised on the basis of the
structure of intermediate π-allylnickel species.

New steroidal based diphosphine ligands have been prepared which lead to
the formation of (+)-2-cyanonorbornane from norbornene in moderate (40-45%)
optical yield.

The addition of hydrogen cyanide to organic molecules is a reaction of considerable commercial importance and is the basis of the Dupont company's large scale preparation of adipo-
nitrile from butadiene (Ref. 1). In this paper we discuss work carried out in our
laboratories which is mainly concerned with the nickel- and palladium-catalysed addition of
hydrogen cyanide to alkenes, alkynes, and alkadienes.

ALKENES

Our interest in the hydrocyanation of alkenes arose from a desire to initiate studies into
asymmetric syntheses involving carbon-carbon bond formation. It had been shown previously
that good yields from hydrocyanation reactions can be obtained when terminal alkenes or
reactive di-substituted alkenes were used (Refs. 1 and 2). Reaction of norbornene with
hydrogen cyanide was found to give exclusively 2-exo-cyanonorbornane in yields of up to 85%
(distilled product) when Pd(diop)$_2$ was used as a catalyst in the presence of excess diop*
ligand when (+)-diop was used both to prepare the palladium complex (1) and as the excess
free ligand, (+)-exo-2-cyanonorbornane was obtained with an optical yield of 36%. It was
established that the most efficient catalysts for hydrogen cyanide addition to alkenes
involved diphosphine ligands which were capable of forming seven-membered chelate rings
with a palladium atom, e.g. (1 and 2). 1,2-Diphosphines, such as prophos and chiraphos,

(1)

(2)

which contain at least one phosphorus atom directly bound to an asymmetric centre have
served as excellent ligands for catalytic asymmetric hydrogenation (Ref. 3). The zero-
valent palladium and nickel complexes of these ligands were however not active as catalysts
for hydrogen cyanide addition. We thus decided to prepare the 3α- and 3β-epimers of the
steroidal ligand (3) which are 1,4-diphosphines with one of the phosphorus atoms directly
linked to an asymmetric carbon atom. The corresponding 1,3-diphosphine (4) was also
prepared. It was found that each of the three diphosphines was a reasonable ligand for

---

*diop ≡ (2,3-isopropylidenedioxybutane-1,4-diyl)bis(diphenylphosphine)

asymmetric hydrogenation of cinnamic acid derivatives using rhodium based catalysts (see Table 1). However, although the 3β-diphenylphosphine form of (3) gave a higher yield than diop when used as a ligand with palladium (optical yields 45 and 36% respectively) the chemical yield was significantly lower (see Table 1). The 1,3-diphosphine (4) gave only a modest optical yield in the hydrocyanation reaction and a poor chemical yield.

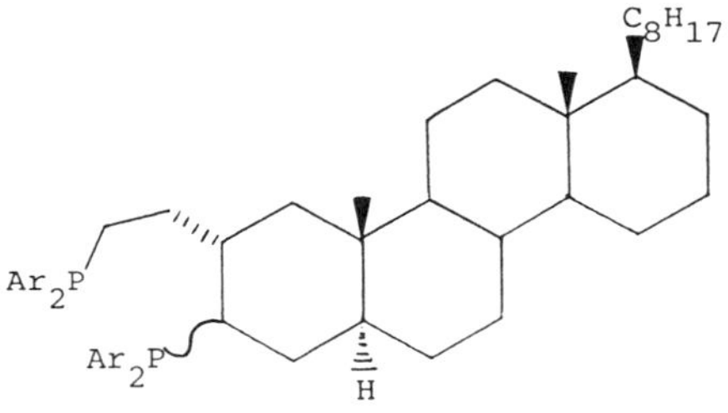

(3) 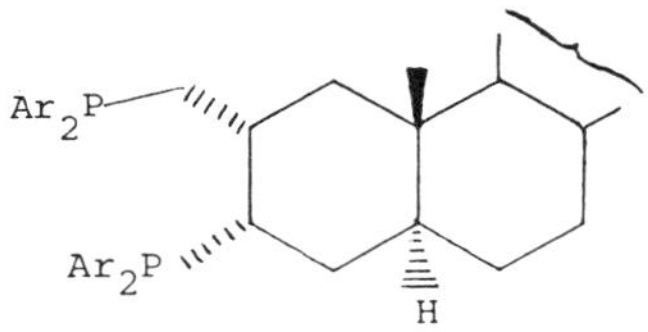   (4)

TABLE 1. Use of steroidal diphosphines as ligands in asymmetric reactions

Reaction	Ligand	Metal	Yield %	Optical yield %
Hydrogenation				
α-Acetamidocinn	3β-(3)	(cod)RhCl	50	80 (+)
amic acid + H$_2$	3α-(3)	(cod)RhCl	60	90 (−)
	(4)[†]	(cod)RhCl	100	82 (−)
Hydrocyanation				
	3β-(3)	(dba)$_2$Pd	23	44 (+)
norbornene + HCN	(4)	(dba)$_2$Pd	<10	14 (−)
	(+) diop	(dba)$_2$Pd	80	36 (+)

[†]Containing *ca*. 10% of the 3β-epimer

The stereochemistry of nickel-catalysed hydrocyanation of alkenes was shown by us (Ref. 4) and by Bäckvall (Ref. 5) to be stereospecifically syn. Thus, for example, we showed that the terminal alkene (5) gave only the product of syn-addition and reaction of 4,t-butyl-cyclohex-1-ene with deuterium cyanide gave only the products of syn addition (6) and (7).

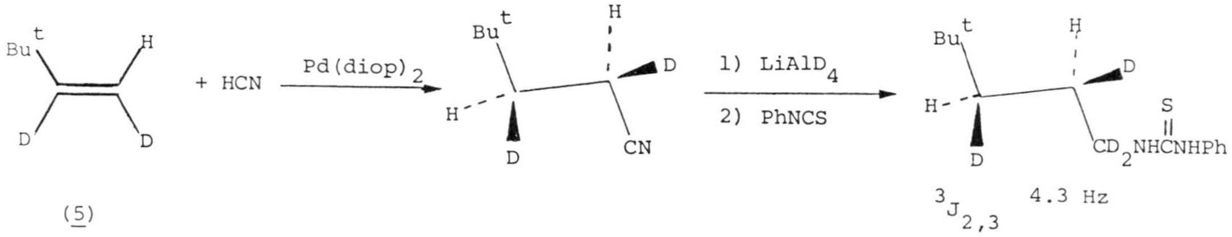

(5)

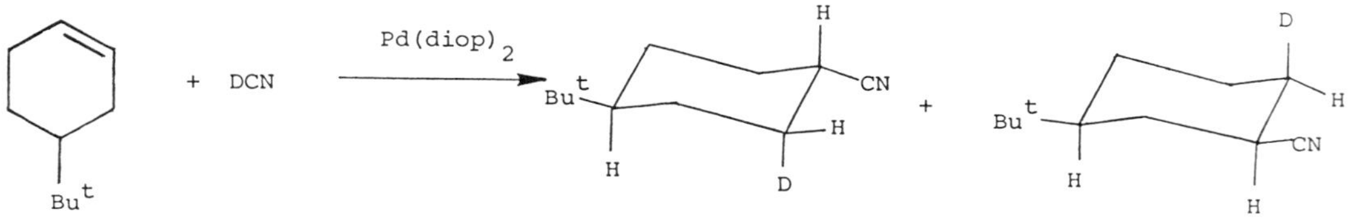

(6)   (7)

The syn-stereospecificity is identical to that shown by metal-catalysed hydrogenation reactions, commonly used for asymmetric hydrogenation studies and thus suggests that similar discrimination between diastereoisomeric transition states may be possible. However, although optical yields greater than 90% are common for asymmetric hydrogenations of cinnamic acid derivatives (Ref. 6) the same ligands only gave moderate (ca. 40%) optical yields in hydrocyanation reactions of norbornene. We, and others (Ref. 6), suspected that the energy difference between the diastereoisomeric transition states arising from chiral ligand-prochiral substrate interactions was increased when the mobility of the transition state geometry was reduced by further coordination to another donor atom or group remotely substituted in the substrate. An example of such coordination is depicted in the Figure for a hydrocyanation reaction.

Figure

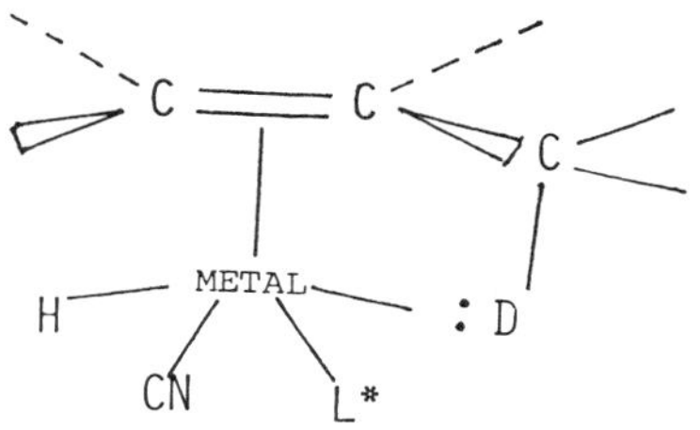

L* = optically active ligand

D = donor atom or group

This suggestion led us to examine the hydrocyanation of molecules containing suitable donor groups and we first encountered evidence for such interactions in the hydrocyanation reactions of alkynes.

ALKYNES

Hydrocyanation of alkynes can be achieved in good yields by heating the alkyne with hydrogen cyanide and nickel tetrakistriphenylphosphite together with excess phosphite ligand in benzene solution (Ref. 7). The substrate to catalyst ratio in this work was ca. 1:200 and recent work by us has shown that if this ratio is increased to ca. 1:90 hydrogen cyanide can be replaced by acetone cyanohydrin without any decrease in yield (see Tables 2 and 3).

Regioselectivity

The regioselectivity of the reaction has been found to be controlled by electronic, steric, and the sought after chelative effects. Terminal, straight chain alkynes such as pent-1-yne or hex-1-yne ($\underline{8}$; R^1 = Pr, Bu, R^2 = H) give predominantly the branched chain nitrile ($\underline{10}$), the product of Markovnikov addition (see Table 2, reactions 1,2,3). When the alkyne contains a single bulky substituent, steric effects dominate the regioselectivity and the terminal nitrile is the preferred product (reactions 4,5,6).

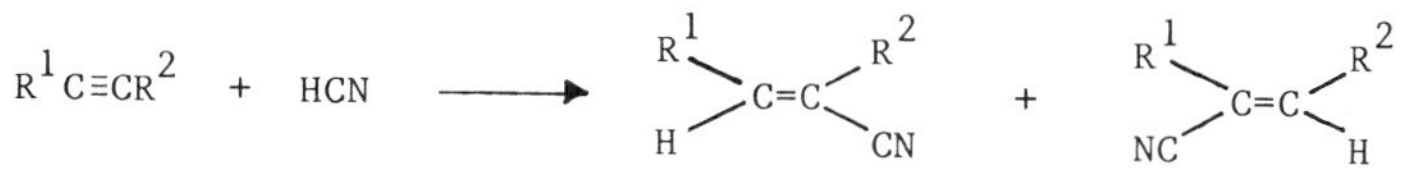

$$R^1 C{\equiv}CR^2 + HCN \longrightarrow$$

$$(\underline{8}) \qquad (\underline{9}) \qquad (\underline{10})$$

Reactions of silylalkynes (Table 3) showed that the regioselectivity could be controlled by varying the size of the groups substituted on silicon. Thus with 1-trimethylsilyl-1-alkynes bearing a bulky group attached to C2 (e.g., Ph reaction 10) only the α-cyano-α,β-unsaturated silane ($\underline{10}$) was obtained. Increasing the size of the silyl substituent by replacing one of the methyl groups with a t-butyl group resulted in high yields of the β-cyano isomer ($\underline{9}$) e.g., reactions 14, 15 and 16.

Reaction of tributylstannylacetylene, also containing an sp-heteroatom bond, gave a 9:1 preference for the α-isomer in high yield. This isomer had previously been prepared by hydrostannylation of cyanoacetylene (Ref. 8) but extension of the hydrocyanation reaction to disubstituted acetylenes should lead to the opposite geometric isomer to that obtained from the hydrostannylation reaction.

TABLE 2   Steric and electronic directing effects in hydrocyanation of alkynes

Reaction	Alkyne R^1	R^2	Conditions	Catalyst:alkyne Ratio	Ratio (9):(10)	Yield %
1	Pr	H	A, HCN	1:185	12:88	60
2	Bu	H	A, HCN	1:185	14:86	73
3	Bu	H	B, HCN	1:40	4:96	33
4	Bu^t	H	B, HCN	1:40	80:20	45
5	Ph	H	A, HCN	1:185	98:2	48
6	Bu^t	Me	B, HCN	1:40	90:10	78

A Reaction for 18 h at 120° in autoclave.

B Slow addition over 18 h of a benzene solution of alkyne and hydrogen cyanide to catalyst plus zinc chloride at 60°.

C Reaction for 18 h in refluxing toluene.

TABLE 3   Regiocontrol in the hydrocyanation of silylalkynes

Reaction	Alkyne R^1	R^2	Conditions[†]	Ratio Catalyst:alkyne	Ratio (9):(10)	Yield %
7	Me_3Si	H	A, HCN	1:185	25:75	26
8	Me_3Si	H	A, HCN	1:90	25:75	74
9	Bu^tMe_2Si	H	A, HCN	1:45	35:67	57
10	Me_3Si	Ph	A, HCN	1:90	0:100	80
11	Me_3Si	Me	A, HCN	1:90	80:20	90
12	Me_3Si	Me	A, $Me_2C(OH)CN$	1:90	80:20	82
13	Me_3Si	Bu	C, $Me_2C(OH)CN$	1:45	72:28	90
14	Bu^tMe_2Si	Me	A, HCN	1:45	98:2	88
15	Bu^tMe_2Si	Bu	A, HCN	1:45	97:3	62
16	Bu^tMe_2Si	Ph	A, HCN	1:45	90:10	73
17	Ph_3Si	Bu	A, HCN	1:18	100:0	75
18	Bu_3Sn	H	A, HCN	1:45	10:90	90

[†] A and C refer to conditions described in Table 2

Stereoselectivity

The addition was shown to be stereospecifically syn for all alkynes bearing aryl, alkyl, and silyl substituents. An X-ray single crystal structure determination on the adduct from 1-triphenylsilylhex-1-yne confirmed the structure (9; R^1 = Ph_3Si, R^2 = Bu). Only in the reaction of dimethyl acetylenedicarboxylate (8; R^1 = R^2 = CO_2Me), an alkyne bearing two electron-withdrawing groups, is there a suspicion that anti-addition may have occurred (Ref. 9). The structure of the product awaits a single crystal X-ray structure determination.

Chelation effects

The first evidence for a contribution to regioselectivity by chelation effects came from a reaction of the alkynol (11) with hydrogen cyanide. In contrast to the reaction of t-butylacetylene (12), the branched chain isomer predominated. In view of the similar size

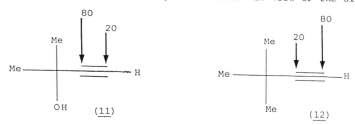

of the groups it would appear that some chelation between the oxygen atom and the nickel atom in the transition state may be occurring. The yields of product in these reactions were all below thirty percent and phenol was formed as a by-product. Reactions of alkynes containing ether functionality are summarised in Table 4. The ethers derived from prop-2-yn-1-ol and but-3-yn-1-ol both gave almost exclusively the product with the nitrile substituted on the carbon bearing the oxygen-bearing substitutent (reactions 19-22). Almost complete regiospecificity was also shown for those ethers bearing a bulky substituent on the other alkyne carbon atom (e.g., Ph or SiMe₃, reactions 23, 24 and 25). The chelative effect alone did not have a dramatic influence on the regioselectivity when steric and electronic effects were similar at each end of the alkyne. Thus the MEM-ether of hex-2-yn-1-ol gave only a small preference for the nitrile which would be preferred by oxygen coordination to the nickel catalyst (reaction 26).

TABLE 4   Hydrocyanation of acetylenic ethers

Reaction	Alkyne R^1	R^2	Conditions†	Ratio (9):(10)	Yield %
19	CH_2OMe	H	A	1:99	70
20	CH_2CH_2OMe	H	A or B	3:97	70
21	CH_2CH_2OMe	H	C	3:97	35
22	$CH_2CH_2OCH_2Ph$	H	A	10:90	78
23	CH_2OSiMe_3	Ph	B	0:100	83
24	CH_2OMe	Ph	A	0:100	66
25	CH_2OMe	$SiMe_3$	A	10:90	40
26	CH_2OCH_2OMe	Bu	A	40:60	90

† Catalyst to alkyne ratio 1:55; 18h reaction time.
A Reactions in autoclave at 120°.
B Reactions with slow addition of reagents to catalyst at 60°
C Reaction in autoclave using $Me_2C(OH)CN$.

The potential coordination ability of groups containing donor atoms other than oxygen is being explored. An initial reaction with N-acetyl-1-amino-prop-2-yne (13; R^1=H, X=NHAc) gave the desired isomer (14) regiospecifically in good yield. Thus the results so far obtained suggest facile routes to the stereospecifically substituted methylene-β-lactones (15, Y=O) and methylene-β-lactams (15; Y=NH).

$$R^1C{\equiv}CCH_2X \ + \ HCN \ \xrightarrow{Ni[0]} \ \underset{H}{\overset{R^1}{>}}C{=}C\underset{CN}{\overset{CH_2X}{<}} \ \longrightarrow \ \underset{H}{\overset{R^1}{>}}C{=}C\overset{CH_2}{\underset{\underset{O}{\overset{\|}{C}}}{<}}Y$$

(13)   (14)   (15)

Experimental

An example of the experimental simplicity of this chemistry is illustrated by the preparation of Z-3-phenyl-2-trimethylsilylprop-2-ene nitrile (10; R[1]=Me$_3$Si, R[2]=Ph).

A solution of 1-(trimethylsilyl)-2-phenylethyne (1.75g, 10 mmol), tetrakis(triphenyl-phosphite)nickel(0)$_3$ (0.26g, 0.22 mmol), triphenylphosphite (0.7 ml, 2.7 mmol) and acetone cyanohydrin (1.8 cm^3, 20 mmol), in toluene (30 cm^3) was heated under reflux for 18 h. The solution was allowed to cool and pentane (10 cm^3) added. The precipitated catalyst was filtered off and washed with more pentane. The solvent was removed under reduced pressure and the residue distilled to give Z-3-phenyl-2-trimethylsilylprop-2-ene nitrile (10; R[1]=Me$_3$Si, R[2]=Ph) (1.4 g, 81%) b.p. 115° (oven) at 0.5 mm of Hg, m.p. 76-77°. ν_{max} (Nujol) 2220s. Mass spectrum m/e 201(50), 186(100), 170(30), 84(40), 73(100). [1]H n.m.r. (90 MHz) δ0.18, s, 9H, SiMe$_3$; 7.3, m, 5H, Ph, 8.06, s, 1H, H3 (Found: C, 71.3; H, 7.3; N, 7.0%. Calc. for C$_{12}$H$_{15}$NSi, C, 71.6; H, 7.5; N, 7.0%).

DIENES

Although the reaction of buta-1,3-diene with hydrogen cyanide is the basis of a large scale industrial process few reports of reactions of other dienes with hydrogen cyanide have appeared. We began to explore the reaction of 1,4- and 1,5-dienes with hydrogen cyanide with the aim of using the second alkene as a chelating function for metal-catalysed reaction of the other alkene. High to moderate yields of β,γ-unsaturated nitriles were obtained when aliphatic dienes were reacted with hydrogen cyanide in the presence of Ni(diop)$_2$ with some free diop and hydroquinone for 18 hours at 120°C with a substrate to catalyst ratio of 1:400. Lower yields were obtained for dienes other than butadiene when diop was replaced by the 1,4-diphosphine, α,α¹-diphenylphosphino-o-xylene but good yields could be obtained if the substrate to catalyst ratio was increased to 1:100. Using this catalyst ratio. It was found that once again hydrogen cyanide could be replaced by acetone cyanohydrin without loss of yield if a catalyst ratio of 1:50 was used.

Reaction of penta-1,4-diene gave E-2-methylpent-3-enenitrile (16) and reaction of either hexa-1,5- or 2,4-diene gave the same equimolar mixture of β,γ-unsaturated nitriles (17) and (18). Unreacted diene from an incomplete reaction of hexa-1,5-diene was shown to be a complex mixture of dienes showing that facile isomerisation occurs under the reaction conditions. The product nitriles (17) and (18) were shown not to isomerise under the reaction conditions, neither to each other nor to the thermodynamically more stable α,β-unsaturated nitriles.

Whilst this work was in progress, similar products were reported for reactions catalysed by Ni[P(OPh)$_3$]$_4$ (Ref. 10). In our hands much lower yields of nitriles were obtained when this catalyst system was used.

Reaction of octa-1,7-diene gave a mixture of four nitriles which were converted to three saturated methyl esters by methanolysis and hydrogenation. The nitrile group was shown to be substituted on carbon atoms 2, 3, and 4 in the chain. No methyl nonanoate, the product of C1-substitution, was present (see Scheme 1). Similar reaction of deca-1,9-diene gave a mixture of unsaturated nitriles which by methanolysis and hydrogenation were shown to have the nitrile group substituted at carbon atoms 2, 3, and 4 in the chain. Again no product of C1-substitution was detected.

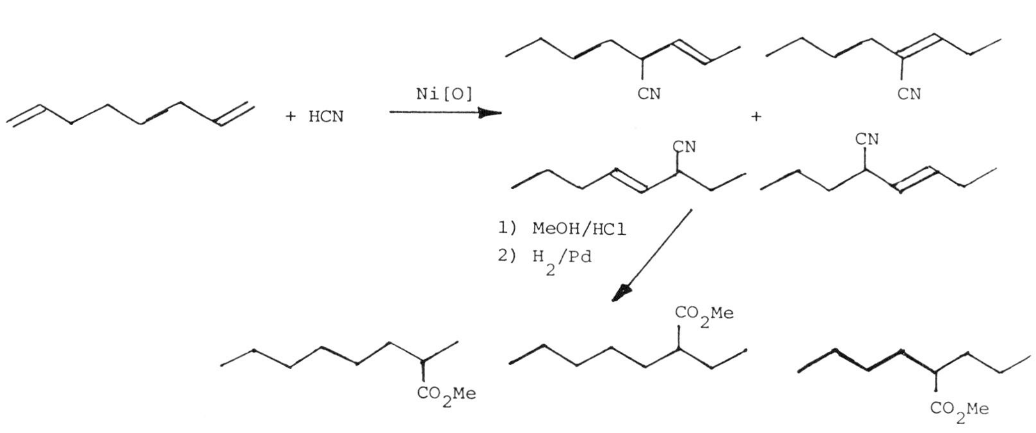

SCHEME 1

A mechanism which is compatible with the above results and with published data (Refs. 1,10,11) involves initial rapid isomerisation of the dienes. The terminal 1,3-diene in the isomeric mixture is almost certainly the most reactive giving a η^3-syn,syn-cyanoallylnickel species (19) (Ref. 12) (see Scheme 2). This can irreversibly transfer cyanide to C2 or C4 giving the E-unsaturated nitriles (20 or 21) or rearrange to give the isomeric allyl species (22) which in turn can lead to the formation of the C3- or C5-substituted products (23 or 24) or isomerise further. The absence of terminal nitrile in all but the reaction of buta-1,3-diene is almost certainly associated with the low equilibrium concentration of the mono-substituted, terminal allyl species (25) in reactions of longer chain dienes. The lack of isomerisation of the product $\beta\gamma$-unsaturated nitriles to the thermodynamically more stable α,β-unsaturated nitriles is possibly due to bidentate coordination of β,γ-unsaturated nitriles to the nickel catalyst which discourages isomerisation.

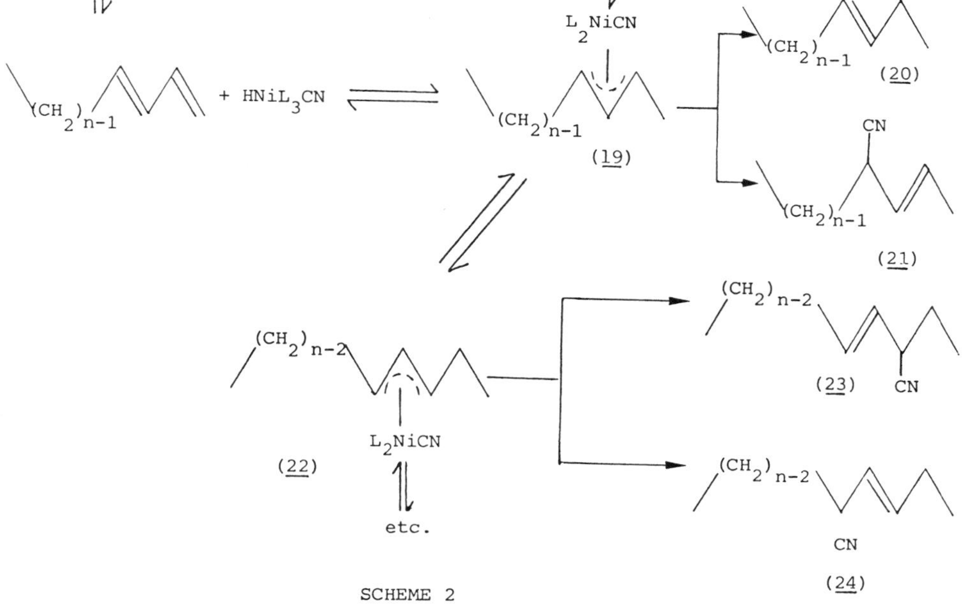

$$L_4Ni \rightleftharpoons L + L_3Ni + HCN \rightleftharpoons HNiL_3CN$$

SCHEME 2

This mechanism suggests that the products from reactions of complex dienes or even diene mixtures may be controlled by the thermodynamic stability of intermediate allylnickel species and kinetics of cyanide transfer to either terminal of such a species. We are exploring the potential of such control and have received encouragement from a reaction of 1-phenylbuta-1,3-diene which gives E-4-phenyl-2-methylbut-3-enenitrile (26) as the sole product in good yield (76%). The allyl species (27) is thermodynamically preferred and kinetically controlled cyanide transfer to give (26) would be expected. Further reactions of dienes are under investigation.

Conclusions

Hydrocyanation reactions of alkenes when carried out with catalysts containing optically active ligands show potential for the preparation of optically active nitriles. Reactions of dienes involve complex rearrangements both of the initial dienes and allylnickel intermediates and can give mixtures of several products. A knowledge of the mechanism suggests that single products may be obtained in high yields from appropriately substituted dienes or even mixtures of dienes. Alkynes can be readily hydrocyanated with complete stereoselectivity and the regioselectivity can be controlled by steric and chelative effects of substituents. Promising new routes to methylene β-lactones and methylene β-lactams are being developed from this chemistry.

The replacement of hydrogen cyanide by acetone cyanohydrin means tha hydrocyanation reactions in many cases can give high yields of useful synthetic intermediates using cheap, easily prepared, non-hazardous reactants and catalysts.

Acknowledgements – We thank the Australian Research Grants Scheme for support, the Australian Government for the award of Commonwealth Postgraduate Research Awards, and Johnson and Matthey Ltd. for a loan of metals.

REFERENCES

1. Chemistry and Engineering News, p.30, April 26th (1971); B.R. James in "Comprehensive Organometallic Chemistry", G. Wilkinson, F.G.A. Stone and E.W. Abel Eds., Vol. 8, p.353, Pergamon Press, N.Y. (1982).
2. P.S. Elmes and W.R. Jackson, Aust. J. Chem., 35, 2041 (1982).
3. M.D. Fryzuk and B. Bosnich, J. Amer. Chem. Soc., 99, 6262 (1977); M.D. Fryzuk and B. Bosnich, J. Amer. Chem. Soc., 100, 5491 (1978).
4. W.R. Jackson and C.G. Lovel, Aust. J. Chem., 35, 2053 (1982).
5. J.E. Bäckvall and O.S. Andell, J. Chem. Soc. Chem. Comm., 1098 (1981).
6. H.B. Kagan in "Comprehensive Organometallic Chemistry", G. Wilkinson, F.G.A. Stone, and E.W. Abel Eds., Vol. 8, p.471, Pergamon Press, N.Y. (1982)
7. W.R. Jackson and C.G. Lovel, Aust. J. Chem., 36, 1975 (1983).
8. A.J. Leusink, H.A. Budding, and J.W. Marsman, J. Organometal Chem., 9, 285 (1967).
9. M.I. Bruce, personal communication.
10. W. Klein, A. Behr, H-O. Luhr, and J. Weisser, J. Catalysis, 78, 209 (1982).
11. C.A. Tolman, W.C. Seidel, J.D. Druliner, and P.J. Domaille, Organometallics,
12. C.A. Tolman, J. Amer. Chem. Soc., 92, 6785 (1970).

Magnesium-anthracene as a source of solubie, active magnesium

Borislav Bogdanović

Max-Planck-Institut für Kohlenforschung, Kaiser-Wilhelm-Platz 1,
D-4330 Mülheim a.d. Ruhr, Germany

Abstract - Magnesium powder reacts with anthracene in THF to give an orange complex having the composition Mg(anthracene).3THF. The preparation, structure and some reactions are discussed as well as it's use in the preparation of organotransition metal complexes and as a catalyst component in the preparation of magnesium dihydride.

The room temperature addition of anthracene to a THF suspension of commercial magnesium powder activated with ethyl bromide leads to the formation of a bright orange complex having the composition Mg(anthracene).3THF. This material was first reported in a patent in 1967 (1) but its significance as a reagent has only recently been realized (2).

The reaction is not confined to anthracene and analogous behaviour has been observed with methyl- and dimethyl-anthracenes as well as with the bis-tri-methylsilyl derivative (3). These, however, offer no advantage over magnesium-anthracene as a reagent. 9,10-Diphenylanthracene, in contrast, reacts with electron transfer from the metal to the substituted anthracene to give a blue radical anion.

The interaction of the magnesium with the 9,10-positions of the anthracene is indicated by protonolysis which gives exclusively 9,10-dihydroanthracene as well as by comparison

$$Mg(anthracene) \cdot 3THF + 2H^+ \longrightarrow \text{[9,10-dihydroanthracene]} + Mg^{2+} + 3THF$$

of the ^{13}C-NMR chemical shifts and C-H coupling constants with those of anthracene and dihydroanthracene.

Compound	δC_1	$^1J_{CH}$	δC_2	$^1J_{CH}$	δC_9	$^1J_{CH}$	δC_{9a}
anthracene	128.13	161.6	125.32	160.5	126.20	158.8	131.68
dihydroanthracene	127.6	–	126.7	–	36.7	–	137.7
Mg(anthracene)·3THF	114.1	148.6	118.1	153.4	57.7	138.0	145.9

Although the oligomeric structure 1 cannot be ruled out, a monomeric structure 2 is made probable by an X-ray determination of the product of the reaction

1

2

of Mg-anthracene with Et$_2$AlH (3) while a monomeric structure has recently been confirmed crystallographically for Mg[9,10-(Me$_3$Si)$_2$anthracene]·2THF (4). A trigonal bipyramidal arrangement around the magnesium is also found in magnesium-diphenylbutadiene and in this case 3 THF molecules are interacting with the metal (5).

Mg-anthracene shows bivalent reactivity: it can dissociate homolytically to give zerovalent magnesium or it can react with transfer of 1 or 2 electrons from the metal to the anthracene to give the blue radical anion or dianion.

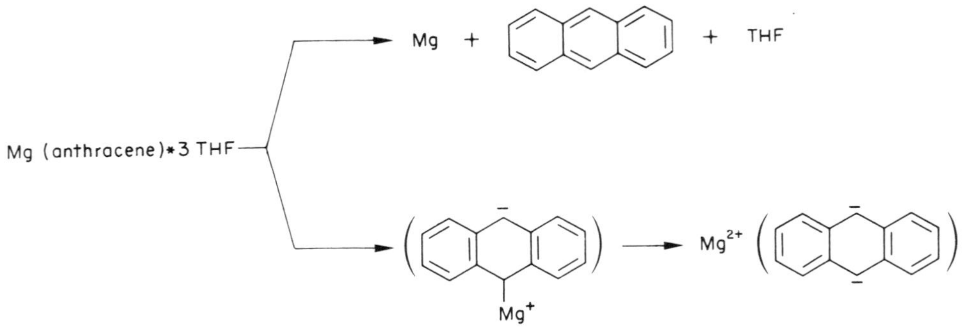

The electron transfer pathway is of interest as a method of preparing substituted 9,10-anthracene derivatives.

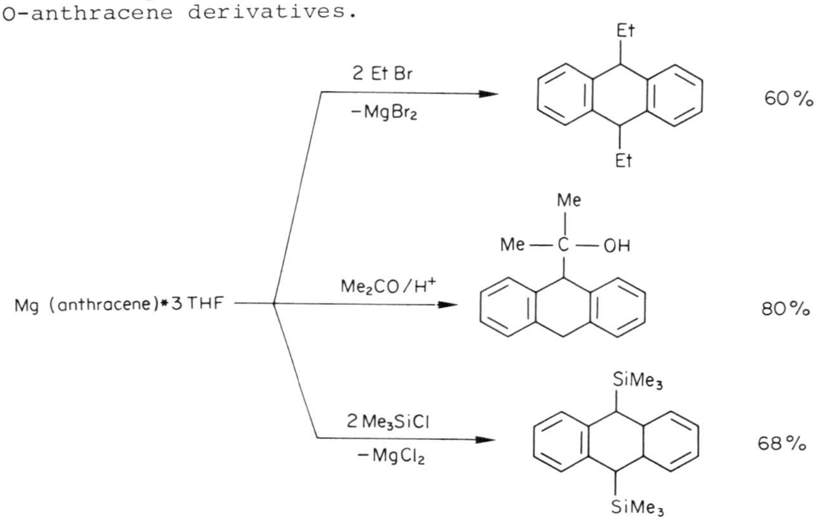

The first indication that Mg-anthracene could be used as a source of atomic magnesium came from kinetic measurements: at 25°, 97-99 % of the anthracene is bonded to the metal while at 60° dissociation occurs and only 60-65 % interacts with the metal. This process is reversible. In some reactions Mg-anthracene does indeed behave as a source of atomic magnesium. For example, reaction with cyclooctatetraene results in exchange (6) while some organic halides react to give Grignard reagents or coupling products.

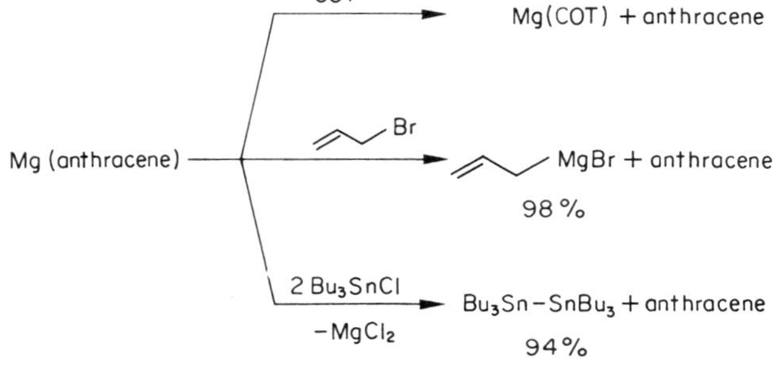

In other cases, a mixture of products resulting from both reaction types is obtained.

The most useful application of Mg-anthracene as a source of magnesium is in the synthesis of organotransition metal complexes which has been developed in collaboration with H. Bönnemann (7). A typical example is the reduction of Co(acac)$_3$ in the presence of cyclopentadiene and 1,5-cyclooctadiene to give $\left[Co(\eta^{2,2}-cod)(\eta^{5}-C_5H_5)\right]$.

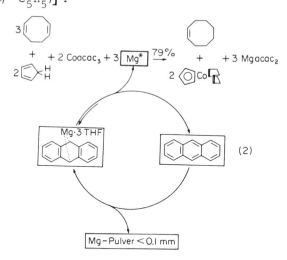

(2)

The use of commercial magnesium powder for this reaction results in only a 10 % yield of the product. The synthesis of cobalt complexes has been studied in detail since these are highly effective catalysts for the cyclocotrimerization of alkynes and nitriles to give pyridine derivatives (8).

This approach has been used to synthesize a whole range of transition metal complexes and typical examples are shown below.

$$PtCl_2 \quad + \quad 3\ Mg^* \quad + \quad 2\ cod \quad \xrightarrow[79^{\circ}C]{45\ \%} \quad \left[Pt(cod)_2\right]$$

$$MoCl_5 \quad + \quad 4\ Mg^* \quad + \quad n\ C_4H_6 \quad \xrightarrow[-30^{\circ}C]{40\ \%} \quad \left[Mo(C_4H_6)_3\right]$$

$$Ni\ acac_2 + Mg^* + 2\ P(O-C_6H_4-CH_3)_3 \quad \xrightarrow[0^{\circ}C,\ C_2H_4]{65\ \%} Ni(C_2H_4)\left[P(O-C_6H_4-CH_3)_3\right]_2$$

Magnesium-anthracene is also involved in the conversion of magnesium into magnesium hydride (9). Commercial magnesium powder reacts only reluctantly with hydrogen (10). The direct hydrogenation of magnesium-anthracene is also unsatisfactory - hydrogenation of the anthracene as well as the magnesium occurs. However, if a catalytic amount of a transition metal halide (e.g. TiCl$_4$, CrCl$_3$) is added then a facile reaction occurs under mild conditions (60°, 60 atm).

$$Mg\ (anthracene)\ *3THF + MX_n \quad \longrightarrow \quad \left[MMgX\right] + anthracene$$

$$Mg\ (anthracene)\ +\ H_2 \quad \xrightarrow{\left[MMgX\right]} \quad MgH_2 + anthracene$$

Initially the magnesium-anthracene reacts with the metal halide to form the

true catalyst releasing anthracene. This method of preparing MgH_2 is recommended for large scale preparations.

The resulting MgH_2 hydride is highly reactive and is pyrophoric both as the
isolated powder or as a suspension in THF. A selection of its reactions are
shown below. The reaction with olefins has been investigated in some detail
(11).

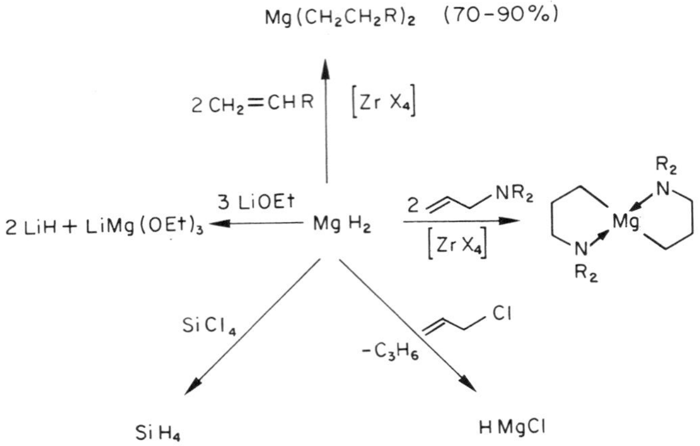

This reaction is catalysed by transition metal halides - $ZrCl_4$ is preferred -
and is limited to 1-alkenes. The reaction is regiospecific, the Mg adding
to C_1 of the olefin, and the resulting magnesium dialkyls can either be
used in situ or can be isolated by removing the solvent (THF). The reaction
with allyl- or homoallyl-amines or -ethers leads to monomeric species in
which the O- or N-donor atom interacts with the metal. Organic halides are,
in general, reduced to saturated hydrocarbons while silicon tetrachloride
is converted into silane in high yield. Metal hydrides are also the product
of the reaction of MgH_2 with alkali metal alkoxides (12). This is in contrast to the behavior of the magnesium hydride prepared by reacting $MgEt_2$
with $LiAlH_4$: this reacts with lithium alkoxides with addition of the alkoxide
to the magnesium to give, for example, $[H_2MgOEt]^- Li^+$ (13). If the transition
metal catalysed reaction of Mg with H_2 is carried out in the presence of an
equivalent of $MgCl_2$ at 0° (the temperature is critical) then HMgCl is formed
(14). This species which was first reported by Ashby (15) is also the

$$Mg + H_2 + MgCl_2 \xrightarrow[\text{anthracene}]{[Cr-Mg]} 2\,HMgCl$$

product of the reaction of MgH_2 with one equivalent of allylchloride and is
being studied further: in the presence of Ti or Zr catalysts it has been
shown to add to olefins to give Grignard reagents.

A highly active, pyrophoric form of magnesium can be prepared by heating
Mg-anthracene in a vacuum at ca. $150-170^\circ$. However, a more convenient method
for preparing

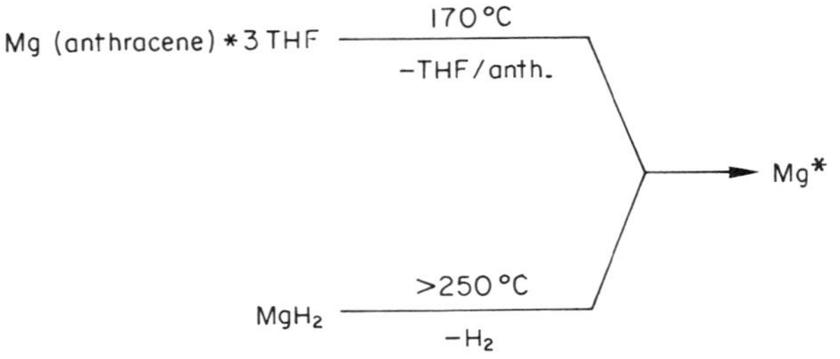

a similar material is the thermolysis of MgH_2 at ca. 250°. The resulting magnesium has an activity similar to that of the material isolated by Rieke (16)
from the reduction of magnesium halide with alkali metals and it can, for
example, be used to prepare Grignard reagents from less reactive organic
halides. Of considerable interest is the fact that this active magnesium
reacts readily with hydrogen at normal pressure and $150-220^\circ$. This reaction
forms the basis of the use of magnesium hydride as a hydrogen storage system

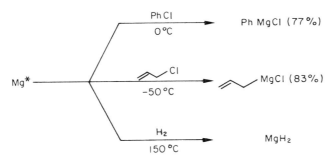

and is discussed in detail in reference 9.

Acknowledgement - The results described here are based upon investigations carried out by E. Bartmann, N. Janke, G. Koppetsch, S.T. Liao, M. Maruthamuthu, K. Schlichte, M. Schwickardi, P. Sikorsky, B. Spliethoff and U. Westeppe and the author would like to express his thanks to these coworkers for their dedicated and enthusiastic engagement.

REFERENCES

1. H.E. Ramsden (Esso Research) U.S. 3,354,190 (1967); Chem. Abs. 68, 114744 (1968). See also P.K. Freeman and L.L. Hutchinson, J. Org. Chem. 48, 879 (1983)
2. B. Bogdanović, S.T. Liao, R. Mynott, K. Schlichte and U. Westeppe, Chem. Ber. 117, 1378 (1984)
3. H. Lehmkuhl, K. Mehler, R. Benn, A. Rufinska, G. Schroth and C. Krüger, Chem. Ber. 117, 389 (1984)
4. H. Lehmkuhl, K. Mehler, A. Shakoor, C. Krüger and Y.H. Tsay, unpublished results (1984)
5. Y. Kai, N. Kanehisa, K. Miki, N. Kasai, K. Mashima, H. Yasuda and A. Nakamura, Chem. Letters 1277 (1982)
6. J. Richter, unpublished results (1982)
7. H. Bönnemann, B. Bogdanović, R. Brinkmann, D.W. He and B. Spliethoff, Angew. Chem. 95, 749 (1983); H. Bönnemann and B. Bogdanović (Studiengesellschaft Kohle mbH) Ger. Offen. DE 3,205,550 (1983); Chem. Abs. 100, 22805 (1984)
8. For a recent review see H. Bönnemann and W. Brijoux, Aspects Homogen. Cat. 5, 000 (1984)
9. B. Bogdanović, S.T. Liao, M. Schwickardi, P. Sikorsky and B. Spliethoff, Angew. Chem. 92, 845 (1980). This aspect is the subject of a recent review article by B. Bogdanović, Angew. Chem. in press (1985)
10. See for example G. Brendel in Ullmanns Encykl. techn. Chem. 4 Edition Vol 13, Verlag Chemie, Weinheim 1977, p. 116
11. B. Bogdanović, M. Schwickardi and P. Sikorsky, Angew. Chem. 94, 206 (1982)
12. B. Bogdanović and E. Bartmann, unpublished results (1983)
13. E.C. Ashby, S.A. Noding and A.B. Goel, J. Org. Chem. 45, 1028 (1980)
14. B. Bogdanović and M. Schwickardi, Z. Naturforsch. 39b, 1001 (1984)
15. E.C. Ashby and A.B. Goel, Inorg. Chem. 16, 2941 (1977)
16. See for example R.D. Rieke, Acc. Chem. Res. 10, 301 (1977)

Stereo- and regioselective organic transformations by transition metal catalysis

Jan-E. Bäckvall

Department of Organic Chemistry, Royal Institute of Technology,
S-100 44 Stockholm, Sweden

Abstract – Palladium-catalyzed oxidation of 1,3-dienes in acetic acid in the presence of LiCl and LiOAc produces 1-acetoxy-4-chloro-2-alkenes with high stereo- and regioselectivity. The chloroacetate products are useful synthons in organic transformations. Sequential substitution of the chloro and acetoxy groups allows a regiochemical choice, and the fact that the chloro group can be sustituted with either retention or inversion offers a stereochemical choice. Synthetic applications to the carpenter bee pheromone and to 1-phosphoryl- and 1-sulfonyl-1,3-dienes are described. A related nickel-catalyzed addition to 1,3-dienes is presented.

Nucleophilic additions to unsaturated hydrocarbons coordinated to a transition metal are important reactions in organic synthesis. In these reactions the question concerning the stereo- and regioselectivity plays a central role. For example, the nucleophile may attack the hydrocarbon on the face opposite to the metal, or it may attack on the same face as the metal via a migration from metal to carbon (Fig. 1). Furthermore, in each case there are several possible regiochemical outcomes depending on the number of carbons that are coordinated to the metal.

Fig. 1

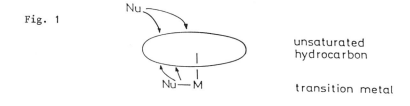

unsaturated
hydrocarbon

transition metal

stereochemistry ?
regiochemistry ?

We have been particularly interested in nucleophilic additions to (π-olefin)- and (π-allyl)-palladium complexes (1), and we recently made a theoretical analysis of the factors governing the stereochemistry of nucleophilic attack on (π-olefin)palladium complexes (2).

During the past three years we have discovered several palladium-catalyzed 1,4-additions to conjugated dienes (1,3). These oxidation reactions involve regio- and stereoselective nucleophilic additions to (π-olefin)- and (π-allyl)palladium intermediates. Three principal reactions, 1,4-diacetoxylation (3a,3b), 1,4-acetoxychlorination (3c,3d), and 1,4-acetoxy-trifluoroacetoxylation (3e) have been developed, which all have their synthetic advantages. These reactions are illustrated on 1,3-cyclohexadiene in Scheme 1. They all proceed via a common intermediate, and the stereochemical outcome depends on the nucleophilic attack on this (π-allyl)palladium intermediate. This paper will deal mainly with the 1,4-acetoxychlorination reaction and the synthetic applications of the chloroacetate products.

Scheme 1

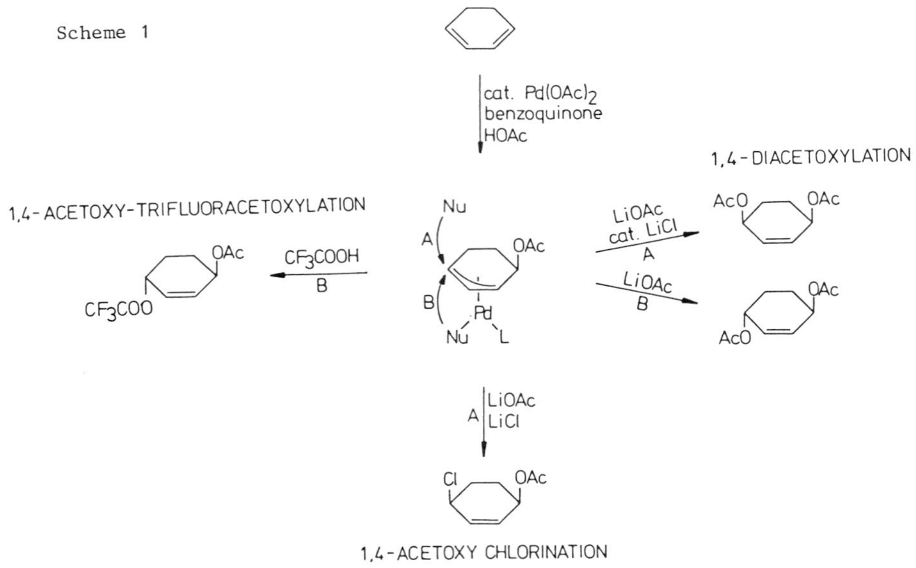

cat. Pd(OAc)$_2$
benzoquinone
HOAc

1,4 - DIACETOXYLATION

1,4 - ACETOXY-TRIFLUORACETOXYLATION

CF$_3$COOH

B

CF$_3$COO   OAc

Nu
A
OAc
B
Pd
Nu   L

LiOAc
cat. LiCl
A

LiOAc
B

AcO   OAc

AcO   OAc

A │ LiOAc
  │ LiCl

Cl   OAc

1,4 - ACETOXY CHLORINATION

1,4-FUNCTIONALIZATIONS OF 1,3-DIENES VIA THE ACETOXYCHLORINATION REACTION

1,4-Acetoxychlorination of conjugated dienes takes place in good yield with high regio- and stereoselectivity (eqs 1 and 2) (3c,3d). The reaction proceeds with overall cis-stereo-

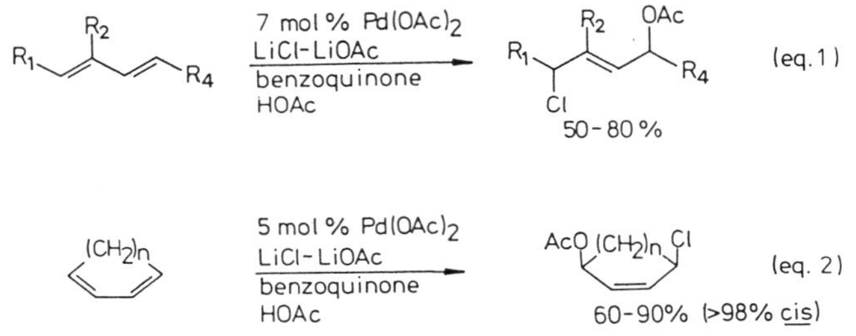

R$_1$ R$_2$ R$_4$ →   7 mol % Pd(OAc)$_2$
LiCl-LiOAc
benzoquinone
HOAc
→   R$_1$ R$_2$ OAc   R$_4$
Cl
50-80 %   (eq.1)

(CH$_2$)$_n$ 5 mol % Pd(OAc)$_2$
LiCl-LiOAc
benzoquinone
HOAc
→   AcO (CH$_2$)$_n$ Cl
60-90% (>98% cis)   (eq. 2)

chemistry with cyclic dienes. Also, for acyclic dienes, it was possible to control the 1,4-relative stereochemistry as shown by the oxidation of E,E- and E,Z-2,4-hexadiene to chloroacetates R*R*-1 and R*S*-1 respectively (eqs 3 and 4).

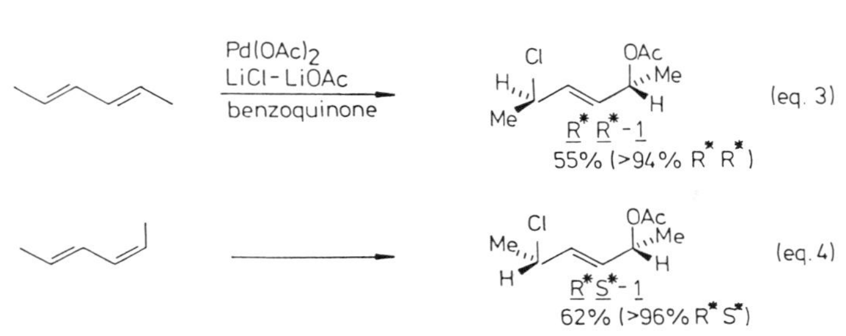

Pd(OAc)$_2$
LiCl-LiOAc
benzoquinone
→   Cl   OAc
H,,,   ,,,Me
Me   H
R*R*-1
55% (>94% R*R*)   (eq. 3)

→   Cl   OAc
Me,,,   ,,,Me
H   H
R*S*-1
62% (>96% R*S*)   (eq. 4)

Regiochemical choice in the 1,4-funtionalizations

An important aspect of the palladium-catalyzed acetoxychlorination reaction is that the chloro and acetoxy groups can be sequentially substituted by two different nucleophiles (Scheme 2). After a completely chemoselective nucleophilic substitution of the chloro group (3c,3d), the acetoxy group can be substituted in a transition metal catalyzed (Pd, Cu,

Scheme 2

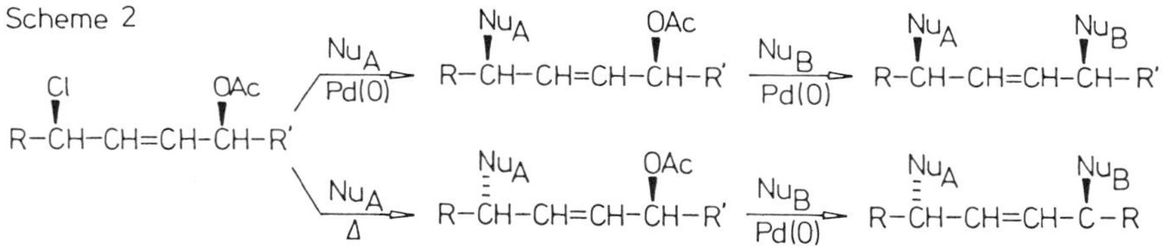

Ni, Mo, Fe) reaction (3,4,5). If the order of the nucleophiles is reversed, this allows a regiochemical choice, which is illustrated in Scheme 3.

Scheme 3

A. Nu_A = Me_2NH, Nu_B = $NaCH(COOMe)_2$:

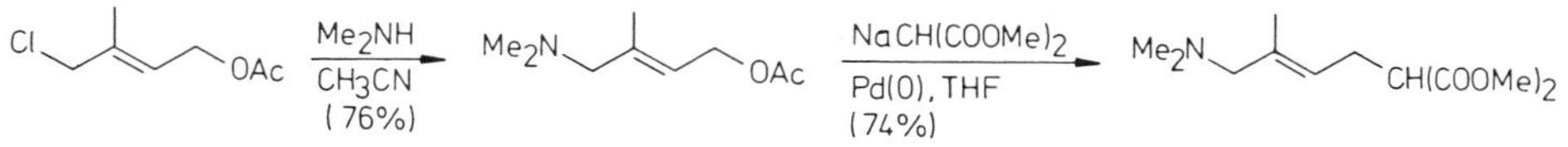

B. Nu_A = $NaCH(COOMe)_2$, Nu_B = Me_2NH :

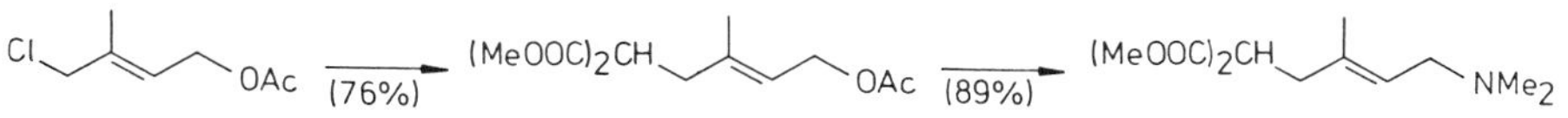

This principle was recently applied to the synthesis of the Monarch butterfly pheromone 4 using the chloroacetate 2 from isoprene (6). Regioselective carbon-carbon bond formation leads to the ketoester 3, which is readily transformed to 4. In this sense the chloroacetate products can be considered as multiple coupling reagents (MCR) (7) with two electrophilic sites.

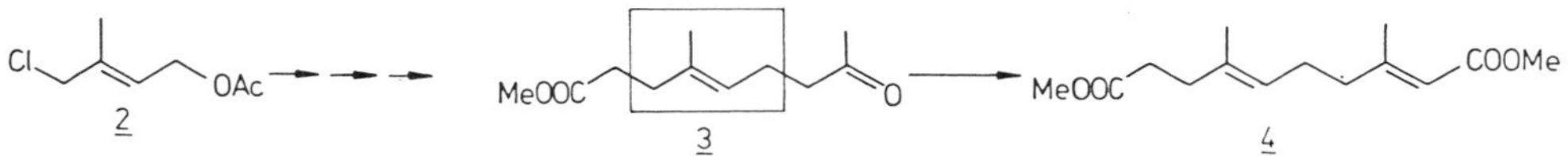

Stereochemical choice in the 1,4-functionalizations

Importantly, the substitution reactions in Scheme 2 are stereospecific, which allows the creation of new carbon-carbon bonds with control of the 1,4-relative stereochemistry. Furthermore, by applying either a metal catalyzed reaction or a classical S_N2-reaction the chloro group can be displaced with either retention or inversion of configuration at carbon (3c,3d). This offers a stereochemical choice in the 1,4-functionalization and is demonstrated in Scheme 4 for 1,3-cyclohexadiene.

Scheme 4

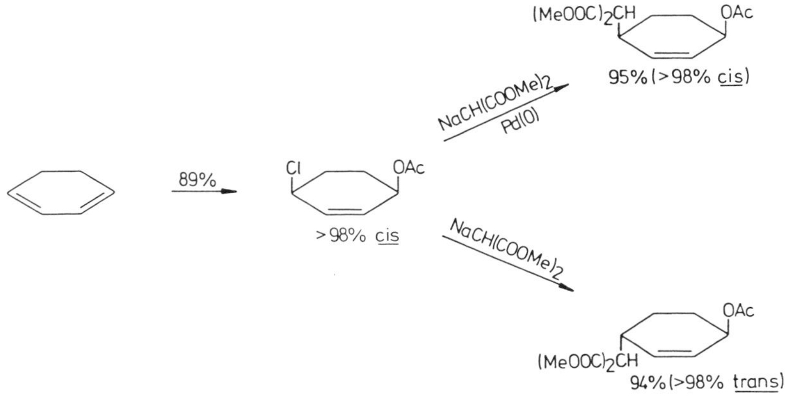

The highly stereospecific palladium(0)-catalyzed substitution reaction of the allylic
chloride, that takes place with complete retention, proceeds via the mechanism outlined in
equation 5 in analogy with the palladium(0)-catalyzed substitution of allylic acetates (4).

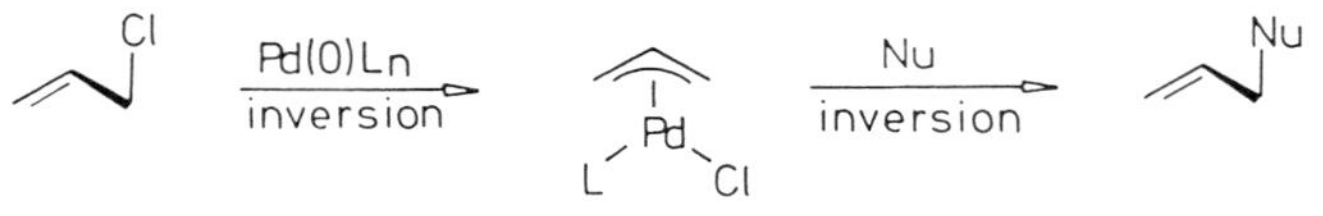

Also in the acyclic systems it was possible to obtain a similar control of the 1,4-relative
stereochemistry in the 1,4-functionalizations (Scheme 5). The chloroacetates R*S*-1 and
R*R*-1 were transformed to both possible diastereoisomers R*S*-5 and R*R*-5 using a number

Scheme 5

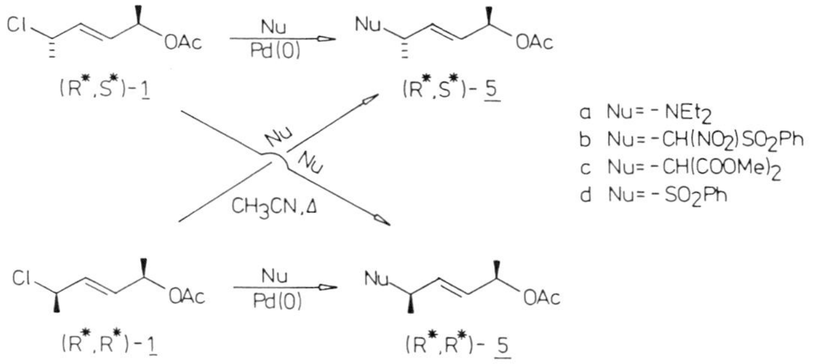

a Nu= -NEt$_2$
b Nu= -CH(NO$_2$)SO$_2$Ph
c Nu= -CH(COOMe)$_2$
d Nu= -SO$_2$Ph

of different nucleophiles. Since the allylic acetoxy group can be stereospecifically substituted by carbon or nitrogen nucleophiles using either palladium (4) or copper (5b,5f,5g)
catalysis, a great number of both cyclic and acyclic derivatives with a defined relative
stereochemistry between carbons are available by this methodology. Such a control of the
relative stereochemistry between distant carbons in acyclic systems is of great importance
in organic synthesis (8).

CARPENTER BEE PHEROMONE SYNTHESIS

The lactone cis-6 was isolated in 1976 and found to be the key component of the sex
attractant of the carpenter bee (9). Since then several syntheses of the lactones cis-6 and
trans-6 have been reported (10). None of the previous methods, however, allow the stereo-

cis-6   trans-6

selective synthesis of both isomers cis-6 and trans-6, but give either only one of the
isomers or a mixture of both isomers that must be separated. By applying the
1,4-functionalization methodology, E,Z- and E,E-2,4-hexadiene were transformed in a
stereospecific manner to the R*S*- and R*R*-isomers respectively of 5-hydroxy-2-methyl-
hexanoic acid lactone (cis-6 and trans-6, Scheme 6) (11). The approach to lactones cis-6
and trans-6 requires the displacement of the chloro group in chloroacetates R*S*-1 and
R*R*-1 by a carboxy anion equivalent and subsequent reduction of the double bond. Attempts
to replace the chloro group with cyanide failed and resulted in elimination of the acetoxy
group. Other methods for stereospecifically introducing a carboxy anion equivalent, such
as metal-catalyzed carbonylation and lithiumdithiane, were also tried without success.

We found that (phenylsulfonyl)nitromethyl anion, which was used by Wade et al (12) as a
carboxy anion equivalent, worked excellently in our system for introducing a masked -COOH
group. Palladium-catalyzed substitution of the chloro group in the chloroacetate R*S*-1
and subsequent hydrolytic workup afforded the alcohol R*S*-7 in 81% isolated yield. This
substitution reaction is completely stereospecific and takes place with retention of

Scheme 6

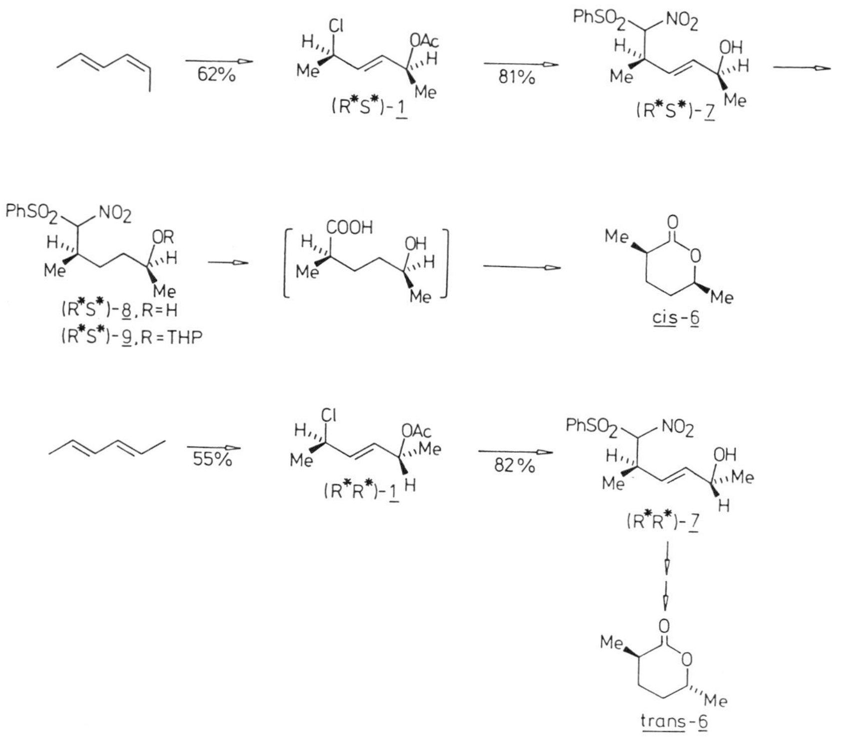

configuration at carbon. It is noteworthy that the (phenylsulfonyl)nitromethyl anion was unable to displace chloride in an ordinary S_N2-reaction (no reaction at 80 °C for several days), whereas the palladium-catalyzed substitution occurs under very mild conditions (7 h, 25 °C).

The hydrogenation of the double bond in R^*S^*-7 using Wilkinson's catalyst, $RhCl(PPh_3)_3$, occurs with complete conservation of the stereochemistry at the allylic carbons. The alcohol R^*S^*-8 obtained was protected as the THP derivative R^*S^*-9. Oxidation of the THP derivative R^*S^*-9 according to Wade ($KMnO_4/NaOH$) (12) followed by acidic workup gave R^*S^*-5-hydroxy-2-hexanoic acid, which spontaneously cyclized to the lactone cis-6. Application of the same 1,4-functionalization sequence to E,E-2,4-hexadiene afforded the trans isomer of the lactone (Scheme 6).

SYNTHESIS OF 1-FUNCTIONALIZED 1,3-DIENES

Nucleophilic substitution of the chloro group in a 1,4-chloroacetate and subsequent elimination of the acetoxy group constitute a potential general route to 1-functionalized 1,3-dienes (eq 6). Using this principle, a number of 1-phosphoryl- and 1-sulfonyl-1,3-dienes were prepared (13). These dienes are important synthetic intermediates as components in Diels-Alder reactions (14)

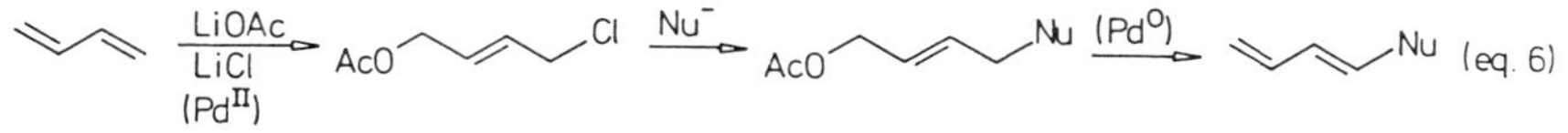

The chloroacetates were smoothly transformed into phosphoryl acetates by employing an Arbuzov reaction. Subsequent palladium-catalyzed elimination (15) of acetic acid produced the desired 1-phosphoryl dienes (eqs 7 and 8).

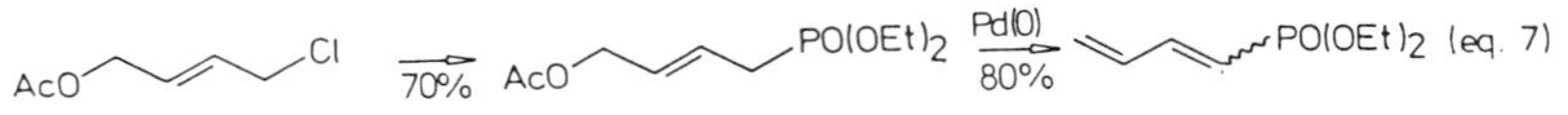

$$AcO\diagup\diagdown\diagup Cl \xrightarrow[70\%]{} AcO\diagup\diagdown\diagup PO(OEt)_2 \xrightarrow[80\%]{Pd(0)} \diagup\diagdown\diagup PO(OEt)_2 \quad (eq.\ 7)$$

$$\underline{E}/\underline{Z}=4/1$$

$$AcO\diagup\diagup\diagdown Cl \xrightarrow[74\%]{} AcO\diagup\diagup\diagdown PO(OEt)_2 \xrightarrow[79\%]{Pd(0)} \diagup\diagup\diagdown PO(OEt)_2 \quad (eq.\ 8)$$

$$>97\%\ \underline{E}$$

In an analogous manner, sulfonyl acetates were prepared by the reaction of the chloroacetates with sodium benzenesulfinate in the presence of a palladium(0) catalyst. This reaction is rapid at room temperature. Subsequent 1,4-elimination of acetic acid either by palladium(0)-catalysis or by treatment with sodium hydride afforded the sulfonyldienes (eqs 9 and 10).

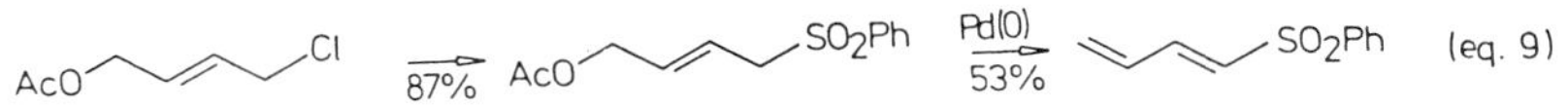

$$AcO\diagup\diagdown\diagup Cl \xrightarrow[87\%]{} AcO\diagup\diagdown\diagup SO_2Ph \xrightarrow[53\%]{Pd(0)} \diagup\diagdown\diagup SO_2Ph \quad (eq.\ 9)$$

$$>95\%\ \underline{E}$$

$$AcO\diagup\diagdown\diagup Cl \xrightarrow[76\%]{} AcO\diagup\diagdown\diagup SO_2Ph \xrightarrow[85\%]{NaH} \diagup\diagdown\diagup\diagdown SO_2Ph \quad (eq.\ 10)$$

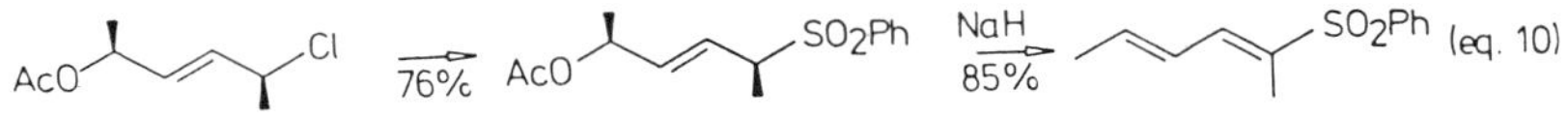

$$>98\%\ \underline{E},\underline{E}$$

NICKEL-CATALYZED ADDITIONS TO 1,3-DIENES

Addition of nucleophiles to 1,3-dienes can be promoted by nickel(0) complexes. We have found that hydrogen cyanide adds very selectively to 1,3-cyclohexadiene in almost quantitative yield on a preparative scale (X=CN, eq 11) (16). Amines (16) and stabilized carbanions (17)

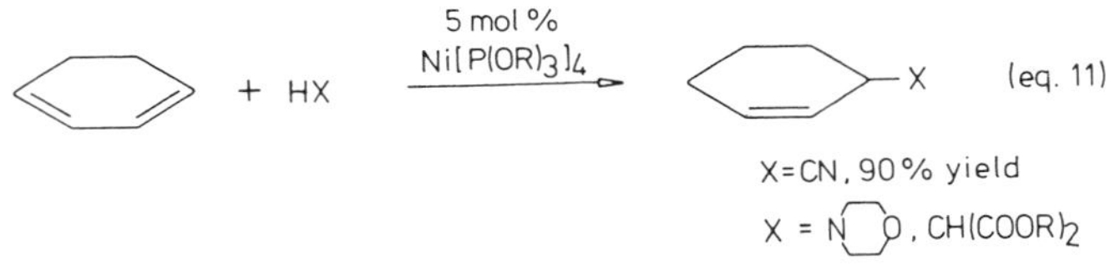

$$\bigcirc + HX \xrightarrow[\quad]{\substack{5\ mol\ \% \\ Ni[P(OR)_3]_4}} \bigcirc\!\!-X \quad (eq.\ 11)$$

$$X=CN,\ 90\%\ yield$$

$$X = N\bigcirc O,\ CH(COOR)_2$$

also add in an analogous manner to 1,3-cyclohexadiene (eq 11). In an effort to elucidate the mechanism of the HCN addition to conjugated dienes, we studied the addition of DCN to 1,3-cyclohexadiene. The results (Scheme 7) indicate that the carbon-carbon bond formation takes place via a _cis_-migration of cyanide from the metal to carbon (18).

Scheme 7

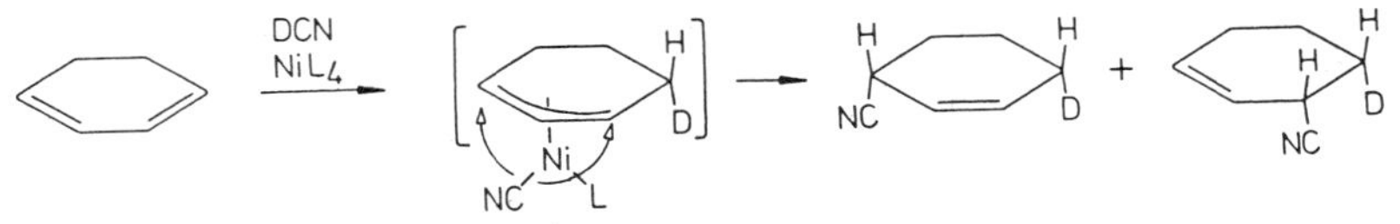

Acknowledgment - I wish to express my sincere appreciation to my collaborators, whose names appear in the references, for their efforts in exploring the chemistry outlined in this report. Financial support from the Swedish Natural Science Research Council and the Swedish Board for Technical Development is gratefully acknowleded.

REFERENCES

1. J.E. Bäckvall, Acc. Chem. Res., 16, 335 (1983); Pure Appl. Chem., 55, 1669 (1983).
2. J.E. Bäckvall, E.E. Björkman, L. Petterson, and P. Siegbahn, J. Am. Chem. Soc., 106, 0000 (1984).
3. (a) J.E. Bäckvall and R.E. Nordberg, J. Am. Chem. Soc., 103, 4959 (1981). (b) J.E. Bäckvall, S.E. Byström, and R.E. Nordberg, J. Org. Chem., in press. (c) J.E. Bäckvall, R.E. Nordberg, and J.E. Nyström, Tetrahedron Lett., 23, 1617 (1982). (d) J.E. Bäckvall, J.E. Nyström, and R.E. Nordberg, manuscript to be submitted. (e) J.E. Bäckvall, J. Vågberg, and R.E. Nordberg, Tetrahedron Lett., 25, 2717 (1984).
4. (a) B.M. Trost, Acc. Chem. Res., 13, 385 (1980). (b) J. Tsuji, "Organic Synthesis with Palladium Compounds", Springer Verlag, Berlin (1980). (c) B.M. Trost and T.R. Verhoeven, in "Comprehensive Organometallic Chemistry", G. Wilkinson Ed.; Pergamon, Oxford, 1982, vol 8, p. 799.
5. (a) R.M. Magid, Tetrahedron, 36, 1901 (1980). (b) Y. Yamamoto, S. Yamamoto, H. Yataga, K. Maruyama, J. Am. Chem. Soc., 102, 2318 (1980). (c) T. Yamamoto, J. Ishizu, A. Yamamoto, ibid., 103, 6863 (1981). (d) B.M. Trost and M. Lautens, ibid., 5543 (1983). (e) J.L. Roustan, J.Y. Marour, and F. Houlihan, Tetrahedron Lett., 3721 (1979). (f) H.L. Goering, E.P. Seitz, and C.C. Tseng, J. Org. Chem., 46, 5304 (1981). (g) E. Erdik, Tetrahedron, 40, 641 (1984).
6. J.E. Nyström and J.E. Bäckvall, J. Org. Chem., 48, 3947 (1983).
7. D. Seebach and P. Knochel, Helv. Chim. Acta., 67, 261 (1984).
8. P.A. Bartlett, Tetrahedron, 36, 2 (1980).
9. J.W. Wheeler, S.L. Evans, M.S. Blum, H.H.V. Velthius, J.M.F. de Camargo, and R.P. Brazil, Tetrahedron Lett., 4029 (1976).
10. (a) W.H. Pirkle and P.E. Adams, J. Org. Chem., 44, 2169 (1979). (b) R. Bacardit and M. Moreno-Manas, Tetrahedron Lett., 21, 5551 (1980). (c) S. Hanessian, G. Demailly, Y. Chapleur, and S. Leger, J. Chem. Soc. Chem. Comm., 1125 (1981).
11. J.E. Bäckvall, S.E. Byström, and J.E. Byström, Tetrahedron, accepted for publication.
12. P.A. Wade, H.R. Hinney, N.V. Amin, P.D. Vail, S.D. Morrow, S.A. Hardinger, and M.S. Saft, J. Org. Chem., 46, 765 (1981).
13. B. Åkermark, J.E. Nyström, T. Rein, J.E. Bäckvall, P. Helquist, and R. Aslanian, submitted for publication.
14. (a) S. Danishefsky, Acc. Chem. Res., 14, 400 (1981). (b) M. Petrzilka and J.I. Grayson, Synthesis, 753 (1981)
15. (a) J. Tsuji, T. Yamakawa, M. Kaito, and T. Mandai, Tetrahedron Lett., 2075 (1978). (b) B.M. Trost, T.R. Verhoeven, and J.M. Fortunak, ibid., 2301 (1979)
16. O.S. Andell and J.E. Bäckvall, unpublished results.
17. C. Moberg, unpublished results.
18. J.E. Bäckvall and O.S. Andell, J. Chem. Soc. Chem. Comm., 260 (1984).

New applications of organometallic derivatives of Li, Mg, B, Al, Si, Ti & V in selective syntheses

Dieter Seebach

Laboratorium für Organische Chemie, Eidgenössische Technische Hochschule,
CH-8092 Zürich (Switzerland)

Abstract - Dramatic changes in regio-, diastereo- and enantio-selectivity in organic transformations which take place when switching metals in the organometallic compounds investigated are described. Some general rules are proposed governing the stereoselective courses of the reactions investigated; these have been supported by x-ray structural studies.

INTRODUCTION

With a few types of functional groups, organic chemists manage to build almost any natural or unnatural product. Representative functional groups are shown in the center box of Scheme 1. Structures which can be assembled by using the characteristic reactivities of these functional groups are also included in this scheme. Very often, the conditions under which the transformations are carried out do not exhibit satisfactory regio-, diastereo-, and enantio-selectivity. It is the purpose of the present lecture to describe some cases of dramatic changes in selectivity brought about by changing the metal of organometallic reagents, with emphasis on our recent work with derivatives of the early transition metals (Ti, Zr, V).

Scheme 1

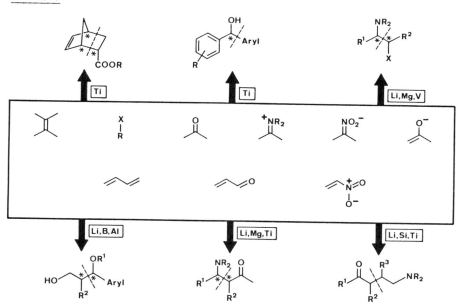

Chemists devoted to organic synthesis have in the last decade turned more and more to increasing the selectivity of known reactions, rather than inventing even more "new novel", and sometimes weird reactions. The main themes of the conference are evidence for this development, and so is my lecture.

TRITYL KETONES IN ALDOL SYNTHESIS

Aldol additions leading to α-substituted β-hydroxycarbonyl derivatives are the more diastereoselective the more bulky the substituent on the carbonylgroup is[1]. The actual diastereoselectivity of a reaction is not nearly as important if crystalline products result as it is
with liquids, because simple recrystallization will usually allow to isolate pure single diastereomers in the first case. Both of these advantageous conditions are fulfilled with tritylketones: they are converted exclusively to Z-enolates by treatment with butyllithium[2,3],
see the silylenolether crystal structure[3] in Scheme 2; these in turn add to aldehydes preferentially with relative topicity[4] ul, see Scheme 3, the selectivities being in excess of
95 % before crystallization. The trityl group can be reductively removed by treatment of the
protected aldols with lithium triethylborohydride. Note that there is a striking difference
between butyllithium, which acts exclusively as a base deprotonating the trityl ketones to

Scheme 2

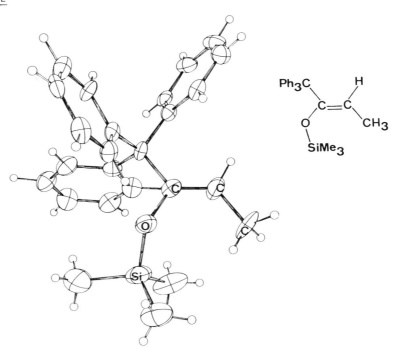

$$Ph_3C \quad\quad H$$
$$\underset{\underset{\displaystyle SiMe_3}{|}}{\overset{\displaystyle |}{C}}=\underset{CH_3}{C}$$

enolates, and the "superhydride"[5,6], which acts solely as a nucleophile adding to the highly
hindered tritylketone carbonyl group - without epimerization at the asymmetric carbon atom in
the α-position[3].

Aluminium enolates of tritylketones are generated with trimethylaluminium[7] in toluene. These
add to aromatic aldehydes and to enolizable ketones with diastereoselective formation of aldols as well. In this case, the selectivity with which the aryl aldols are obtained, results
from equilibration between the stereoisomeric aluminium aldolates in solution, coupled with
the fact that one isomer precipitates and can be filtered off. Luckily, the aldols thus obtained are the products of combination of the aldehyde carbonyl group and the trityl ketone

Scheme 3

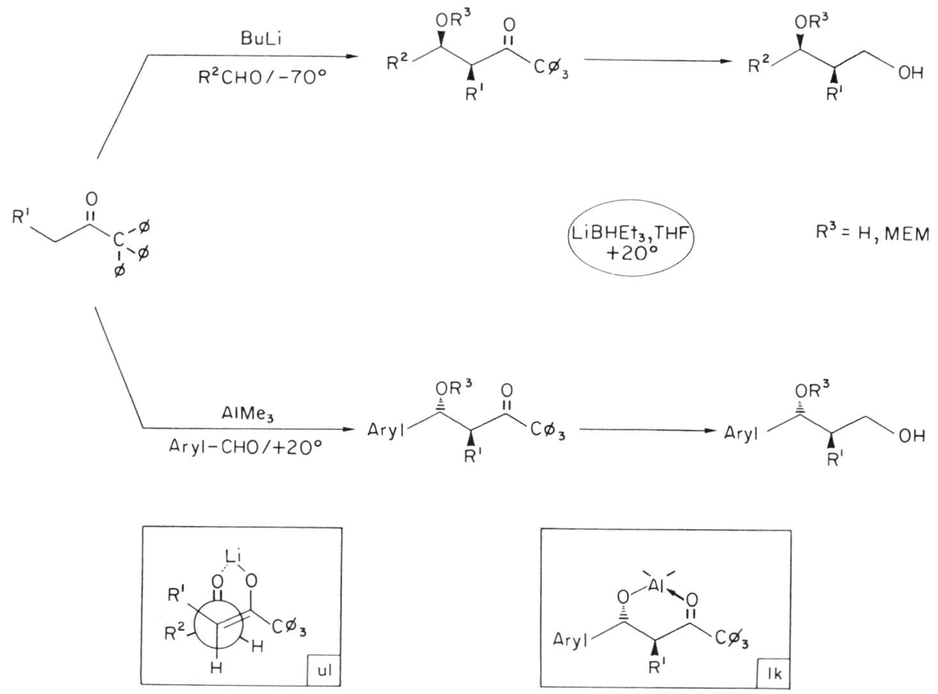

with relative topicity 1k, so that either pure diastereomer is readily accessible by the trityl-ketone method[6], see Scheme 4. Some representative 2-substituted 1.3-diols, with protected secondary and free primary OH-group, are shown in Scheme 5, including the yields in the three step sequence and the diastereoselectivities (without crystallization of the tritylaldol!).

Obviously, the tritylgroup serves several purposes in this method: (i) steric protection of the carbonyl group to furnish enolates with butyllithium, (ii) steric bulk to secure formation of Z-enolates, (iii) steric bulk to provide for high ul-selectivity in the kinetically controlled addition to aldehydes, (iv) anionic leaving group in the cleavage step, and last not least (v) crystallizability of the intermediate aldols.

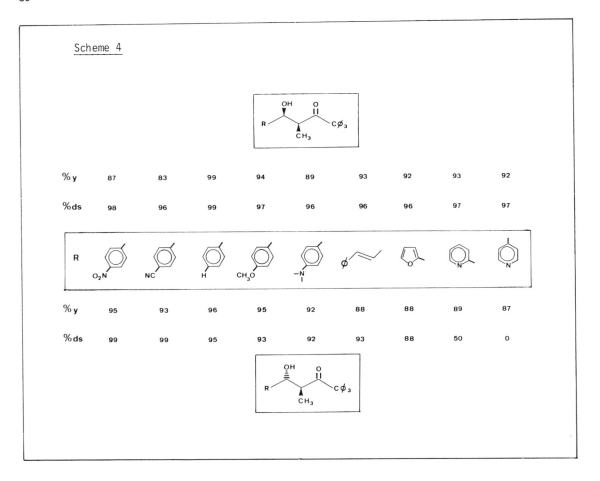

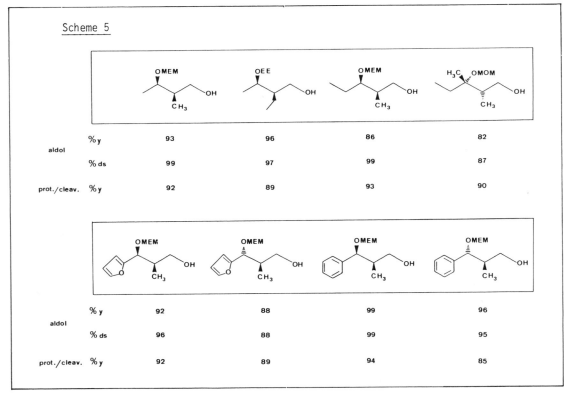

u-1-HYDROXYALKYL-TETRAHYDRO-ISOQUINOLINES

There are numerous alkaloids which are formally derived from the title compounds of this section[8]. While the enantioselective simple alkylation of the THIQ skeleton can be achieved through suitable chiral amidines[9], diastereoselective hydroxyalkylations in the 1-position of THIQ were not possible until we found most recently[10], that replacement of the lithium in the non-selective 2-pivaloyl derivatives by magnesium bromide renders highly selective reagents, see Scheme 6. The products were shown[10,11] to have u-configuration[4], resulting from practically exclusive carbon-carbon bond formation with relative topicity ul. Since the pivaloyl group can be shifted from nitrogen to oxygen, and subsequently removed, either with retention or with inversion of configuration, both possible diastereomers are available by this method[10], and several highly substituted hydroxybenzylated intermediates for THIQ-al-

Scheme 6

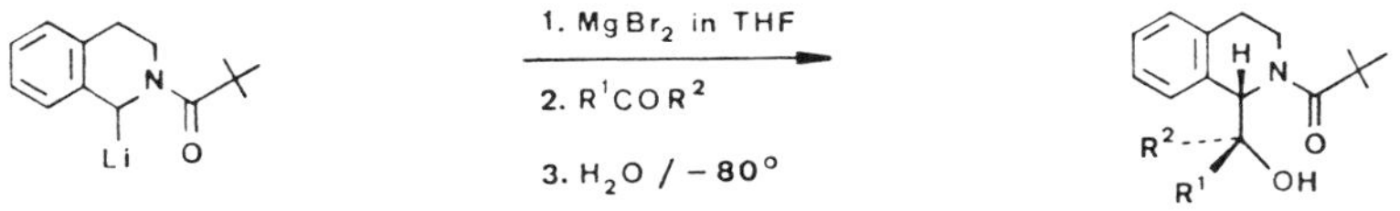

R^1	H	H	H	H	H	C$_6$H$_5$
R^2	C$_2$H$_5$	CH$_2$C$_6$H$_5$	CH=CHC$_6$H$_5$	C$_6$H$_5$	3.4-(OCH$_2$O)C$_6$H$_3$	CH$_3$
% d s	> 97	>97	>97	> 97	>97	96

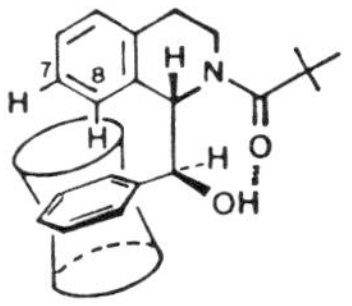

kaloid synthesis have been obtained[12]. Although it was pleasing to be able to carry out this transformation so selectively, it was not satisfying that we had no explanation whatsoever.

Being well aware of possible pitfalls in drawing conclusions about mechanisms from crystal structures of intermediates[13] we undertook an x-ray analysis of the magnesium-organic reagent which might be involved in this reaction. This appeared to us as the only way to at least initiate a discussion of the mechanism. The result[11] is shown in Scheme 7: <u>A</u> is the first structure of a metal derivative of a so-called dipol-stabilized carbanion[14]. The magnesium is octahedrally coordinated; it occupies an equatorial position on the THIQ-1-position; the most electronegative ligands Br and O=C are in an anti-position; two of the three tetrahydrofuran (solvent) ligands form ca. 90° angles with the carbon atom on magnesium. If we assume, (i) that the isolated Mg-complex is also the species responsible for the reaction in solution, (ii) that the THF molecule in the less hindered exo-position - which happens to be the one with the longest Mg,O-bond - is replaced by the aldehyde electrophile in the first

Scheme 7

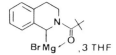

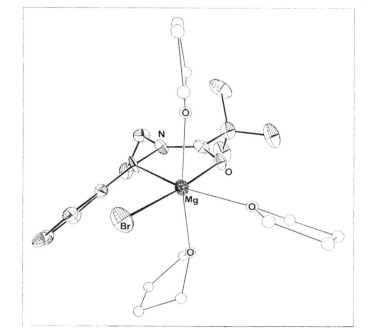

A

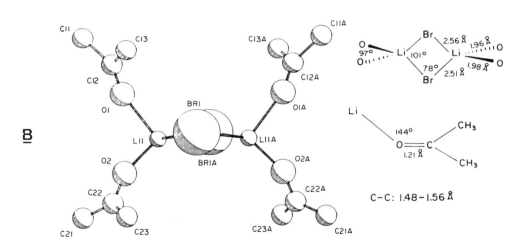

B

C−C: 1.48−1.56 Å

C

ul   lk

step of the reaction, and (iii) that the complex of an aldehyde with a magnesium has the same non-linear geometry as that with a lithium[15,16], see <u>B</u>, we arrive at the following conclusions: The aldehyde molecule experiences steric hindrance when moving from its complexation site towards the carbon atom in the THIQ-1-position with relative topicity lk, see Scheme 7 <u>C</u>, while there is free space for an ul-approach-leading to the observed product. In addition, the more crowded octahedral geometry with 90° angles between ligands on magnesium may be responsible for the higher selectivity as compared with lithium, which is usually tetrahedrally coordinated (ca. 109° angles) - the bond lengths being almost identical.

DIASTEREOSELECTIVE PREPARATION OF NITROALCOHOLS AND NITROKETONES

Nitroalkanes are highly versatile reagents for synthesis, due to their reactivity in carbon carbon bond-forming processes, as well as to the multitude of transformations of the nitro-group to other functional groups[17]. In the past five years, our main efforts in this area were directed towards the stereoselective preparation of 2-nitro-alcohols and of γ-nitro-ketones by nitroaldol additions and by Michael additions of enolates to nitroolefins, respectively.

In an attempt to obtain more highly functionalized products, we found that THP-protected nitroethanol as well as 2-alkoxynitroalkanes can be doubly deprotonated[18] to chelation-stabilized lithium α-lithio-nitronates[19]. These add to aldehydes to give, after diastereoselective protonation and deprotection, 2-nitro-1.3-diols of l-configuration[19], see Scheme 8.

Scheme 8

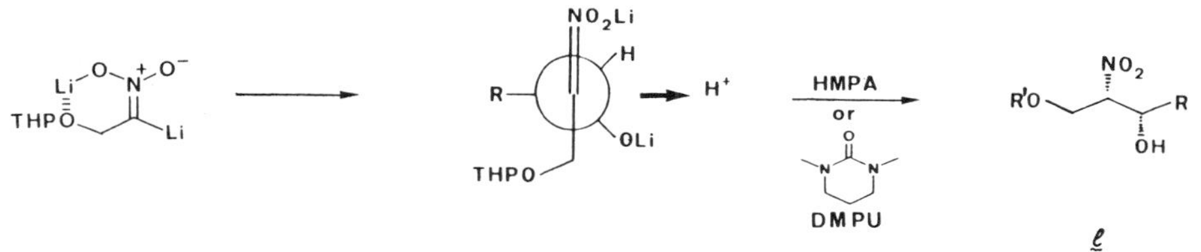

R = CHMe$_2$	93		R = C≡C-C$_{13}$H$_{27}$	88
R = C$_4$H$_9$	85		R = C$_6$H$_5$	>95
R = c-C$_6$H$_{11}$	>95	(% ds)	R = p-CH$_3$O-C$_6$H$_4$	80
R = CH=CH-C$_6$H$_5$	91		R = p-NC-C$_6$H$_4$	75

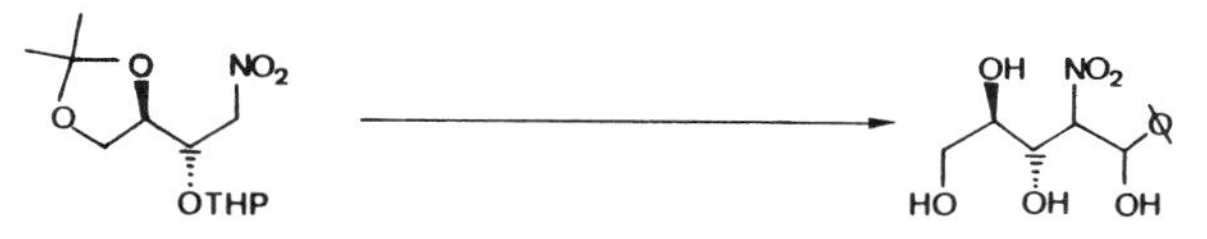

1 diastereomer

73%, mp 141°

In this case, the diastereoselectivity depends upon the presence of a cosolvent (hexamethyl-
-phosphoric acid triamide, HMPTA, or dimethyl-propylene urea, DMPU[20]). An extensive discussion of the mechanism of this reaction is given in a full paper to be published shortly[19].

The coupling between the two trigonal centers of an enamine (or an enolate) and a nitroolefin
takes place with relative topicity lk[21]. This steric course of reaction could be reversed
by employing z- instead of E-nitropropene in additcns of lithium enolates[22], while both E-
and z-nitrostyrene combine with Li-cyclohexenolate and with morpholino-cyclohexene with relative topicity lk[23,24], see Scheme 9. We have now found that switching to the trimethyl-

Scheme 9

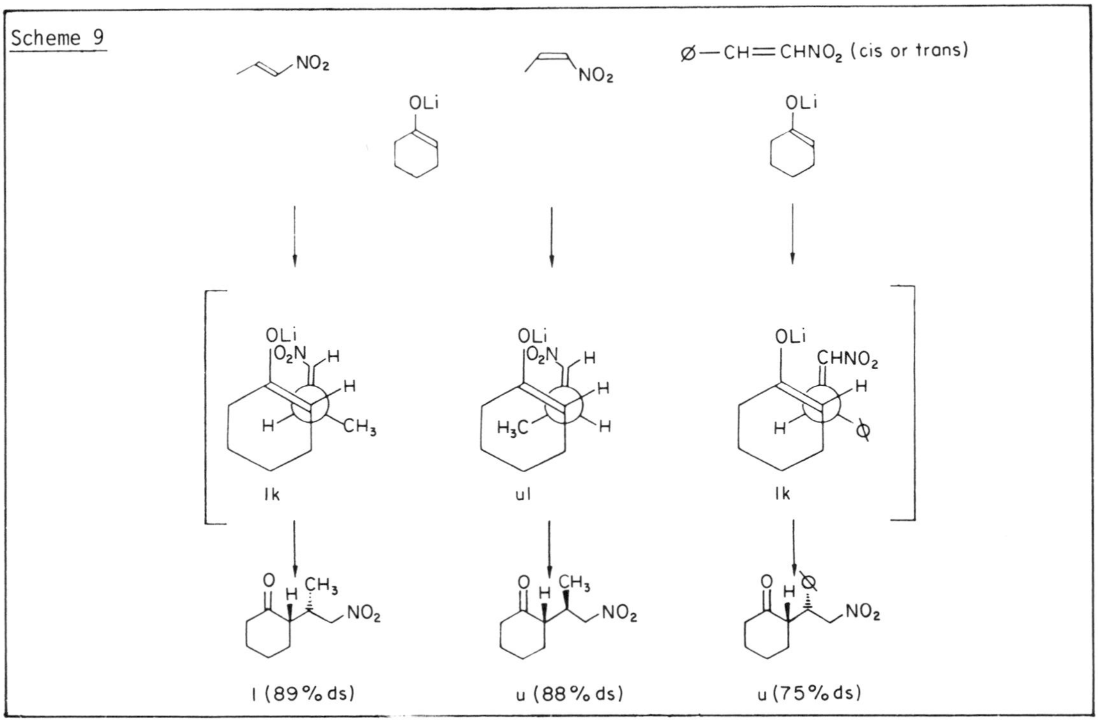

Scheme 10

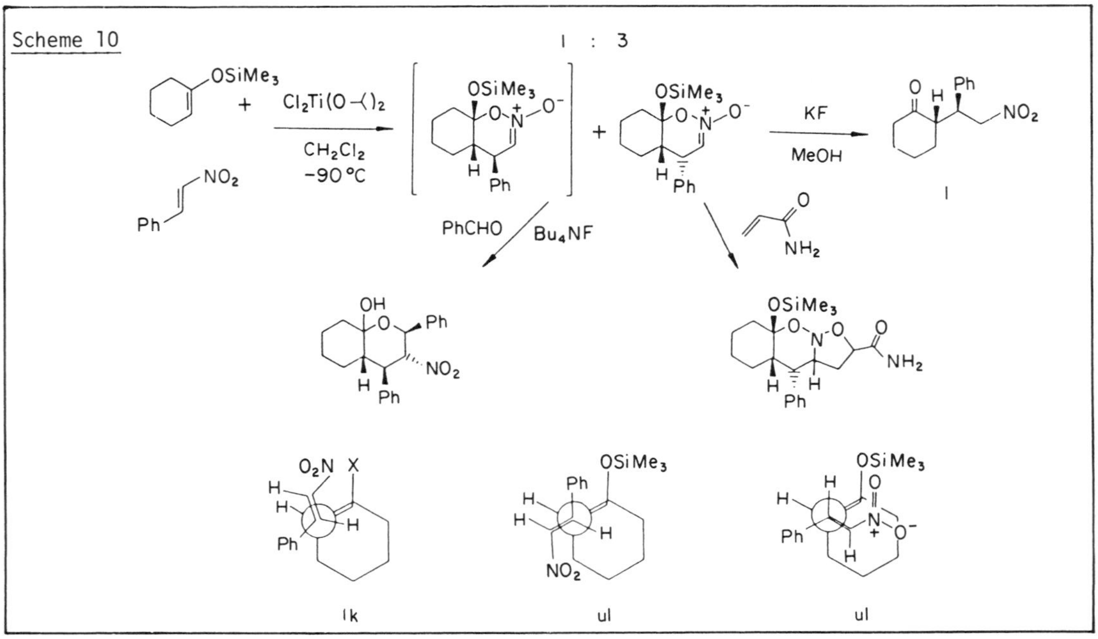

silyl-enolether of cyclohexanone and to Ti(IV) Lewis acid mediated conditions leads to a reversal of the steric course, also with nitrostyrenes[25], see Scheme 10. The reaction is carried out in methylene chloride at low temperature, with dichloro-diisopropoxy-titanium[26], it furnishes a mixture of (4+2)-cycloadducts with a 4:1 preference of the one formed by C,C bond formation with relative topicity ul. A discussion and leading references to analogous effects of TiX_4 on related reactions are given in a forthcoming paper[25]. The cycloadduct can be used for further elaboration of the carbon skeleton, for instance by 1.3-dipolar cycloadditions or silyl nitroaldol additions[27].

SELECTIVITIES ACHIEVED WITH Ti- AND Zr-REAGENTS

In three recent review articles[28a-c], we have given a survey of the striking functional--group-, regio-, diastereo-, and enantioselectivities observed with organotitanium and -zirconium reagents $(RTiX_3, TZrX_3)$[28d]. The accompanying five Schemes 11-15 give some examples. Thus, nitro group-containing substrates, which are not compatible with the conventional polar RLi and RMgX-derivatives, can be employed (Scheme 11)[29,30]. The trichlorotitanium iminoacylchlorides from $TiCl_4$ and isocyanides add to aldehydes and ketones to give, after aqueous workup, α-hydroxy-carboxylic acid amides containing additional functional groups such as bromoaryl, morpholino, or diethylphosphonate[31,32] (Scheme 12). Lithiated derivatives of allylmercaptanes can be rendered highly regioselective (α- vs. γ-reactivity) by addition of chloro-triisopropoxy-titanium prior to the electrophiles[33-35] (Scheme 13). Triphenoxy-crotyltitanium adds to aliphatic and aromatic aldehydes[36] and ketones[37] preferentially with relative topicity lk (Scheme 14). Examples of asymmetric addition of aryl groups to aromatic aldehydes[28c,38] to give chiral benzhydrols of high enantiomeric excesses (at least after recrystallization!) are given in Scheme 15; the reagent is derived from binaphthol; the relative topicity of the reaction is lk. The examples in Schemes 11-15 demonstrate the great potential of the titanate- and zirconate-derived reagents for selective syntheses of multifunctional products - with a minimum of protections and detours!

Scheme 11

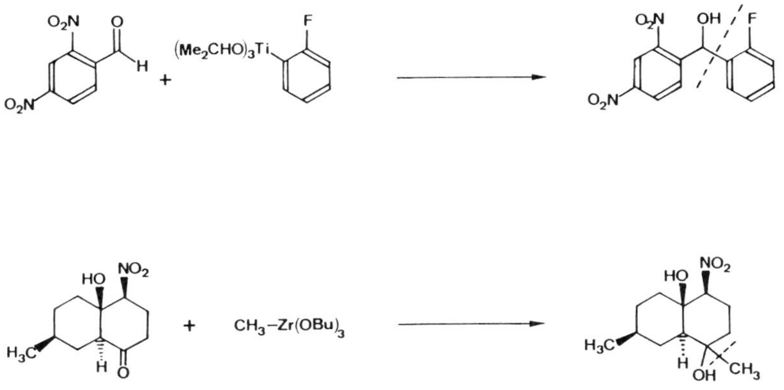

Scheme   12

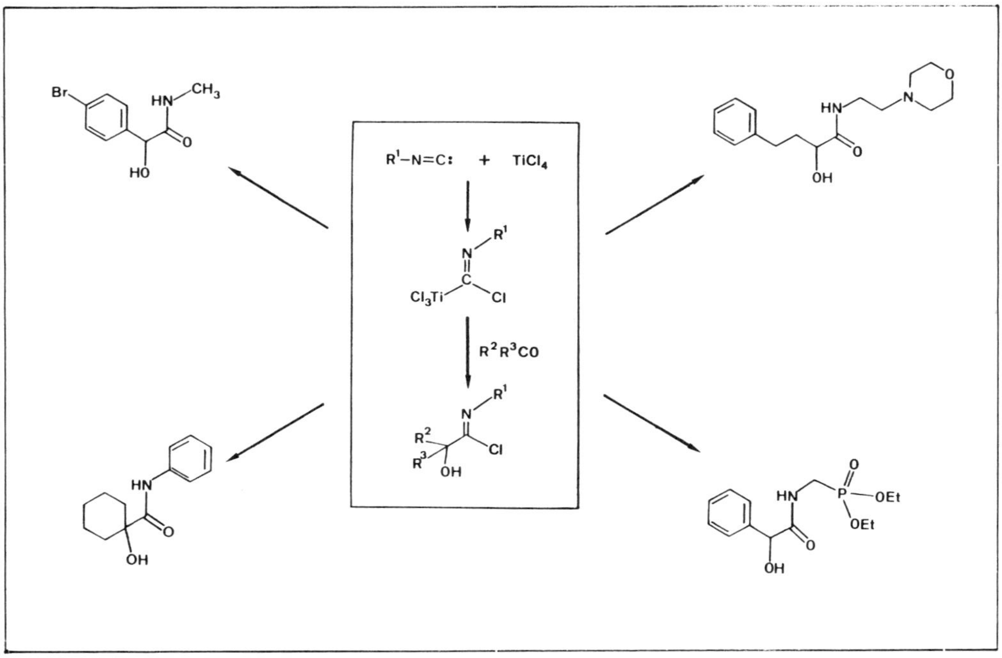

Scheme   13

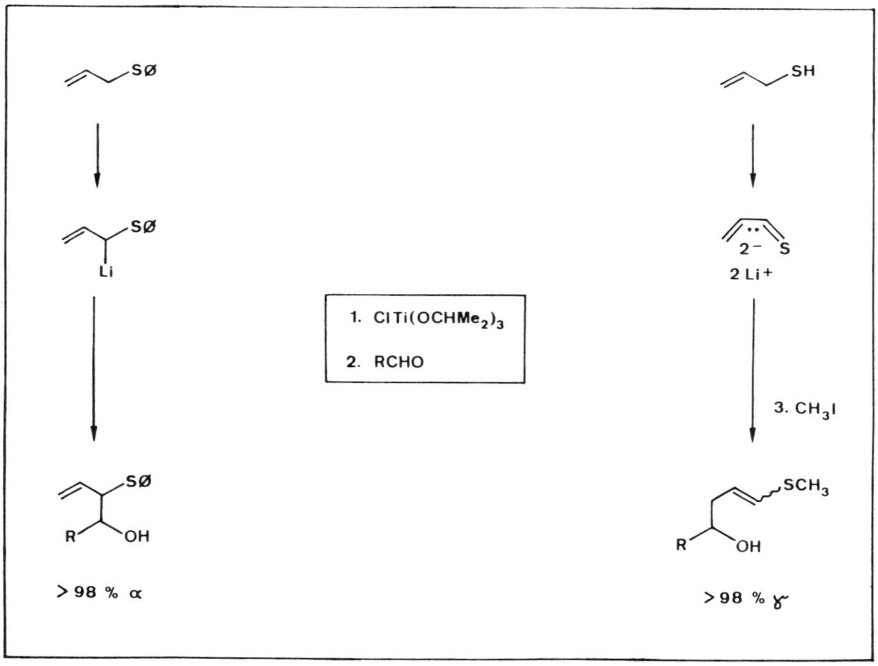

Scheme 14

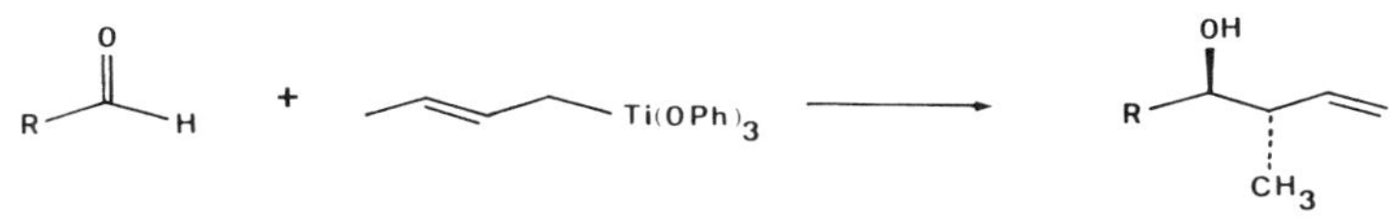

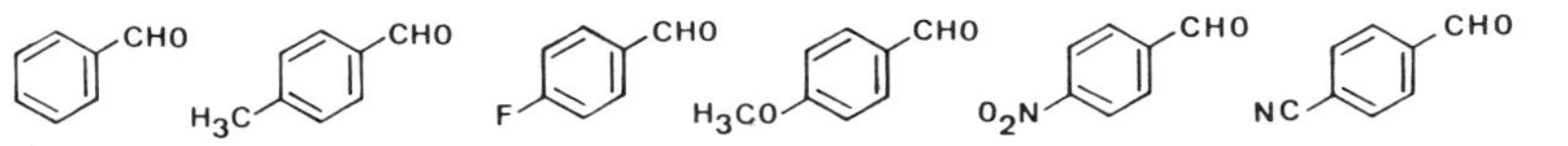

| % ds | 85 | 93 | 93 | 98 | 81 | 80 |

| % ds | 96 | 99 | 93 | | 95 | 93 |

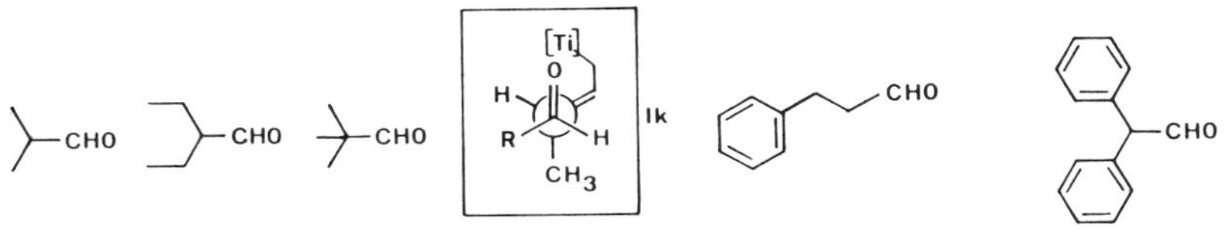

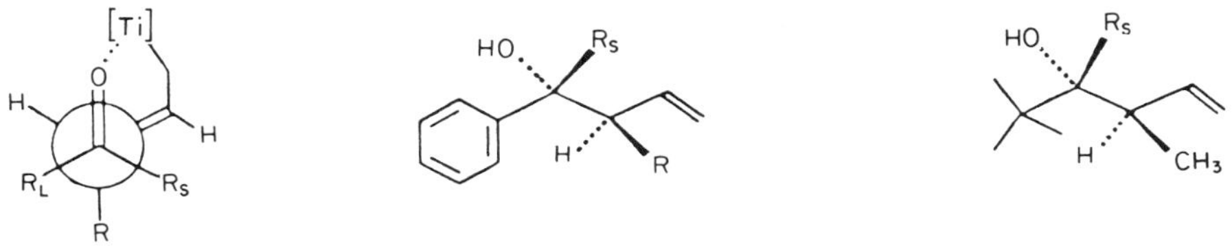

lk (RL>RS)
ul (RS>RL)

RL	RS	% ds with R		
		CH$_3$	C$_4$H$_9$	CH(CH$_3$)$_2$
O	H	85	93	98
O	CH$_3$	88	87	87
O	C≡C—CH$_3$	72	77	77
Me$_3$C	H	>98		
Me$_3$C	CH$_3$	>98		
Me$_3$C	O	96		

Scheme 15

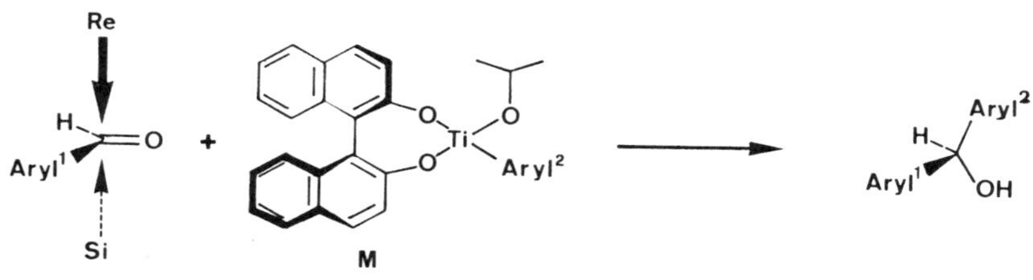

relative topicity   lk

$\left(\text{M from Re}-,\ \text{P from Si}-\text{face}\right)$

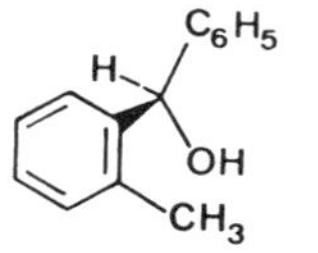

85 % ee
$[\alpha]_D^{25} = +0{,}67°$

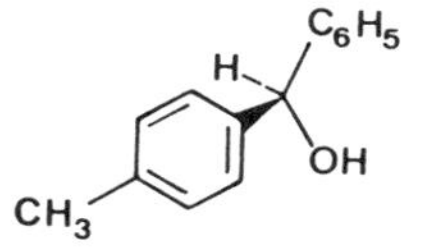

86 % ee
$[\alpha]_D^{25} = +10{,}3°$

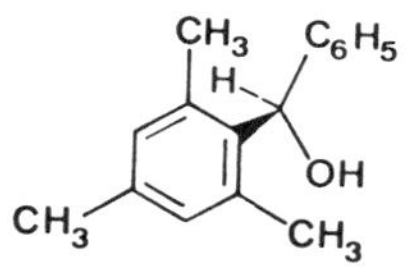

60 % ee
$[\alpha]_D^{25} = +148{,}8°$

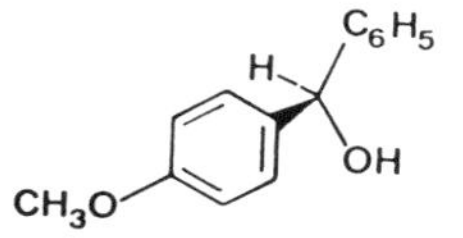

82 % ee
$[\alpha]_D^{25} = +18{,}8°$

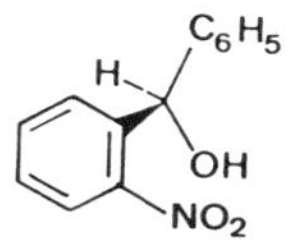

39 % ee
$[\alpha]_D^{25} = +79{,}8°$

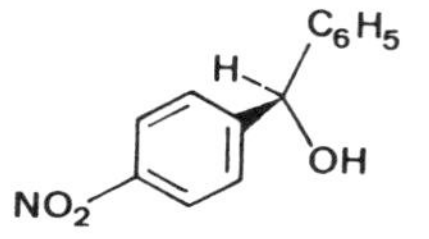

76 % ee
$[\alpha]_D^{25} = -74{,}9°$

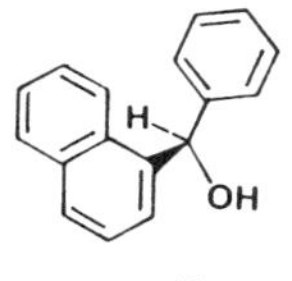

> 98 % ee
$[\alpha]_D^{25} = +59{,}5°$

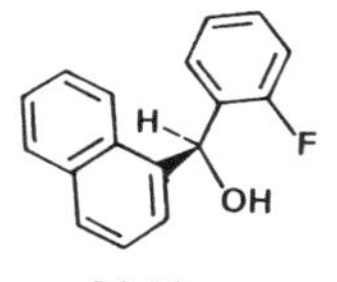

89 % ee
$[\alpha]_D^{25} = +112°$

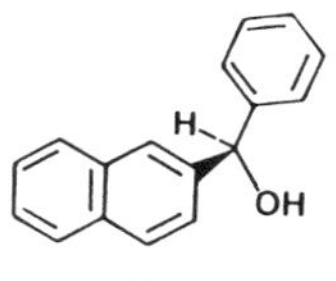

> 98 % ee
$[\alpha]_D^{25} = +7{,}4°$

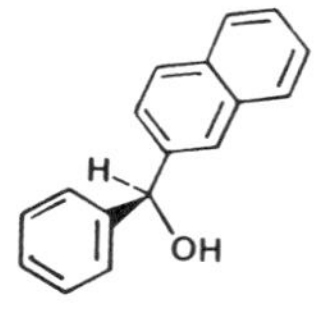

46 % ee
$[\alpha]_D^{25} = -7{,}4°$

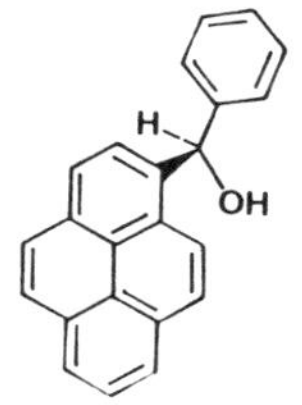

95 % ee
$[\alpha]_D^{25} = +264°$

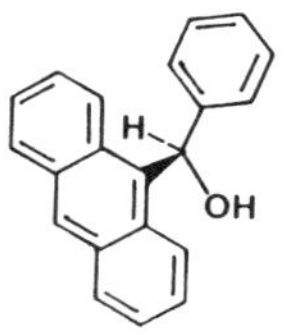

69 % ee

TiCl$_4$-MEDIATED, GENERALIZED MANNICH REACTIONS

The alkyl and aryl organotitanium reagents with three alkoxy or aryloxy groups on the metal were found[28] to add to aldehydes and ketones with overall transfer of the organic group to give secondary and tertiary alcohols, respectively. When the RO-ligands on titanium were replaced by the more electronegative chlorines[28d] or by the less electronegative dialkylamino groups[28,39], different reactions took place with the corresponding reagents. In the first case (MeTiCl$_3$)[40], a geminal dimethylation was observed, in the second case [MeTi(NEt$_2$)$_3$][39], an aminative alkylation, i.e. replacement of the carbonyl oxygen by an alkyl and a dialkyl-amino group, resulted, see Scheme 16. Obviously, the diethylaminosubstituted reagent is

Scheme 16

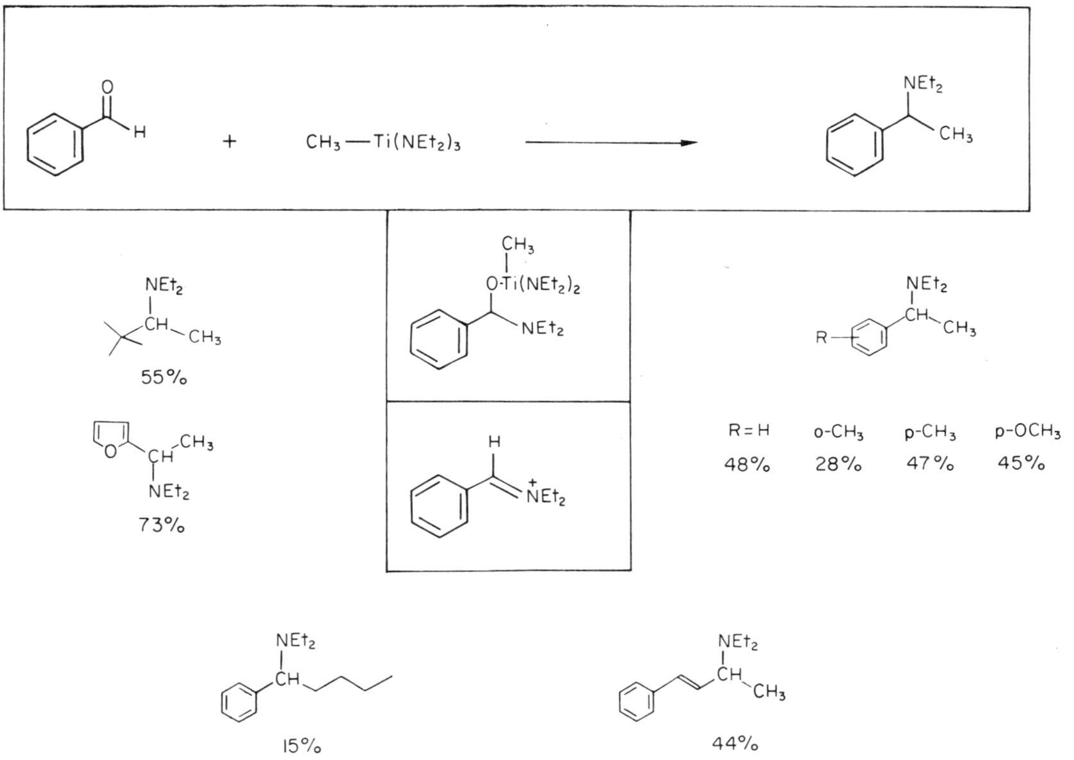

transferring an amino group first. The adduct is a source of an iminium ion, or at least is synthetically equivalent to one, and the OTi moiety is substituted by an alkyl group. The yields of the original procedure were not satisfaying, and the method was only applicable to non-enolizable aldehydes. We sought for alternative routes of generating geminal dialkyl-amino-titanoxy derivatives of the general type invoked as intermediates in the direct aminative alkylation. We found[28c] that the adducts of lithium dialkylamides to aromatic alde-hydes and to cinnamaldehyde could be "titanated" with tetrachlorotitanium, and that the re-agents thus formed also act as electrophilic α-aminoalkylating reagents. The yields obtained by this variation were much higher, see the numbers in parentheses, Scheme 17. The adducts of organolithium reagents to formamides, to benzamides, but also to the enolizable N-methyl--pyrrolidone, could be used as well for "titanation" and subsequent reaction with a second organolithium reagent, see Scheme 18. In this way, a geminal dialkylation of the amide car-

Scheme 17

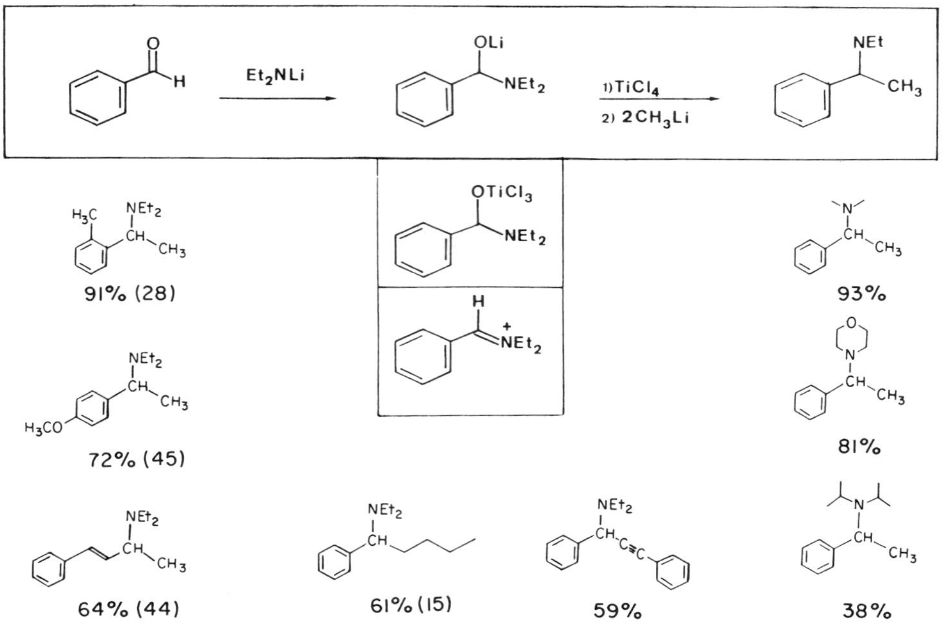

Scheme 18

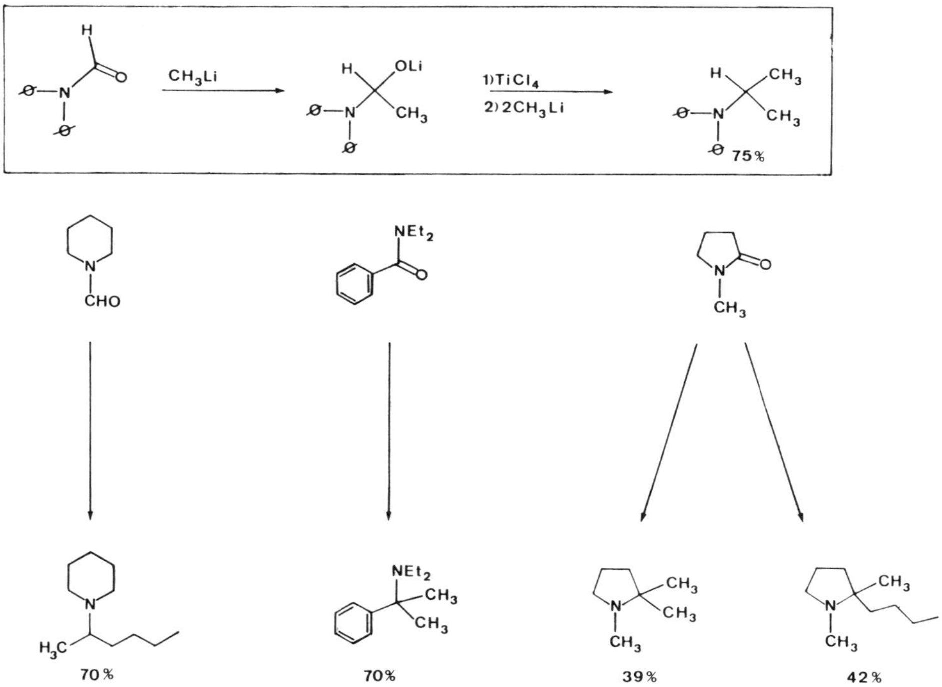

bonyl carbon by two different alkyl groups becomes feasable. Finally, aliphatic aldehydes can be employed for aminative alkylations if a dialkylamino-trichloro-titanium is first added, followed by an alkyllithium reagent, see Scheme 19.

Scheme 19

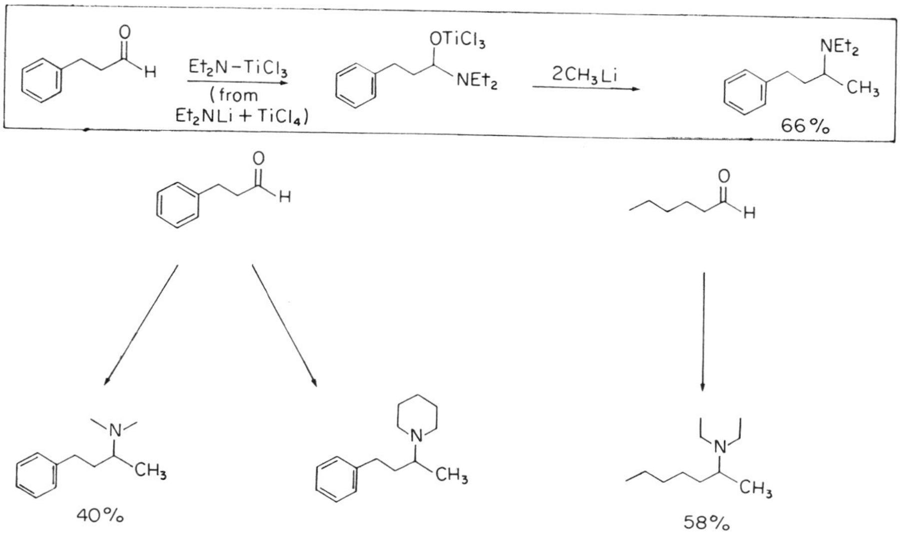

In all three cases, the trichlorotitanium α-dialkylamino-alkoxide intermediates are present as highly insoluble precipitates. It can be shown by filtration or decantation that the supernatant liquid does not contain any reagent. The insolubility in hexane, toluene, or diethyl ether might be taken as evidence for the iminium-salt character of these reagents. In one case, we have filtered off the solid, dried it in vacuo (oil pump), and stored it under an inert atmosphere for a day at up to 40 °C - without deterioration: reaction with a nucleophile still gave the expected product. When combined with the organolithium nucleophiles, the resulting reaction mixtures may slowly change color and macroscopic appearance, they usually look ugly and are heterogeneous until the workup procedure commences. Vigorous stirring is necessary throughout.

Thus, the same type of reagent is generated by three different routes, and used for aminoalkylations of nucleophiles, see Scheme 20, a highly versatile method of preparing tertiary amines[28c,32]. Primary and secondary amines may be available also by this route, if N-benzyl- or N-silylamines are applied[32].

Scheme 20

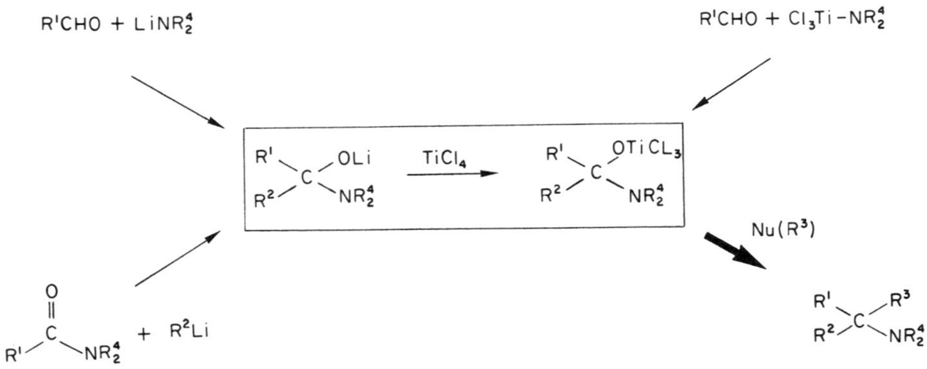

Mannich bases are β-dialkylamino ketones[41]. They are usually made from secondary amines, an aldehyde, mostly formaldehyde, and a ketone component. A recent modification is the dimethyl-aminomethylation of lithium enolates by the so-called Eschenmoser salt[42,43]. In those cases, in which two new asymmetric carbon atoms are generated intermolecularly, the two diastereo-meric Mannich bases are formed with poor stereoselectivity.

We hoped to be able to extend the scope of the classical Mannich reaction, and maybe perform it diastereoselectively, by employing our α-trichloro-titanoxy amines as sources of iminium salts[43]. Since silylenolether did not combine directly with these aminoalkylating re-agents[32], lithium enolates were used as nucleophiles. These had to be free of amines, and therefore had to be generated from silylenolethers and methyllithium, or else the diisopro-pylamine, cogenerated with the enolates from carbonyl compounds and the lithium amide LDA, had to be removed by evaporation. The β-amino-ketones and esters shown in Scheme 21 were ob-tained in good yields[32,44,45]. The ketones were extremely labile towards elimination of amines, with formation of α.β-unsaturated ketones, while the esters were quite stable. All three routes (Scheme 20) to aminoalkylating reagents, as outlined above, can be used for the preparation of these Mannich bases.

To test the stereoselectivity of the reaction, we used cyclohexanone enolate and the iminium salts derived from aromatic aldehydes, see Scheme 22 and Figure 1. Aminobenzylated cyclo-hexanones were formed, and we were pleased to find, that one of the two possible diastereo-mers prevailed to the extend given by the % ds numbers in the scheme (percentages of the major diastereomers in the crude product mixtures[44]). We do not as yet know the configuration of the main products. In the series investigated, however, the same configurational isomer appears to arise preferentially (α-N-CH-signal at lower field in the NMR spectrum, cf. Fig.1). [In the meantime, we have been able to determine the x-ray crystal structure[46] of the adduct from benzaldehyde, piperidine, and cyclohexanone. The configuration is u. Thus, the relative topicity of the addition is lk - as is the one of the corresponding aldol addition.]

Scheme 21

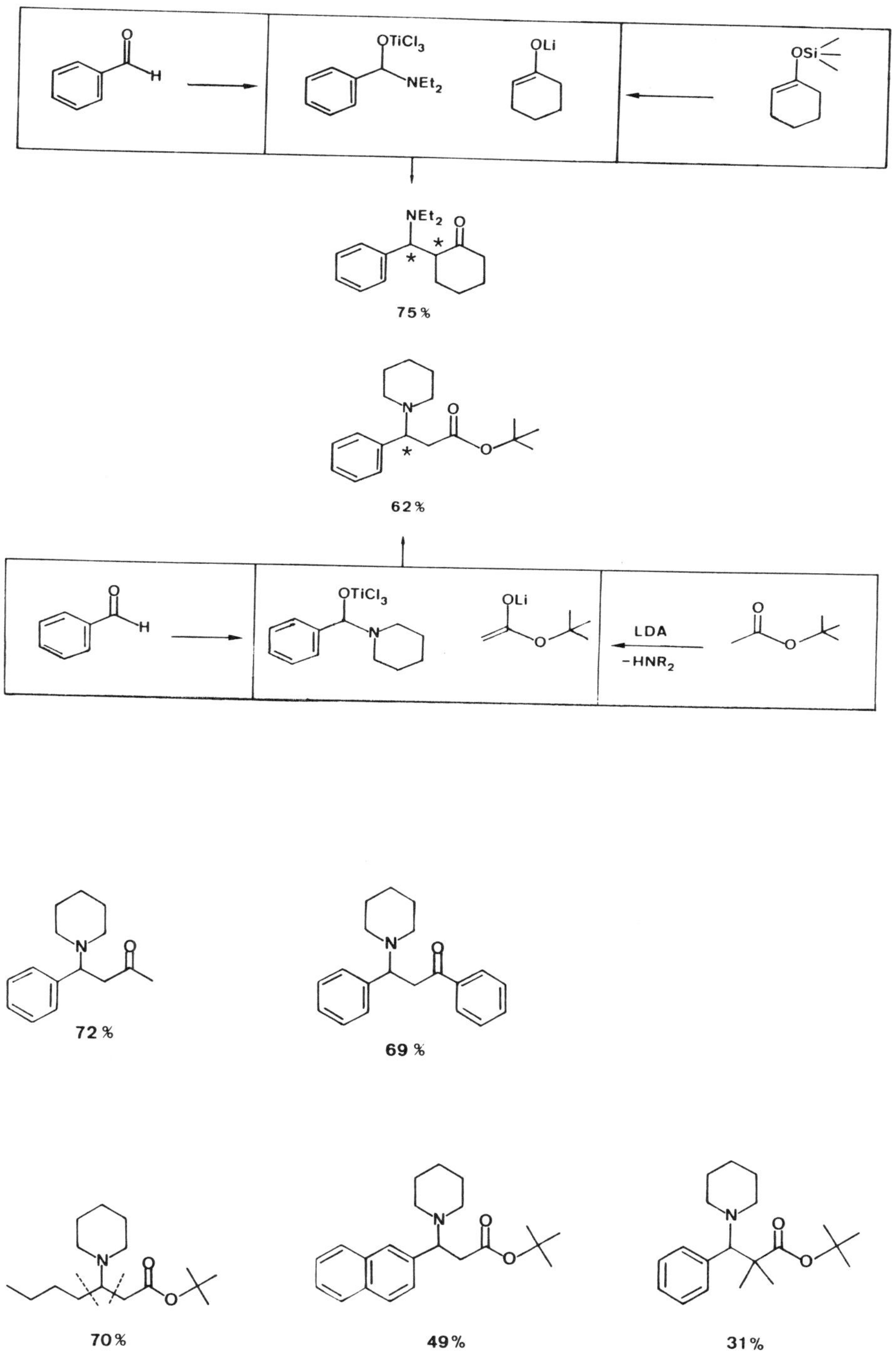

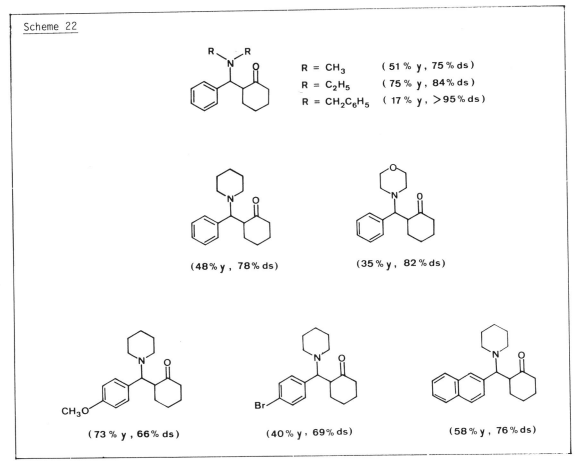

Scheme 22

R = CH$_3$ (51 % y , 75 % ds)
R = C$_2$H$_5$ (75 % y , 84 % ds)
R = CH$_2$C$_6$H$_5$ (17 % y , >95 % ds)

(48 % y , 78 % ds)

(35 % y , 82 % ds)

(73 % y , 66 % ds)

(40 % y , 69 % ds)

(58 % y , 76 % ds)

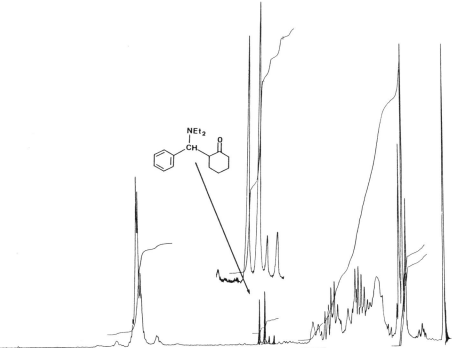

Fig. 1. ^{1}H-NMR spectrum (CDCl$_3$, 90 MHz) of the crude product from benzaldehyde, lithium diethylamide, titanium tetrachloride, and lithium cyclohexenolate.

As demonstrated in Scheme 23, the experiments described here make the Mannich reaction, i.e. the combination of a carbonyl compound, an amine, and a CH-acidic component to give a tertiary amine with formal elimination of water, a much more general process. In principle, the CH-acidic component can be any compound of which a lithium derivative is available - including methane! There are still some drawbacks to overcome, such as the necessity of using two equivalents of the lithium derivative per equivalent of aminoalkylating reagent. It must also be pointed out that carefully controlled reaction conditions are necessary in the multi-step one-pot procedures. Finally, after we will hav
will be necessary to study the mechanism of these reactions in more detail.

Scheme 23

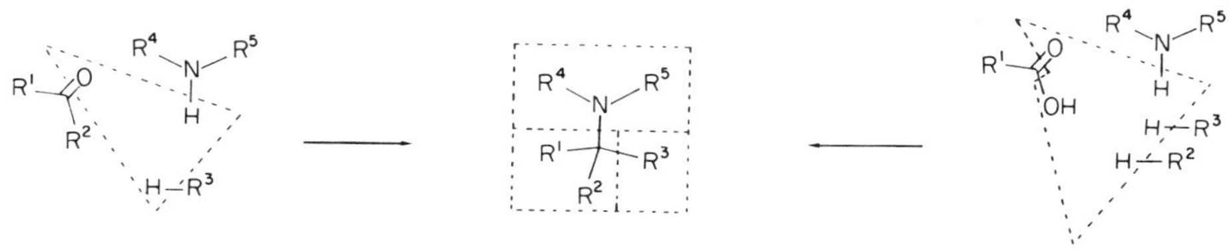

AMINATIVE REDUCTIVE COUPLING OF AROMATIC ALDEHYDES WITH $CH_3V(NR_2)_3$

After having found the aminative alkylation of aromatic aldehydes by alkyl-tris(dialkyl-amino)-titanium, we decided to investigate and compare the reactivity of analogous vanadium derivatives[47]. We hoped that, due to the many stable oxidation states of this transition metal, we would observe fundamentally different reactions.

Indeed, when the methyl-tris(diethylamino)-vanadium was allowed to react with benzaldehyde in diethylether, a ca. 50 % yield of N.N.N'.N'-tetraethyl-1.2-diphenyl-ethylene diamine was formed, see Scheme 24. Thus, reductive coupling of the carbon atoms of the aldehyde with re-

Scheme 24

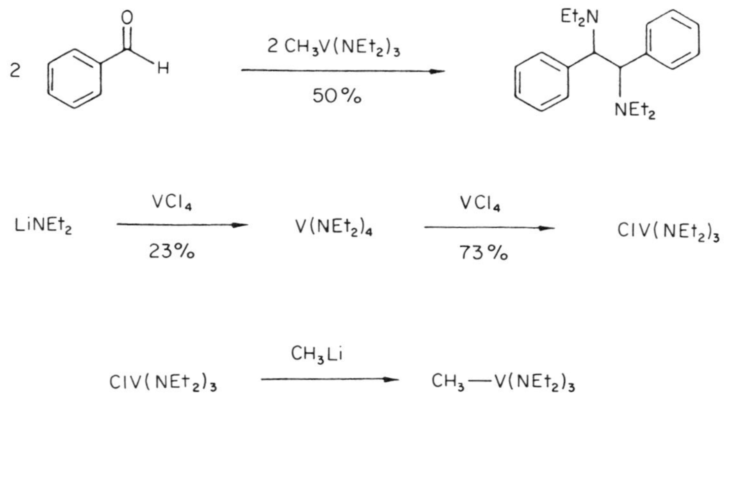

placement of the oxygen by a diethylamino group occurs; the vanadium is oxidized from V(4)
to V(5). The reagent effecting the coupling process is obtained by treatment of tetrakis(diethylamino)-vanadium first with tetrachloro-vanadium (ratio 3:1) and then with methyllithium.
Alternatively, coupling reagents of this type can also be generated in situ by combining
tetrachlorovanadium first with three equivalents of a lithium dialkylamide and then with one
equivalent of methyllithium. The yields of diaryl ethylene diamines obtained after subsequent
addition of aromatic or heteroaromatic aldehydes were somewhat lower than with the reagent
from isolated tetrakis(dialkylamino)-vanadium, see Scheme 25. In all cases ca. 1:1 mixtures
of meso and racemic diamines were isolated and separated chromatographically. The meso-iso-
mers migrate faster on Al_2O_3 columns ($C_nH_{2n+2}/Et_2O/Et_3N$), melt at higher temperature, show
the signals of the benzylic α-N-CH hydrogens at higher field in the NMR spectra, and hardly
give rise to europium shift reagent-induced changes of the ^{1}H-NMR spectra (cf. bottom part of
Scheme 25). The different reactions taking place with analogous titanium and vanadium derivatives can be explained with the same primary reaction step, see Scheme 26: an adduct is first
formed with transfer of a dialkylamino group; formally, this gives rise to an iminium ion in
the case of titanium, and to an α-amino-benzyl radical in the case of vanadium. With the titanium reagent a methyl group is transferred in a subsequent step, while the radical dimerizes to the ethylene diamine, the aminative coupling product observed with the vanadium reagent. It seems that the vanadium-bound methyl group takes the role of a "spectator". The
fact, that the diamine from benzaldehyde is formed with $V(NEt_2)_4$, $VCl(NEt_2)_3$ and $MeVCl(NEt_2)_2$

Scheme 25

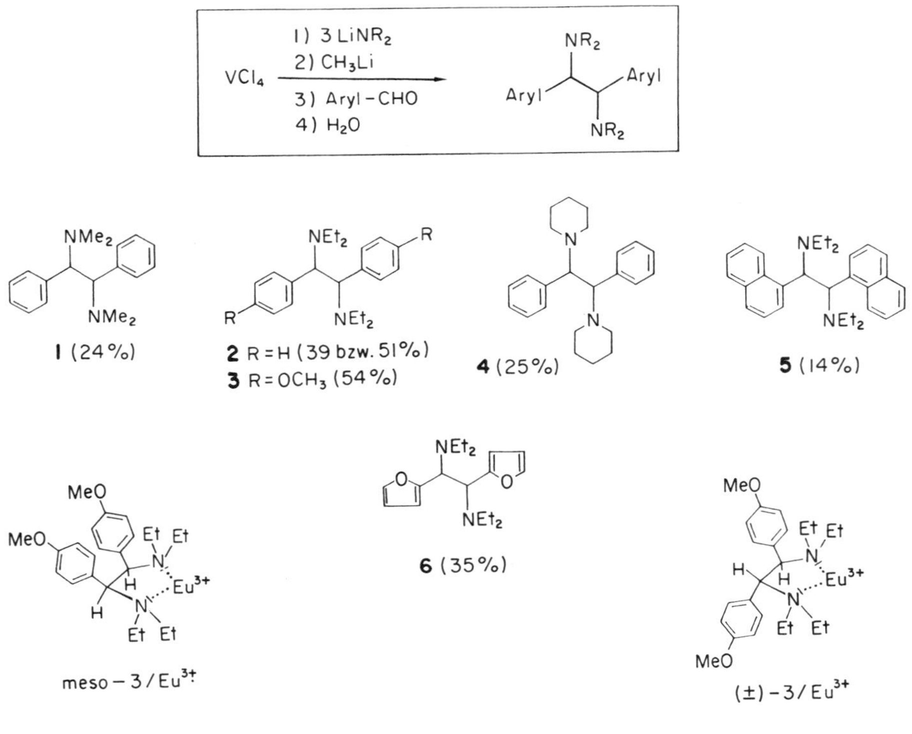

meso − 3 / Eu³⁺

6 (35%)

(±) − 3 / Eu³⁺

Scheme 26

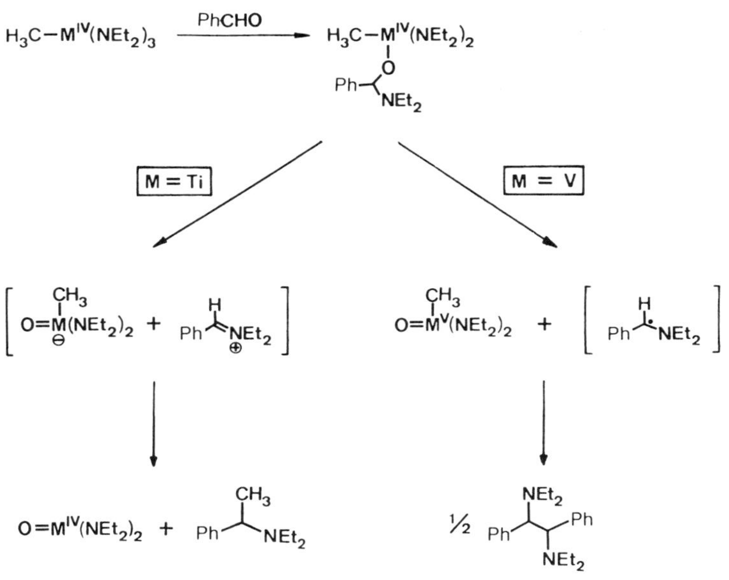

in only 15, 10, and 8 % yield, respectively, shows, however, how important the ligand sphere, and thus the redox potential of the vanadium is for the coupling reaction to take place.

A full account with experimental details and characterization of twelve diaryl ethylene di-amines has appeared in the meantime[48]. We have also found[32] that diaminoethanes can be prepared by adding potassium metal to certain α-dialkylamino trichlorotitanium alkoxides (see Schemes 17, 19, 20).

CONCLUSION

The examples described here demonstrate the powerful effects of different metals on the selectivity of organometallic reagents. Much work is needed in order to be able to choose the "best" metal not so much by trial and error, but by understanding employment. A possible route to this desirable situation is the x-ray structure determination of intermediates, as used in the case of the 1-magnesio-tetrahydroisoquinoline derivative described above.

ACKNOWLEDGEMENT

I am grateful to Silvia Sigrist, Martin Schiess, and Albert K. Beck for their help in preparing the manuscript. My special thanks go to the coworkers who have done the experiments, and in part the thinking, which led to the results described here; their names are given in the list of references. Generous support by the SANDOZ AG (Basel) and by the Schweizerischer Nationalfonds zur Förderung der wissenschaftlichen Forschung is gratefully acknowledged.

LIST OF REFERENCES

1) C.H. Heathcock, Asymmetric Synthesis, Vol. 3, Academic Press, New York, 1984.

2) R. Locher, Dissertation No. 6917, ETH Zürich.

3) D. Seebach, M. Ertas, R. Locher and W.B. Schweizer, Helv. Chim. Acta 68 (1985), to be published in issue No. 1.

4) D. Seebach and V. Prelog, Angew. Chem. 94, 696 (1982); Angew. Chem., Int. Ed. Engl. 21, 654 (1982).

5) H.C. Brown and S. Krishnamurthy, J. Am. Chem. Soc. 95, 1669 (1973); H.C. Brown, S.C. Kim and S. Krishnamurthy, J. Org. Chem. 45, 1 (1980).

6) M. Ertas, ETH Zürich, 1984, hitherto unpublished results.

7) A. Meisters and T. Mole, Inst. J. Chem. 27, 1655 (1974); E.A. Jeffery and A. Meisters, J. Organomet. Chem. 82, 307 (1974).

8) G. Grethe: "Isoquinolines, Part 1": The Chemistry of Heterocyclic Compounds, Vol. 38, J. Wiley & Sons, New York, 1981; T. Kametani: The Chemistry of Isoquinoline Alkaloids, Hirokawa-Elsevier, Tokyo-Amsterdam, 1969.

9) A.I. Meyers and L.M. Fuentes, J. Am. Chem. Soc. 105, 117 (1983); A.I. Meyers, L.M. Fuentes and Y. Kubota, Tetrahedron 40, 1361 (1984).

10) D. Seebach and M.A. Syfrig, Angew. Chem. 96, 235 (1984); Angew. Chem., Int. Ed. Engl. 23, 248 (1984); M.A. Syfrig, Dissertation No. 7574, ETH Zürich, 1984.

11) D. Seebach, J. Hansen, P. Seiler and J.M. Gromek, J. Organomet. Chem. (1985), in press.

12) J. Huber, ETH Zürich, hitherto unpublished results.

13) A.S.C. Chan, J.J. Pluth and J. Halpern, J. Am. Chem. Soc. 102, 5952 (1980).

14) D. Seebach and D. Enders, Angew. Chem. 87, 1 (1975); Angew. Chem., Int. Ed. Engl. 14, 15 (1975); P. Beak et al., Chem. Rev. 78, 275 (1978); P. Beak et al., Chem. Rev. 84, 471 (1984); D. Seebach, J.J. Lohmann, M.A. Syfrig and M. Yoshifuji, Tetrahedron 39, 1963 (1983) [Tetrahedron Symposia-in-Print No. 9].

15) R. Amstutz, Dissertation No. 7210, ETH Zürich, 1983.

16) R. Amstutz, J.D. Dunitz, Th. Laube, W.B. Schweizer and D. Seebach, Chem. Ber. 118 (1985), in press.

17) D. Seebach, E.W. Colvin, F. Lehr and Th. Weller, Chimia 33, 1 (1979).

18) D. Seebach and F. Lehr, Angew. Chem. 88, 540 (1976); Angew. Chem., Int. Ed. Engl. 15, 505 (1976).

19) M. Eyer and D. Seebach, J. Am. Chem. Soc. 107 (1985), in press.

20) T. Mukhopadhyay and D. Seebach, Helv. Chim. Acta 65, 385 (1982); D. Seebach, A.K. Beck, T. Mukhopadhyay and E. Thomas, Helv. Chim. Acta 65, 1101 (1982); D. Seebach, R. Henning and T. Mukhopadhyay, Chem. Ber. 115, 1705 (1982).

21) A. Risaliti, M. Forchiassin and E. Valentin, Tetrahedron Lett. 1966, 6331; D. Seebach and J. Goliński, Helv. Chim. Acta 64, 1413 (1981).

22) R. Häner, Th. Laube and D. Seebach, Chimia 38, 255 (1984).

23) D. Seebach, A.K. Beck, J. Goliński, J.N. Hay and Th. Laube, Helv. Chim. Acta 68 (1985), in press.

24) Th. Laube, Dissertation No. 7649, ETH Zürich, 1984.

25) D. Seebach and M.A. Brook, Helv. Chim. Acta 68 (1985), in press.

26) T. Mukaiyama, Angew. Chem. 89, 585 (1977); Angew. Chem., Int. Ed. Engl. 16, 817 (1977); T. Mukaiyama, Organic Reactions 28, 203 (1982).

27) M.A. Brook, ETH Zürich, hitherto unpublished results.

28) a) "Organometallverbindungen von Titan und Zirconium als selektive nucleophile Reagentien für die Organische Synthese", B. Weidmann and D. Seebach, Angew. Chem. 95, 12 (1983); Angew. Chem., Int. Ed. Engl. 22, 31 (1983).
b) "Titanium and Zirconium Derivatives in Organic Synthesis. A Review with Procedures". D. Seebach, B. Weidmann and L. Widler, Modern Synthetic Methods 1983, Vol. 3, 217-353, Salle + Sauerländer (Aarau, Switzerland) and J. Wiley & Sons (New York, USA), 1983.
c) "Some recent advances in the use of titanium reagents for organic synthesis",

D. Seebach, A.K. Beck, M. Schiess, L. Widler and A. Wonnacott, Pure Appl. Chem. 55, 1807 (1983).
For related work see also the review:
d) "Organotitanium Reagents in Organic Synthesis. A Simple Means to Adjust Reactivity and Selectivity of Carbanions", M.T. Reetz, Top. Curr. Chem. 106, 1 (1982).

29) B. Weidmann, L. Widler, A.G. Olivero, C.D. Maycock and D. Seebach, Helv. Chim. Acta 64, 357 (1981).

30) B. Weidmann, C.D. Maycock and D. Seebach, Helv. Chim. Acta 64, 1552 (1981).

31) M. Schiess and D. Seebach, Helv. Chim. Acta 66, 1618 (1983).

32) M. Schiess, ETH Zürich, hitherto unpublished results.

33) L. Widler, Dissertation No. 7371, ETH Zürich, 1983.

34) L. Widler, Th. Weber and D. Seebach, Chem. Ber. 118 (1985), in press.

35) Y. Ikeda, K. Furuta, N. Meguriya, N. Ikeda and H. Yamamoto, J. Am. Chem. Soc. 104, 7663 (1982).

36) L. Widler and D. Seebach, Helv. Chim. Acta 65, 1085 (1982).

37) D. Seebach and L. Widler, Helv. Chim. Acta 65, 1972 (1982).

38) D. Seebach, A.K. Beck, S. Roggo and A. Wonnacott, Chem. Ber. 118 (1985), in press.

39) D. Seebach and M. Schiess, Helv. Chim. Acta 65, 2598 (1982).

40) M.T. Reetz, J. Westermann and R. Steinbach, J. Chem. Soc., Chem. Commun. 1981, 237.

41) H. Hellmann and G. Opitz: "α-Aminoalkylierungen", Verlag Chemie, Weinheim, 1960; B. Reichert: "Die Mannich-Reaktion", Springer-Verlag, Berlin, 1959; F.F. Blicke, Org. Reactions 1, 303 (1942); J.H. Brewster, Org. Reactions 7, 99 (1953); M. Tramontini, Synthesis 1973, 703.

42) J. Schreiber, H. Maag, N. Hashimoto and A. Eschenmoser, Angew. Chem. 83, 355 (1971); Angew. Chem., Int. Ed. Engl. 10, 330 (1971).

43) H. Böhme and H.G. Viehe, "Iminium Salts in Organic Chemistry", Part 1 and 2, J. Wiley & Sons, New York, 1976.

44) D. Seebach, M. Schiess and C. Betschart, Helv. Chim. Acta 67, 1593 (1984).

45) C. Betschart, Diplomarbeit, ETH Zürich, 1984.

46) W.B. Schweizer, ETH Zürich, hitherto unpublished experiments, 1984/85.

47) R. Imwinkelried, part of the projected Ph.D. thesis, ETH Zürich.

48) R. Imwinkelried and D. Seebach, Helv. Chim. Acta 67, 1496 (1984).

Asymmetric syntheses via heterocyclic intermediates

Ulrich Schöllkopf

Institut für Organische Chemie der Universität
Tammannstraße 2, D-3400 Göttingen, FRG

Abstract - Metalated Bis-lactim ethers **3** of 2,5-diketopiperazines **1** react with electrophiles highly diastereoselectively to give the adducts **4**. $E^{\oplus}$ enters trans to the inducing chiral center C-6. Acid hydrolysis of **4** liberates the optically actice amino acid ester **6** and the amino acid ester **5** that functions as the chiral auxiliary in the synthesis of **1**. Examples are described with $E^{\oplus}$ = alkyl halides, carbonyl compounds, epoxides, α,ß-unsaturated esters and thioketones. In many cases the amino acid methyl esters **6** are obtained in essentially optically pure form. - After exchange of lithium in **3** for tris(dimethylamino)titanium aldehydes react with 28 with exceedingly high diastereoselectivity to give essentially enantiomerically and diastereomerically pure products **26,32**. - 3-Metalated dyhydrooxazinones (type **49,52**) can also be utilized for the asymmetric synthesis of amino acids.

1 INTRODUCTION

Optically active, non-proteinogenic amino acids deserve attention because of their documented or potential biological activity. Some are valuable pharmaceuticals, such as L-Dopa, (S)-α-Methyldopa. D-Penicillamine or D-Cycloserine. Others are components of pharmaceuticals, for instance D-phenylglycine or D-(p-hydroxyphenylglycine) in the semisynthetic penicillines Ampicillin or Amoxycillin. - In biochemistry, they are valuable tools to investigate the mechnism of enzyme reactions (1). In fact, enzyme inhibition studies with non-proteinogenic amino acids have furnished valuable information about the mode of action of certain enzymes.

Obviously, there is a demand for optically active - if possible optically pure - uncommon amino acids both for pure applied organic or bioorganic chemistry. Since asymmetric synthesis is - at least in principle - the shortest and most economic way to optically compounds, it is a challenge for the synthetic organic chemist, to develop asymmetric synthesis of amino acids.

2 STRATEGY

Our approach is based on heterocyclic chemistry and on the following concept. 1. From a racemic lower amino acid and a chiral auxiliary an heterocycle is built up, that is CH-acidic adjacent to the potential amino group and that contains two sites susceptible to hydrolysis. 2. An electrophile is introduced diastereoselectively via the anion of the heterocycle. 3. Subsequently the heterocycle is cleaved by hydrolysis to liberate the chiral auxiliary and the new optically active amino acid. This lecture deals with the use of metalated bis-lactim ethers **3** of 2,5-diketopiperazines **1** according to scheme 1 (2). In addition, the utilization of 3-metalated dihydrooxazinones (type **49,52**) is briefly discussed.

The bis-lactim ether **2** reacts with butyllithium or LDA (THF, -70°C) to give the lithium compounds **3**. Electrophiles react with **3** to give the adducts **4**, whereby chirality is transfered from C-6 to C-3. $E^{\oplus}$ enters at C-3 trans to R^1 at C-6. The degree of asymmetric induction (=de=$(D_1 - D_2)/D_1 + D_2) \bullet$ 100) exeeds in many cases 95 % and reaches up to 99 %. The products **4** can be hydrolyzed liberating the optically active target molecules **6** and the esters **5** (the chiral auxiliaries). The esters **5** and **6** are separable either by fractional distillation or - eventually after further hydrolysis to the amino acids - by chromatography.

Scheme 1

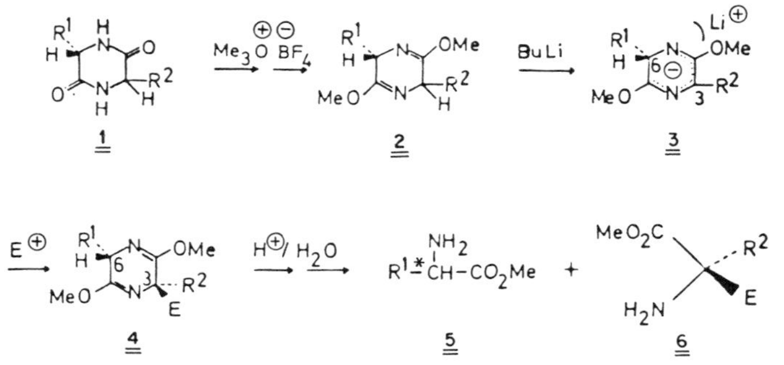

1,2,3	a	b	c	d	e
R^1	iPr	iPr	tBu	iC_4H_9	CH_3
R^2	Me	H	H	H	H

3 (R)-α-METHYL AMINO ACID ESTERS 9 FROM 3a AND ALKYL HALIDES

The bis-lactim ether **2a** yields with butyllithium (or LDA) regiospecifically **3a**. This reacts with alkyl halides with virtually complete asymmetric inductions to give the (3R)-addition products **7**. In the ^{1}H-NMR only (6S,3R)-diastereomers are detectable. Acid hydrolysis of **7** liberates (besides methyl L-valinate **8**) the α-methyl amino acid methyl esters **9** which are enantiomerically pure by ^{1}H-NMR-standard (scheme 2) (3).

Scheme 2

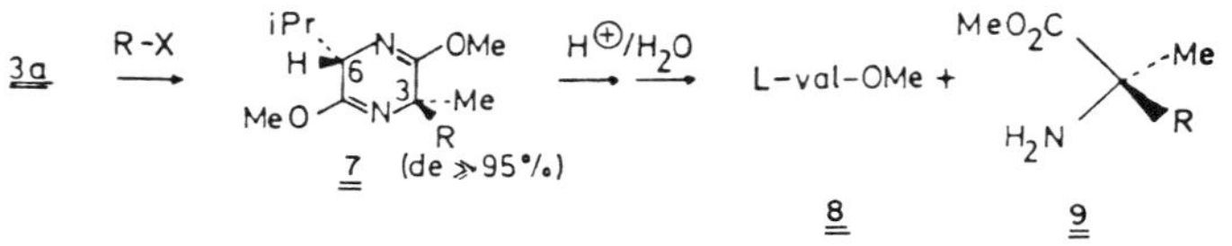

R = alkyl-,benzyl-,allyl- and proparayl

We assume that the ion pairs **3** contain a planar anion with the lithium cation situated near N-1. Furthermore, we postulate a mobile equilibrium between two diastereomeric ion pairs **10** ans **11**, which lies far on the left side because of steric reasons. Due to attractive complexation between Li$^⊕$ and X-R , **10a** reacts via **12a**‡ to (3R,6S)-**7** and **11a** via **13a**‡ to (3S,6S)-**7**. **10a** reacts faster than **11a**, since **12a**‡ is relatively strain free, wheras **13a**‡ is strained due to steric congestion "at the bottom side" (scheme 3).

Scheme 3

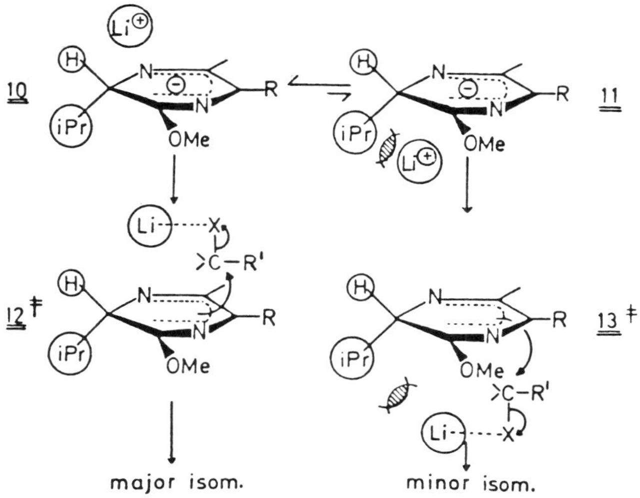

	a	b
R	Me	H

4 (R)-α-ALKENYL ALANINE METHYL ESTERS 16 FROM 3a AND KETONES

Ketones add to **3a** with exceedingly, high diastereoface selection to give the (3R)-adducts **14** (d.e.>95%). With acetophenone C-3 becomes also a chiral center, although the enantioface selection at the carbonyl group is relatively poor. After dehydratation **14→15**, hydrolysis of **15** yields the (R)-α-alkenyl alanine esters **16**. These are enantiomerically pure by ^{1}H-NMR-standard (4) (scheme 4).

Scheme 4

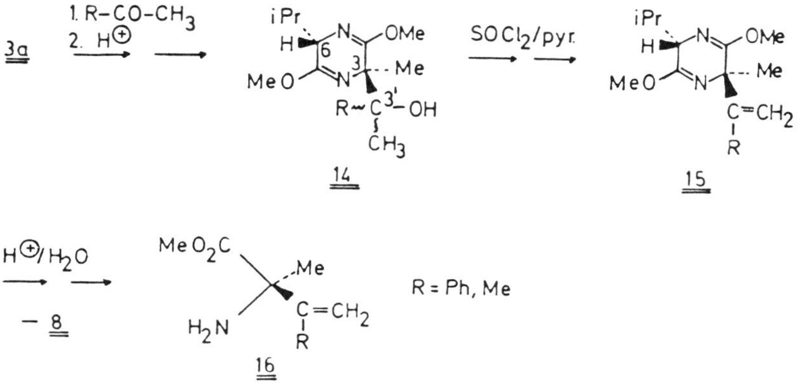

The diastereofacial bias of **3a** toward carbonyl compounds can be explained by a model concept analogous to the one put forward in 3.2. TS **17a**‡, leading to the major isomer, is of lower energy than TS **18a**‡ which is strained due to steric hindrance at "the bottom side" (scheme 5).

Scheme 5

major isom.   minor isom.

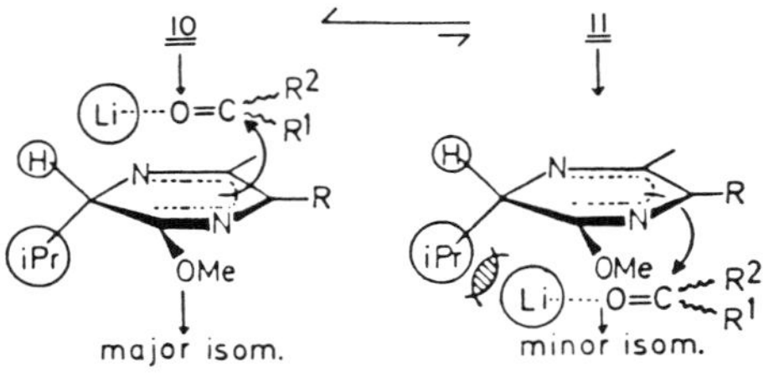

5 (α-UNSUBSTITUTED) AMINO ACID METHYL ESTERS **20** FROM **3b** AND ALKYL HALIDES

As expected, the bis-lactim ether **2b** of cyclo(L-val-gly) **1b** is lithiated by butyllithium regiospecifically in the glycine part to give **3b**. This reacts with alkyl halides to afford the (3R)-products **19** with de-values from 70 -> 95 % (scheme 6) (5). On hydrolysis, the products **19** are cleaved to methyl L-valinate **8** and (R)-amino acid methyl esters **20** (5). A comparison of the results depicted in scheme 6 with those reported in 3.1 reveals that a methyl group at the prochiral center C-3 in **3** is beneficial to the degree of asymmetric induction. The results are rationalized on the basis of the TSs **12b**‡ and **13b**‡ (scheme 3), although it is hard to explain, why the induction is in general higher with R = Me than with R = H.

Scheme 6

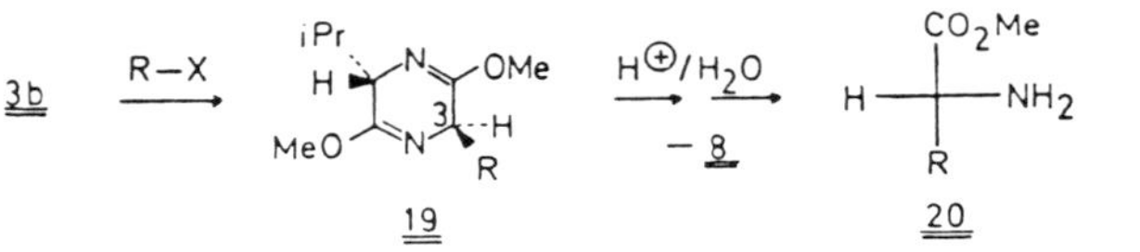

19,20, R	Bz	CH$_2$CH=CHPh	CH$_2$C≡CPh	CH$_2$OBz	n-C$_7$H$_{15}$
% de. ee	92	>95	>95	94	70

6   AMINO ACID METHYL ESTERS **20** FROM **3c** AND ALKYL HALIDES

Bulkier than iso-propyl is tert-butyl. Hence it is not surprising, that **3c** reacts with all alkyl halides, tried so far -apart from methyl iodide - with de> 95 % (6), i.e. with essentially complete asymmetric induction (scheme 7).

Scheme 7

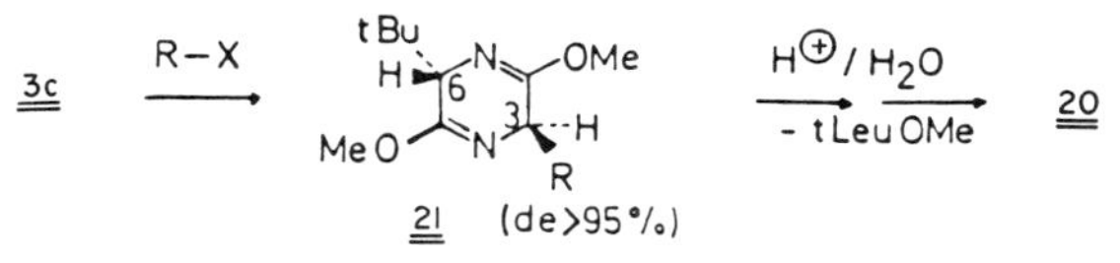

7   (R)-ß-HYDROXY VALINE **23** AND (R)-ß-METHYLENE PHENYLALANINE ESTER **25** FROM **3b** AND ACETONE, RESP. ACETOPHENONE

Ketones such a acetone and acetophenone afford with **3b** the (3R)-adducts **22** with de > 95 %; only (3R)-diastereomers are detectable in the ^{1}H-NMR (scheme 8).From **22a** practically optically pure (R)-ß-hydroxy valine **23** is obtainable (7), from **22b** (R)-ß-methylene phenylalanine methyl ester **25** (via the olefin **24**) (8) (scheme 8).

Scheme 8

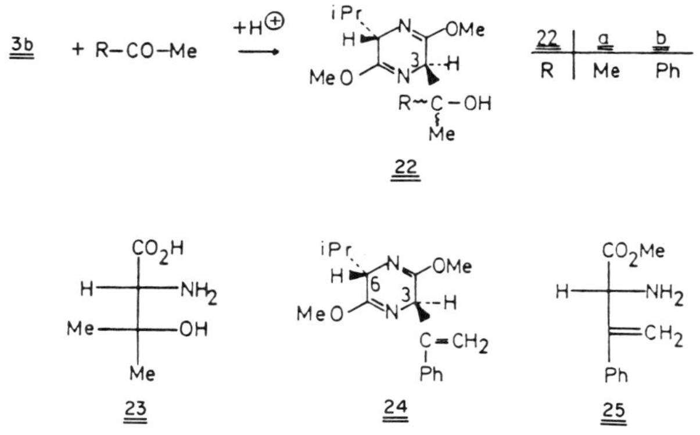

The high diastereoselectivity observed in the addition of ketones to **3b** can be rationalized on the basis of the model concept depicted in scheme 5. TS **17b**‡ is of considerably lower energy than **18b**‡.

8 (2R)-3-SUBSTITUTED SERINES 27

8.1 Addition of Aldehydes to the Lithium Compound 3b

Compared with ketones (cf.7), aldehydes react with the lithium compound 3b with somewhat lower diastereoselectivity (9). The asymmetric induction at C-3 (de at C-3) are listed in scheme 9 as well as the (3R,3'S) : (3R,3'R)-ratios.

Scheme 9

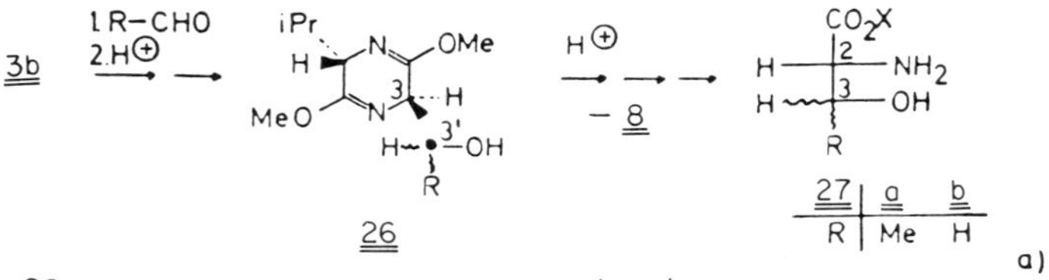

26	R	(3R,3'S)	(3R,3'R)	(3S,3'S)	(3S,3'R)	% de C-3	u : l
a	Ph	16	13	1	1	86	1,2:1
b	iPr	55	10	1	1	94	5,5:1
c	tBu	35	18	2	1	85	2:1
d	Me	19,5	4,7	1	1	85	4:1

a) (3R,3'S):(3R,3'R)

The diastereoface selection with regard to the anion of 3b is best explained on the basis of the TS 17b‡ and 18b‡ (scheme 5). The enantioface selection at the carbonyl group can be rationalized on the basis of the chair like TSs 29a‡ and 30a‡ (9) (scheme 11). The (3R,3'S)-epimers are formed predominantly, probably because the 1,3-diaxial R↔OMe - and the R↔Li-repulsion in 3oa‡ outweighs the 1,2 R↔H-repulsion in 29a‡ (scheme 11). - As described in ref. (9), (2R)-3-substituted serine methyl esters 27a - or serines 27b - can be obtained from the compounds 26.

8.2 Addition of Aldehydes to a Titanium Derivative of 2b

All factors, that render the TSs 17b‡, 18b‡, 29‡ and 30‡ more compact should enhance both the diastereoface selection with respect to the anion and the enantioface selection with regard to the carbonyl group. Consequently, exchange of lithium for metals with shorter metal-oxygen- and metal-nitrogen-bonds should lead to an higher degree of de at C-3 and to an higher (3R,3'R)-ratio in 26. This working hypothesis seems to be correct. Exchange of lithium for tris(dimethyl-amino)titanium - for example - has a dramatic effect as can be seen by comparing the data in scheme 9 with those in scheme 10. The titanium compounds 28 yields with aldehydes essentially diastereomerically pure (3R,3'S)-adducts 26.

Scheme 10

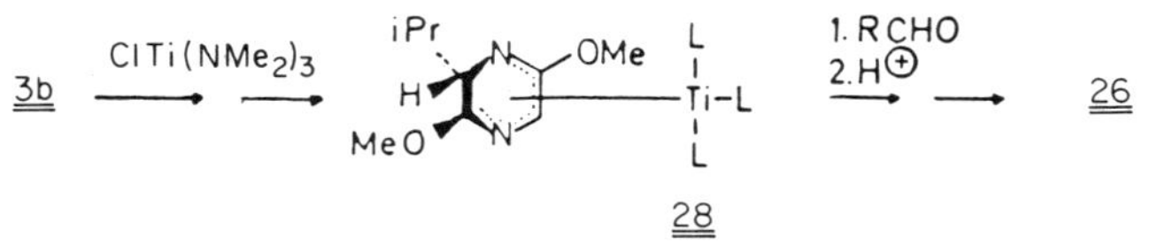

26	R	(3R,3'S)	(3R,3'R)	(3S,3'S)	(3S,3'R)	% de C-3	u : l
a	Ph	32	1	_a)	_a)	>99	32:1
b	iPr	151	2,3	1	1	97	65:1
c	Me	88	0,6	1	1	95,6	146:1

a) Not detectable any more with capillary GLC. b) (3R,3'S):(3R,3'R)

Scheme 11 depicts the TSs **29**‡ and **30**‡ . With M = Ti(NMe_2)$_3$, the TSs are more compact than with M = Li. Hence, **29b**‡ and **30b**‡ differ more in energy than **29a**‡ and **30a**‡ .

Scheme 11

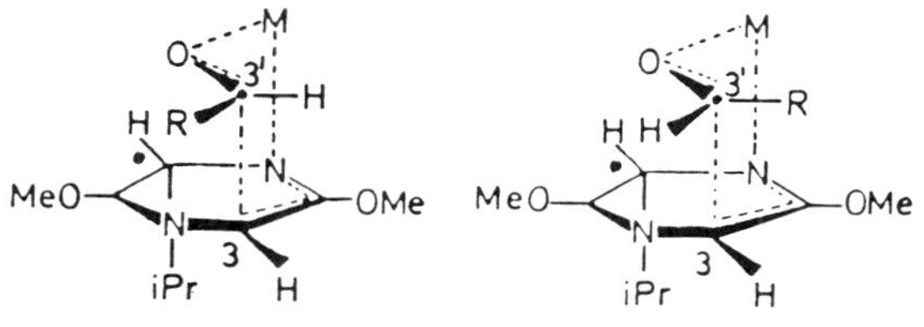
M
O
3'
H
R
MeO
N
OMe
iPr
3
H
M
O
3'
R
H
MeO
N
OMe
iPr
3
H

29‡ 30‡

29,30	a	b
M	Li–	Ti(L_3)

L = NMe_2

Scheme 10 contains examples with achiral aldehydes. With chiral aldehydes the problem of double stereodifferentiation comes into play. Scheme 12 depicts results obtained with **28** and (R)- and (S)-glyceraldehyde (cyclohexylidene protected). As expected, based on the TS-models **29b**‡ and **30b**‡ - whereby **29b**‡ is the overall dominating low energy TS - and on the Felkin-Anh-model of carbonyl addition, (S)-glyceraldehyde reacts with exceedingly high diastereoselectivity in any respect (scheme 12) to form **32**, but (R)-glyceraldehyde reacts with high de at C-3 (>96 %) to form **31**, but with a lower threo/erythro ratio (3R,1'R) : (3R,1'S) of ca. 19, compared with ca. 128 for (S)-glyceraldehyde. With (S)-glyceraldehyde, TS **29b**‡ is a totally matched pair - arrangement of the glyceraldehyde moiety with respect to the heterocyclic anion is favorable and carbonyl attack is Felkin - whereas TS **30b**‡ is a totally mismatched pair- arrangement of the aldehyde moiety with respect to the heterocyclic anion is poor and carbonyl attack is anti-Felkin. Hence, the energy difference of the two transition states is relatively large.

Scheme 12

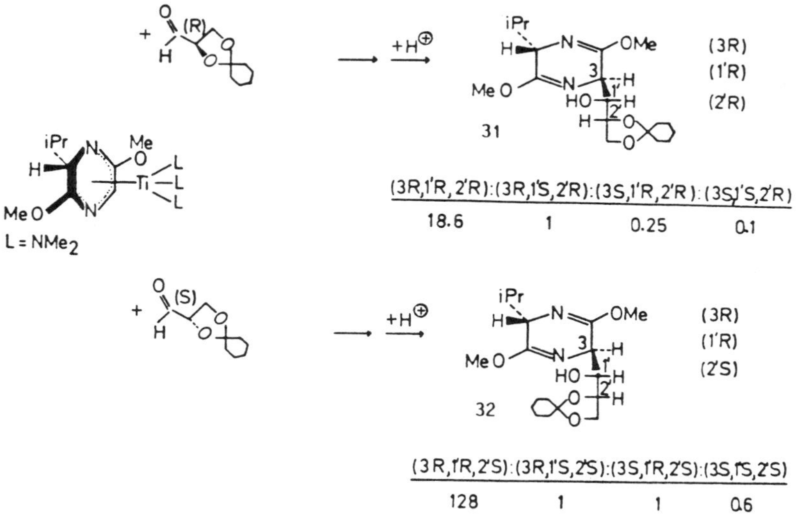
(R)
+H⊕
iPr
OMe
MeO
HO
(3R)
(1'R)
(2'R)
31
iPr
Me
Ti
L
MeO
L = NMe2
(3R,1'R,2'R):(3R,1'S,2'R):(3S,1'R,2'R):(3S,1'S,2'R)
18.6 1 0.25 0.1
(S)
+H⊕
iPr
OMe
MeO
HO
(3R)
(1'R)
(2'S)
32
(3R,1'R,2'S):(3R,1'S,2'S):(3S,1'R,2'S):(3S,1'S,2'S)
128 1 1 0.6

On hydrolysis (subsequent to O-acetylation (10))of **31** and **32** followed by removal of **8** by fractional distillation the derivatives **33** of D-2-amino-2-deoxy-xylonic acid and **34** of L-2-amino-2-deoxy-arabinoic acid are obtained. Both have the threo-relation between C-2 and C-3.

Scheme 13

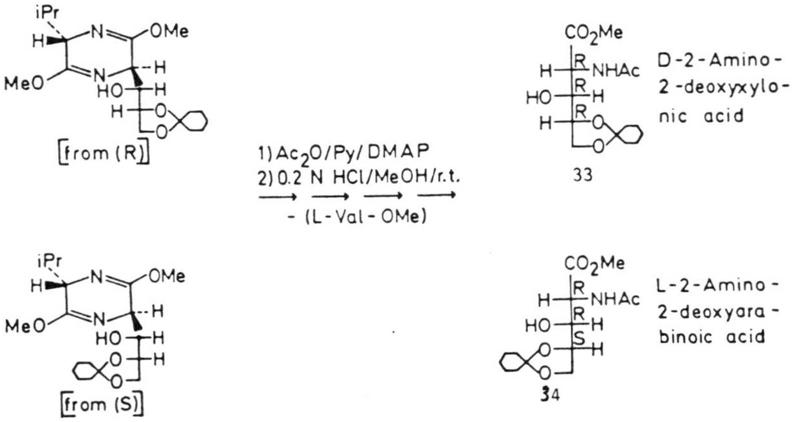

9   (R)-HOMOSERINES FROM **3b** AND EPOXIDES

Epoxides (type **35**) react with **3b** after addition of BF$_3$-etherate to give the addition products of type **36**. The de-values are in the range of 91 - 97 % (11). Prior to hydrolysis the hydroxy group of the adducts has to be Mem-protected, then hydrolysis proceeds smoothly (0.1 N HCl, ca. 16 h) to give the O-Mem-protected (R)-homoserine methyl esters of type **37**. Scheme 14 depicts one typical example with propene oxide. The amino acid ester **37** is essentially configurationally pure at C-2.

Scheme 14

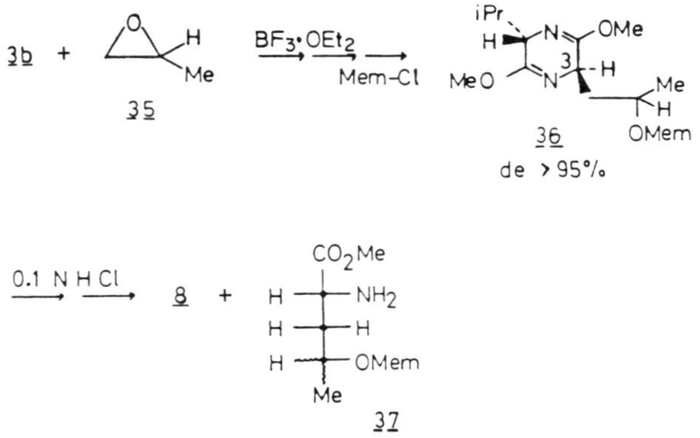

10 DERIVATIVES **40** OF (R)-GLUTAMIC ACID FROM **3b** AND METHYL ACRYLATES (MICHAEL-ADDITION)

As shown in scheme 15 Michael-addition of **3b** to methyl acrylates (type **38**) takes place with exceedingly high diastereoselectivity to give the precursors **39** of (D)-glutamic acids. This is an intriguing results since, for instance, methyl acrylate is a relatively small electrophile. Hence, the size of the electrophile can't be the sole decisive factor determining the degree of asymmetric induction. Probably, in the transition state of the Michael-addition the π-system of the heterocyclic anion an the π-system of the α,ß-unsaturated ester are arranged parallel to each other forming a kind of π-complex in which the ester moiety is turned inside (cf.**41** and **42** in scheme 16). In this arrangement the electrophile comes close to the chiral inducing center, rendering the energy difference of the relatively strainfree "topside" TS and of the strained "bottomside" TS relatively large (scheme 16.)

Scheme 15

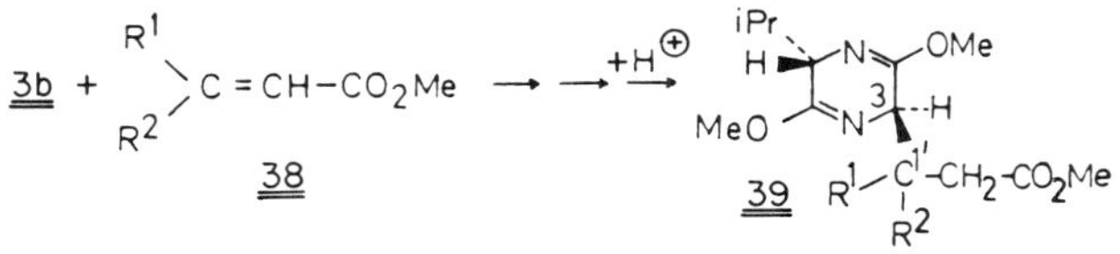

R^1	R^2	(3R):(3S)	(1'S):(1'R)
H	H	200 : 1	
Me	Me	140 : 1	
H	Ph	225 : 1	19 : 1

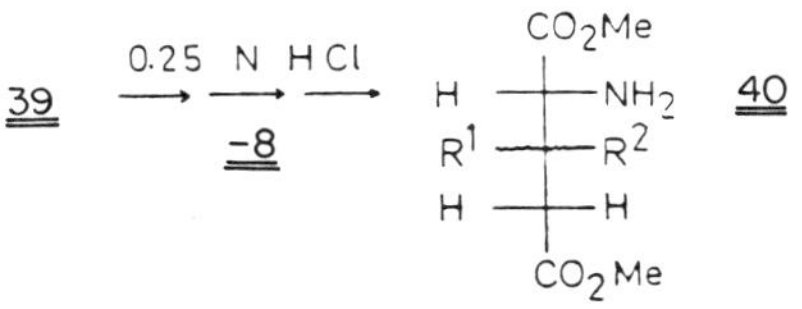

Scheme 16

41
Strainfree TS‡

42
Highly strained TS‡

11 (R)-CYCLOHEXENYL GLYCINE **46** FROM **3b** AND CYCLOHEXANETHIONE

Like ketones, thioketones (type **43**) also react with **3b** with virtually complete asymmetric induction. For instance, thiocyclohexanone **43** gives the addition product **44** (after S-methylation) with ca. 97 % asymmetric induction (scheme 17). Surprisingly, when treated with Raney-Ni, **44** undergoes a regioselective elimination of methanethiol forming the Hofmann-olefin **45** exclusively. From **45**, (R)-(-)-cyclohexenyl glycine **46** can be obtained , which is enantiomerically pure by NMR-standard (12) (scheme 17).

Scheme 17

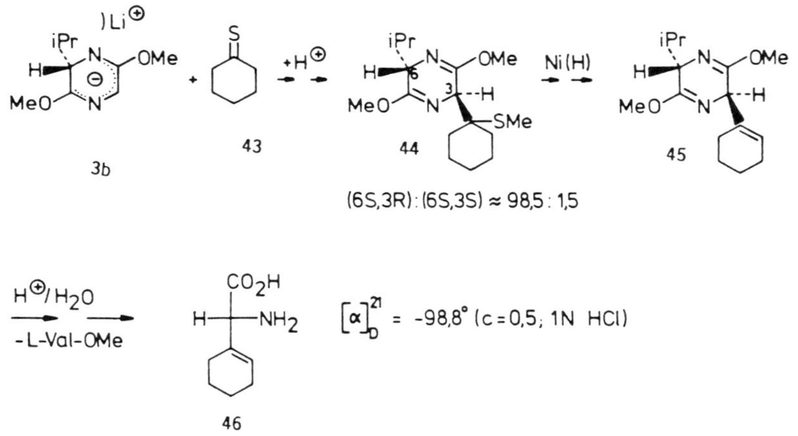

$(6S,3R):(6S,3S) \approx 98,5 : 1,5$

$[\alpha]_D^{21} = -98,8° \ (c = 0,5; \ 1N \ HCl)$

12 α-ALKYL-α-[PHENYL OR (2-FURYL)]-GLYCINES OF TYPE **56** FROM DIHYDROOXAZINONES **48** AND **51** (CHIRALLY SUBSTITUTED AT C-6)

Dihydrooxazinones of type **48** or **51** are also useful as vehicles for the asymmetric synthesis of non-proteinogenic amino acids, although they are not as versatile as the bislactim ethers **2**. An optically active α-hydroxy carboxylic acid of type **47** serves as the chiral auxiliary. Starting from these auxiliaries the 3-substituted dihydrooxazinones with a chiral residential center at C-6 are built up as depicted in scheme 18 for the 3-phenyl derivative **48** (13,14).

Scheme 18

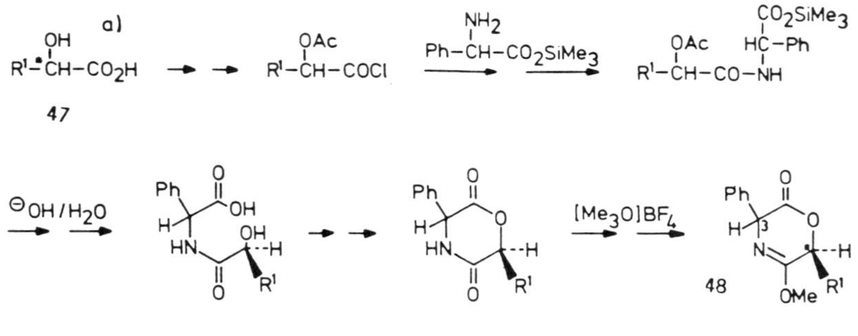

a) Chiral auxiliary: R¹=iPr, from L-Valine; R¹= tBu, hard to prepare optically pure,

R¹= Ph, partial racemisation during oxazinone synthesis

Scheme 19 shows results of alkylation experiments performed with the potassium derivative **49** of **48**. With R^1 = tert-butyl the degree of asymmetric induction during alkylation **49→50** is rather high, even with the relatively small methyl iodide (13). However, with R^1 = iso-propyl only benzyl bromide reacts with de >95 %. The 3-(2-furyl)-substituted dihydrooxazi-none **51** behaves similarly (scheme 20) (14). Again, with benzyl type halides the degree of asymmetric induction is exceptionally high, higher than with the bulky cyclohexyl methyl iodide.

Scheme 19

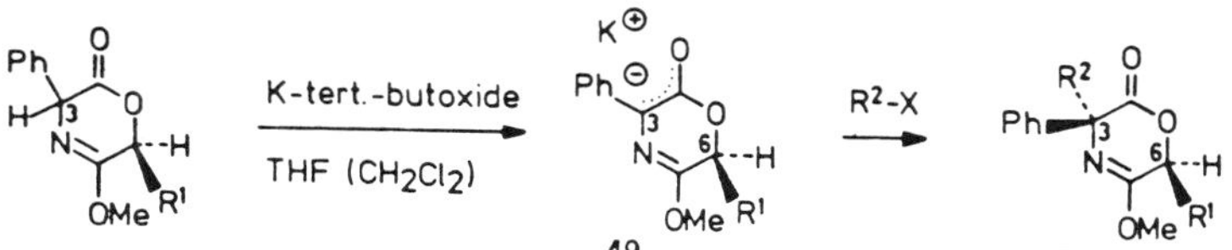

R^1	R^2	C-3 %de.	Config.
iPr	$CH_2C_6H_5$	>95	S
iPr	$CH_2CH=CH_2$	83	S
iPr	CH_3	60	S
tBu	$CH_2C_6H_5$	>95	S
tBu	CH_3	92	S

Scheme 20

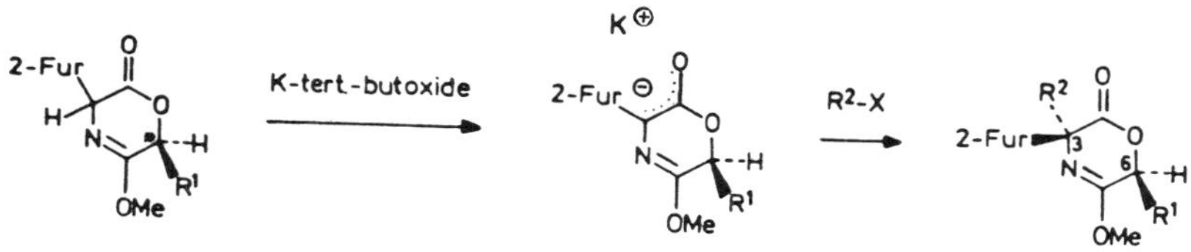

2-Fur:

53	R^1	R^2	% de. C-3	Config.
		$CH_2C_6H_5$	>95	S
		CH_2-4-pyr.	>95	S
	iPr	CH_2-2-thien.	>95	S
		$CH_2CH=CH_2$	70	S
		CH_3	58	S
		CH_2-cyclohex.	62	S
		$CH_2C_6H_5$	>95	S
	tBu	$CH_2CH=CH_2$	>95	S
		CH_3	>95	S

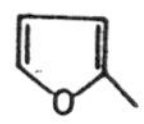

The relatively high diastereoface selectivities observed with benzyl type halides (scheme 19,20) can be rationalized on the basis of the "folded" transition states of type **54** and **55** (scheme 21) in which the aryl ring (heteroaryl ring) faces the heterocyclic anion. This brings the incoming group close to the chiral inducing center C-6 rendering the energy difference of the two transition states relatively large. The "aryl-inside" conformations might be stabilized by some kind of charge-transfer- or HO-LU-interactions between the heterocyclic anion and the aryl or heteroaryl ring.

Scheme 21

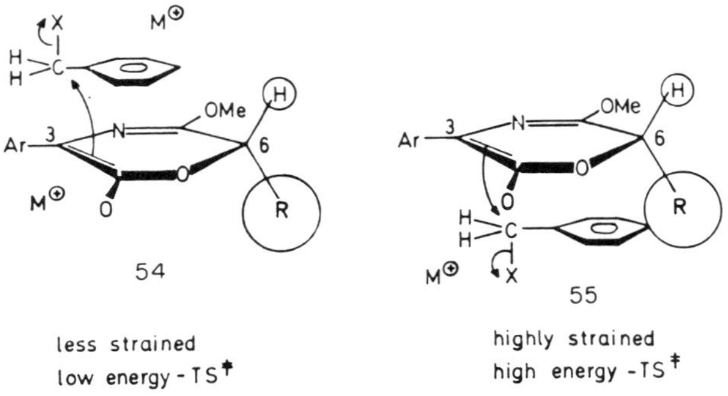

54

less strained
low energy - TS‡

55

highly strained
high energy - TS‡

Ar= C_6H_5, 2-Furyl

Hydrolysis of the alkylation products **50** or **53** followed by removal of the chiral auxiliary **47** gives the target molecules, the optically active α-alkyl-α-[phenyl or (2-furyl)]-glycines **56** (scheme 22) (13,14).

Scheme 22

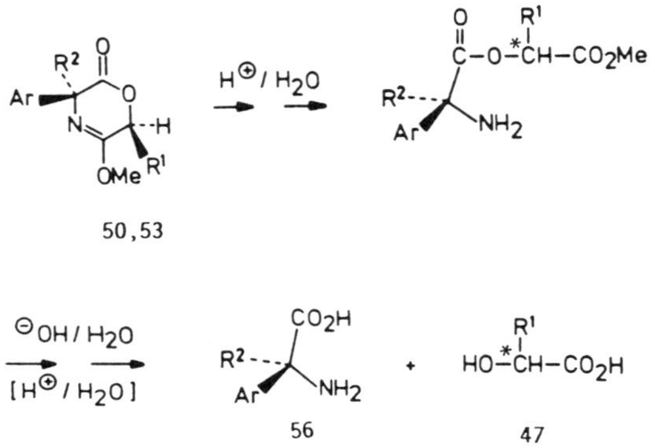

50,53

56: Ar=phenyl, 2-furyl

56   47

Acknowledgement - Thanks for enthusiastic cooperation are due to W.Hartwig, who carried out the initial cruzial experiments, but also to U.Busse, Y.Chiang,C.Deng, M.Grauert, U.Groth, R. Lonsky, H.-J.Neubauer, J.Nozulak, D.Pettig, K.-H.Pospischil, R.Scheuer, K.-O.Westphalen. Thanks for supporting these studies are due to the Deutsche Forschungsgemeinschaft, the Fonds der Chemischen Industrie, the BASF AG and the Degussa AG.

REFERENCES
(1) G.Nass, K.Poralla and H.Zähner, Naturwissenschaften **58**,603, (1971);W.Trowitzsch and H.Sahm, Z.Naturforschung, Teil C 32,78 (1977).
(2) Review on previous results: U.Schöllkopf, Top.Curr.Chemistry **109**.65 (1983); Tetrahedron **39**, 2085 (1983); Pure & Appl. Chem. **55**, 1799 (1983).
(3) U.Schöllkopf, U.Groth, K.-O.Westphalen and C.Deng, Synthesis **1981**,696; U.Groth, Y. Chiang and U.Schöllkopf, Liebigs Ann.Chem.**1982**,1756; U.Groth and U.Schöllkopf,Synthesis **1983**, 37; U.Schöllkopf and R.Lonsky, Synthesis **1983**, 675.
(4) U.Schöllkopf, U.Groth and Y.Chiang, Synthesis **1982**, 864.
(5) U.Schöllkopf, U.Groth and C.Deng, Angew.Chem.,Int.Ed.Engl. **20**, 798 (1981).
(6) U.Schöllkopf and H.-J.Neubauer, Synthesis **1982**, 861.
(7) U.Schöllkopf, J.Nozulak and U.Groth, Synthesis **1982**, 868.
(8) U. Schöllkopf and U.Groth, Angew.Chem.,Int.Ed.Engl. **20**, 977 (1981).
(9) U.Schöllkopf, U.Groth, M.-R.Gull and J.Nozulak, Liebigs Ann.Chem. **1983**, 1133.
(10) M.Grauert, Diplomarbeit, Univ. Göttingen 1984.
(11) M.-R. Gull, Dissertation, Univ.Göttingen 1985.
(12) U. Schöllkopf, J.Nozulak and U. Groth, Tetrahedron **40**, 1409 (1984).
(13) W. Hartwig and U. Schöllkopf, Liebigs Ann.Chem. **1982**, 1952.
(14) U.Schöllkopf and R.Scheuer, Liebigs Ann. Chem. **1984**, 939.

Stereoselectivity and stereospecificity: new tools in industrial organic chemistry

Jacques MARTEL

Centre de Recherches Roussel Uclaf, 102 route de Noisy,
93230 Romainville, France

Abstract - Taking into account industrial constraints requires
an increased implementation of stereoselective and stereospecific
methods either in the preparation or in the recovery of optically
active compounds. Examples of stereoconversion of one enantiomer
into the other are given as illustrations of this general trend.

INTRODUCTION

The introduction of the concepts of stereoselectivity and stereospecificity, clearly defined in the sixties (1) has strongly influenced the evolution of many sectors in industrial organic chemistry, especially in the
field of fine chemicals to which I am by now committed for quite a few
years. From the standpoint of economy it would be hard to imagine today how
we could do without having recourse to a high stereoselectivity and even
more to the stereospecificity of reactions.

ECONOMY AND ITS IMPLICATIONS

I want also to underline the economy standpoint since it is at the very
heart of most of our industrial constraints and successes. To sum it up
economy is always the ultimate target.

As it is usually defined economy is any act or method undertaken to keep
down expenses or, alternatively, any arrangement or organization for
efficient operations.

Both views of the definition are characteristic of various aspects of our
industrial research either at the conceptual or at the implementation level
as it will be shown in a moment. Furthermore trying to keep down expenses
does not appear only as constraint but also as an incentive to look for new
short pathways, new methodologies, new technologies or new concepts. When
applied to syntheses this last point of view has far reaching pratical
consequences and it is worth mentioning them here.

For instance a convergent overall sequence of reactions has generally to
be preferred to a linear one in synthetic work (2).

An example of convergent synthesis is shown in Figure 1 : it is the well
known stereoselective synthesis of dl-trans-chrysanthemic acid from readily
available starting materials and solvents (3).

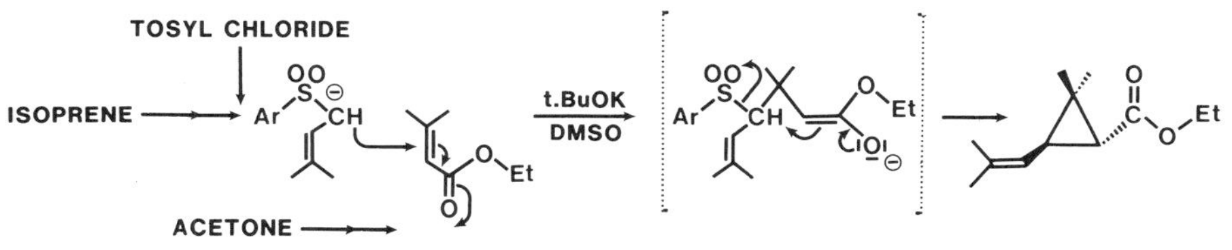

FIG. 1

Needless to say safe reagents and solvents, as cheap as possible are always
selected. We rigorously exclude not only potentially dangerous reactions
but any reaction that could not be controlled at some stage. Again to stay
thrifty we try to avoid any type of reaction yielding wastes or side
products the disposal of which would lead to water or land pollution.

In the realm of fine chemicals most of the time we are dealing with intermediates and final products that are optically active and consist of only
one pure isomer. This is especially true for the pyrethroid field from
which are drawn most of the examples I am going to give you in this lecture. With respect to optical purity it has become more and more necessary to
aim directly at the desired optically active compound, implementing highly
stereoselective or better stereospecific methods for bond formation and
avoiding a resolution step. In this way, the amount of unwanted isomer is
reduced as much as possible and therefore a series of tedious separations
and puriifcations, that are both time, energy and money consuming, are
avoided.

However it is not always possible to reach directly the desired optically active isomer and for this reason the resolution of a racemic compound
and the recovery of the unwanted isomer are important problems that often
rise in the industry.

EXAMPLES OF RESOLUTIONS AND RECOVERIES

Let me give you a few examples and tell you how we cope with such situa-
tions.

Let us start with dl-trans-chrysanthemic acid the synthesis of which was
shown in the previous slide; this racemic acid which has two asymmetric
centers is conveniently resolved, in a common fashion by cristallisation of
its diastereoisomeric salts with an optically active amine, which in this
precise case is the D(-) threo precursor base of chloramphenicol (4) shown
in Figure 2.

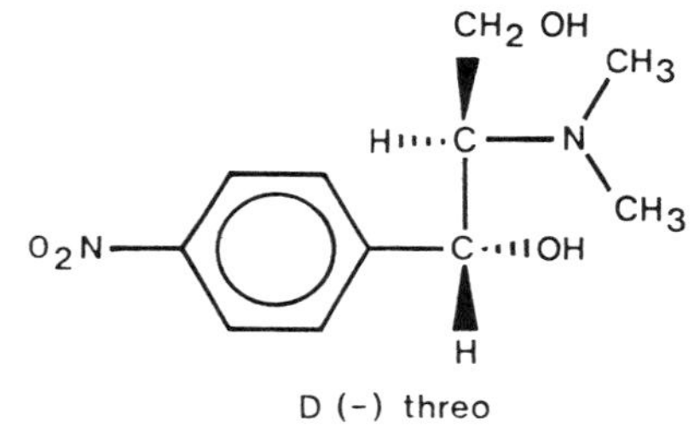

D (-) threo

FIG. 2

The optically active (1R, 3R) trans-chrysanthemic acid, thus obtained,
corresponds to the natural product and since its enantiomer is also recovered the problem is now what to do with this unwanted isomer, the esters
of which are devoid of pesticidal activity. We could have tried to racemize
it but we choose, instead, to turn it into its biologically active enantiomers by successive inversions of configurations at the centers 1 and 3.

The trans (1S, 3S) isomer of chrysanthemic acid, like its antipode, is
thermodynamically more stable than the corresponding cis compound. Therefore, in order to effect the inversion of one of the asymmetric centers,
some trick has to be found to reverse or at least alter the unfavorable
thermodynamic stability. Formation of a cyclic intermediate is precisely
the trick we took advantage of in the recovery of the unwanted isomer. This
could be performed along two lines which I am going to discuss now.

In the first method (5), shown in Figure 3 the trisubstituted double bond
is first hydrated with sulfuric acid to the tertiary alcohol. This tertiary
alcohol is now able to trap the carboxylic ester group into a six membered
lactone provided the ester is given an opportunity to equilibrate. Strongly
basic treatment of the ester gives rise to the enolate, the selective

protonation of which yields the expected δ-lactone. Thus the inversion ot
the asymmetric center at 3 has been successfully carried out. To reconsti-
tute the unsaturated side chain the δ-lactone is treated with Lewis acids
under carefully controlled conditions yielding nearly exclusively the (1R,
3S) cis chrysanthemic acid (5).

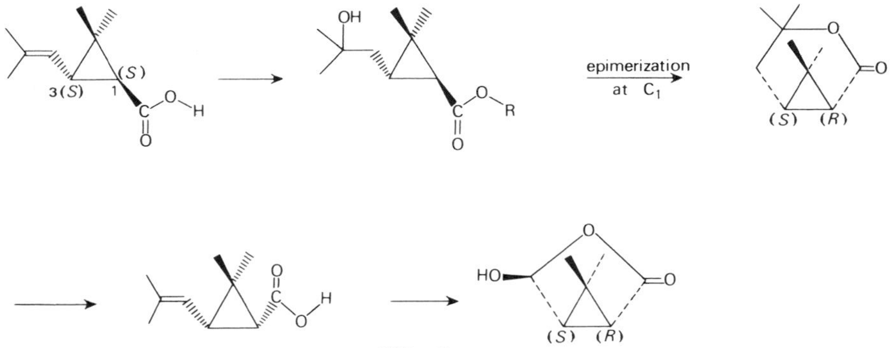

FIG. 3

This series of reactions opens a new access to the biologically important
derivatives of (1R, 3S) cis chrysanthemic acid (6). Indeed ozonolysis of
the double bond, followed by reduction of the resulting ozonide gives a key
lactolic intermediate (7), the relative stability of which may be used to
prepare derivatives of either cis or trans chrysanthemic acids of the
proper natural configurations as it will be shown in a moment.

In the series of reactions just described, the recovery started by configurational inversion of the asymmetric center at 1 but it is possible to
invert first the center at 3 as shown in Figure 4. In this second method
(7) the olefinic side chain of the antipodal (1S, 3S) trans-chrysanthemic
methyl ester is cleaved by ozonolysis ; subsequent reduction of the ozonide
yields the caronaldehyde that still belongs to the trans series.

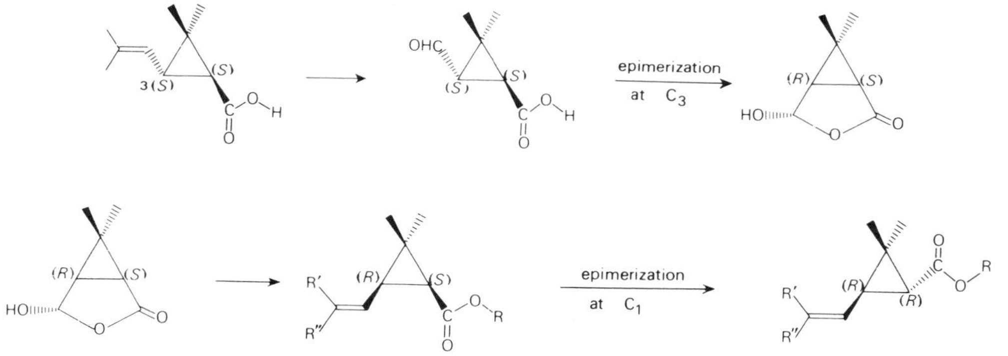

FIG. 4

Taking advantage of the stabilisation imparted by the cyclic lactolic form,
I mentioned previously, it is possible now to induce the inversion of
configuration at the aldehyde center by mild basic treatment followed by
mild acidic work-up (8). Now a Wittig reaction (9) on the free cis lactolic
form gives back the required unsaturated side chain, this time with the
desired 3 (R) configuration. The corresponding (1S, 3R) cis carboxylic
ester is now isomerized under basic conditions to its stable trans (1R, 3R)
isomer.The overall result of this series of reactions is the complete
stereoconversion of the unwanted isomer into the desired one.

It may be worthwhile to spend a few minutes on the mechanisms of lactol
formation. As I told you earlier, the trans-caronaldehyde arising from
ozonolysis of the unsaturated side chain of trans-chrysanthemic acid and
subsequent reduction of the ozonide, is treated by sodium methylate in
methanol under reflux for a few hours and then the mixture is hydrolyzed by
refluxing in aqueous acidic dioxane to the hydroxylactone, the lactolic
form of cis-caronaldehyde. Since sodium methylate in methanol is also able

to induce the nearly complete isomerization of the formyl group of the free cis-caronaldehyde into the corresponding trans derivative it is clear that under our experimental conditions there is a subtle interplay of several competing reactions. A detailed investigation of the isomerization mechanism in this series has been performed by American Chemists (10). Taking into account their results it is possible to propose the following scheme (Fig. 5).

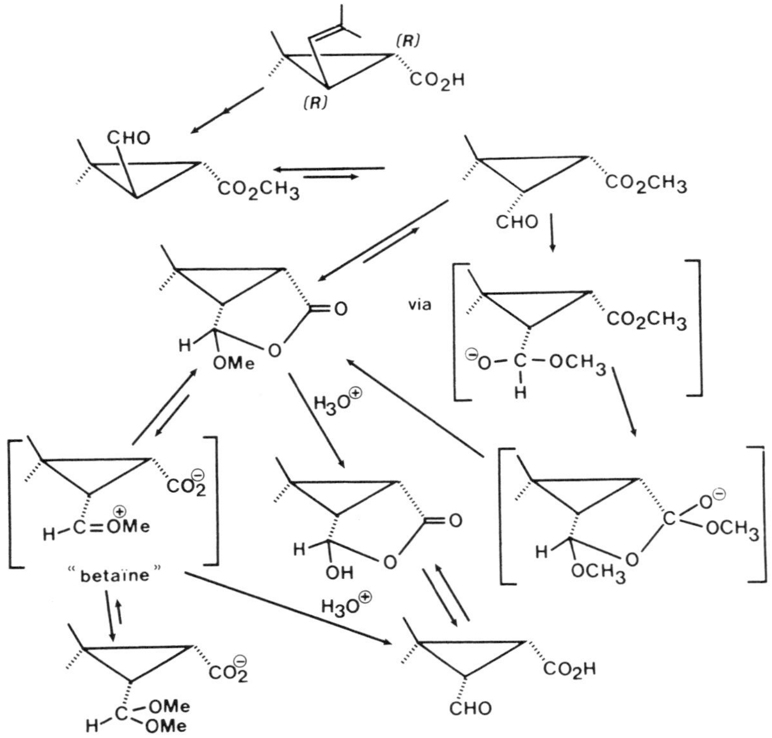

FIG. 5

The ester of trans caronaldehyde arising from the cleavage of the double bond is partly isomerized to its cis isomer. The latter is trapped as the methylether of the lactolic form. Although endowed with a relative stability, this methyl ether appears too labile under the conditions of the reaction to be the form that drives the equilibrium towards the cis derivative. What seems likely is the existence of a series of equilibria between this lactolic ether form, the corresponding betaine and the dimethylketal that results from methylate addition to the betaine. This last dimethylketal appears stable enough to displace this series of equilibria towards the cis derivative. Acidic hydrolysis then yields cis caronaldehyde either in its cyclic lactol form or its tautomeric form with the free formyl group.

Thus our method of successive stereoconversion of the asymmetric centers of trans-chrysanthemic acid initially designed to recover one enantiomer led us to new interesting derivatives in the biologically active cis series readily available through the intermediate lactol.

Interestingly enough this intermediate free lactol, although fairly labile, may be readily converted into its ketolic ether by mild treatment with an alcohol under slightly acidic conditions.

We took advantage of this interesting property to resolve racemic alcohols. We were led to this discovery in the following way. At the time we wanted to market a pure, biologically active, product in the allethrin series (Fig. 6).

As some of you probably know, allethrin is the mixture of 8 diastereoisomers resulting from the esterification of dl-allethrolone with both
cis and trans dl-chrysanthemic acids. In this mixture the biologically most
active component is the ester of (1R, 3R) trans-chrysanthemic acid with the
4(S)-enantiomer of allethrolone and this ester called S-bioallethrin was
our target (Fig. 7).

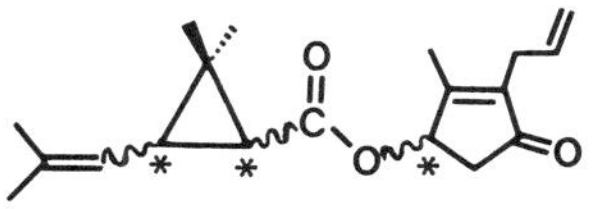

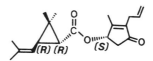

ALLETHRIN
3 unresolved asymmetric centres*
8 stereoisomers

FIG. 6

S-BIOALLETHRIN
S allethronyl *1R* <u>trans</u> chrysanthemate
(single isomer)

FIG. 7

Since we had already in hand the optically pure (1R, 3R) trans-chrysanthemic acid we had to resolve racemic allethrolone into the desired 4(S)
isomer and its 4(R) enantiomer and this last compound had to be recovered.

Let us note at this point (Fig. 8), that resolution of alcohols, generally
requires the preparation of intermediate hemi-succinic esters or equivalent
hemi-esters which are then resolved by conventional means through the
separation of diastereoisomeric salts with opticallyactive amines such as
ephedrine (11). Once the separation of the diastereoisomeric salts has been
performed, one has to get rid of the hemisuccinic moiety of the optically
active ester by hydrolysis.

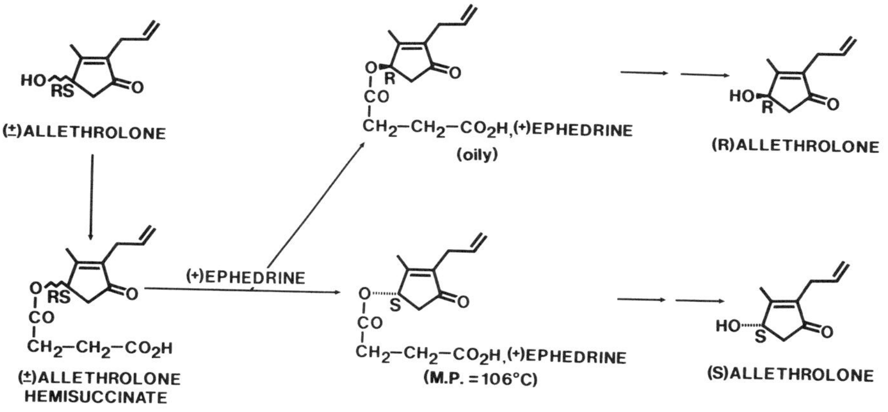

FIG. 8

The overall process is fairly lengthy. In order to shorten this procedure
and in connection with the properties of sugar hemi-ketals we tried to
directly condense dl-allethrolone with the optically active chrysanthemic
lactol or its ether in order to obtain separable diastereoisomers. As a
matter of fact, under controlled acidic conditions the condensation of the
two molecules yields nearly only the mixture of exo-derivatives which are
separated either by crystallisation or by usual physical methods (chromatography etc...) (Fig. 9). After separation of the enantiomers, the resolved
alcohol is recovered from the cyclic hemiketolic derivative by mild hydrolysis in aqueous acid. These conditions are particularly well suited to
cyclopentenolone derivatives which are fairly labile under alkaline
conditions (12).

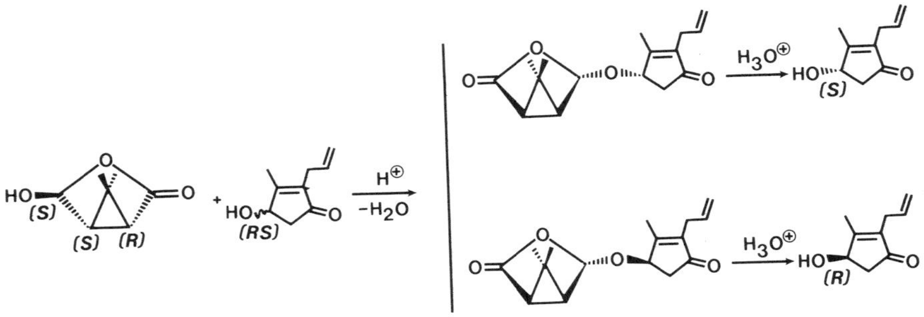

FIG. 9

The principle of this method of resolution has been extended to the other
lactolic compounds shown in Figure 10 among which the anthracene adduct
appear very promising since it gives rise very often to crystalline
derivatives (13).

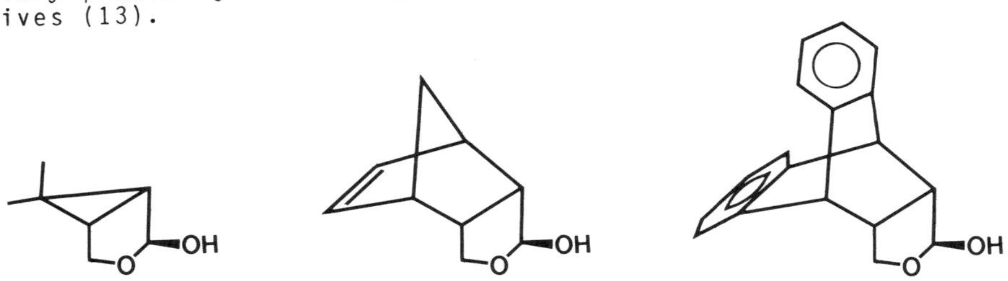

FIG. 10

The same principle has been applied to the resolution of various racemic
amines through the transient formation of hemiaminal derivatives (13).

This new resolution method looks very promising also for alcohols sensitive
to alkaline conditions. For instance it is possible to resolve labile
cyanohydrins into their enantiomers since the experimental conditions
exclude any basic treatment (12). To underline the importance of this new
method or cyanohydrin resolution let us look at Figure 11 which shows the
resolution of the m-phenoxy benzaldehyde cyanohydrin, corresponding to the
alcoholic part of the potent insecticide deltamethrin with the (S)
configuration at the asymmetric center.

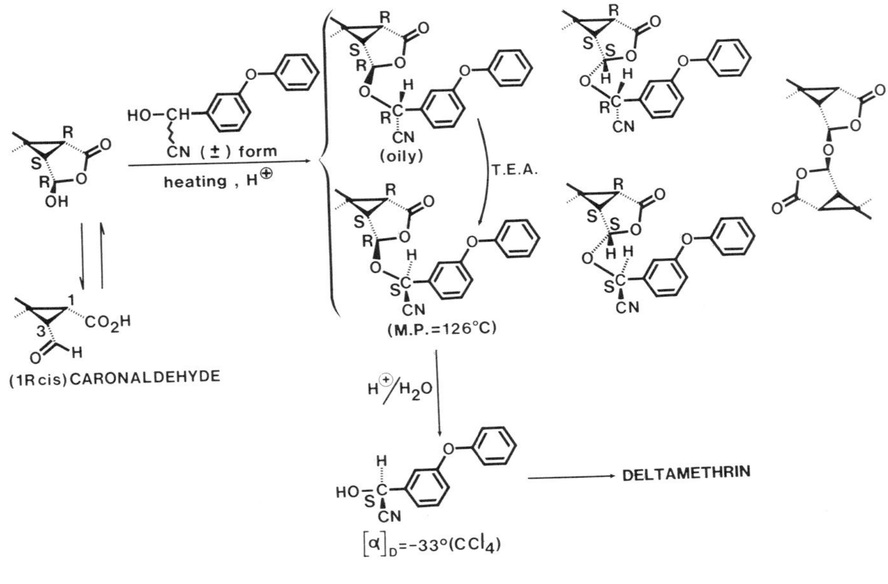

FIG. 11

Heating the optically active caronaldehyde with the racemic cyanohydrin of m-phenoxy benzaldehyde in the presence of a small amount of tosylic acid affords mainly two diastereoisomers. One of them is induced to crystallize out of the solution and through mild acidic hydrolysis it was ascertained that it was the required (S) cyanohydrin enantiomer. The other remaining oily diastereoisomer could also be hydrolyzed into the antipodal (R) cyanohydrin. But owing to the lability of the proton of the asymmetric carbon of the cyanohydrin a stereoconversion of this last diastereoisomer into the other could be successfully carried out under the influence of weak organic bases. In this way it is possible not only to resolve the mixture of cyanohydrin enantiomers but to selectively convert one of the enantiomer into the other and in this case, the desired one.

Now let us turn to the other 4(R) enantiomer of allethrolone : how to recover it ? Several methods of racemization of such allylic alcohols did exist in the literature but the yields in this sensitive series were rather poor. Instead of a racemization it was thought that a stereo-conversion of the unwanted (R) isomer into the desired (S) enantiomer would be the best solution.

From a perusal of the literature at this time it was clear that there were very few authenticated cases of nucleophilic displacements of allylic alcohol derivatives in the cyclopentene series and even less in the cyclopentenolone series. Furthermore the sensitivity of allethrolone to dehydration under acidic or alkaline conditions was well known, leading first to cyclopentadienone and then to dimers or polymers. Nevertheless we decided to study this stereoconversion, the first step of which was the esterification of the hydroxyl of the 4(R) center by mesylic or tosylic acid (Fig. 12).

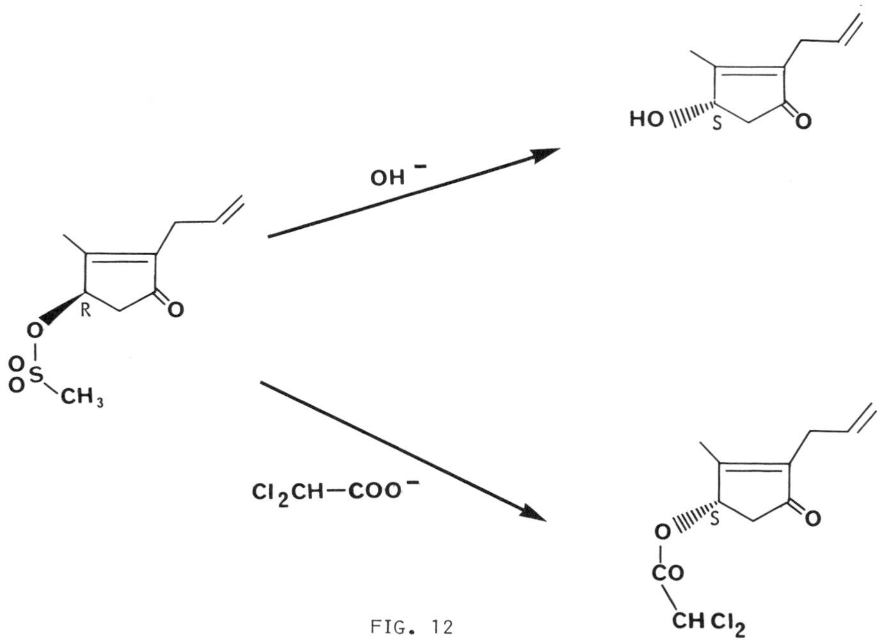

FIG. 12

This step was successfully carried out by the sulfene method (14), yielding the corresponding alcohol mesylate. A thorough study of the various nucleophiles that could perform the required inversion of the allylic mesylate of allethrolone showed that the nucleophile, the solvent and the experimental conditions play a dominant role in the successful outcome of the displacement. A few anions such as hydroxyl or dichloroacetate gave excellent results but, again, to comply with our industrial constraints, it was better to treat allethrolone mesylate with an alkaline salt of d-trans chrysanthemic acid in order to get directly the desired S-bioallethrin, as shown in Figure 13 (12).

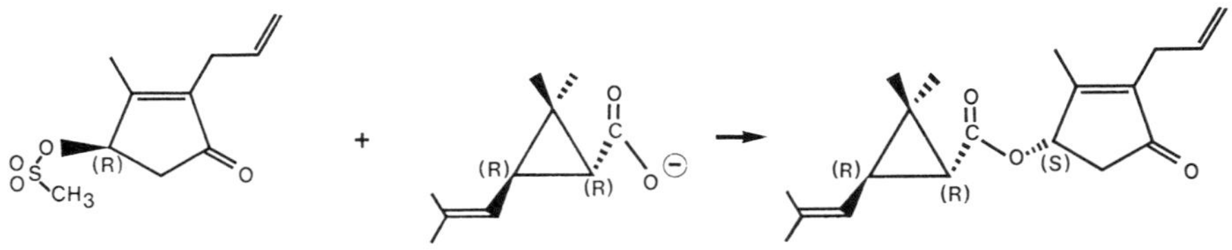

FIG. 13

The economy requirement always forces us to look for simple methods that allow us to do without separations whenever possible. Success is, of course, dependent on highly skilled chemists able to find the proper experimental conditions that give the right transformation. Allow me to give you a final example concerning the industrial preparation of the very potent insecticide Deltamethrin, shown in Figure 14.

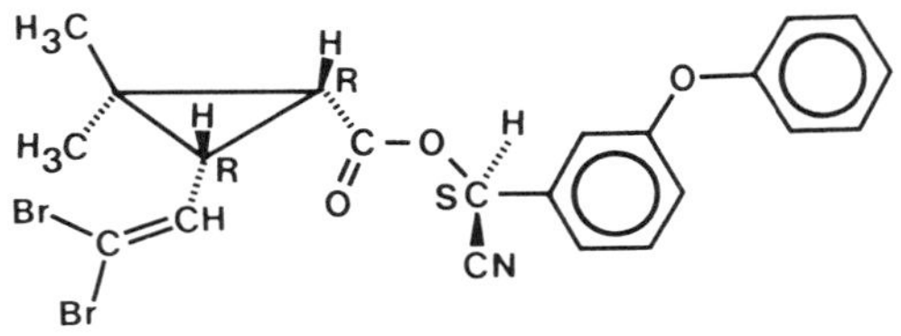

DELTAMETHRIN

FIG. 14

Both components are optically active and therefore one could have thought of separately resolving each component; on the one hand the chrysanthemic part, which incidentally is prepared via our key intermediate the optically active lactolic compound, and, on the other hand, the m-phenoxybenzaldehyde cyanohydrin part. Again our lactolic key intermediate is able to do this kind of resolution as it was just shown.

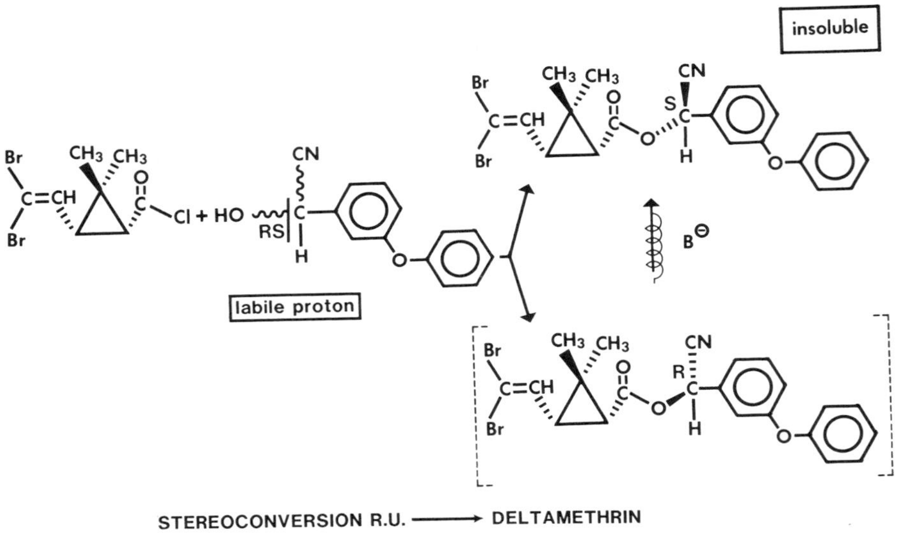

STEREOCONVERSION R.U. ———→ DELTAMETHRIN

FIG. 15

For economic reasons another simple solution was preferred (Fig. 15). The optically active d-cis-dibromo chrysanthemic acid is esterified with the racemic cyanohydrin of m-phenoxy benzaldehyde and the mixture of the two diastereoisomers is heated under alkaline conditions that induce a series of racemizations and deracemizations at the cyanohydrin asymmetric center. Under carefully chosen conditions one of the isomer crystallyzes out of the medium and fortunately the desired one. In this way a kinetically induced second order resolution of the cyanohydrin is carried out yielding, finally, only the desired (S) enantiomer (15). Therefore although at the start there is a mixture of diastereoisomers, at the end there is only a single, pure, optically active compound and in the meantime any separation or purification of isomers and any loss of organic compound has been avoided.

CONCLUSION

I hope I have convinced you about the fundamental role played by the concepts of stereoselectivity and sterospecificity in industrial organic chemistry. As an organic chemist I strongly feel that there are still fine days ahead for experimental organic chemistry but at the same time I feel that we need also a deepening of our understanding of a few fundamental problems such as the role of the solvent and its influence on the outcome of a reaction. The interplay of theoretical concepts and experimental work still remains the basis of scientific progress.

REFERENCES

1. Stereochemistry of Carbon Compounds, E.L. Eliel Ed., McGraw-Hill, New-York, 1962 p. 434

2. L. Velluz, J. Valls and J. Mathieu, Angew. Chem., Int. Ed., Engl. 6, 778 (1967); J.B. Hendrickson, J. Am. Chem. Soc., 99, 5439 (1977); S.H. Bertz, J. Am. Chem. Soc., 104, 5801 (1982)

3. J. Martel and C. Huynh, Bull. Soc. Chim. Fr. 3, 985 (1967)

4. Roussel Uclaf, French Patent 1.536.458 (1967)

5. Roussel Uclaf, French Patent 2.036.088 (1969)

6. M. Elliott, A.W. Farnham, N.F. Janes, P.H. Needham, D.A. Pulman, Nature (London), 248, 710 (1974)

7. Roussel Uclaf, French Patent 1.580.474

8. J. Martel, Roussel Uclaf DOS 1.935.320 (1970), 1.935.321 (1970), 1.935.386 (1970), 1.966.839 (1969)

9. Roussel Uclaf, French Patent 2.032.526; J. Martel DOS 1.807.091 (1969)

10. P.R.O. de Montellano and S.E. Dinizo, J. Org. Chem., 43, 4323 (1978)

11. Roussel Uclaf, French Patent 2.166.503 (1971)

12. J.J. Martel, J.P. Demoute, A.P. Tèche et J.R. Tessier, Pestic. Sci., 11, 188 (1980)

13. J. Martel et al., unpublished results

14. R.K. Crossland and K.L. Servis, J. Org. Chem., 35, 3195 (1970)

15. Roussel Uclaf, French Patent 2, 375, 161 (1976)

New routes to gem dialkylcyclopropane carboxylic acids. Application to the synthesis of chrysanthemic acids

M.J. Devos, L. Hevesi, P. Mathy, M. Sevrin, G. Chaboteaux and A. Krief [*]

Facultés Universitaires Notre-Dame de la Paix, Department of Chemistry
61, rue de Bruxelles, B-5000 - Namur (Belgium)

Abstract - Synthesis of suitable esters of (IR) *trans* chrysanthemic acid and their (IR) *cis* dihalogenovinyl analogues have previously been reported from this laboratory. These compounds were prepared from isopropylidene triphenyl phosphorane and α,β-unsaturated esters with suitable functional groups. The use of α-metallo sulfones and α-metallo nitroalkanes was tried in place of isopropylidene triphenyl phosphorane in order to lower the size of the reagent as compared to the size of the delivered iso-propylidene moiety (M_w = 42). α,β-unsaturated esters are not suitable electrophiles but alkylidene malonates seem to fulfil the requirements. Scopes and limitations of the methods are described.

Suitable esters of (1R)trans chrysanthemic acid 1 and of their (1R) cis dihalogenovinyl analogues 2 are among the most powerful insecticides [1,2] sold respectively for domestic and agricultural uses. Their economical value is very high and for example their world turnover has reached 450 million dollars in 1981.

SCHEME I

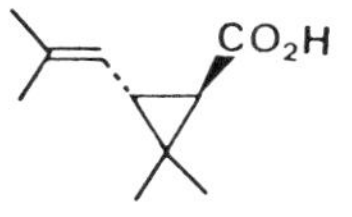

1R,3S Chrysanthemic Acid (trans)

1

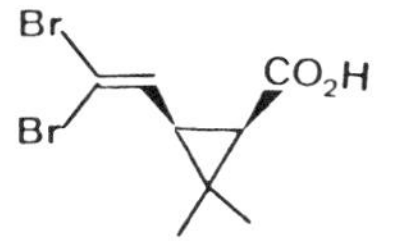

1R,3R Dibromovinyl Chrysanthemic Acid (cis)

2

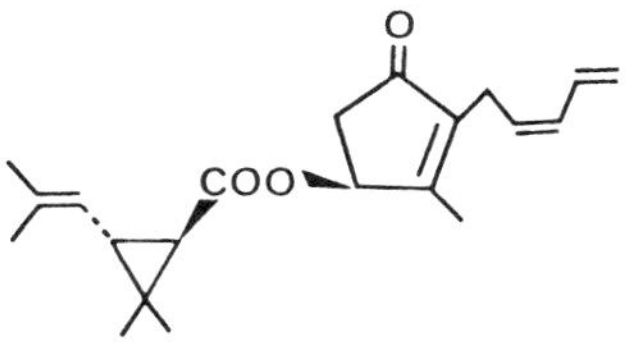

Pyrethrin

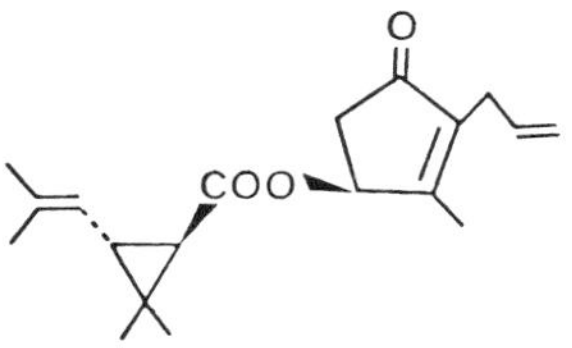

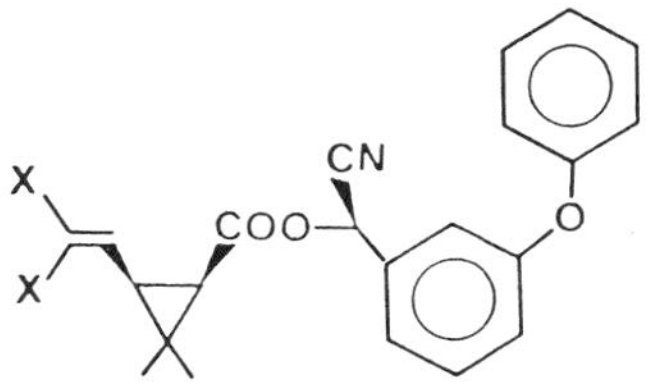

S-Bioallethrin

1R,3R (trans)

High (1)

High

< 1 day

NAME	Cypermethrin X = Cl	Deltamethrin X = Br
Stereochemistry	Mixture	1R,3S (cis)
Houseflies Toxicity	130	1400 (2.10⁻⁹ g/kg)
KD effect to insects	without	
Mammals Toxicity	1	4 (1.3 g/kg)
Photochemical Stability	4 months	

123

Since 1976, we have described several syntheses of such derivatives [3-10] (Schemes II-X).
These include the enantioselective syntheses [5,6] of both (1R)trans chrysanthemic acid and its
((R)cis gem dibromovinyl analogue (Scheme VII-X). The key step in these syntheses is without
contest the construction of the cyclopropane ring possessing the required functionalities.

We have shown that these molecules can be prepared from isopropylidene triphenyl phosphorane and suitably functionalized αβ-unsaturated esters [3-7]. In these reactions the phosphorous ylide has played the role of a carbene, allowing the cyclopropanation by formal delivery
of an isopropylidene moiety.

SCHEME II [3]

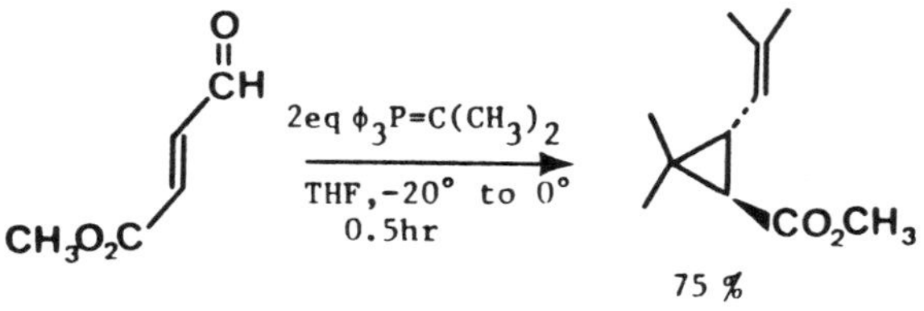

SCHEME III [3]

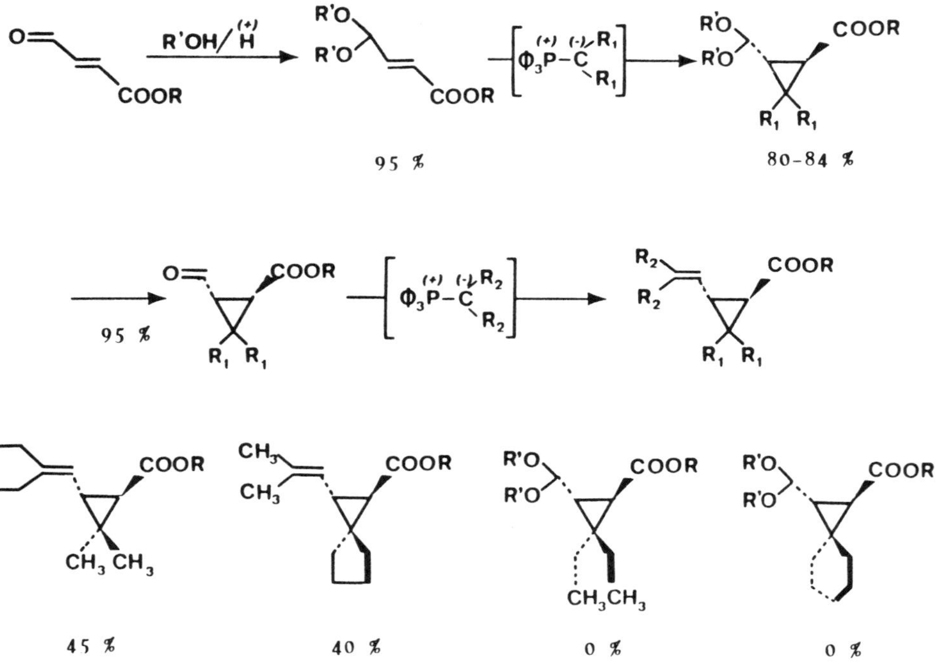

SCHEME IV [4]

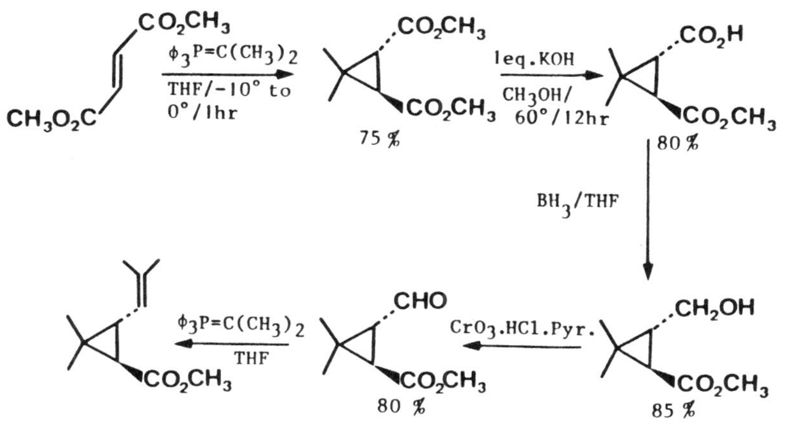

SCHEME V [4]

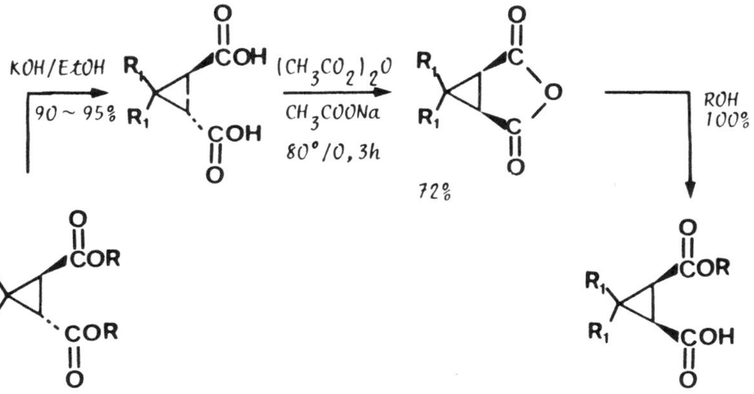

SCHEME VI [4]

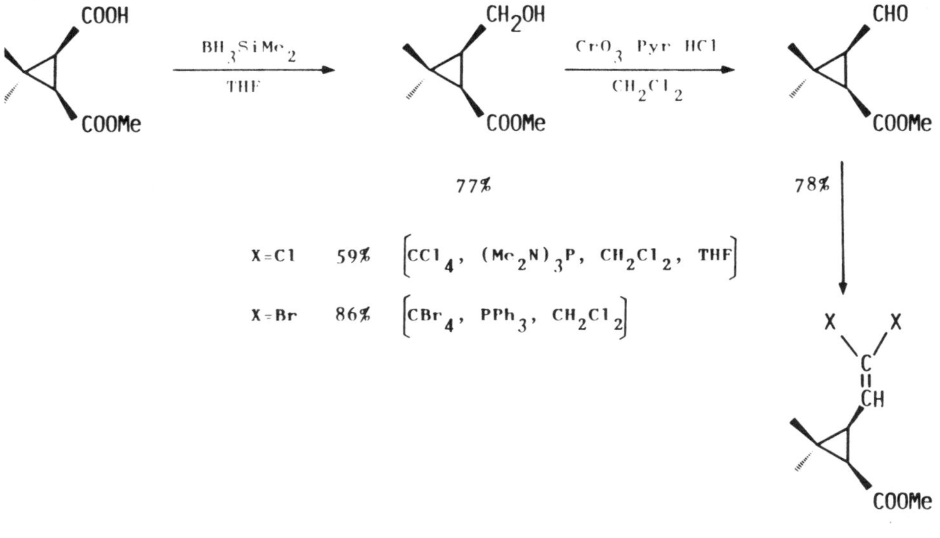

$$X = Cl \quad 59\% \quad \left[CCl_4, \ (Me_2N)_3P, \ CH_2Cl_2, \ THF \right]$$

$$X = Br \quad 86\% \quad \left[CBr_4, \ PPh_3, \ CH_2Cl_2 \right]$$

SCHEME VII [5]

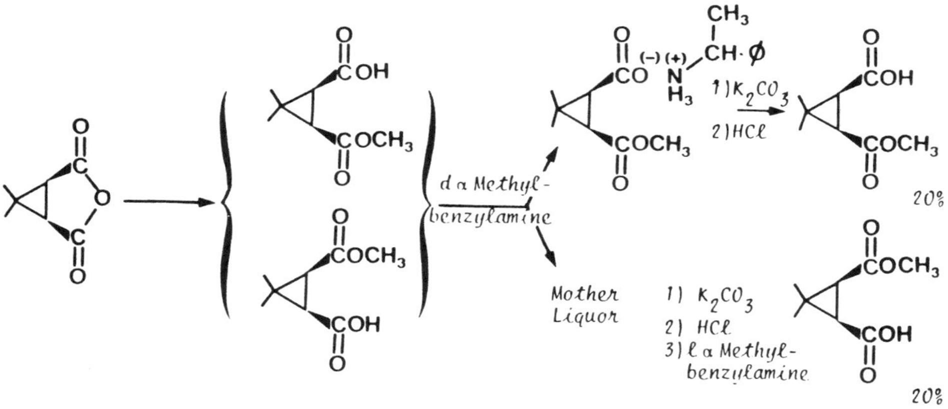

$[\alpha]_{20}^{D} = 30,91$

SCHEME VIII [5]

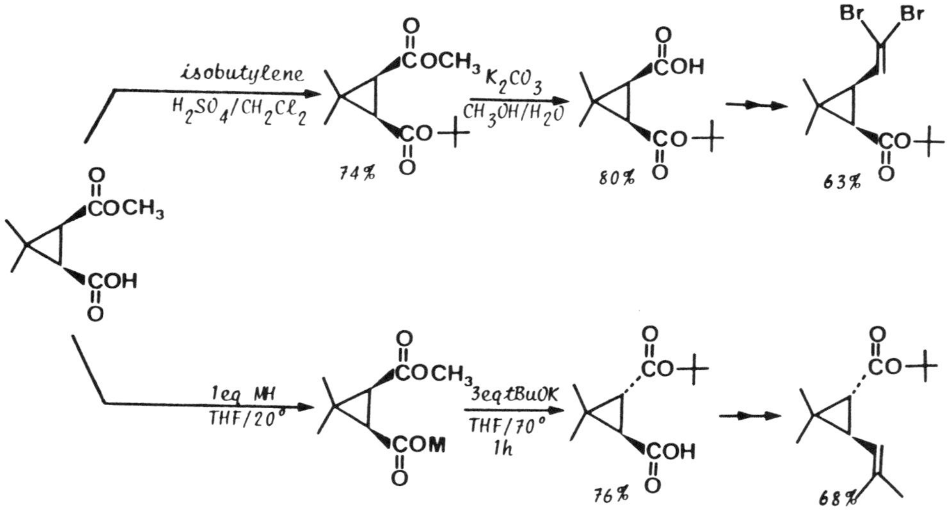

SCHEME IX [6]

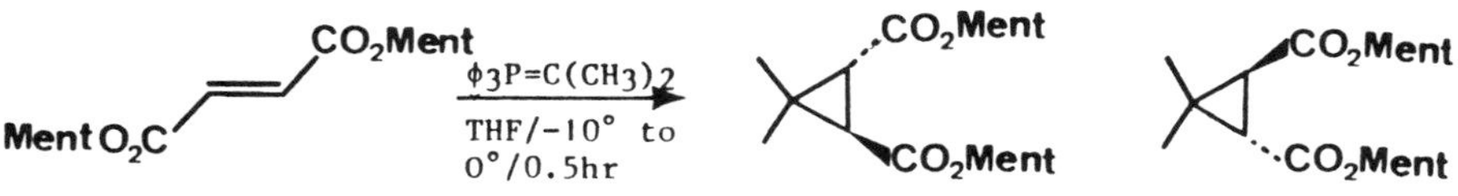

$\geqslant 80\%$ overall yield

Natural l-menthol	0.625 mol/l	28	72
Natural l-menthol	0.125 mol/l	13	87
Unnatural d-menthol	0.125 mol/l	87	13

SCHEME X [6]

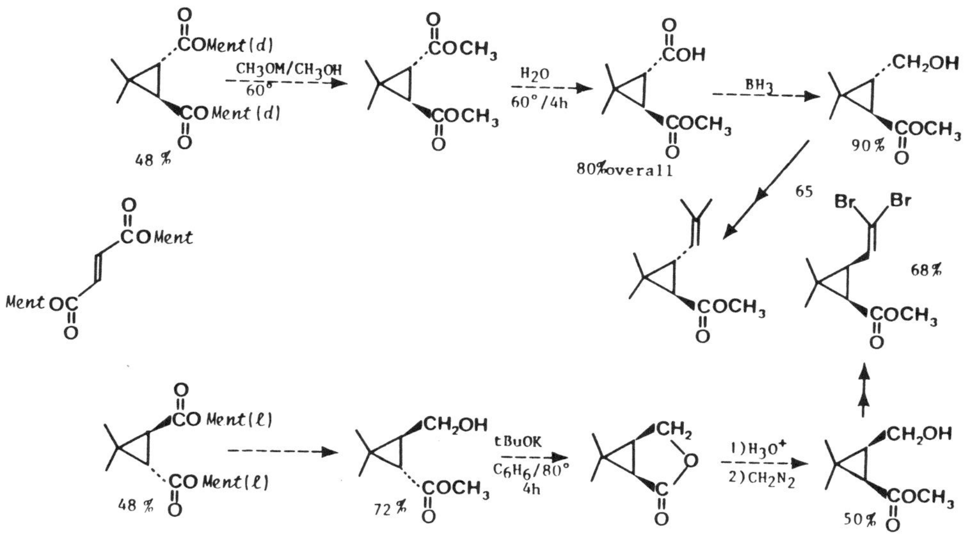

This carbenoid offers the advantage to be readily available, highly reactive, thermally stable and interestingly, the triphenyl phosphine recovered at the end of the cyclopropanation reaction [11] can be recycled.

There remains however still some goal to achieve in that field since the synthesis of isopropylidene triphenyl phosphorane requires a strongly basic system (nBuLi/THF) and another problem arises from the important size of the reagent (Mw 312 for isopropyl triphenyl phosphonium bromide) required for the transfert of the isopropylidene moiety (Mw = 42). We have set up a research program which should involve an identical synthetic strategy but with the objective to replace isopropylidene triphenyl phosphorane by another reagent able to transfer the isopropylidene moiety and able to avoid the problems cited for the phosphorous ylide case.

Several other α-heterosubstituted organometallics [16] bearing two alkyl groups on the carbanionic center are susceptible to react with αβ unsaturated esters to produce gem dialkyl cyclopropane carboxylic esters. Some of them such as sulfur ylides [10,12], diazo alkanes and anions derived from sulfoximes [14,15] have been already successfully used for that purpose. To our opinion they do not offer advantages over the phosphorous ylide method, due to the instability of the formers and the size of the reagent in the latter case. There remain other candidates [16] such as selenonium and arsonium ylides which should have closely related advantages and problems and α-halogeno organometallics whose well-known instability [17] should preclude the use. α-Metallo sulfones [18] and α metallo nitroalkanes [19] retain in contrast our attention. Sulfones [18] and nitro [19] alkanes are both readily available and easily metallated. Isopropyl methylsulfone requires nBuLi or KDA in THF whereas lithium, sodium or potassium methylate in methanol or caesium hydroxyde in water are sufficiently strong to perform the metallation of 2-nitropropane. In the later case, the salts can be taken out from the solvent and stored.

Interestingly, in both cases, comparative size of the starting material and isopropylidene would be in case of successfull cyclopropanation, respectively 284/42 and 89/42.

Although novel, the reactions have few precedent in the literature. It is, for example well known that α metalloallyl sulfones allow the cyclopropanation of β.β dimethyl acrylate [20] and of β,β - dimethyl alkylidene malonate [21] and has produced chrysanthemic acid (Scheme XI). More recently α-metallo nitromethane [22] has been used for the cyclopropanation of alkylidene malonates. However, these reactions have never involved, to our knowledge, the cyclopropanation of electrophilic olefins with a formal monoalkyl or dialkylsubstituted carbene.

Our first trials with α,β unsaturated esters were not successful. In most cases the first step of the reaction(e.i. the conjugate addition of the organometallics across the carbon-carbon double bond of the enoate),took usually place [23] but we do not find yet the right conditions which permit the cyclisation of the resulting γ-heterosubstituted organometallics (Scheme XII). In the special case of maleate and fumarate,another reaction occurs and an α,β unsaturated esters is obtained   The formation can be rationalized by formal 1,2-metal-hydrogen shift followed by β-elimination of the heteroatomic moiety.

We then investigated the reaction of these organometallics with alkylidene malonates or with tri-carbomethoxy-ethylene and were now successful in synthesizing the corresponding cyclopropanes. We observed however a quite important difference of reactivity between the two reagents.

In the reactions involving sulfones, most of the work has been carried out with 2-lithio 2-phenylsulfonyl propane. We found that the solvant plays a crucial role at least for the cases studied.

The addition reaction takes place in THF but not in methanol or DMSO or in a mixture of THF and these solvents. The cyclisation reaction however does not occur in THF, but is efficient in methanol, DMSO or a mixture of these solvents with THF (Scheme XIII).

SCHEME XI

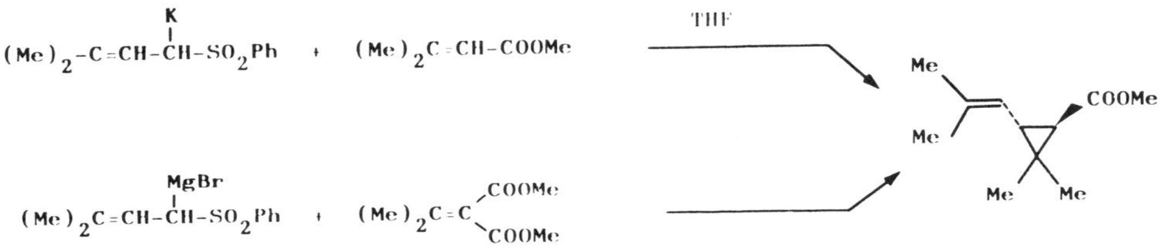

SCHEME XII

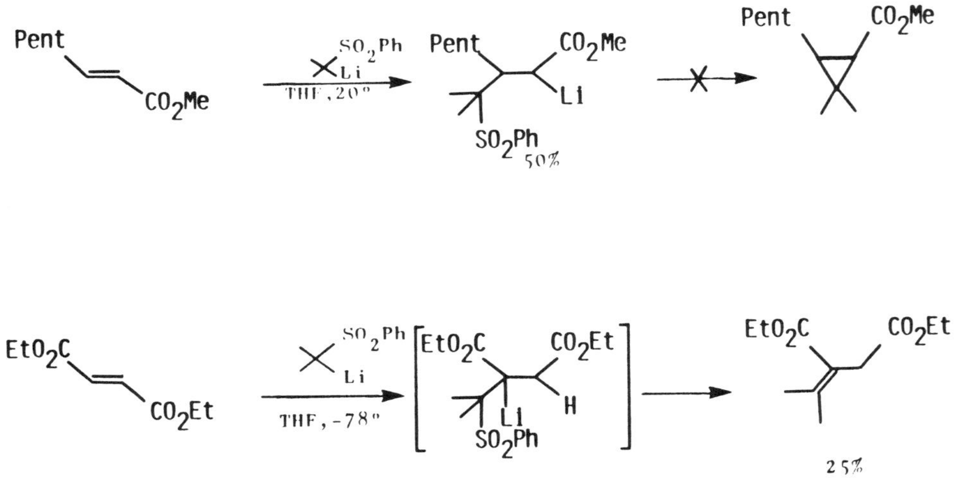

SCHEME XII (suite)

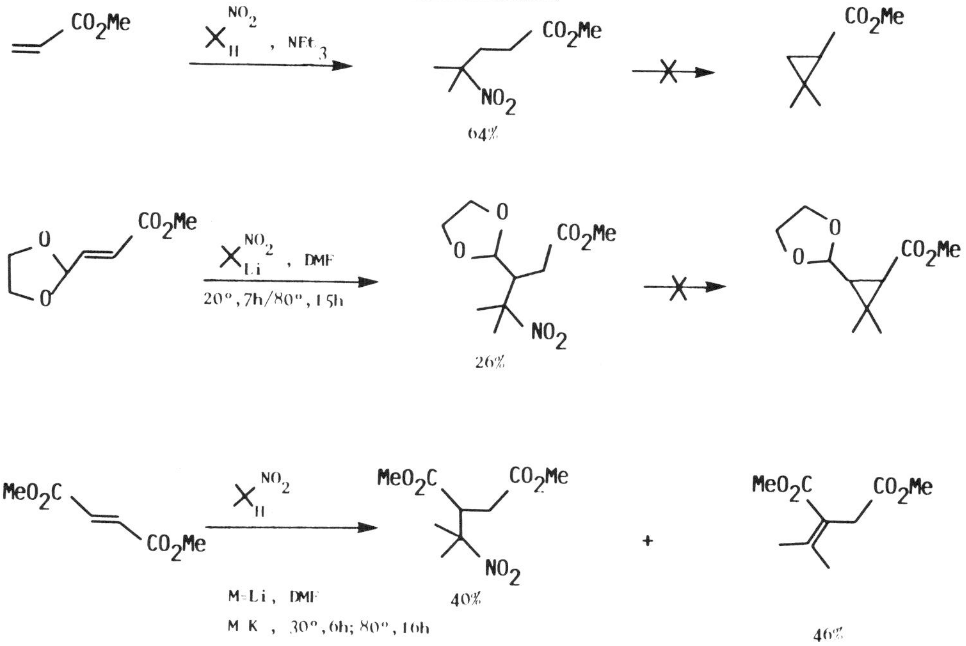

64%

26%

M-Li , DMF   40%

M-K , 30°,6h; 80°,16h   46%

SCHEME XIII

X = SO₂Ph

X = NO₂   M = K

62-80%

50-82%

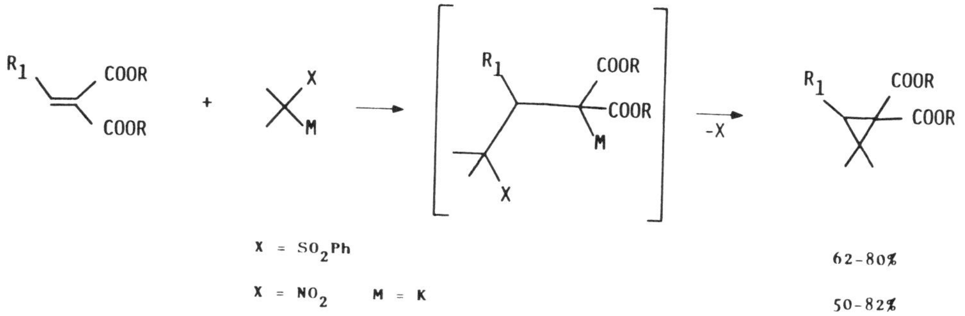

Several 2-metallo 2 nitro propanes (M = Li,Na,K,Cs) were reacted with a series of alkylidene malonates. Best conditions imply the potassium salt in DMSO. The reactions are conducted at 80° for few hrs. and lead to cyclopropane carboxylic esters in 50-80 % yield (Scheme XIII). In some cases (M = Li) and under suitable conditions, the Michael adduct can be trapped before cyclisation occured. Scope and limitations of these reactions will be discussed as well as their use for the synthesis of chrysantemic esters and their analogues.

References

1. M. Elliot, and N.F. Janes, Chem. Soc. Rev. $\underline{7}$, 473 (1978).

2. D. Arlt, M. Jautelat, and R. Lantzsch, Angew. Chem. int. ed. $\underline{20}$, 703 (1981).

3. M.J. De Vos, L. Hevesi, P. Bayet and A. Krief, Tet. Lett. 3911 (1976).

4. J.N. Denis, M.J. De Vos and A. Krief, Tet. Lett. 1847 (1978).

5. M.J. De Vos and A. Krief, J. Amer. Chem. Soc., $\underline{104}$, 4282 (1982).

6. M.J. De Vos and A. Krief, Tet. Lett. 103 (1983).

7. a. M.J. De Vos and A. Krief, Tet. Lett. 1511 (1979).

   b. M.J. De Vos and A. Krief, Tet. Lett. 1515 (1979).

8. M.J. De Vos and A. Krief, Tet. Lett. 1845 (1978).

9. M.J. De Vos and A. Krief, Tet. Lett. 1891 (1979).

10. M. Sevrin, L. Hevesi and A. Krief, Tet. Lett. 3915 (1976).

11. a. R. Mechoulam, and F. Sondheimer, J. Amer. Chem. Soc. $\underline{80}$, 4386 (1958).

   b. J.P. Freedman, Chem. Ind. (London) 1254 (1959).

   c. H.J. Bestmann and F. Seng, Angew. Chem. $\underline{74}$, 154 (1962).

12. E.J. Corey and M. Jautelat, J. Amer. Chem. Soc. $\underline{89}$, 3912 (1967).

13. H.M. Walborsky, T. Sugita, M. Ohno and H. Inouye, J. Amer. Chem. Soc. $\underline{89}$, 3912 (1967).

14. C.R. Johnson and E.R. Janiga, J. Amer. Chem. Soc. $\underline{95}$, 7692 (1973).

15. C.R. Johnson, Acc. Chem. Res. $\underline{6}$, 341 (1973).

16. A. Krief, Tetrahedron $\underline{36}$, 2531 Report 94.

17. G. Köbrich, Angew. Chem. int. ed. Engl., $\underline{6}$, 41 (1967).

   G. Köbrich, Bull. Soc. Chim. Fr., 2712 (1969).

18. a. M. Julia and B. Badet, Bull. Soc. Chim. Fr. 525 (1976) and references cited.

   b. M. Julia and D. Arnould, Bull. Soc. Chim. Fr. 743 (1973).

   c. M. Julia and J.M. Paus, Tet . Lett. 4833 (1973) and references cited.

19. a. O. von Schickh, G. Apel, H.G. Padeken, H.H. Schwarz and A. Segnitz, Methoden der Organi-schen Chemie, Vol 10/1 p.1 (1971). Eugen Müller ed.

   b. K.C. Kerber, G.W. Urry, and N. Kornblum, J. Amer. Chem. Soc. $\underline{87}$, 4520 (1956).

   c. N. Kornblum, Angew. Chem. Int. Ed. Engl. $\underline{14}$, 734 (1975).

20. J. Martel and C. Huynh, Bull. Soc. Chim. Fr. 985 (1967).

21. M. Julia and A. Guy-Roualt, Bull. Soc. Chim. Fr. 1411 (1967).

22. K. Annen, H. Hofmeister, H. Laurent and R. Wiechert, Chem. Ber. $\underline{111}$, 3094 (1978).

23. N. Ono, A. Kamimura, and A. Kaji, Synthesis 226 (1984).

Dien-ketens: key compounds in natural product synthesis

Gerhard Quinkert

Institut für Organische Chemie der Universität, Niederurseler Hang, D-6000 Frankfurt am Main 50, Federal Republic of Germany

Abstract The discovery of the photochemical cleavage of linear conjugated cyclohexadienones[2] was made by accident rather than on purpose[3]. Description of the electronic mechanism of photo-isomerization of cyclohexa-2,4-dienones[4] was instrumental in leading to the notion of surface crossings between singlet states[5]. First hand information of the scope and mechanistic details of a reaction finally transforming phenols into hexa-dienoic acids (or derivatives)[6] paved the way for useful application. Intermolecular addition of protic nucleophiles has already been used in natural product synthesis[7]. Its intramolecular version[8] still lacks utilization.

Dien-ketens have been known ever since the chapter on light-induced reactions of linear conjugated cyclohexadienones was opened[2,9]. Figure 1 introduces various reactions leading to and away from dien-ketens.

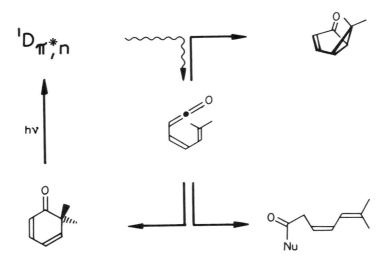

Fig. 1. Dien-ketens conveniently accessible by ring cleavage of $^1(\pi^*,n)$-excited cyclohexa-2,4-dienones[2], partition themselves between one or more of three reaction channels[10]: mono-cyclization (→cyclohexa-2,4-dienones), bicyclization (→bi-cyclo[3.1.0]hex-2-enones) and/or addition of protic nucleophiles (→substituted hexadienoic acids or derivatives).

THE ADDITION OF PROTIC NUCLEOPHILES TO KETENS IS, OF COURSE, A WELL KNOWN RE-
ACTION. THE THOROUGH STUDY OF 1,10-SECO-ANDROSTA-1,3,5(10)-TRIEN-1-ONE (4)[8,11]
GIVES ALREADY AN IDEA HOW COMPLEX THE REACTION ACTUALLY IS AND HOW ADDITION
MAY BE ACHIEVED EVEN IN CASES WHICH AT FIRST SIGHT LOOK RATHER UNPROMISING
(SEE FIG. 2).

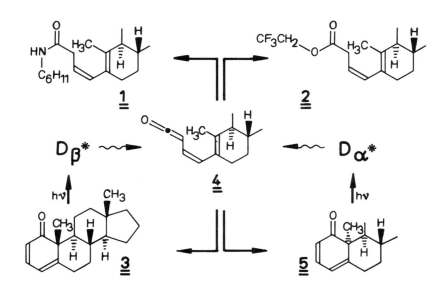

Fig. 2. Overall view of reactions leading to and away from the
dien-keten 4 related to androsta-2,4-dien-1-one (3)[8,11]: UV-
irradiation of 3 or of its 10α-epimer 5 yields the same seco-
isomer 4. The latter adds the strong nucleophile cyclohexylamine (→ 1,6-adduct and/or 1,2-adduct 1, if the concentration
of the amine is low or high, respectively), or the weak nucleophile TFE (→ 2) provided the strong nucleophile DABCO is also
present (if not: monocyclization → 3 and 5).

FIGURE 3 ADUMBRATES UNDER WHAT CONDITIONS WHICH TYPE OF ADDUCT MAY BE EXPECT-
ED: HIGH CONCENTRATION OF A PROTIC NUCLEOPHILE FAVOURS 1,2-ADDITION, WHEREAS
AT LOW CONCENTRATION ESPECIALLY IN CASES WHERE PATTERN OF SUBSTITUTION SUP-
PORTS THE REQUIRED CONFORMATION[12], 1,6- OR 1,4-ADDITION MAY OCCUR AS WELL.
THIS IS ESPECIALLY TRUE FOR STRONG NUCLEOPHILES. WEAK NUCLEOPHILES MAY QUITE
FAIL TO FORM ADDUCTS AT ALL (CF. REVERSIBLE PHOTOEPIMERIZATION OF 3 INTO 5 IN
TFE IN FIG. 2).

FIGURE 4 DISCLOSES IN WHICH WAY AN ADDUCT MAY STILL BE PRODUCED IF THE PROTIC
NUCLEOPHILE PRESENT PROVES TO BE TOO WEAK TO REACT WITH A DIEN-KETEN DIRECT-
LY: BY EXTRA HELP OF DABCO. THE STRONG BUT APROTIC NUCLEOPHILE CAN ONLY TEM-
PORARILY BE BOUND TO THE ELECTROPHILIC CARBON ATOM OF A KETEN GROUP AND AFTER
TRANSFORMATION OF THE INITIALLY FORMED ZWITTERIONIC INTERMEDIATE INTO AN ACYL
AMMONIUM DERIVATIVE HAS TO MAKE WAY FOR THE (DEPROTONATED) NUCLEOPHILE Nu^-.

THE DIEN-KETEN 4 PLAYS A CENTRAL ROLE IN THE PHOTOCHEMISTRY OF THE ANDROSTA-
2,4-DIENONES EPIMERIC AT C-10 (FIG. 2). IT IS SUGGESTED BY THE CHEMIST'S
IMAGINATION AND ESTABLISHED BY DIVERS SPECTROSCOPIC METHODS. A COLLAGE (FIG.

5) DISPLAYS THE VARIOUS TYPES OF SPECTRA RECORDED.

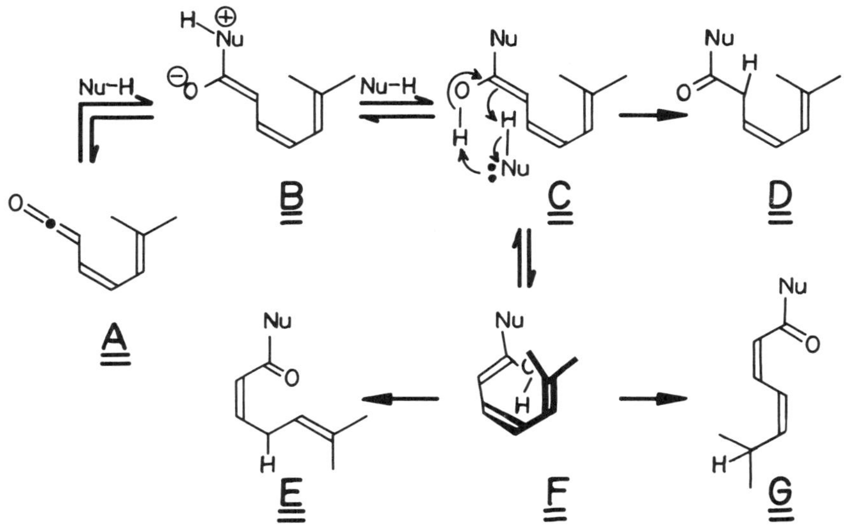

Fig. 3. Addition of a protic nucleophile Nu-H to a dien-keten A leading to zwitterionic intermediate B which by prototropic isomerization affords hexatrienol C. The latter compound gives the 1,2-adduct D by another Nu-H supported prototropic isomerization and affords the 1,4-adduct E or the 1,6-adduct G by 1,5- or 1,7-sigmatropic rearrangement.

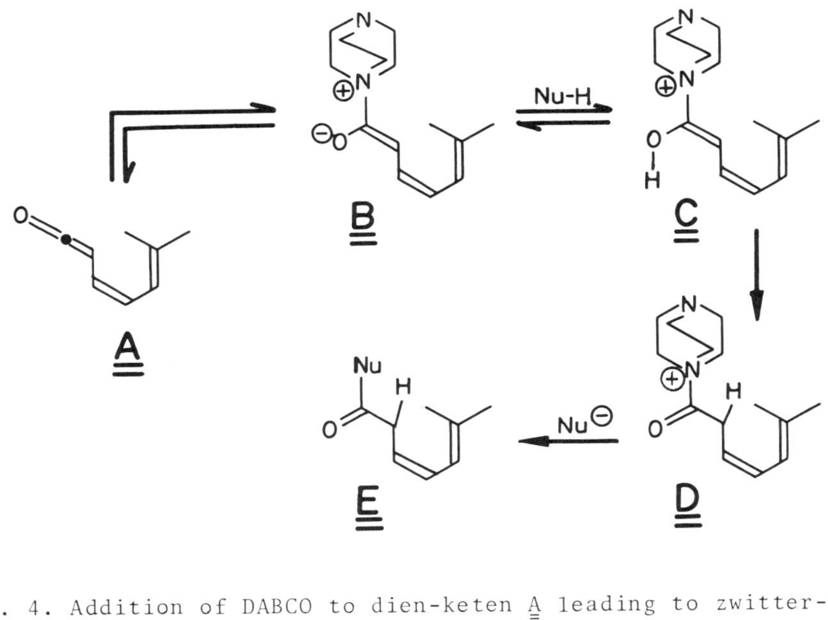

Fig. 4. Addition of DABCO to dien-keten A leading to zwitterionic intermediate B which after abstraction of a proton from the weak nucleophile Nu-H affords the hexatrienol C. The latter compound by prototropic isomerization furnishes the highly reactive acyl ammonium derivative D which is readily substituted by Nu⁻ and gives 1,2- adduct E.

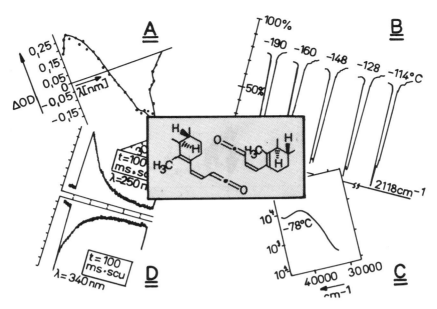

Fig. 5. Spectroscopic evidence for the transient dien-keten
4[11]. A: Spectrum of absorption differences monitored after
flash photolysis of 3 at room temperature in TFE using light
>280 nm; B: IR-spectra in the keten band region monitored aft-
er irradiation of a suspension of 3 at -190°C in nujol with
light >335 nm; C: UV/VIS-absorption spectrum after irradiation
of 3 at -76°C in n-hexane with 313 nm light; D: Transmission/
time plots monitored after flash photolysis of 3 at room temperature in TFE using light >280 nm.

CONVENTIONAL SPECTROSCOPY AT LOW TEMPERATURE[4,14] AND FLASH SPECTROSCOPY AT
ROOM TEMPERATURE[11,15] MATCH WELL. THE IR ABSORPTION IN THE 2100 CM^{-1} REGION
ENABLES KETENS TO BE READILY IDENTIFIED. 4 SHOWS TWO BANDS. THE ONE AT 2118
CM^{-1} INCREASES IRREVERSIBLY AT THE EXPENSE OF THE OTHER ONE AT 2104 CM^{-1}. WE
ATTRIBUTE THESE TWO BANDS TO DIFFERENT ROTAMERS OF THE DIEN-KETEN (B IN FIG.
5). FLASH SPECTROSCOPY SHOWS FIRST ODER DECAY OF THE TRANSIENT DIEN-KETEN AT
250 NM OR FIRST ORDER RISE OF THE TWO EPIMERIC ANDROSTADIENONES 3 AND 5 AT
340 NM: THE HALF LIFE OF THE KETEN AT ROOM TEMPERATURE AMOUNTS TO 0.07S (D IN
FIG. 5). POINT-TO-POINT REGISTRATION OF THE DIFFERENCE IN ABSORPTION CAUSED
BY THE LIGHT FLASH AT ROOM TEMPERATURE LEADS TO SPECTRUM A IN FIG. 5. ITS
MAXIMUM AT ABOUT 245 NM IS CLOSE TO THE ABSORPTION MAXIMUM OF THE KETEN TRAP-
PED AT -78°C (C IN FIG. 5); ITS MINIMUM CORRESPONDS TO THE ABSORPTION MAXIMUM
AT ABOUT 330 NM OF THE DIENONES PRODUCED ON HEAT-INDUCED MONOCYCLIZATION.

OTHER DIEN-KETENS HAVE BEEN SIMILARLY PROVEN. SUCH IS THE CASE FOR THE KIN-
ETICALLY UNSTABLE SECO-ISOMER 9 RELATED TO RAC-8. IN TFE THE BICYCLIC ISOMER
RAC-10 IS FORMED[16]: THE METHYL SUBSTITUENT AT C-2 FAVOURS THOSE CONFORMATIONS
OF 9 WITH THE KETEN GROUP TURNED INSIDE AND HEREBY FACILITATES BICYCLIZATION.
DABCO AGAIN INTERFERES IN SUPPORT OF ADDITION OF THE WEAK PROTIC NUCLEOPHILE
TO THE KETEN AFFORDING RAC-7. 1,2-ADDUCTS ARE FORMED IMMEDIATELY IF STRONG
PROTIC NUCLEOPHILES ARE AVAILABLE (I.E. FORMATION OF RAC-6 IN FIG. 6). HIGH
QUANTUM (~0.6) OR CHEMICAL YIELD (>80%) IS TYPICAL FOR ADDUCT FORMATION[17].

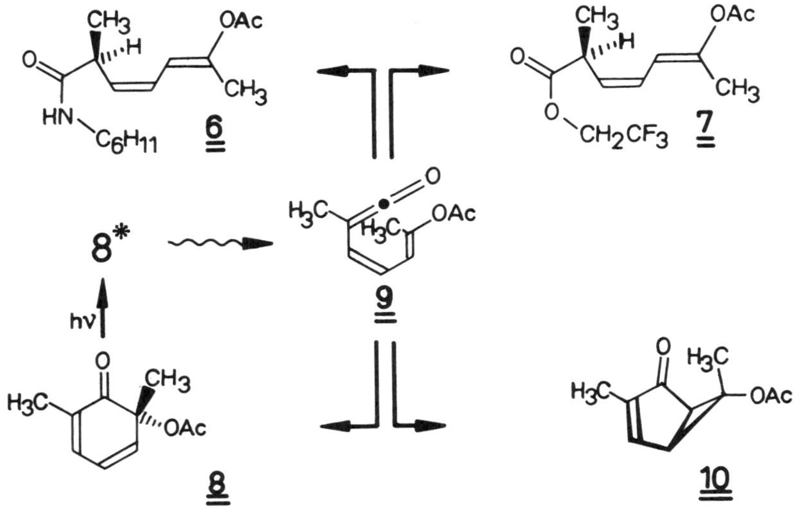

Fig. 6. Overall view of reactions leading to and away from the dien-keten 9 related to rac-8[8,10]: UV-irradiation of rac-8 yields the seco-isomer 9. The latter adds the strong nucleophile cyclohexylamine (→ 1,2-adduct rac-6), or the weak nucleophile TFE (→ rac-7) provided the strong nucleophile DABCO is also present (if not: bicyclization → rac-10).

A COLLECTION OF O-QUINOL ACETATES DIFFERING IN NUMBER AND/OR POSITION OF METHYL SUBSTITUENTS IS PRESENTED IN THE TABLE. THERE INFORMATION IS GIVEN FOR WHICH INDIVIDUAL CYCLOHEXA-2,4-DIENONE WHICH PROTIC NUCLEOPHILE HAS TO BE USED AS SOLVENT (METHANOL, TFE) OR ADDITIVE (CYCLOHEXYLAMINE) IN ORDER TO ACCOMPLISH ADDUCT FORMATION OR ISOMERIZATION TO BICYCLIC KETONE.

TABLE. Quick instructions to use the right conditions: The shaded (dotted) area guarantees adduct formation (bicyclization).

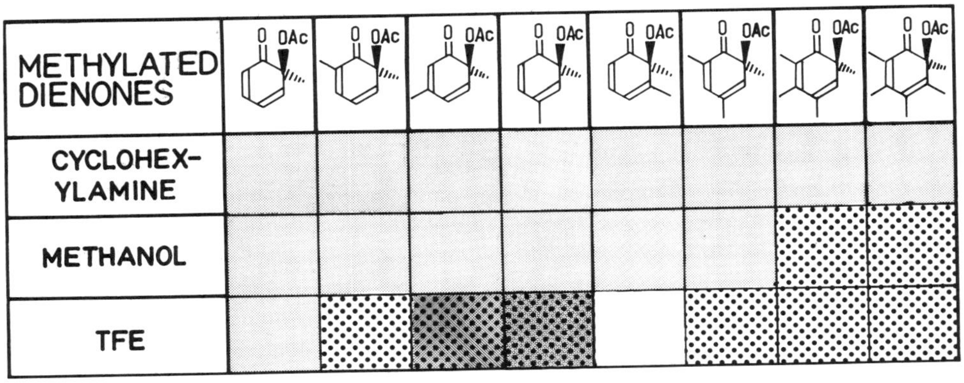

RELATIVE POPULATION OF THE VARIOUS CONFORMERS OF A GIVEN DIEN-KETEN[6] AS WELL AS POLARITY OF THE SOLVENT COME INTO PLAY. BY QUANTITATIVE STUDY[15] IT HAS BEEN SHOWN THAT POLARITY OF THE SOLVENT IS PRACTICALLY WITHOUT INFLUENCE ON THE RATE OF MONOCYCLIZATION. THE RATE OF BICYCLIZATION, HOWEVER, GROWS RAPIDLY WITH INCREASING SOLVENT POLARITY.

O-QUINOL ACETATES, A CONVENIANTLY ACCESSIBLE SUBGROUP OF CYCLOHEXA-2,4-DIEN-
ONES, GENERALLY PRODUCE 1,2-ADDUCTS WITH EXCEEDINGLY HIGH STEREOSELECTION[18].
RAC-6 IS REPRESENTATIVE AND SHOWS 3Z,5E-CONFIGURATION. THIS STRUCTURAL INFOR-
MATION UNAMBIGOUSLY FOLLOWS FROM X-RAY ANALYSIS[13] AND IS SUPPORTED BY IR- OR
NMR-DATA. GROUND STATE CONTROL OF DIEN-KETEN CONFIGURATION HAS BEEN THOUGHT
TO BE THE CAUSE[6] (FIG. 7).

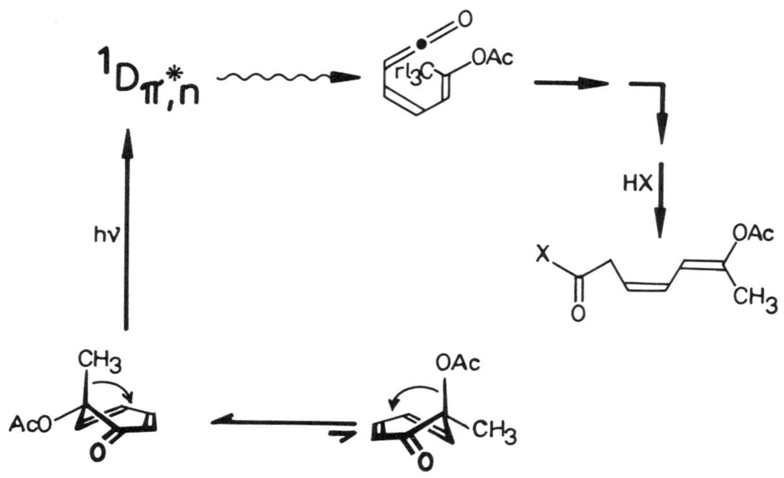

Fig. 7. View of highly efficient stereoselection upon 1,2-ad-
duct formation. It is supposed, first that each of the two
conformers of a linear conugated cyclohexadienone ringopens
stereospecifically[6] (following the direction of the arrow),
and secondly that in the thermodynamically more stable conformer the methyl group occupies the pseudoaxial position[6].

SYNTHESIS OF COMPLEX TARGET COMPOUNDS IS THE SUREST STANDARD BY WHICH TO TEST
THE REAL TENDENCY OF A REACTION MECHANISTICALLY ALREADY KNOWN. FIGURE 8 DIS-
PLAYS COMPOUNDS PRECEDING OR FOLLOWING VARIOUS KETENS FORMED DURING THE SYN-
THESIS OF THE NATURALLY OCCURING FOOD COLOURANT DIMETHYLCROCETIN[7]: PHOTOCHEM-
ICAL RING CLEAVAGE IS THE KEY REACTION OF THE THREE-STEP SYNTHESIS. IT TAKES
PLACE GRADUALLY ON THE STEREOISOMERIC BIS-DIENONES. WITH ONLY A LIMITED A-
MOUNT OF METHANOL ADDED, MONOCYCLIZATION AND NUCLEOPHILE ADDITION STRUGGLE
FOR THE KETENS FORMED. WHEN CIRCULAR POLARIZED LIGHT WAS USED UNDER THOSE
CONDITIONS, THE RECOVERED BIS-DIENONES CONTAINED A FRACTION WHICH HAD BECOME
OPTICALLY ACTIVE. THE OPTICALLY INACTIVE FRACTION WAS, OF COURSE, THE MESO-
COMPOUND 13. AND THIS WAS THE WAY CONFIGURATION WAS ASSIGNED. IN THE COURSE
OF THE SYNTHESIS OF DIMETHYLCROCETIN PHOTOCHEMICAL CLEAVAGE OF LINEAR CONJU-
GATED CYCLOHEXADIENONES THUS HAS BEEN USED TO SOLVE TWO TOTALLY DIFFERENT
PROBLEMS. THIS CANNOT SATISFY THE DESIRE TO PROVE THE USEFULNESS OF A REAC-
TION THE DISCOVERY OF WHICH HAS BEEN CLASSIFIED AS ACCIDENT[3]. AT LEAST THE
INTRAMOLECULAR VARIANT OF THE INTERMOLECULAR ADDITION OF NUCLEOPHILES TO
DIEN-KETENS SHOULD BE DEVELOPED BY CONCEPTION.

THE INTRAMOLECULAR ADDITION OF AN ALCOHOLIC HYDROXY GROUP TO AN APPROPRIATELY
ARRANGED KETEN GROUP PRODUCED PHOTOCHEMICALLY IN THE NOW WELL-KNOWN WAY,

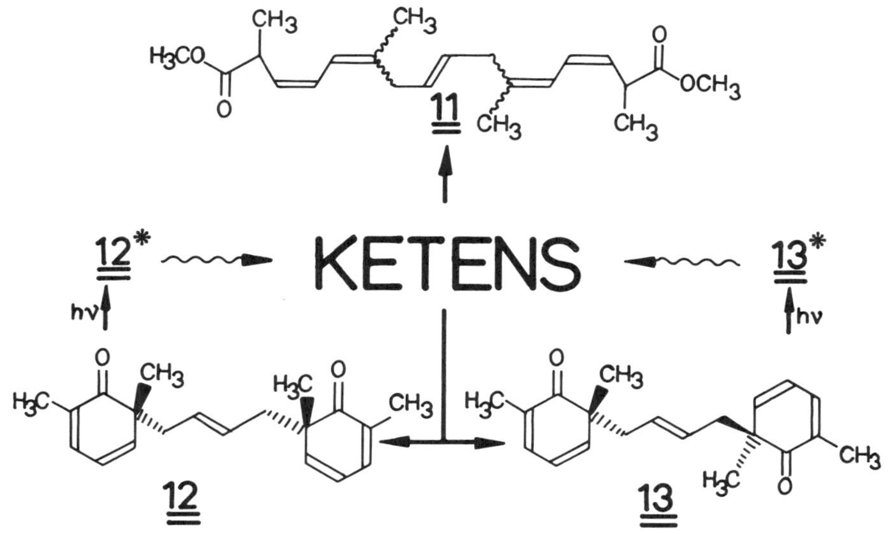

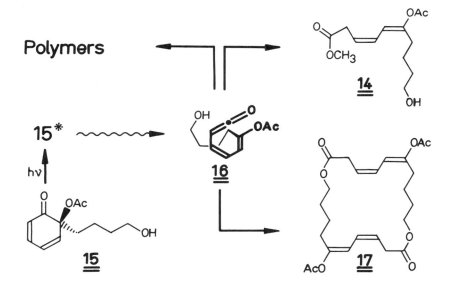

Fig. 8. A group of dien-ketens acting as key compounds during a synthesis of dimethylcrocetin[7]. On irradiation rac-12 and/or 13 (even with sun-light of Frankfurt), in the absence of a protic nucleophile, reversibly photoisomerize. In methanol solution several stereoisomeric tetrahydrodimethylcrocetins (of constitution 11) almost quantitatively are produced, which can easily be transformed into dimethylcrocetin.

Fig. 9. Overall view of reactions leading to and away from dien-keten 16 related to rac-15[8]: UV-irradiation of rac-15 yields the seco-isomer 16. The latter adds the extermal nucleophile methanol (→ 1,2-adduct 14). It dimerizes in ether (→ 17) provided DABCO is also present (if not: only polymeric material is formed).

SHOULD OPEN A NEW AVENUE TO MACROCYCLIC LACTONES. WHEN RAC-15 WAS IRRADIATED
(FIG. 9) IN ETHER SOLUTION ONLY POLYMERIC MATERIAL COULD BE ISOLATED. CHANGING
SOLVENT FROM ETHER TO METHANOL AFFORDED THE INTERMOLECULAR 1,2-ADDUCT 14. THE
BIFUNCTIONAL KETEN 16 OBVIOUSLY HAD BEEN FORMED BUT INTRAMOLECULAR ADDITION
LEADING TO THE 11-MEMBERED LACTONE MUST HAVE FAILED. THE HYDROXYKETEN, THERE-
FORE, HAD TO BE ACTIVATED: DABCO HAS BEEN ABLE TO PROMOTE ADDITION OF A WEAK
PROTIC NUCLEOPHILE TO THE KETEN GROUP OF DIEN-KETENS IN PREVIOUS EXAMPLES
(FIGS. 2 AND 6) AND OUGHT TO BE TRIED AS ACTIVATOR FOR AN INTRAMOLECULAR CASE,
TOO (FIG. 10).

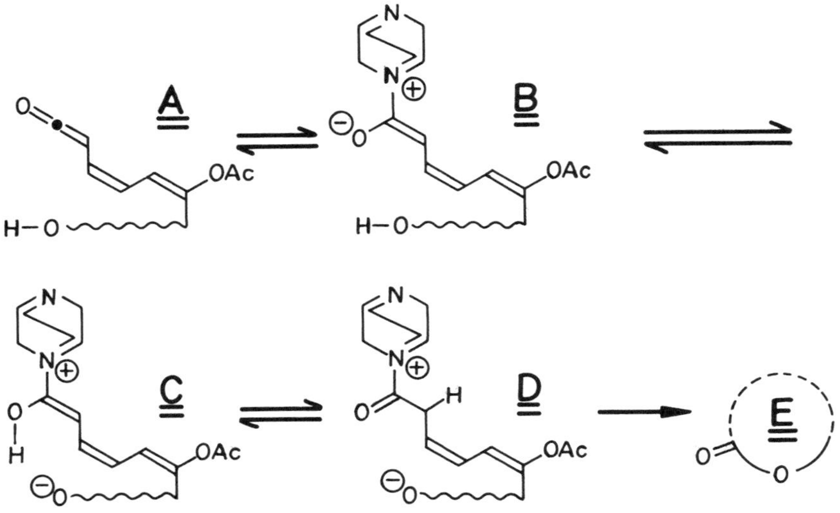

Fig. 10. Addition of DABCO to dien-keten A with an alcoholic
hydroxy group in an appropriate position leading to zwitterionic intermediate B which after abstraction of a proton from
the hydroxy group affords the hexatrienol C. The latter compound by prototropic isomerization furnishes the highly reactive acyl ammonium derivative D which is substituted by RO⁻
and gives lactone E.

IRRADIATION OF AN ETHER SOLUTION OF RAC-15 IN THE PRESENCE OF DABCO GAVE THE
SYMMETRICAL 22-MEMBERED BIS-LACTONE 17 (FIG. 9). ITS CONSTITUTION (1,2-AD-
DUCT), CONFIGURATION (3Z,5E), AND CONFORMATION (TRANS-LACTONE[19]) FOLLOWED FROM
X-RAY STRUCTURE ANALYSIS (FIG. 11): RING STRAIN CAUSED CYCLO-DIMERIZATION TO
OUTWEIGH COMPLETELY CYCLO-MONOMERIZATION. IF ONLY THE SIDE CHAIN OF THE DIEN-
KETEN WOULD BE LONG ENOUGH, THE MONO-LACTONE SHOULD PREDOMINATE OVER THE COR-
RESPONDING BIS-LACTONE. AS A MATTER OF FACT, THE 15-, 16-, 17-, AND 18-MEM-
BERED MONO-LACTONES, WITH A METHYL SUBSTITUENT AT THE CARBON ATOM ADJACENT TO
THE CARBONYL GROUP OF THE LACTONE MOIETY, COULD BE ISOLATED IN 19, 59, 62 OR
80% YIELD, RESPECTIVELY. IN EACH CASE THE PHOTOCHEMICAL REACTION WAS PERFORMED
IN TOLUENE AND, OF COURSE, IN THE PRESENCE OF DABCO. SINGLE-CRYSTAL STRUCTURE
ANALYSIS[8] OF RAC-18 AGAIN PROVED (TRANS-LACTONE[19]) CONFORMATION AND (3Z,5E)
CONSTITUTION OR CONFIGURATION (FIG. 12). DIFFERENT TYPES OF 16-MEMBERED LACTO-
NES WITH ONE OR TWO SUBSTITUENTS ADJACENT TO THE LACTONE GROUP (FIG. 13) SIM-
ILARLY HAVE BEEN PREPARED: IN ORDER TO LEARN HOW TO INTRODUCE, TRANSFORM OR

REMOVE FUNCTIONAL GROUPS WHICH MAY BE SUITABLE FOR SYNTHETIC PURPOSES.

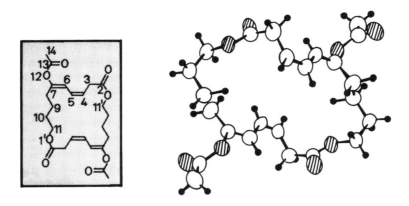

Fig. 11. Conformation of <u>bis</u>-lactone <u>17</u> as observed in the crystalline state[8]. Note S$_2$-symmetry, <u>trans</u>-lactone[19] and 3Z, 5E configuration.

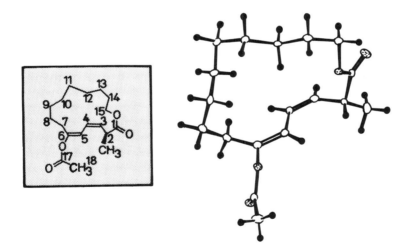

Fig. 12. Conformation of <u>mono</u>-lactone <u>rac-18</u> as observed in the crystalline state[8]. Note the <u>trans</u>-lactone[19] and 3Z,5E configuration.

RECENT ACTIVITY IN MACROLIDE SYNTHESIS[20] HAS STIMULATED SEARCH FOR NEW METHODS TO MAKE LARGE-SIZED LACTONES. AFTER WE HAVE FOUND A PHOTOCHEMICAL ENTRY INTO THIS CLASS OF COMPOUNDS OURSELVES[8] WE ARE NOW TEMPTED TO BECOME DEEPLY INVOLVED IN MACROLIDE SYNTHESIS. WHAT HAS BEEN ACHIEVED SO FAR - THE PREPARATION AND IRRADIATION OF HYDROXYLATED, SUITABLY SUBSTITUTED CYCLOHEXA-2,4-DIENONES (FIGS. 14 AND 15) - IS AN ENCOURAGING BEGINNING.

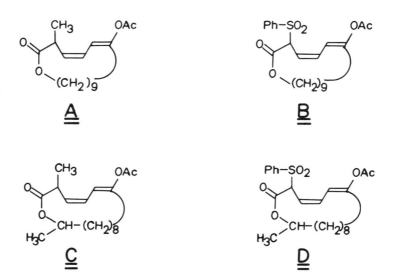

Fig. 13. 16-Membered lactones differing in kind and/or number of substituents obtained by DABCO promoted cyclization of the 2- and/or 15-mono- or disubstituted (3Z,5E)-6-acetoxypenta-deca-1,3,5-trien-15-ol-1-ones.

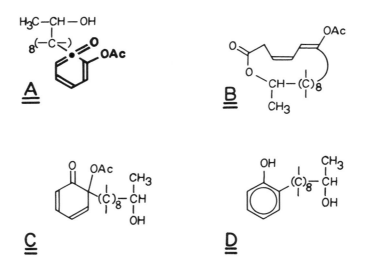

Fig. 14. Scheme that shows how to project and direct the construction of the basic skeleton of large-sized lactones using hydroxylated dien-ketens as key compounds.

DIEN-KETENS WHICH ARE HYDROXYLATED AT THE APPROPRIATE POSITION IN THE SIDE CHAIN (A IN FIG. 14) MAY PLAY THE ROLE OF KEY COMPOUNDS IN MACROLIDE SYN-THESIS. THEIR SUCCESSIVE CYCLO-ISOMERS B ARE POLYFUNCTIONAL BY ORIGIN AND EQUIPPED WITH STRUCTURAL ACCESSORIES THAT WERE ALREADY PRESENT IN THE PRE-CEDING CYCLO-ISOMERS C. THE LATTER COMPOUNDS ARE SATISFACTORILY AVAILABLE BY WESSELY-OXIDATION, IF THE O'-POSITION OF THE FOREGOING PHENOLS D IS SUBSTI-

TUTED BY A DEACTIVATING (I.E. $C_6H_5SO_2-$) GROUP. THE O-SUBSTITUTED PHENOLS D CAN BE EASILY BUILT UP BY O-ALKYLATION OF PHENOLETHERS OR BY WITTIG-REACTION OF SUBSTITUTED SALICYLALDEHYDES AND HYDROGENATION OF THE RESULTING OLEFINS.

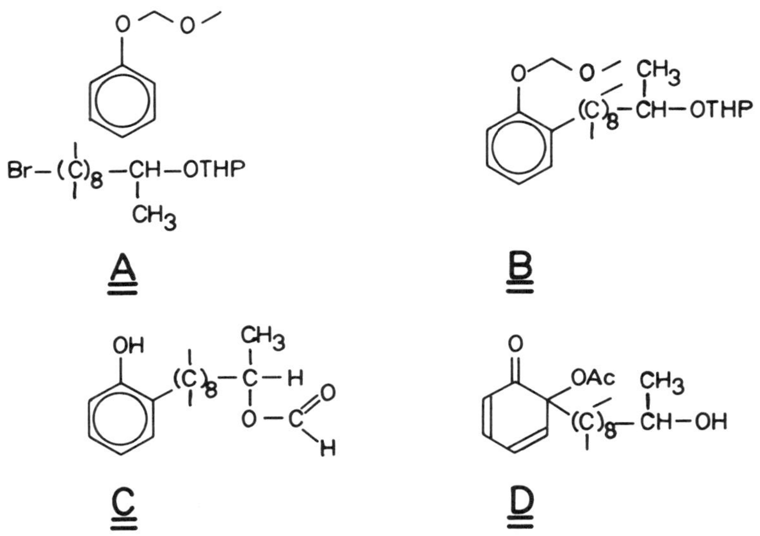

Fig. 15. Scheme that shows how to activate or deactivat functional groups during the preparation of hydroxylated cyclohexa-2,4-dienones used for photoisomerization finally to large-sized lactones.

THE METHOXYMETHOXY SUBSTITUENT (A IN FIG. 15) IS EMPLOYED AS AN EFFECTIVELY ACTIVATING AND O-DIRECTING GROUP IN HYDROGEN/METAL-EXCHANGE REACTIONS[21]. THE TETRAHYDROPYRANYLETHER OF A HALOGENATED ALCANOL SERVES AS ALKYLATING REAGENT. THE ALKYLATION PRODUCT (B IN FIG. 15) AFTER LITHIATION OF THE O'-POSITION READILY REACTS WITH DIPHENYL DISULFIDE. AFTER REMOVAL, BY MILD METHANOLYSIS OF BOTH THE PROTECTING GROUPS, THE REACTIVATED COMPOUND ON TREATMENT WITH H_2O_2 IN METHYLEN DICHLORIDE/FORMIC ACID IS TRANSORMED INTO PHENYLSULFONYLATED FORMATES (C IN FIG. 15). LEAD TETRAACETATE-OXIDATION, CATALYZED BY BORON TRI-FLUORIDE[22], IN METHANOL/ETHYL ACETATE GAVE THE PHENYLSULFONYLATED DIENONES (D IN FIG. 15) IN GOOD YIELD. THE SULFONE SUBSTITUENT, AFTER PHOTOLYSIS OF THE DIENONE AND DABCO PROMOTED CYCLIZATION OF THE RESULTING HYDROXYLATED DIEN-KETEN, MAY BE REMOVED BY 1,6-ELIMINATION ON TRANSESTERFICATION OF THE ENOL ACETATE GROUP.

ACKNOWLEDGEMENTS I would like to thank the coworkers who have made this project possible: B.-J. Freitag, E. Kleiner, H. Englert, and F. Cech (mechanistic part), U.-M. Billhardt, G. Fischer, J. Glenneberg, G. Wess, K. Wagner, N. Heim, and P. Nagler (synthetic part) and especially emphasize the important contribution of Dr. G. Dürner and his crew (analytical and preparative HPLC). I enjoyed a fruitfull cooperation with Prof. D. Rehm (mechanistic part), Professor H. Kessler and Dr. G. Zimmermann (NMR-analysis), Professor H. Fuess, Dr. J.W. Bats, and Dr. E.F. Paulus (X-ray structure analysis).

The project was generously supported for many years by Deutsche Forschungsgemeinschaft, Fonds der Chemischen Industrie, and Hoechst AG.

LIST OF REFERENCES AND FOOTNOTES

1. Abbreviations used: DABCO = 1,4-Diazabicyclo|2.2.2|octane; TFE = 2,2,2-trifluorethanol.

2. D.H.R. Barton and G. Quinkert, Proc. Chem. Soc. 197 (1958); J. Chem. Soc. 1 (1960).

3. D.H.R. Barton, Frontiers in Bioorganic Chemistry and Molecular Biology (Yu. Ovchinnikov and M.N. Kolosov, Eds.), p. 21; Elsevier/North-Holland Biomedical Press, Amsterdam (1979).

4. G. Quinkert, Pure Appl. Chem. 33, 285 (1973).

5. L. Salem, Science 191, 822 (1976).

6. G. Quinkert, Stereoselective Synthesis of Natural Products (W. Bartmann and E. Winterfeldt, Eds.), p. 151; Excerpta Medica, Amsterdam (1979).

7. G. Quinkert, K.R. Schmieder, G. Dürner, K. Hache, A. Stegk, and D.H.R. Barton, Chem. Ber. 110, 3582 (1977).

8. G. Quinkert, G. Fischer, U.-M. Billhardt, J. Glenneberg, U. Hertz, G. Dürner, E. Paulus, and J.W. Bats, Angew. Chem., Int. Ed. Eng. 23, 430 (1984).

9. Racemization of usnic acid had been postulated earlier proceeding through a dien-keten intermediate: G. Stork, Chem. and Ind. 915 (1955).

10. G. Quinkert, F. Cech, E. Kleiner, and D. Rehm, Angew. Chem., Int. Ed. Engl. 18, 557 (1979).

11. G. Quinkert, H. Englert, F. Cech, A. Stegk, E. Haupt, D. Leibfritz, and D. Rehm, Chem. Ber. 112, 310 (1979).

12. (1E,3Z)-1,2,3,4,5-pentamethyl-6-oxohexa-1,3,5-trienyl-acetate is the only dien-keten so far known to give a 1,4-adduct in addition to a 1,2-adduct[13]. For cases where 1,2- and 1,6-adducts were formed side by side see l.c.[11].

13. G. Quinkert, U.-M. Billhardt, E.F. Paulus, J.W. Bats, and H. Fuess, Angew. Chem., Int. Ed. Eng. 23, 442 (1984).

14. G. Quinkert, Photochem. Photobiol. 7, 783 (1968).

15. B.-J. Freitag, Dissertation, Univ. Frankfurt am Main, 1983.

16. Photoisomerization of highly substituted linear conjugated cyclohexadienones into bicyclo|3.1.0|hex-2-enones was firstly reported by J. Griffiths and H. Hart, J. Am. Chem. Soc. 90, 5296 (1968). These authors believed $^3(\pi^*,\pi)$-excited cyclohexa-2,4-dienones to be involved. $^3(\pi^*,\pi)$-excited o-quinol acetates, however, undergo photo dienone/phenol rearrangement affording isomeric phenols[10].

17. Because of photochemical instability chemical yield of bicyclic isomers rarely exceeds 50-60%.

18. 2Z,5E- or 2Z,4Z-configuration stereoselectively is to be expected for 1,4- or 1,6-adducts resulting from o-quinol acetates, if sigmatropic 1,5- or 1,7-hydrogen migration would act as product determining step (Fig. 3).

19. For designation of an ester (lactone) group see J. Dale, Stereochemie und Konformationsanalyse, p. 65; Verlag Chemie, Weinheim 1978.

20. See relevant compounds in previous or forthcoming issues of SYNFORM (Verlag Chemie, Weinheim) or of Natural Product Reports (Royal Society of Chemistry).

21. M.R. Winkle and R.C. Ronald, J. Org. Chem. 47, 2101 (1982).

22. F. Takacs, Monatsh. Chem. 95, 961 (1964).

The "*tert*-amino effect" in heterocyclic synthesis; *in situ* generation of 1,5-dipolar intermediates

D.N. Reinhoudt,* W. Verboom, B.H.M. Lammerink and W.P. Trompenaars

Laboratory of Organic Chemistry, Twente University of Technology, 7500 AE Enschede, The Netherlands

Abstract - The mechanism of the "*tert*-amino effect" has been elucidated and the scope of possible applications for the synthesis of heterocyclic compounds has been investigated. In N,N-dialkylanilines that have a 2π-substituent at the ortho position, two different modes of hydrogen shift are possible. An *antarafacial* [1,6] hydrogen shift takes place in a helical conformation to give a conjugated 1,5-dipolar intermediate that further cyclizes to pyrrolizine derivatives. A *suprafacial* [1,5] hydrogen shift gives a non-conjugated 1,5-dipolar intermediate that reacts further via intramolecular addition of the nucleophilic centre of the dipole to the iminium group. The potential of these reactions for the synthesis of heterocycles is illustrated with the results of our work on the synthesis of Mitomycin type of compounds.

INTRODUCTION

Reviewing the chemistry of ortho-substituted N,N-dialkylanilines reveals a number of unexpected and in most cases unexplained reactions. The first examples were reported by Pinnow (1) at the end of the last century. The common feature in these reactions is the interaction of the dialkylamino group with an unsaturated ortho-substituent in the aromatic ring. Two typical examples given in Scheme 1 show the unexpected reactivity of the dialkylamino group at one of the α-positions (Ref. 1 & 2).

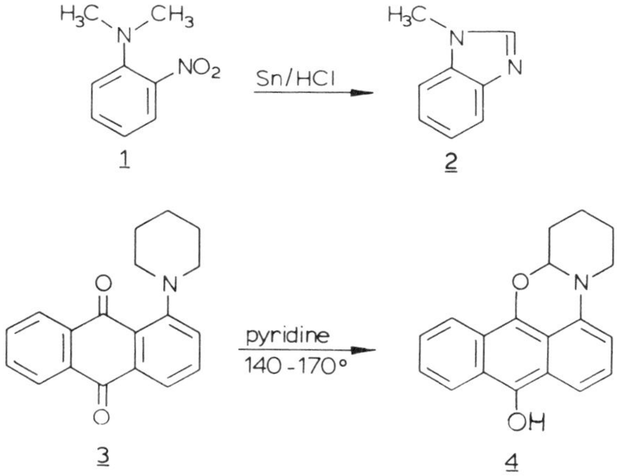

Scheme 1

In 1972 Meth-Cohn and Suschitzky (3) have reviewed these reactions and their potential applications in heterocyclic chemistry.

RESULTS

We have been interested for a number of years in the possible applications of ring-enlargement for the synthesis of medium-sized heterocycles via (2+2)-cycloaddition of unsaturated heterocycles with acetylenes and subsequent ring-opening of the cyclobutene ring in the resulting (2+2)-cycloadducts (Ref. 4-7). We found that enamines of heterocyclic ketones react with electron-deficient acetylenes under mild conditions and also that the cyclobutene rings in such (2+2)-cycloadducts undergo a rapid ring-opening reaction. This rearrangement (e.g. 6 → 7) (Scheme 2) has been discussed in terms of a disrotatory electrocyclic process (Ref. 8)

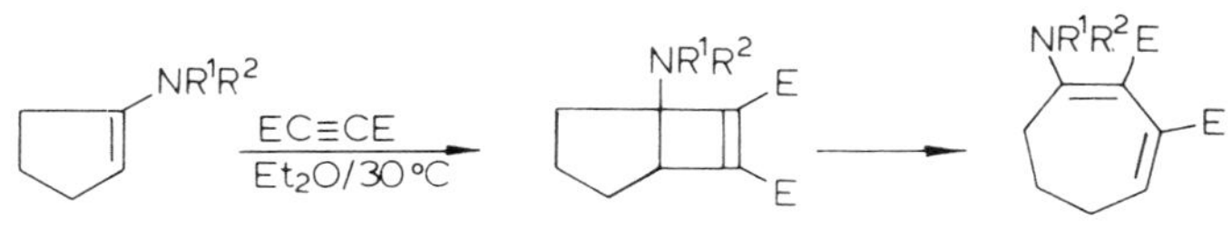

Scheme 2

but we have recently shown that a conrotatory ring-opening is the preferred pathway (Ref. 9).
When we studied the ring-opening in more detail e.g. the effect of the solvent on the rate of
reaction we found that 3-(1-pyrrolidinyl)cyclobutenes (e.g. 8 and 10) rearrange in a polar
solvent in a different way (Ref. 10). X-ray analysis revealed that the reaction products were
pyrrolizines (9 and 11) (Scheme 3).

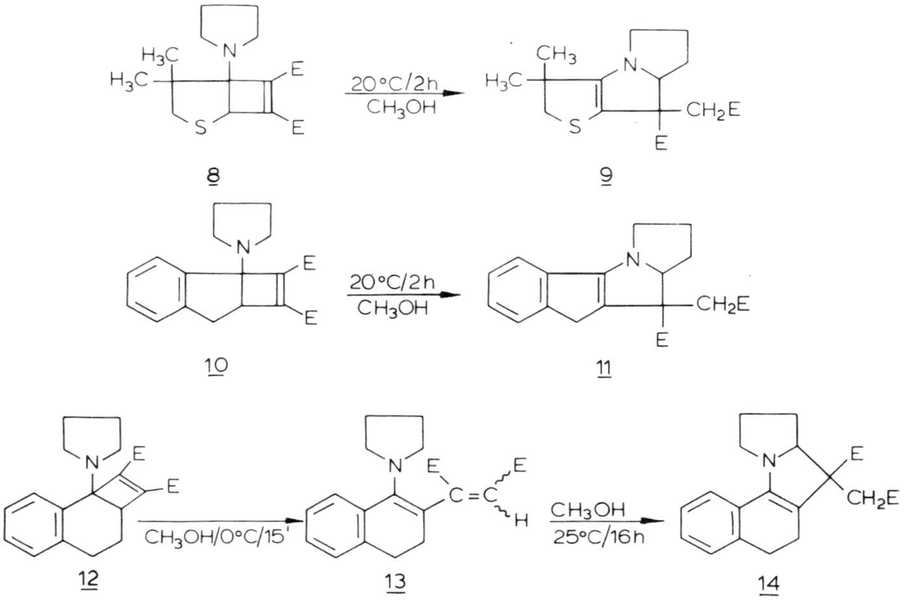

Scheme 3

The same products (e.g. 9 and 11) were obtained from reactions of the appropiate enamines
with dimethyl acetylenedicarboxylate (DMAD) in methanol (Ref. 11 & 12). This formation of
pyrrolizines resembles the reactions of the ortho-substituted N,N-dialkylanilines (1 and 3)
because also in these reactions insertion into the -NCH$_2$-group of the dialkylamino group
takes place. When the isomerization of the (2+2)-cycloadduct of the pyrrolidine enamine of
α-tetralone (12) to the corresponding pyrrolizine 14 was studied in detail by ^{1}H NMR spectroscopy, a transient intermediate was observed with a 1-(1-pyrrolidinyl)-1,3-diene structure
(13) (Scheme 3). Independent synthesis of such 1-(1-pyrrolidinyl)-1,3-dienamines and subsequent thermal, or in some cases Lewis acid-catalyzed, reaction revealed also the formation
of pyrrolizines (Ref. 13).

We have studied the isomerization reaction of 1-(1-pyrrolidinyl)-1,3-butadienes, in which a
hydrogen atom is transferred and a carbon-carbon bond is formed, in detail with the dimethyl
(E)- and (Z)-[2-(1-pyrrolidinyl)benzo[b]thien-3-yl]-2-butenedioates (15) (Ref. 14). These
isomers were prepared by reaction of 2-(1-pyrrolidinyl)benzo[b]thiophene and DMAD in methanol at 20 °C in yields of 64% [(E)-15] and 22% [(Z)-15], respectively. These two compounds
allowed us to study the formation of pyrrolizines in more detail. We found that the conversion of 15 (Scheme 4) takes place at temperatures of 80-120 °C. In solvents of low polarity
such as toluene both the isomers of 15 gave rise to the exclusive formation of the benzo-
thieno[3,2-b]pyrrolizine 17, whereas in a polar solvent such as acetic acid a mixture of 16
and 17 was obtained in a ratio of 94:6. The effect of the solvent on the ratio of 16 and 17
is given in Table 1. Our results also show that the rate of reaction increases with increasing polarity of the solvent. A kinetic study showed that (E)-15 reacts 37 times faster in
acetic acid than in toluene. Besides, the E isomer reacted 5 times faster than the Z isomer
at 110 °C in toluene. Experiments with dimethyl(E)- and (Z)-[2-(1-pyrrolidinyl-$2,2,5,5$-d_4)-
benzo[b]thien-3-yl]-2-butenedioate (18) showed two important results. Firstly, they proved

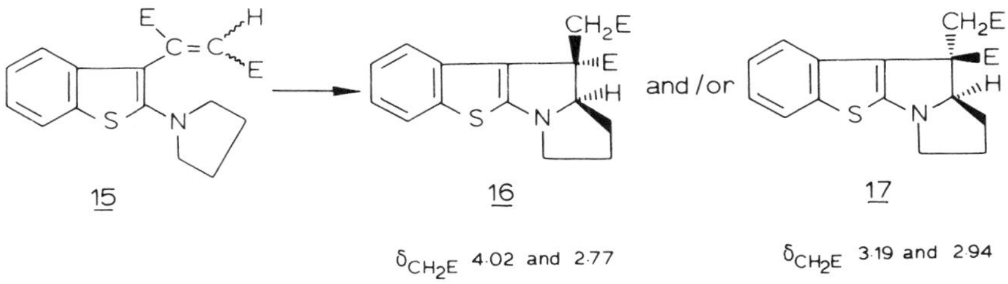

δ_{CH_2E} 4.02 and 2.77 δ_{CH_2E} 3.19 and 2.94

Scheme 4

that the hydrogen transfer is exclusively an intramolecular process even in protic solvents such as 1-butanol. Secondly, we found that (E)- and (Z)-18 gave two different sets of dia-stereomeric pyrrolizines. In the reaction products of (E)-18 one proton (H^b) is replaced by

TABLE 1. Reaction of (E)-15 in solvents of different polarity[a]

Solvent	Temp, ^0C	Time, h	Ratio of 16 : 17[b]
Toluene	110	15	0 : 100
Toluene/acetic acid[c]	110	15	15 : 85
Acetonitrile	81	48	33 : 67
Acetonitrile (ZnCl$_2$)	81	16	67 : 33
1-Butanol	118	15	67 : 33
Acetic acid	100	15	∿ 94 : 6

[a]Reactions have been carried out to complete conversion.
[b]Ratios determined by ^1H NMR spectroscopy and confirmed by isolation.
[c]Molar ratio of (E)-15 to acetic acid (5:1).

deuterium whereas in the products of (Z)-18 the other proton (H^a) is replaced by deuterium.

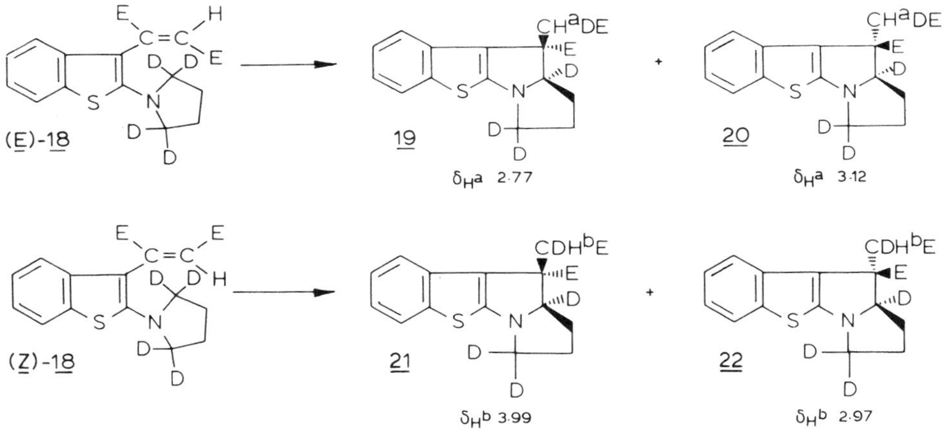

(E)-18 ⟶ 19 δ_{H^a} 2.77 + 20 δ_{H^a} 3.12

(Z)-18 ⟶ 21 δ_{H^b} 3.99 + 22 δ_{H^b} 2.97

Scheme 5

All these observations are in agreement with a mechanism in which the first step is an antarafacial [1,6] hydrogen shift in 15 to give a 1,5-dipolar intermediate 23 which may ei-ther cyclize directly to 17 or may undergo stereomutation to 24 which subsequently cyclizes to 16. Compared to 23 the conformer 24 will be more stable because the steric interaction of the more bulky CH$_2$E-group at the "inner" side in 23 is replaced by the more planar ester function in 24. However, the conversion of 23 into 24 requires a polar transition state in which the conjugation of the 6π-electron system of the 1,5-dipole is lost. As a consequence the activation energy of the stereomutation will be lower in polar solvents because the di-polar transition state will be better stabilized. This explains our observation that the ra-tio of 16:17 increases with increasing polarity of the solvent. The results of our experi-ments with the deuterated compounds (E)- and (Z)-18 confirm that the [1,6] hydrogen shift occurs in a *helical conformation*. Moreover, these results prove that the 1,5-dipolar cycli-zation must occur from the same helical conformation in which the hydrogen is transferred and also that stereomutation occurs without equilibration of the two mirror images of

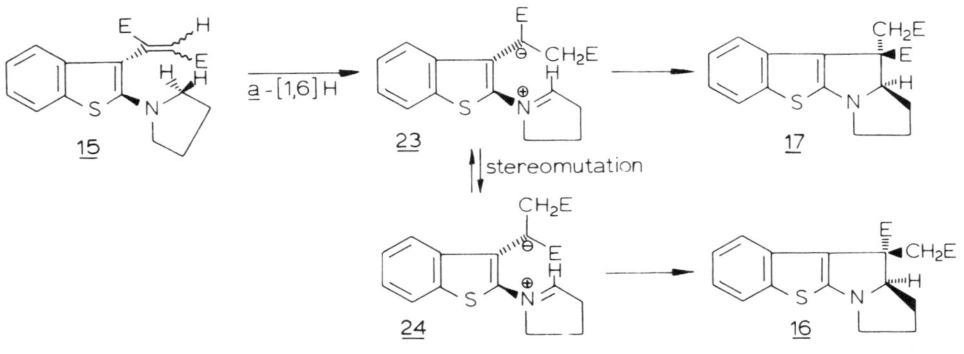

Scheme 6

the helix. In Scheme 7 the two possible ways of reaction of the E isomer, either via a clockwise or an anti-clockwise helix, are depicted. Only under the above mentioned conditions we explain the exclusive formation of (RRR)-20 and (SSS)-20. Because these are enantiomers the

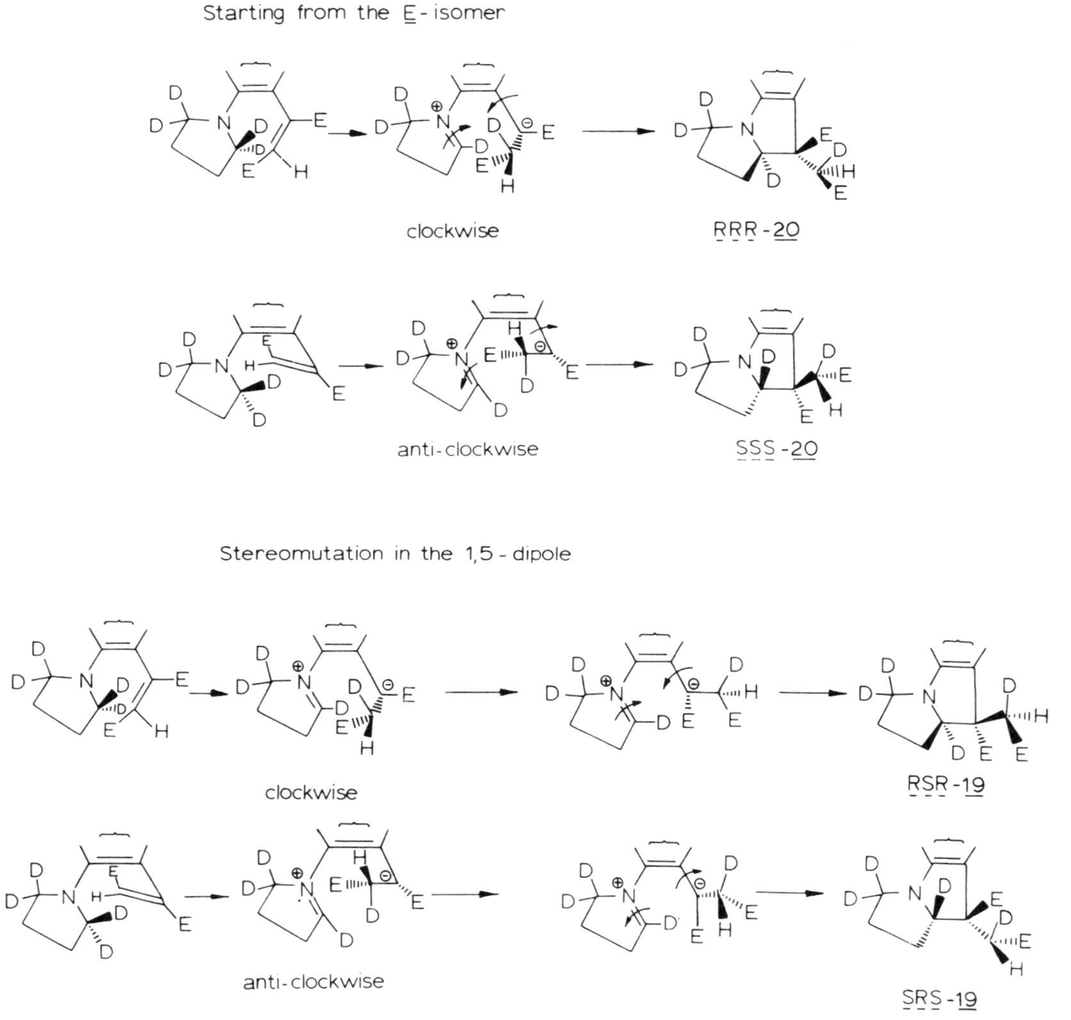

Starting from the E-isomer

Stereomutation in the 1,5-dipole

Scheme 7

^{1}H NMR spectra are identical which means that the CHDE-group will show only one singlet for the H$^{\alpha}$-proton. Stereomutation prior to 1,5-dipolar cyclization leads to the exclusive formation of (RSR)-19 and (SRS)-19. Reaction of (Z)-18 gives the (SRR)-22 and (RSS)-22, in which the H^{b}-proton is present, upon direct cyclization but (SSR)-21 and (RRS)-21 after stereomutation of the 1,5-dipolar intermediate followed by 1,5-dipolar cyclization. Our conclusion

is that compound 17 is formed from an intermediate with a short lifetime (23) and 16 from 24 the formation of which is favoured by more polar solvents. This was confirmed by reaction of (E)- and (Z)-15 in the presence of chiral catalysts. Whereas the product from direct cyclization (17) showed no enantiomeric excess (e.e.), we found e.e.-values of ∿ 20% for 16.

TABLE 2. Influence of a chiral catalyst on the cyclization of (E)- and (Z)-15

Solvent	Chiral catalyst	Ratio 16 : 17	e.e. (%) 16	17
Toluene	5 eq (−)-Menthol	0 : 100	–	0
	5 eq (+)-Cinchonine	8 : 92	0	0
	15 eq (+)-Camphoric acid	64 : 36	22	0
1-Butanol	5 eq (+)-Quinidine	77 : 23	18	0
	15 eq (+)-Camphoric acid	76 : 24	20	0
Acetic acid	1 eq (+)-Quinidine	90 : 10	18	0

In the compounds that undergo this combination of [1,6] hydrogen transfer and 1,5-dipolar cyclization, one of the double bonds of the reacting 1-(1-pyrrolidinyl)-1,3-butadiene moiety constitutes part of a heteroaromatic ring. In view of the similar reaction pattern with that commonly observed in ortho-substituted *tert*-anilines, we have also investigated the thermal reactions of 2-*vinyl-N,N*-dialkylanilines (Ref. 15).

Dimethyl (Z)-2-[2-(1-pyrrolidinyl)phenyl]butenedioate (25) was obtained by reaction of DMAD with 2-(1-pyrrolidinyl)phenylcopper in a yield of 75%. When 25 was heated in toluene or in 1-butanol the corresponding pyrrolo[1,2-*a*]indole derivatives 26 and 27 were obtained (Scheme 8). In 1-butanol the reaction is stereospecific, only 26 was obtained. This result definitely

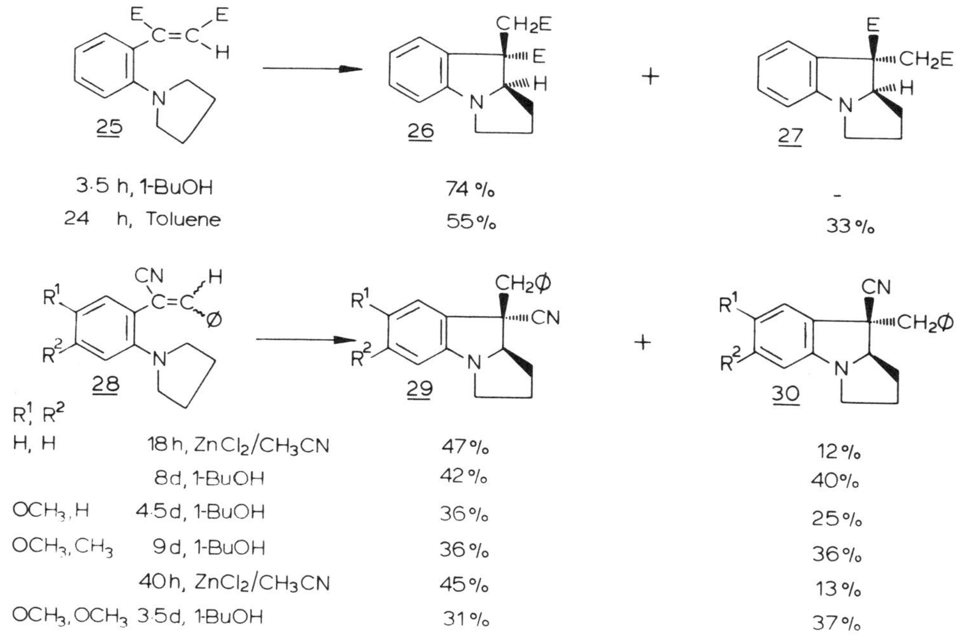

Scheme 8

relates the formation of pyrrolizines with the *tert*-amino effect observed in other *N,N*-dialkylanilines. A systematic study with variation in the structure of the vinyl and the dialkylamino moiety revealed that the electron-withdrawing group at the α-position of the vinyl group is essential. Compounds 28 rearrange upon heating to the corresponding pyrrolo-[1,2-*a*]indoles 29 and 30 in yields of 60–80%. 2-Vinyl-*N,N*-dialkylanilines in which no cyano or ester function is present at the α-position of the vinyl moiety did not react. Furthermore, the pyrrolidinyl moiety in 25 or 28 can not be replaced by other dialkylamino groups.

A more general but different type of thermal rearrangement occurs when two electron-withdrawing substituents are present at the β-carbon atom of the vinyl moiety. Such compounds

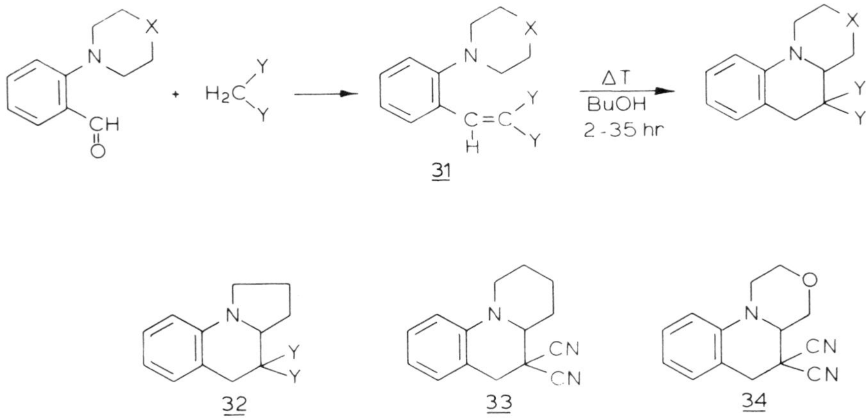

Y = COOMe
Y = CN   Scheme 9

(31) are easily accessible by condensation of the corresponding aldehydes with either dimethyl malonate or malonitrile in yields of ⁿ 90%. When compounds 31 (X = - and Y = COOCH₃
or CN) were heated in 1-butanol the corresponding pyrrolo[1,2-a]quinolines 32 were obtained
in yields of 67 and 82%, respectively. The piperidinyl (31, X = CH₂ and Y = CN) and morphol-
inyl (31, X = O and Y = CN) derivatives reacted in a similar way to give 33 (78%) and 34
(84%), respectively. When these reactions were carried out in 1-deuterio-1-butanol we did not
observe incorporation of the label in the products (32-34) which means that also in these
reactions the hydrogen atom is transferred, in this case via a [1,5] hydrogen shift, intra-
molecularly.

Mechanism

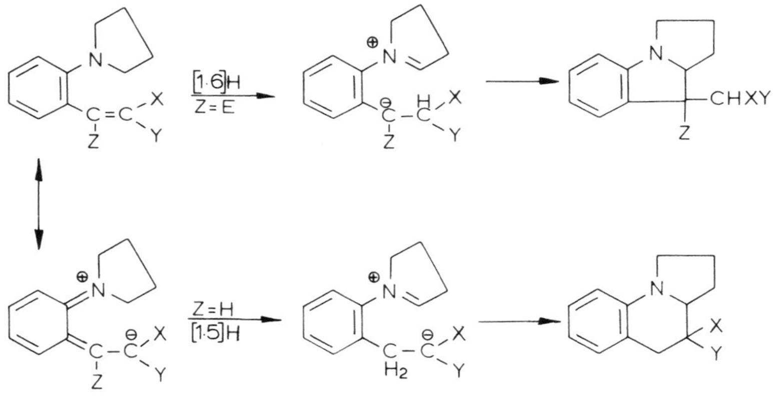

Scheme 10

In Scheme 10 the two principally different modes of reaction of 2-vinyl-N,N-dialkylanilines
are summarized. The type of reaction depends on the stabilization of the negative charge in
the dipolar intermediate that is formed by intramolecular hydrogen transfer. In 2-vinyl-N,N-
dialkylanilines that have an electron-withdrawing group at the α-position of the vinyl group
the [1,6] hydrogen shift is preferred whereas in compounds with two electron-withdrawing
groups at the β-carbon atom of the vinyl group a [1,5] hydrogen shift takes place.

These experiments not only reveal the mechanism of the tert-amino effect in our systems but
it is not unlikely now that other reactions of the same type are possible. For the synthesis
of heterocyclic compounds our results provide a new methodology because it allows carbon-
carbon bond formation by insertion into a non-activated NCH₂-group.

The first possible enlargement of the scope that we have investigated were reactions of 2-
acyl-N,N-dialkylanilines because we had previously found that reactions of enamines with
trifluoroacetic anhydride yielded in some cases not the expected trifluoroacylated enamines
but 1,3-oxazines (Ref. 16). The formation of these 1,3-oxazines may be rationalized by a
[1,5] hydrogen transfer and subsequent ring-closure. A number of 2-(trifluoroacetyl)-N,N-di-
alkylanilines were converted by heating in 1-butanol, or in acetonitrile with a catalytic
amount of zinc chloride, into pyrrolo[1,2-a][3,1]benzoxazine [36, 37; X = (CH₂)₂] or pyrido-
[1,2-a][3,1]benzoxazine [36, 37; X = (CH₂)₃] derivatives (Ref. 17).

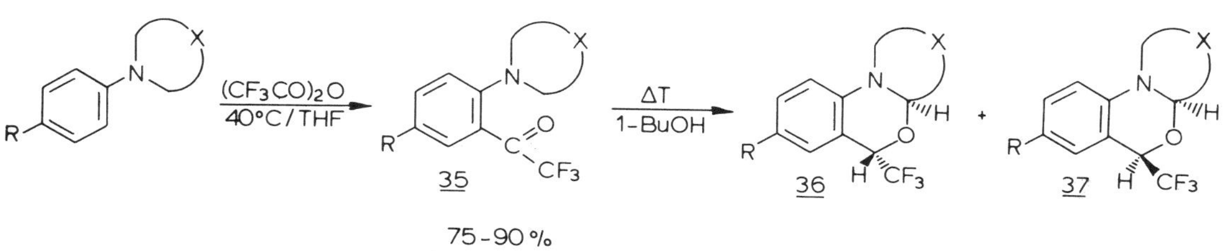

75-90%

Scheme 11

R	X	Conditions	Yield	36 : 37
CH_3	$(CH_2)_2$	45 h 1-BuOH	91%	4.5 : 1
CH_3	$(CH_2)_3$	90 h 1-BuOH	95%	1 : 1
OCH_3	$(CH_2)_2$	20 h 1-BuOH	77%	2.5 : 1
OCH_3	$(CH_2)_3$	20 h 1-BuOH	90%	2 : 1
CH_3	–	17 h $ZnCl_2/CH_3CN$	51%	– –
H[a]	$(CH_2)_2$	115 h 1-BuOH	85%	3.5 : 1

[a] 35 prepared otherwise.

The electron-withdrawing CF_3 group adjacent to the carbonyl moiety is crucial for a rapid hydrogen transfer; the corresponding 2-acetyl-N,N-dialkylanilines did not react. Similar N,N-dialkylanilines ortho-substituted with an imino or a thiocarbonyl group (38 and 40) were prepared by reaction of 35 with the anion of 1,1,1-trimethyl-N-(4-methylphenyl)silanamine [$H_3C-C_6H_4-NHSi(CH_3)_3$] (Ref. 19) or with Lawesson reagent (Ref. 20), respectively. The reaction of the imine 38 required 5 days at 118 °C in 1-butanol to give the pyrrolo[1,2-a]quinazoline derivative 39 in 66% yield. The trans stereochemistry of 39 was determined by using

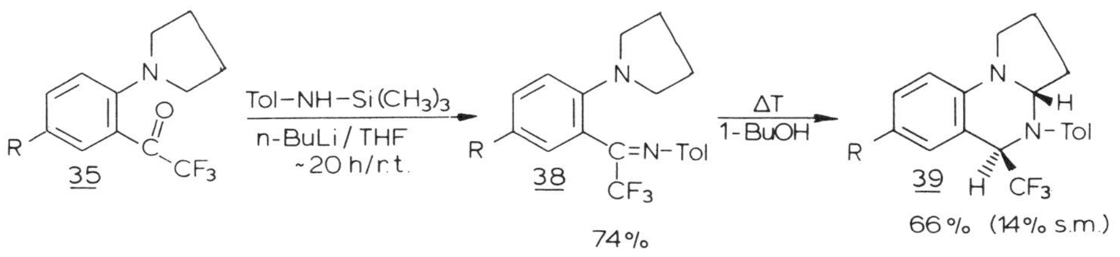

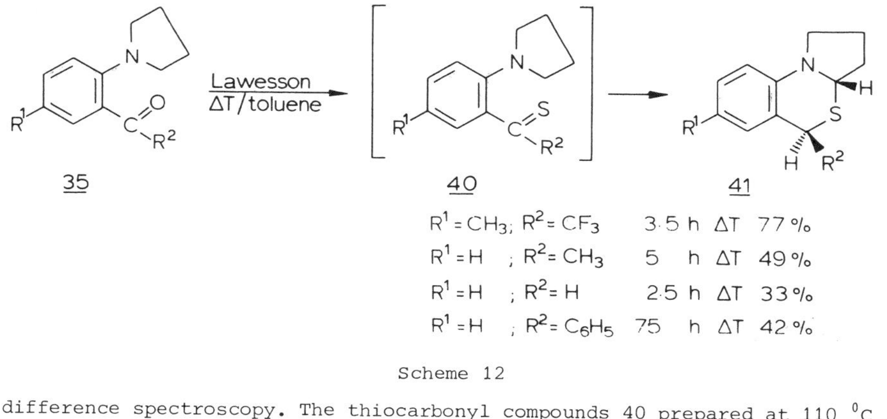

74%   66% (14% s.m.)

$R^1 = CH_3;\ R^2 = CF_3$ 3.5 h ΔT 77%
$R^1 = H\ ;\ R^2 = CH_3$ 5 h ΔT 49%
$R^1 = H\ ;\ R^2 = H$ 2.5 h ΔT 33%
$R^1 = H\ ;\ R^2 = C_6H_5$ 75 h ΔT 42%

Scheme 12

[1]H NOE difference spectroscopy. The thiocarbonyl compounds 40 prepared at 110 °C in toluene by reaction of 35 with Lawesson reagent could not be isolated. Instead we obtained the corresponding pyrrolo[1,2-a][3,1]benzothiazines 41. It should be noted that the formation of the compounds 41 in which R^2 = H, CH_3 or C_6H_5 represents the first examples in our work where the ortho 2π-moiety of the phenyl ring is not substituted with a strongly electron-withdrawing group. Obviously the stabilization of the negative charge in the dipolar intermediate at the sulphur atom does not require such an assistance for the [1,5]hydrogen shift to take place (see further Scheme 10) (Ref. 18).

SYNTHESIS OF MITOMYCIN C TYPE OF ANTI-TUMOR ANTIBIOTICS

A possible application of the *tert*-amino effect for the synthesis of heterocyclic compounds can be illustrated by our current work on the synthesis of the skeleton of Mitomycin C with functional groups that are supposed to be required for biological activity (Ref. 21).

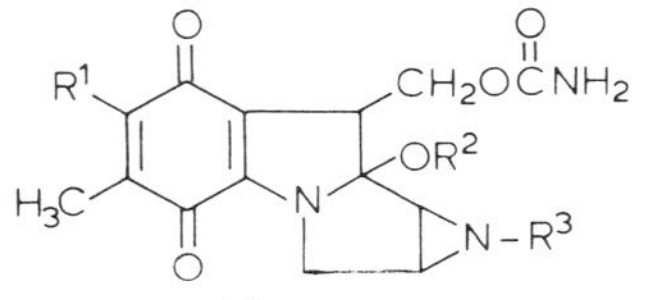

compound	R^1	R^2	R^3
mitomycin A	H_3CO	CH_3	H
mitomycin B	H_3CO	H	CH_3
mitomycin C	H_2N	CH_3	H
porfiromycin	HO	CH_3	CH_3

The pyrrolo[1,2-*a*]indole skeleton of the Mitomycins may be obtained by a route analogous to the thermal rearrangement of 25 to 26 and 27 (see Scheme 8). However, four additional functional groups are most likely necessary for biological activity:
  - The urethane group
  - The aziridine or another alkylating function
  - The quinone function
  - The methoxy group or another leaving group

Our approach outlined in Scheme 13 is based on the introduction of the (potential) urethane function in the substituted vinyl group, the functionalized 1-pyrrolidinyl group and a

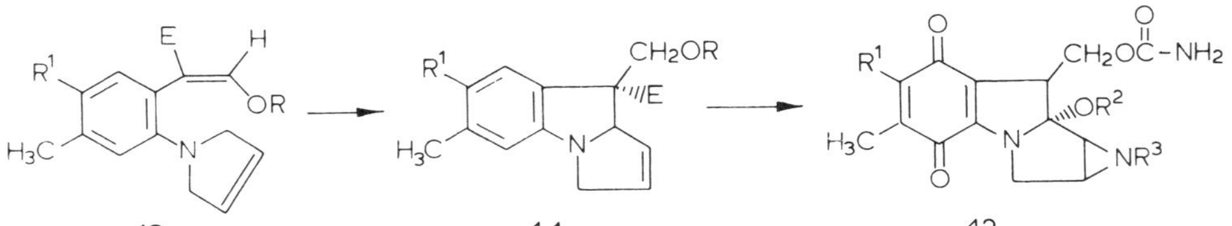

Scheme 13

potential quinone function in a precursor of type 43 prior to cyclization. This approach introduces the maximum complexity in the molecule and allows simple modification in all three building blocks, prior to [1,6] hydrogen shift and 1,5-dipolar cyclization.

The potential urethane group was introduced via an OR substituent at the β-carbon atom of the vinyl moiety (Ref. 22). Condensation of 2-(1-pyrrolidinyl)benzeneacetonitrile (45) with ethyl formate gave the sodium salt of α-(hydroxymethylene)-2-(1-pyrrolidinyl)benzeneacetonitrile (46) in 90% yield. Alkylation of this salt with several alkylating agents afforded the enol ethers 47 in yields of 80-90%. Subsequent thermal rearrangement gave the corresponding *cis*- and *trans*-pyrrolo[1,2-*a*]indoles 48 and 49, respectively (Scheme 14) in 50-80% yield. The *O*-protecting methyl group in 48 or 49 (R=CH$_3$) could be cleaved by boron tribro-

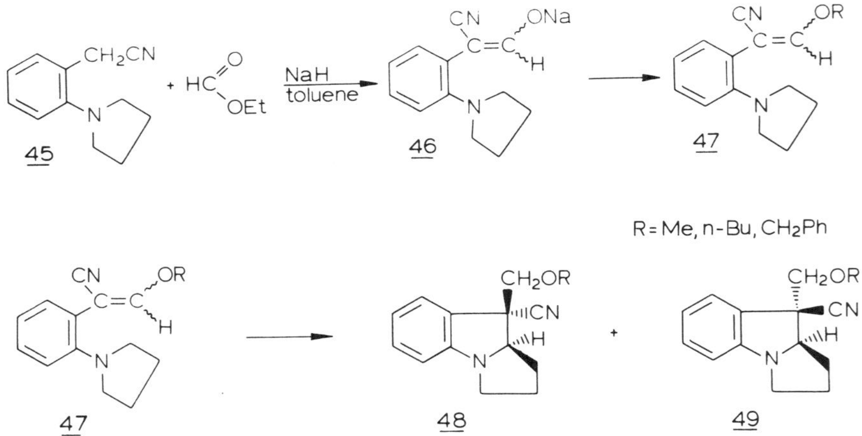

R = Me, n-Bu, CH₂Ph

Continued at next page

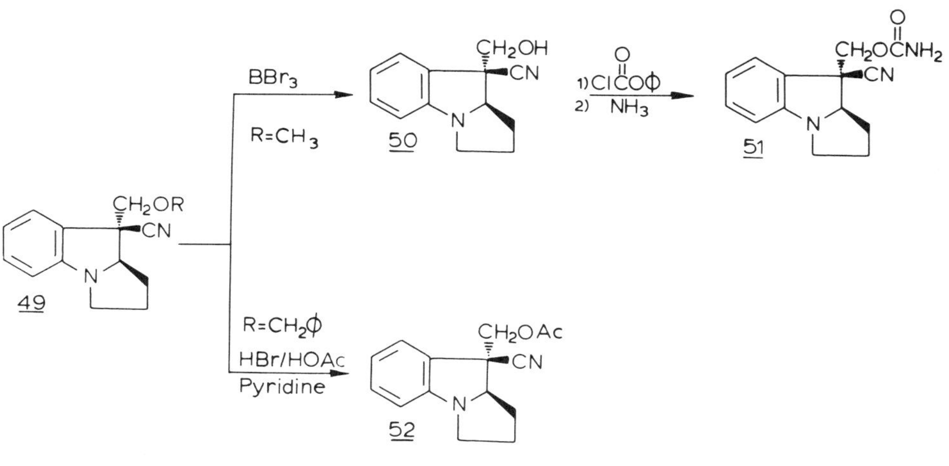

Scheme 14

mide to give 9-(hydroxymethyl)-2,3,9,9a-tetrahydro-1H-pyrrolo[1,2-a]indole-9-carbonitrile (50) that could be converted into the urethane derivative 51 by standard procedures. As an alternative the benzyl ether moiety in 48 or 49 (R = CH$_2$Ph) can be cleaved in a hydrobromic acid/acetic acid mixture. The resulting acetate 52 was obtained in a yield of 61%.

The second functional group required for biological activity of the Mitomycins is the aziridine or another alkylating function. In our approach the pyrrolo[1,2-a]indole skeleton can easily be functionalized in this position because this only requires the use of substituted 1,4-dihalo-2-butenes in the synthesis of 43. We have used a 2,5-dihydropyrrole moiety as a potential alkylating group. Compounds 53 were prepared starting from the appropriate (substituted) 2-nitrobenzeneacetonitrile, reduction of the nitro group and subsequent N,N-dialkylation with 1,4-dibromo-2-butene. Condensation with the appropriate carbonyl compound gave 53.

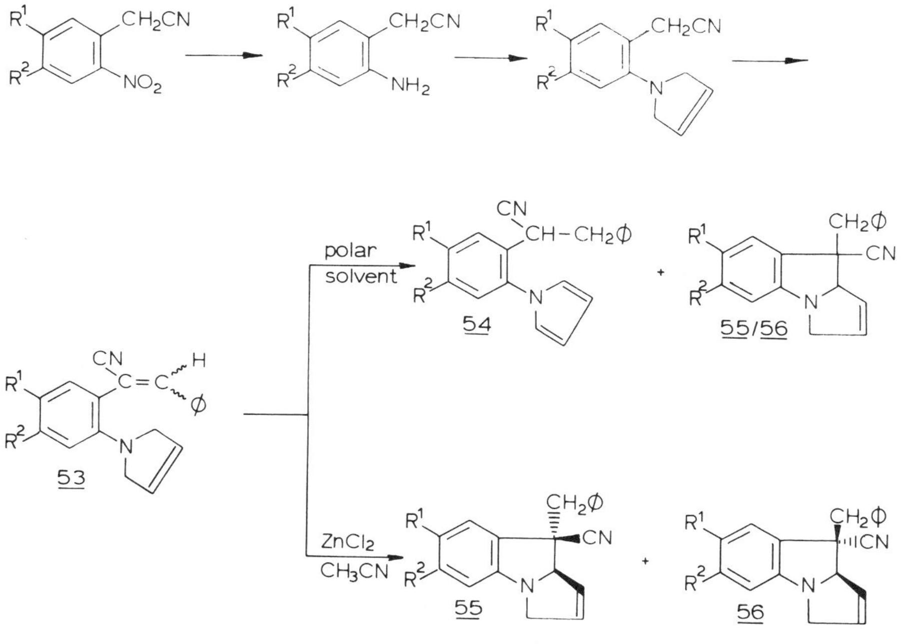

Scheme 15

Heating of 53 in a polar protic solvent like 1-butanol gave the corresponding pyrrolo[1,2-a]-indoles 55 and 56 in low yield. The major product was the pyrrole derivative 54 that is formed by a deprotonation-protonation of the 1,5-dipolar intermediate generated by the [1,6] hydrogen transfer (Scheme 10). However, when the reaction was carried out in acetonitrile in the presence of zinc chloride the pyrrolo[1,2-a]indoles 55 and 56 (R^1=R^2=OCH$_3$) were obtained in yields of 19 and 34%, respectively. Further functionalization of the double bond in 55 (or 56) by the usual reagents for aziridine formation resulted in oxidative elimination to the pyrrole derivative 57. Reaction with osmium tetroxide gave the corresponding *cis*-diol 58

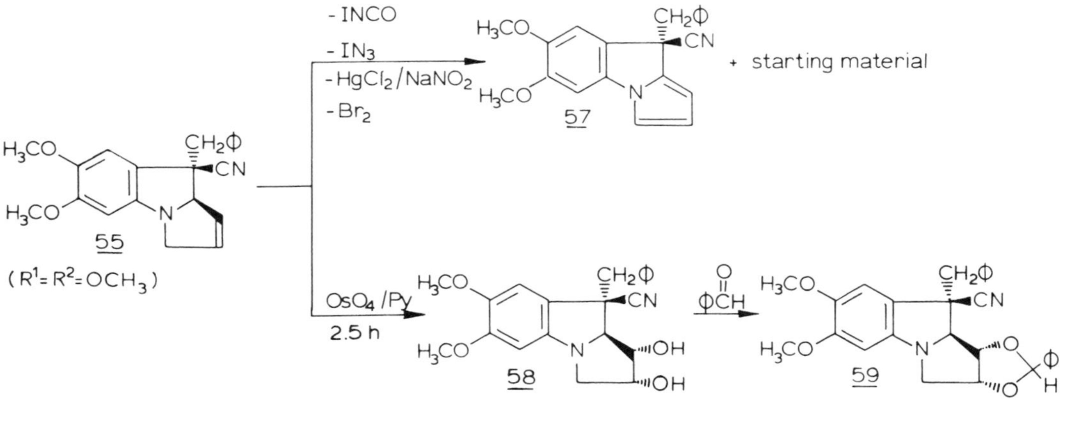

Scheme 16

that was further transformed into the acetal 59 by reaction with benzaldehyde. An X-ray structure determination of 59 revealed that the *cis*-hydroxylation had taken place stereose-lectively with the reagent approaching from the least hindered side.

Introduction of a quinone function can in principle be carried out by nitration of a suita-ble substituted aromatic compound, reduction of the nitro group and subsequent oxidation of the resulting aniline derivative.

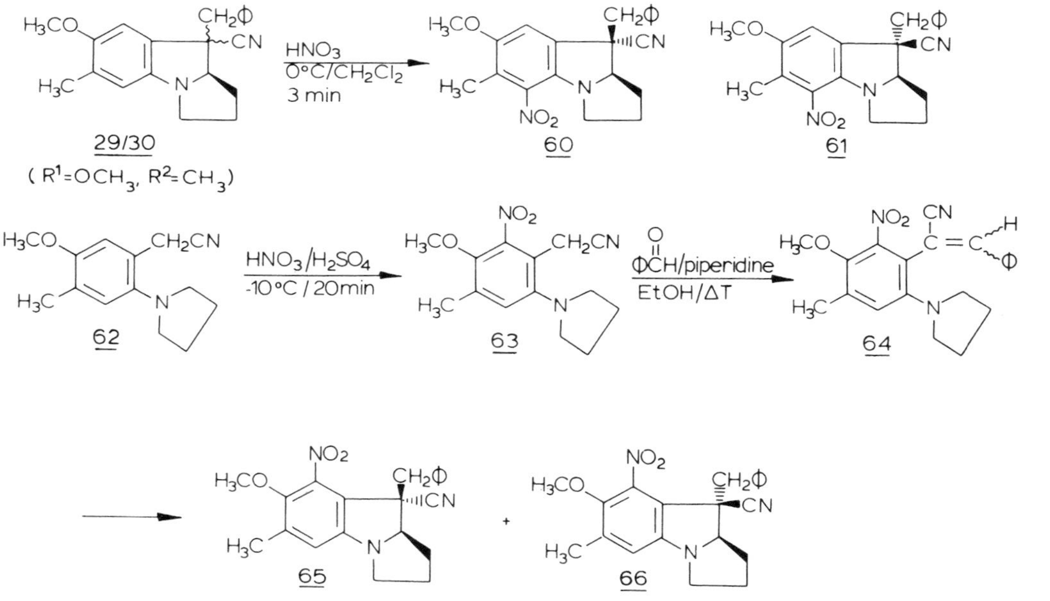

Scheme 17

In our approach the nitro groups can be introduced at different stages of the synthesis. The most simple would be nitration of the pyrrolo[1,2-*a*]indoles (29, 30, 55, or 56) (Ref. 23). We found that nitration of 29 or 30 (R^1=OCH$_3$, R^2=CH$_3$) required very specific conditions and even then the yield was only 20% of 60 or 61. As an alternative 62 was nitrated to give 63 in a higher yield (36%). The condensation with benzaldehyde yielded 64 that could be con-verted by heating into the corresponding nitropyrrolo[1,2-*a*]indoles 65 and 66. The presence of the nitro group substantially reduces the rate of cyclization. Finally the reduction of the nitro group in compounds 60 and 65 gave no problems but the subsequent oxidation with

Fremy's salt did not produce the corresponding *para*-quinones. In low yield we could identify the compounds 67 and 68, respectively (Scheme 18).

Further work on the synthesis of Mitomycin C analogues is in progress.

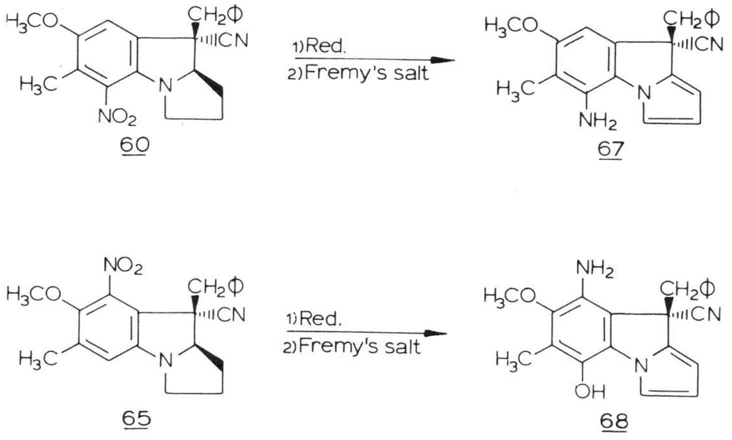

Scheme 18

Acknowledgements - We are grateful for the financial support of this work by the "Koningin Wilhelmina Fonds" and by the Netherlands Foundation for Chemical Research (SON) with financial aid from the Netherlands Organization for Advancement of Pure Research (ZWO).

REFERENCES

1. J. Pinnow, *Chem. Ber.* 32, 1666-1669 (1899).

2. E.P. Fokin and V.V. Russkikh, *J. Org. Chem. (USSR)* 2, 902-906 (1966).

3. O. Meth-Cohn and H. Suschitzky, *Adv. Heterocycl. Chem.* 14, 211-278 (1972).

4. D.N. Reinhoudt, *Adv. Heterocycl. Chem.* 21, 253-321 (1977).

5. D.N. Reinhoudt and C.G. Kouwenhoven, *Recl. Trav. Chim. Pays-Bas* 93, 129-132 (1974).

6. D.N. Reinhoudt and C.G. Kouwenhoven, *Tetrahedron* 30, 2093-2098 (1974).

7. D.N. Reinhoudt and C.G. Kouwenhoven, *Tetrahedron* 30, 2431-2436 (1974).

8. D.N. Epiotis, *Angew. Chem., Int. Ed. Engl.* 13, 751-780 (1974).

9. D.N. Reinhoudt, W. Verboom, G.W. Visser, W.P. Trompenaars, S. Harkema and G.J. van Hummel, *J. Am. Chem. Soc.* 106, 1341-1350 (1984).

10. G.W. Visser, W. Verboom, W.P. Trompenaars and D.N. Reinhoudt, *Tetrahedron Lett.* 23, 1217-1220 (1982).

11. D.N. Reinhoudt, J. Geevers, W.P. Trompenaars, S. Harkema and G.J. van Hummel, *J. Org. Chem.* 46, 424-434 (1981).

12. W. Verboom, G.W. Visser, W.P. Trompenaars, D.N. Reinhoudt, S. Harkema and G.J. van Hummel, *Tetrahedron* 37, 3525-3533 (1981).

13. G.W. Visser, W. Verboom, P.H. Benders and D.N. Reinhoudt, *J. Chem. Soc., Chem. Commun.* 1982, 669-671.

14. D.N. Reinhoudt, G.W. Visser, W. Verboom, P.H. Benders and M.L.M. Pennings, *J. Am. Chem. Soc.* <u>105</u>, 4775-4781 (1983).

15. W. Verboom, D.N. Reinhoudt, R. Visser and S. Harkema, *J. Org. Chem.* <u>49</u>, 269-276 (1984).

16. W. Verboom, D.N. Reinhoudt, S. Harkema and G.J. van Hummel, *J. Org. Chem.* <u>47</u>, 3339-3342 (1982).

17. W. Verboom, B.G. van Dijk and D.N. Reinhoudt, *Tetrahedron Lett.* <u>24</u>, 3923-3926 (1983).

18. W. Verboom, M.R.J. Hamzink, D.N. Reinhoudt and R. Visser, *Tetrahedron Lett.*, accepted for publication.

19. P.A.T.W. Porskamp and B. Zwanenburg, *Synthesis* <u>1981</u>, 368-369.

20. B.S. Pedersen, S. Scheibye, N.H. Nilsson and S.-O. Lawesson, *Bull. Soc. Chim. Belg.* <u>87</u>, 223-228 (1978).

21. H.W. Moore, K.F. West, K. Srinivasacher and R. Czerniak, in "Structure-activity relationships of anti-tumour agents", D.N. Reinhoudt, T.A. Connors, H.M. Pinedo and K.W. van de Poll (Eds), Martinus Nijhoff, The Hague, 1983, p. 93-110.

22. W.C. Dijksman, W. Verboom, D.N. Reinhoudt, C.G. Hale, S. Harkema and G.J. van Hummel, *Tetrahedron Lett.* <u>25</u>, 2025-2028 (1984).

23. M. Ihara, K. Takahashi, Y. Kigawa, T. Ohsawa, K. Fukumoto and T. Kametani, *Heterocycles* <u>6</u>, 1658-1665 (1977).

New n-ylids and their synthetic utility

M. Baumann, H. Bosshard, W. Breitenstein and H. Greuter*

Central Research Laboratories, CIBA-GEIGY AG, CH-4002 Basel, Switzerland

Abstract. - An efficient new synthesis of dimethylmaleic anhydride by decarboxylative dimerisation of maleic anhydride in the presence of 2-aminopyridine was
found several years ago in our laboratories. It has now been established that bicyclic pyridinium ylids are key intermediates in this unusual transformation. Generalisation of the reaction principle has provided a new, versatile synthesis of
2,3-disubstituted maleic anhydrides. Their preparative value as synthetic building
blocks is exemplified by natural product syntheses, some technical uses and a versatile one-pot-synthesis of multifunctional aromatic compounds via intramolecular
condensation reactions.

INTRODUCTION

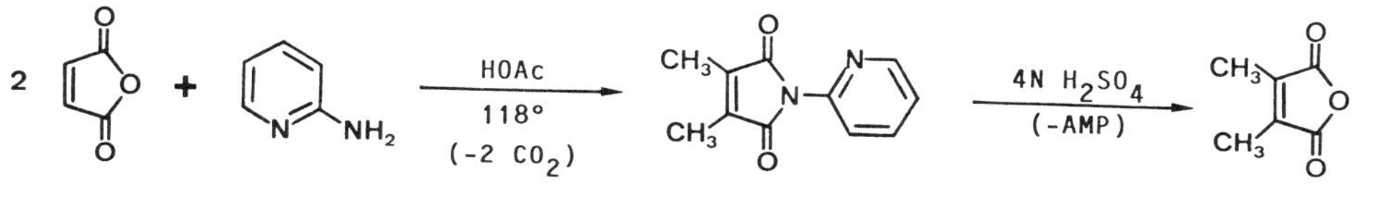

On heating 2-aminopyridine with two equivalents of maleic anhydride in acetic acid, rapid
evolution of carbon dioxide indicates that an unexpected reaction is taking place. Indeed,
on evaporation of the solvent, the imide 1 of dimethyl maleic acid is obtained as the major
product. Direct hydrolysis of the residue with 4N sulfuric acid without isolating the intermediates gives dimethyl maleic anhydride (DMA) in 75 % yield (Ref. 1).

Intrigued by this unique reaction we embarked on a systematic study of its mechanism. We
hoped that by understanding the reaction we might be able to generalize it.

MECHANISM OF THE DECARBOXYLATIVE DIMERISATION OF MALEIC ANHYDRIDE TO DIMETHYL
MALEIC ANHYDRIDE

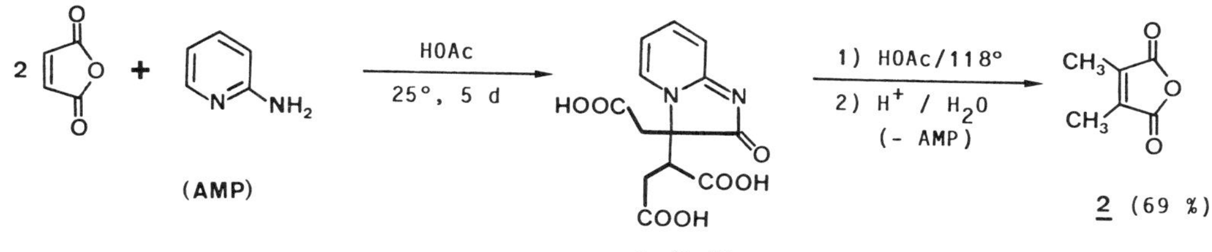

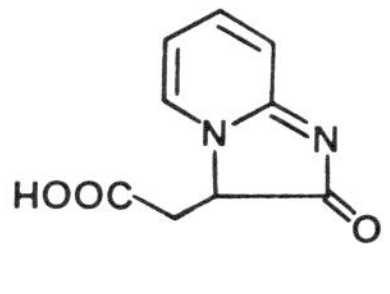

We speculated that some of the reaction intermediates might be isolable if the reaction between 2-aminopyridine and maleic anhydride in acetic acid were conducted at lower temperature. Indeed, reaction at room temperature for five days gave a complex mixture, from which the crystalline tricarboxylic acid 3 could be isolated in 8 % yield. 3 is a likely intermediate: on heating in refluxing acetic acid, 3 is transformed into DMA in 69 % yield. The structure of 3 clearly shows the presence of two C_4-dicarboxylic acids; it therefore indicates how maleic anyhdride may have entered the reaction. The structure also shows that acetic acid is not incorporated into the molecule. We speculated that 4 is the precursor of 3. In our discussion therefore we shall first concentrate on the preparation of 4 and its reactions; we then will return and discuss how 3 is transformed into 2.

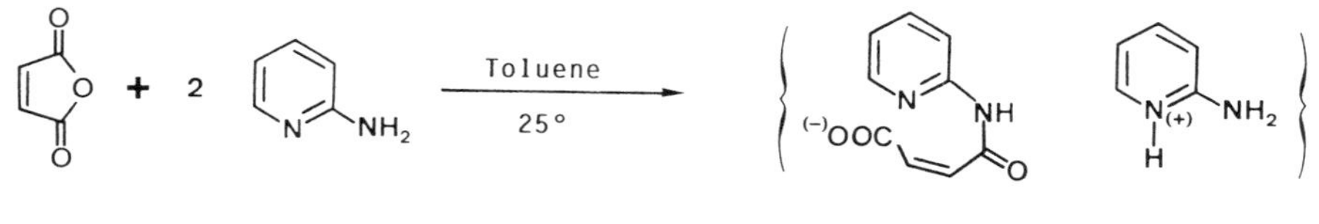

5 (95 %)

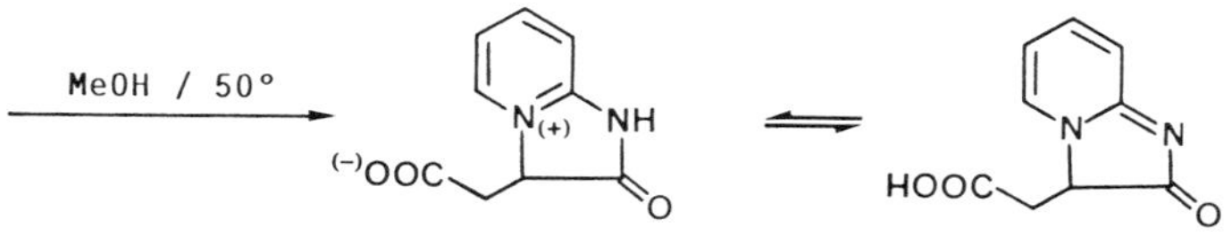

4a (64 %)   4b

When maleic anhydride is added to two equivalents of 2-aminopyridine, dissolved in toluene, the aminopyridinium salt of the maleamic acid 5 precipitates in almost quantitative yield. On heating in methanol it rapidly cyclises to the desired imidazo[1,2-a]pyridine compound 4. In D_2O 4 exists mainly as the pyridinium salt 4a, in DMSO the tautomer 4b can also be detected.

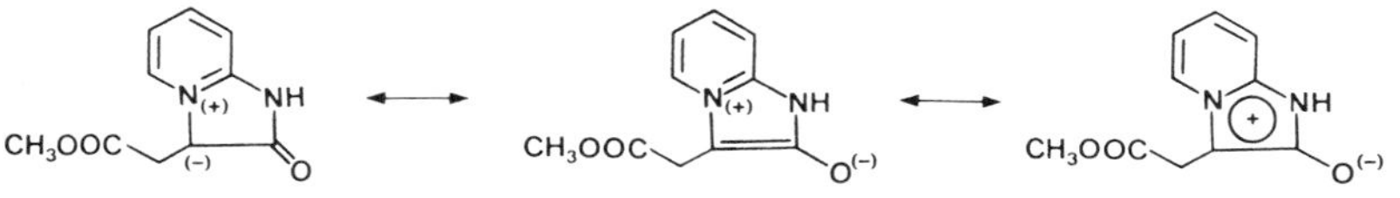

6a   6b   6c

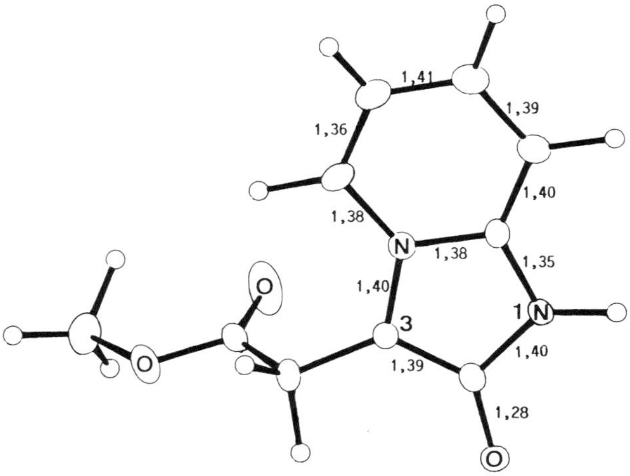

6

On esterification with methanol, 4 is transformed into the methyl ester 6. The X-ray analy-

sis showed to our surprise that 6 possesses the structure shown: The molecule is completely planar, a hydrogen atom can be localised at N-1 but not at C-3. 6 therefore has the ylid
structure shown. The fact that all C-C and C-N bonds have almost equal length indicates
that the charges are highly delocalised. This means that 6 is a mesoionic compound which is
probably best described by formula 6c (Ref. 2). (We shall keep this fact in mind even
though in our further discussion we shall use the term "ylid" in order to indicate more
clearly at which atom reaction takes place.)

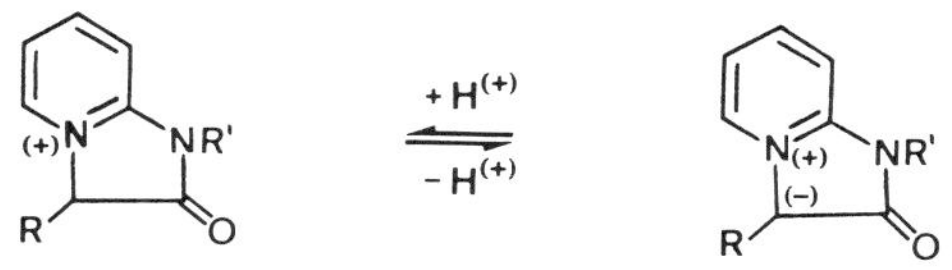

	R	R'	pK*$_{MCS}$
6	CH$_2$COOCH$_3$	H	3,6
4	CH$_2$COOH	H	3,5; 5,2
7	CH$_2$COOH	CH$_3$	3,5; 5,5

The ester 6 which we just discussed belongs, when protonated, to a class of imidazo-
[1,2-a]pyridinium salts of type A which have acidities in the range of mono- and dicarboxy-
lic acids. For instance, a value of 3,6 has been measured for the pK* of protonated 6 in
methylcellosolve 80 % / water. The pK* values of the acid 4 are 3,5 and 5,2. Very similar
values are measured for the N-methylated compound 7. It is clear, therefore, that under the
reaction conditions these compounds exist to a considerable degree in the deprotonated
form.

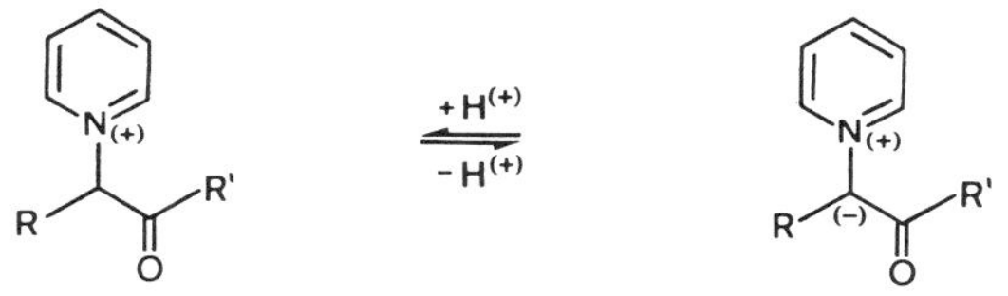

R	R'	pK*$_{MCS}$
CH$_3$	OCH$_3$	9,2
H	OCH$_3$	9,4
H	C$_6$H$_5$	10,4 (10,5[1])

1) R.G. Pearson et al.,
JACS 70, 1933 (1948).

By contrast, pK* values of monocyclic pyridinium salts are about 6 units higher. The work
of Kröhnke has shown that these ylids may undergo some very interesting and useful transformations (Ref. 3). However, DMA and other disubstituted maleic anhydrides which, as we
shall see, are readily available from our bicyclic pyridinium ylids, cannot be obtained
from monocyclic pyridinium ylids.

These bicyclic pyridinium salts are excellent Michael donors, even under acidic conditions.
Reaction of isolated 4 with maleic anhydride in acetic acid at room temperature gives 3 in
59 % yield. Further transformation of 3 leads to 2 as we have already discussed. Similarly
the ester 6, when reacted with maleic anhydride in refluxing acetic acid gives the anhy-
dride-ester 9 in 32 % yield. We assume by analogy that compound 8 is an intermediate in
this reaction. The formation of 9 is our first example of how the reaction principle may be
generalised to create new disubstituted maleic anhydrides. We shall return to this point
later.

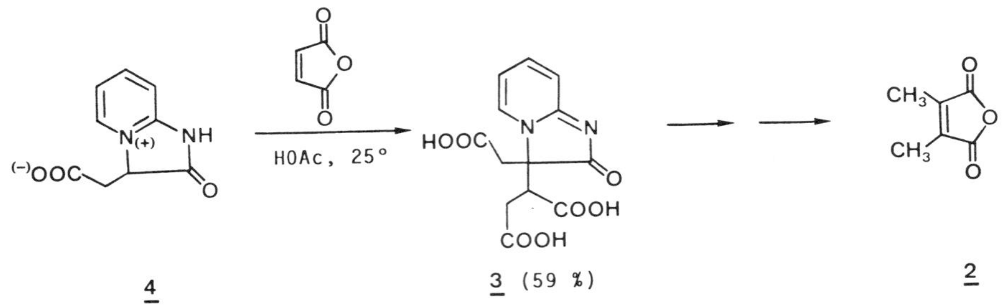

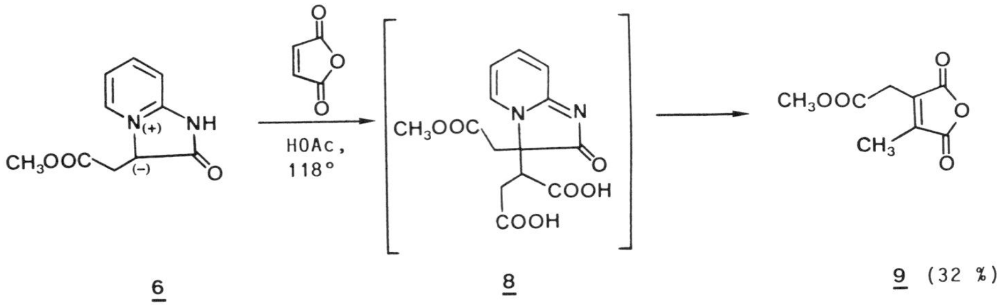

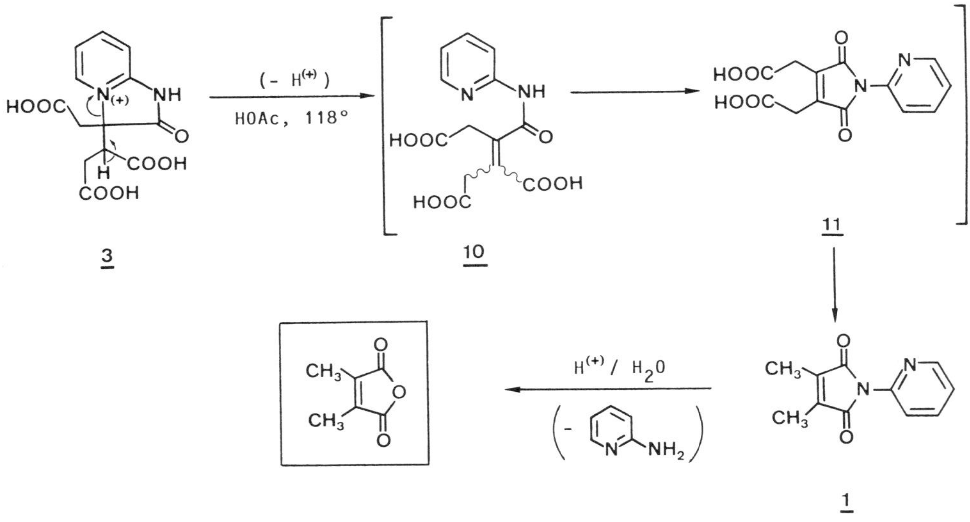

It has already been mentioned that compound 3 on heating in refluxing acetic acid forms the imide 1 with loss of CO$_2$. Our postulated mechanism is as shown: Loss of a proton and elimination of pyridine as internal leaving group leads to the postulated amide 10 which is easily cyclised to the imide 11. This imide, by double decarboxylation, is finally transformed into 1. From 1, DMA is prepared by hydrolysis. This procedure also allows the 2-aminopyridine to be recycled.

We have recently found that 2-aminopyridine may be replaced by catalytic amounts of N-substituted heterocycles of the general formula shown. Examples are shown: with N-methyl-, N-t-butyl- and N-phenyl 2-aminopyridine the yields are in the 60 % range. We believe that the catalytic effect is due to the fact that these amines cannot form imids.

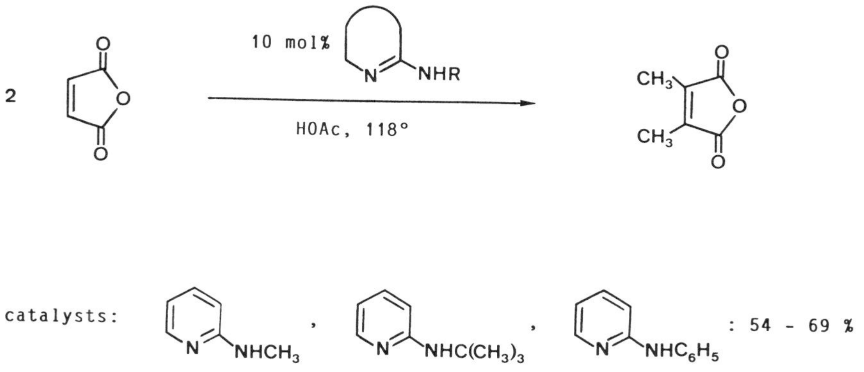

GENERAL SYNTHESIS OF DISUBSTITUTED MALEIC ANHYDRIDES

We have already discussed the formation of 9 from precursor 6 as our first example of the generalisation of the reaction principle. It is conceivable that any pyridinium salt 13 may be transformed into an anhydride of type 14 in analogy to the reactions discussed before.

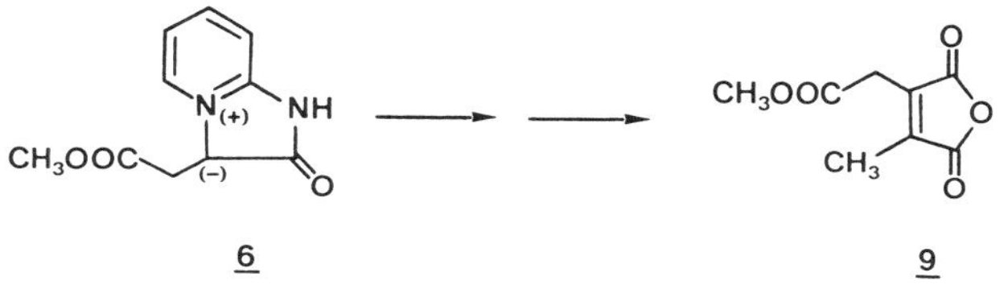

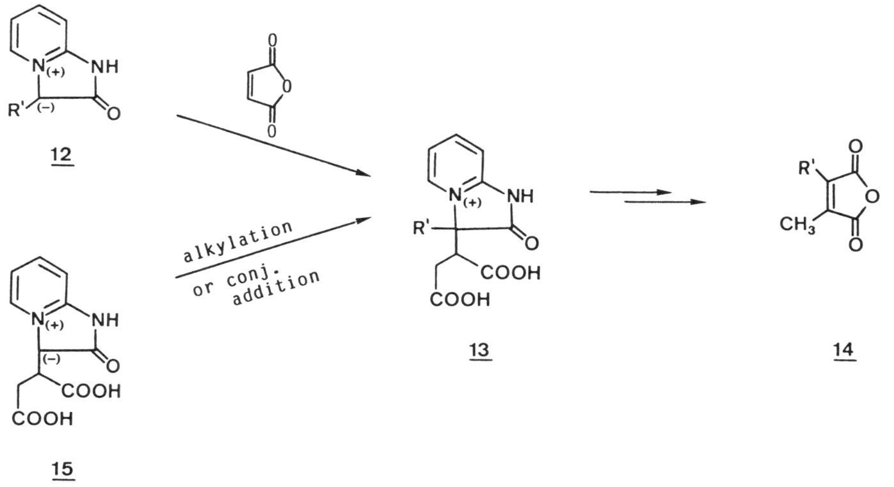

There are two approaches to generate 13: 13 may be formed from a precursor 12 by addition of maleic anhydride as in the case 6 → 9; on the other hand, 13 may also be formed by al-kylation of a precursor 15, for example by conjugate addition of 15 to a suitable acceptor. We have experimentally confirmed that these postulated transformations are indeed of general applicability. Let us examine two examples.

Reaction of 2-bromohexanoic acid chloride with 2-aminopyridine leads via the N-acylated intermediate to the pyridinium bromide 16 in 48 % yield. Heating 16 in acetic acid with maleic anhydride, followed by hydrolysis of the intermediate imide gives butyl-methylmaleic anhydride 17 in 60 % yield.

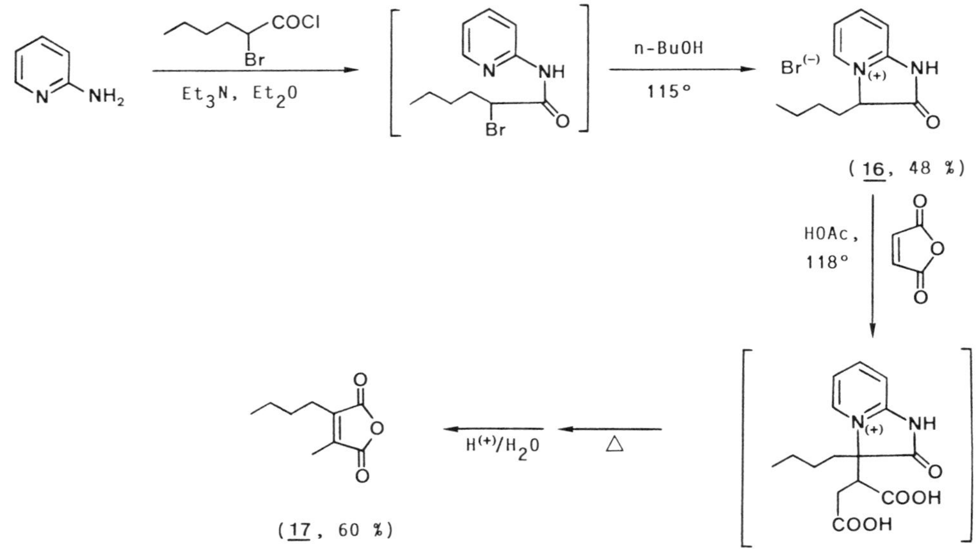

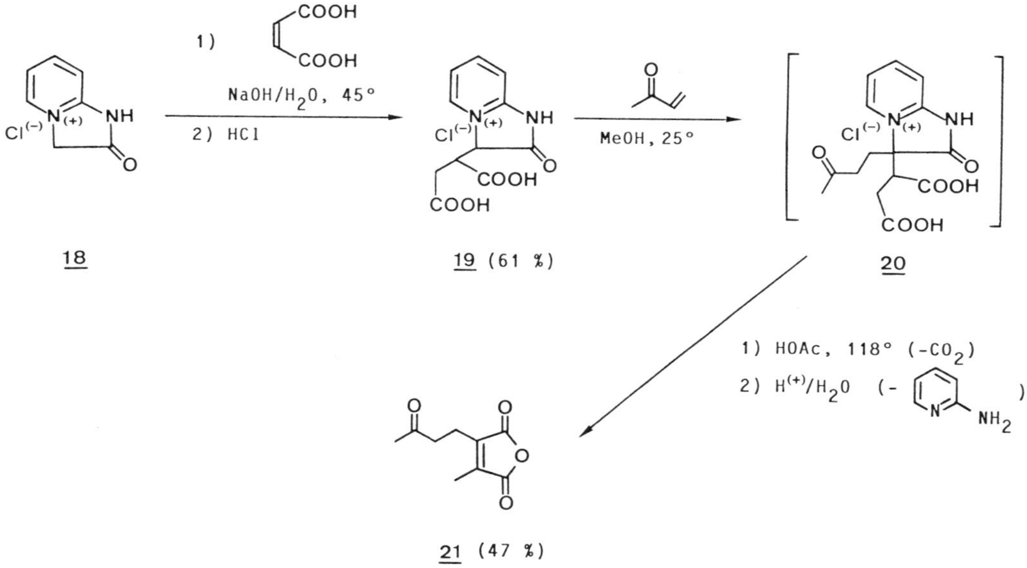

(**16**, 48 %)

(**17**, 60 %)

18   **19** (61 %)   **20**

21 (47 %)

Pyridinium salt **18** proved to be a key intermediate. It is easily obtained from chloroacetic acid and 2-aminopyridine in 60 % yield. On deprotonation in water it reacts with maleic acid to give the salt **19** after reprotonation. **19** is a very good Michael donor: it adds readily to methyl vinyl ketone to give salt **20**. On refluxing the crude adduct **20** in acetic acid, the anhydride **21** is obtained in 47 % overall yield after aqueous hydrolysis. This last step liberates 2-aminopyridine, which in a formal sense may be considered a catalyst: it induces the crucial reactivities of all the reaction intermediates without being destroyed in the overall process. In this, and in the cases discussed before, 2-aminopyridine may be recovered almost quantitatively.

SYNTHETIC APPLICATIONS

There are several natural products which by retrosynthetic analysis contain a DMA structural unit. Cantharidine is certainly the best known of all compounds whose synthesis has been attempted starting from DMA. After unsuccessful attempts by Diels and Alder to synthesise the compound by direct addition of DMA to furan (Ref. 4), cantharidine was first synthesised by Ziegler by a multistep synthesis (Ref. 5). Forty years after the first synthesis Dauben successfully used high pressure to achieve a simple synthesis of cantharidine. Although attempts to add DMA to furan at pressures up to 40 kbars failed, a bicyclic derivative of DMA as dienophile proved to be successful (Ref. 6). DMA, however, can be added to cyclopentadiene. Using this reaction, Kreiser (Ref. 7) prepared isoalbene and was able to correct the initially proposed structure of the nature product which possesses endo-methyl groups.

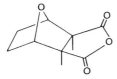

Cantharidine

O. Diels & K. Alder (1929)

K. Ziegler et al. (1942)

W. Dauben et al. (1980)

Albene

Isoalbene

W. Kreiser et al. (1978)

Bovolide

G. Lardelli et al. (1966)

J.N. Schumacher et al. (1966)

Spiroxabovolide

E. Demole et al. (1973)

(onions, leeks)

M. Albrand et al. (1980)

The presence of the so-called bovolide has been established in milk (Ref. 8), it is also found in tobacco leaves (Refs. 9 & 10), together with spiroxabovolide (Ref. 11). Both compounds have been successfully synthesised by addition of the corresponding Grignard reagents to DMA (Refs. 9 & 11). It may be interesting to note that even DMA is a component of cigarette smoke (Ref. 12). One may calculate that tons of it are vaporised worldwide each year! Finally, a rather unpleasantly smelling thio analogue of DMA has been found in onions and leeks (Ref. 13).

Another derivative of DMA is the natural product calythrone (Ref. 14). As calythrone was claimed to have insecticidal properties (Ref. 15), we were interested in the synthesis of calythrone and calythrone derivatives. In the following synthesis of calythrone analogues we have adopted and mixed steps from synthetic work of Vandewalle in Belgium (Ref. 16), Pattenden in Great Britain (Ref. 17) and Massy-Westropp in Australia (Ref. 18). This almost international synthesis was governed by our need as Swiss industrial chemists to create as many derivatives as possible from a common intermediate late in the synthesis. A few of our examples are shown; unfortunately none of the calythrone derivatives has exhibited insecticidal activity superior to that of the natural product.

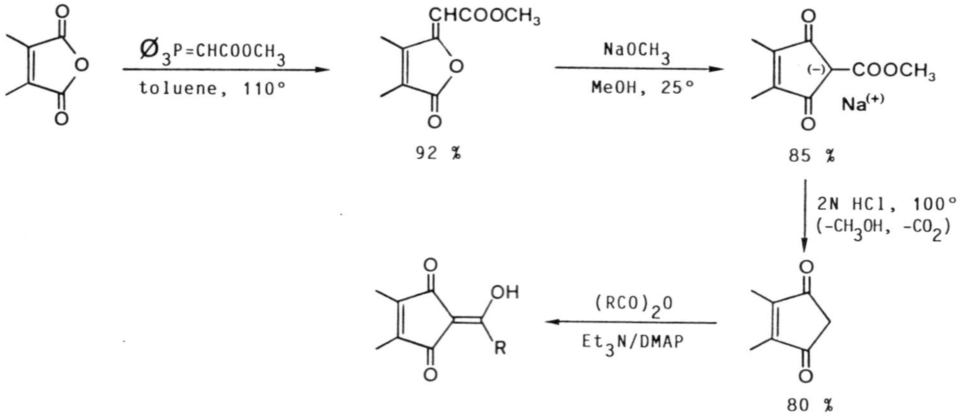

$\emptyset_3P=CHCOOCH_3$
toluene, 110°

92 %

NaOCH$_3$
MeOH, 25°

85 %

2N HCl, 100°
($-CH_3OH$, $-CO_2$)

80 %

$(RCO)_2O$
Et$_3$N/DMAP

R = $CH_2CH(CH_3)_2$ (Calythron, 20 %)
  = C_2H_5 (25 %)
  = C_4H_9 (45 %)
  = C_6H_5 (60 %)

An extremely valuable use of DMA was found, however, when the photochemical properties of imids of DMA were exploited. It is known that such imids undergo [2+2]-cycloadditions with high quantum yield (Refs. 19 & 20). Polymerisable monomers are obtained, if suitable derivatives (i.e. methacrylates) are prepared from these imids. Polymerisation induced by radical initiators leads to linear polymers with pendant DMA groups. No crosslinking is observerd under these conditions as the sterically hindered double bond in DMA is totally inert towards radical polymerisation. On irradiation in the presence of a sensitizer, however, crosslinking occurs efficiently leading to highly insoluble polymers (Ref. 22). Thus, photoreactive polymers containing pendant DMA groups find important industrial uses, for example in photolithographic applications and in photoresists.

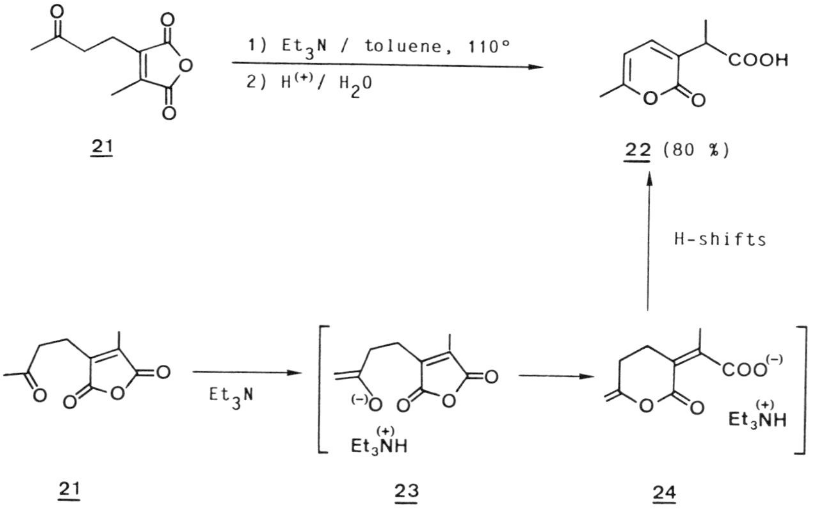

21

1) Et$_3$N / toluene, 110°
2) H$^{(+)}$/ H$_2$O

22 (80 %)

H-shifts

Et$_3$N

21   23   24

Disubstituted maleic anhydrides of type 21 are polyfunctional molecules which undergo interesting transformations. One is shown: treatment of 21 with one equivalent of triethylamine or any other tertiary amine leads to the α-pyrone 22 in 80 % yield. We assume that the mechanism of this unique rearrangement is as follows: attack of the enolate oxygen of 23 on the anhydride carbonyl leads to the acid salt 24 which then undergoes two hydrogen shifts to give, after protonation, the α-pyronylpropionic acid 22.

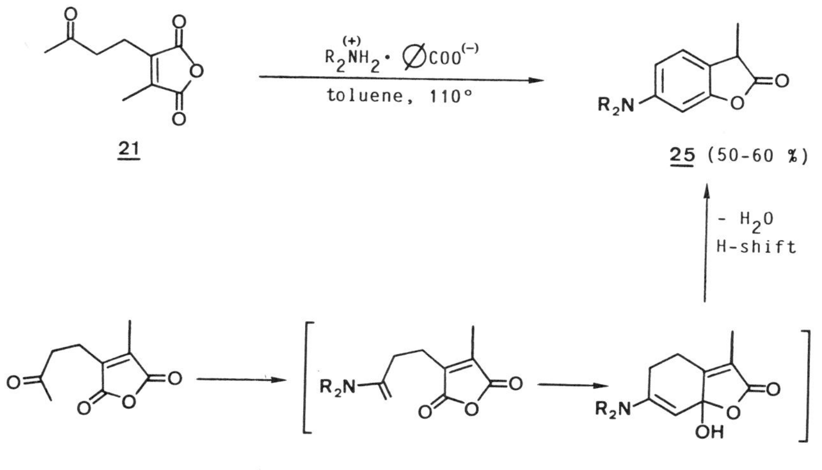

If, on the other hand, 21 is treated with a secondary amine salt, e.g. pyrrolidinium benzoate, another surprising reaction takes place: under these conditions benzofuranones of type 25 are obtained in 50-60 % yield. The formation of these compounds may best be explained by assuming an intermediate enamine 26, which then cyclises to the tetrahydrobenzene derivative 27. Aromatisation is achieved by loss of water and a hydrogen shift.

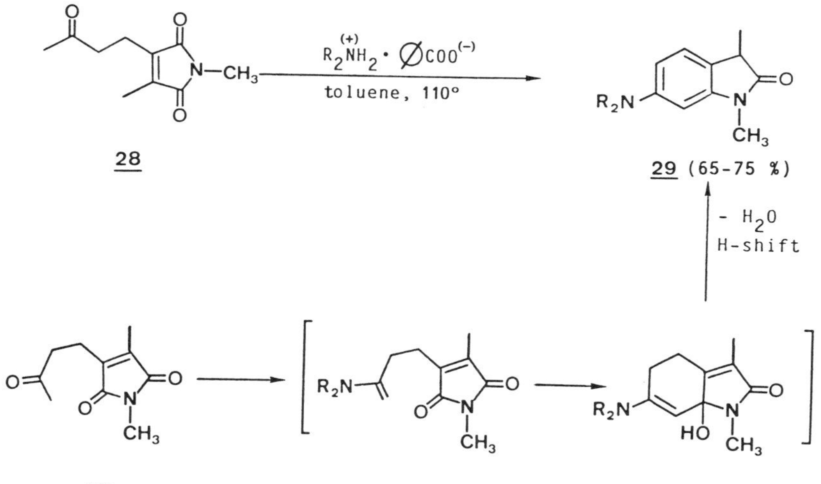

A very similar reaction is observed when the imide 28 is treated under the same conditions. This reaction leads to indolinone derivatives of type 29. In analogy to the mechanism discussed before, we assume that an enamine 30 cyclises to 31, which, as in the case of the benzofuranone, aromatises by a hydrogen shift and loss of water.

CONCLUSION

Imidazo[1,2-a]pyridinium compounds are valuable synthetic intermediates due to their high nucleophilicity even in acidic medium. The exploration of their reactivity has led to unexpected, but highly useful results. We believe that further efforts to exploit their synthetic potential will continue to be fruitful.

Acknowledgement - The authors wish especially to thank Dr. T. Winkler (NMR) and Mrs. G. Rihs (X-ray) for valuable contributions to the structure eludidations. Many valuable discussions were held with Dr. Hp. Sauter, Dr. G. Rist, and especially Dr. D. Bellus.

REFERENCES

1. M.E. Baumann and H. Bosshard, Helv. 61, 2751 (1978).
2. W.D. Ollis and C.A. Ramsden, Adv. Heterocycl. Chem. 19, 1 (1976).
3. F. Kröhnke and W. Zecher, Angew. Chem. 74, 811 (1962).
4. O. Diels and K. Alder, Ber. 62, 554 (1929).
5. K. Ziegler, G. Schenck, E.W. Krockow, A. Siebert, A. Wenz and H. Weber, Ann. Chem. 551, 1 (1942).
6. W.G. Dauben, C.R. Kessel and K.H. Takemura, J. Amer. Chem. Soc. 102, 6893 (1980).
7. W. Kreiser, L. Janitschke and L. Ernst, Tetrahedron 34, 131 (1978); W. Kreiser and L. Janitschke, Tetrahedron Letters, 601 (1978).
8. G. Lardelli, G. Dijkstra, P.D. Harkes and J. Bolding, Recueil 85, 43 (1966).
9. J.N. Schumacher and D.L. Roberts, US Pat. 3,251,366 (1966).
10. E. Demole and D. Berthet, Helv. 55, 1866 (1972).
11. E. Demole, C. Demole and D. Berthet, Helv. 56, 265 (1973).
12. J.N. Schumacher, C.R. Green, F.W. Best and M.P. Newell, J. Agric. Food Chem. 25, 310 (1977).
13. M. Albrand, P. Dubois, P. Etievant, R. Gelin and B. Tokarska, J. Agric. Food Chem. 28, 1037 (1980).
14. A.R. Penfold and J.L. Simonsen, J. Chem. Soc., 412 (1940).
15. M. Elliott and K.A. Jeffs, Proc. Chem. Soc., 374 (1961).
16. M. Vandewalle, L. van Wijnsberghe and G. Witvrouwen, Bull. Soc. Chim. Belg. 80, 39 (1971).
17. N.G. Clemo, D.R. Gedge and G. Pattenden, J. Chem. Soc., Perkin Trans 1 1448 (1981).
18. R.A. Massy-Westropp, Austr. J. Chem. 34, 2369 (1981).
19. G.O. Schenk, W. Hartmann, S.P. Mannsfeld, W. Metzner and C.H. Krauch, Chem. Ber. 95, 1642 (1962).
20. F.C. De Schryver, N. Boens and G. Smets, J. Polym. Sci. Part A 10, 1687 (1972).
21. J. Berger and H. Zweifel, Angew. Makromol. Chem. 115, 163 (1983).
22. H. Zweifel, Photogr. Sci. Eng. 27, 114 (1983).

Stereoselection at 3-carbon units. Recent studies and applications in natural product synthesis

Günter Helmchen

Institut für Organische Chemie der Universität, Am Hubland, D-8700 Würzburg, Federal Republic of Germany

Abstract – Given a sufficient set of practical highly stereoselective CC-bond forming asymmetric syntheses, a large variety of natural product molecules may be constructed by straightforward combination of 2- and 3-carbon units. Presented in this communication are: (i) Lewis acid promoted Diels-Alder reactions with acrylates of lactic acid esters, and (ii) homoaldol additions of metallated N-allyl urea derivatives. Stereoselectivity achieved for these reactions is discussed by considering (a) the relation of the allylic unit to the adjacent heteroatom, and (b) the interaction between the allylic unit and the chiral inductor group R* causing diastereoface-differentiation.

During the last decade the science - or art - of asymmetric synthesis, in the sense of achieving stereoselection under control of an external chiral auxiliary, has born several practical and broadly applicable methods. In natural product syntheses, externally controlled asymmetric reactions are usually applied at an early stage, for construction of one out of several stereogenic units. One established, others are then created via internally controlled diastereoselection. This approach works well with polycyclic compounds but it needs further development for applications to acyclic molecules (Ref. 1). Some of these, notably some macrolide antibiotics, have in fact been synthesized using external control for construction of almost each and every chirality center (Ref. 2). However, the arsenal of desirable methods still being small, usually reactions of only one type have been employed in an iterative mode. It appears worth-while to extend this approach by devising strategies wherein several asymmetric syntheses of different types are combined. For illustration of this concept, a total synthesis of (-)(S,S)-α-bisabolol will be described in the following.

A TOTAL SYNTHESIS OF (-)-α-BISABOLOL: STRATEGY

(-)-α-Bisabolol (1a, Scheme 1) is one of the physiologically active constituents of camomile oil. Not unusual with mono- and sesquiterpenes, all of its stereoisomers have been isolated from various plants (1b-d, Ref. 3). Accordingly, our synthesis scheme, presented as retro-synthesis below, provides for a route that can yield each isomer with equal facility.The scheme starts with cleavage of the side chain double bond, a Wittig reaction being, of course, the corresponding forward operation. Next, construction of one of the tertiary centers is envisaged by homoaldol reaction of the methyl ketone 2 with a propionaldehyde homoenolate 3. It has been demonstrated experimentally that no stereocontrol is exerted by the chirality center of 2 (Ref. 4). Therefore, a chiral homoenolate equivalent (indicated by the star) was required in order to obtain stereoselection in this step. Further retrosynthesis leads to the carboxylic acid 4 which is considered to be obtained by asymmetric Diels-Alder reaction from isoprene (5) and a chiral derivative of acrylic acid (6). A variety of efficient chiral auxiliaries being known, no difficulty was envisaged for this step. However, there was a challenge involved. The methyl ketone 2 has been prepared by Tamm and co-workers at Bâle in 36 % yield from readily available cheap (-)(S)-limonene (7)(Ref. 4). To be competitive, a route via asymmetric synthesis required a chiral auxiliary priced at less than ca. 10 US$/kg.

Scheme 1

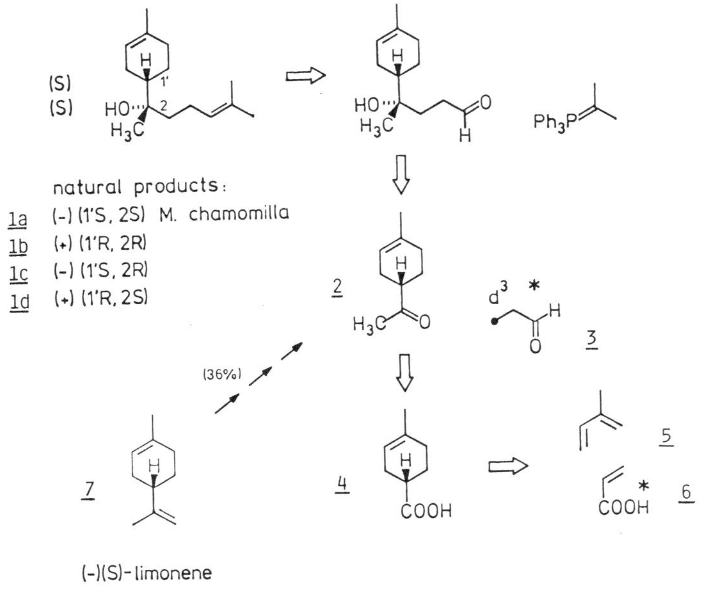

In summary, our objective was a total synthesis of (-)-α-bisabolol using as key steps two asymmetric syntheses, the first of which had yet to be devised; for the second a new chiral auxiliary had to be found.

A CHIRAL PROPIONALDEHYDE HOMOENOLATE EQUIVALENT (H. RODER)

Of three known types of homoenolate equivalents (Ref. 5), heterosubstituted allylic metal-organic compounds are probably most generally applicable (Ref. 6). Such compounds (Scheme 2) have been prepared from a variety of allylic derivatives 8 by metalation with butyllithium and subsequent transmetalation by addition of metal salts MY. Depending on metal and solvent, carbon metal bond character ranging from essentially ionic to largely covalent is displayed. Compounds (9) of the latter type, because of low basicity and a high degree of structural or-ganization, are best suited for addition reactions to aldehydes and ketones. Many of these reactions appear to be of S_E2'-type or to be chelation controlled. As a consequence, location of the metal controls α,γ-regioselectivity. Therefore, metallatropic equilibria (9a/9b) are an issue of major concern (Ref. 7). Fortunately, work of Tischler, Seebach, Hoppe and others (Ref. 8) has shown that an acyl group within X of 9 can effectively lock certain metal ions by coordination into the α-position. Corresponding reagents display excellent γ-selectivity and, furthermore, yield addition products 10 of Z-configuration. This indicates that substi-tuents X of 9 prefer the cis disposition with respect to the double bond ("cis-effect"). Hy-drolysis of addition products 10 gives γ-hydroxy-aldehydes 11 which spontaneously cyclize. Resultant lactols are usually oxidized to yield γ-lactones which are characterized.

Scheme 2

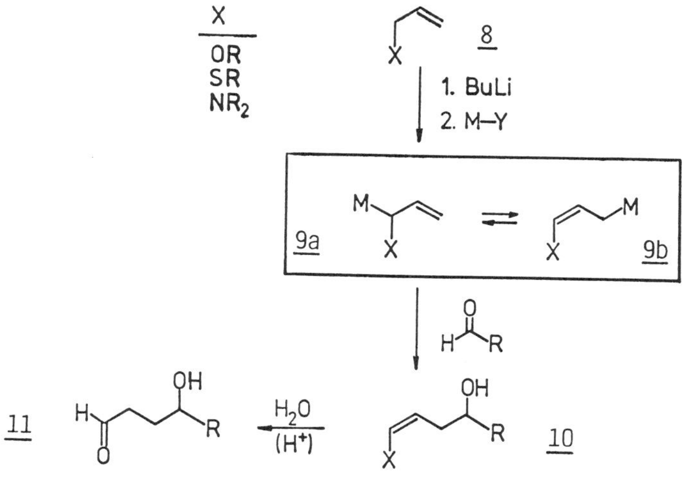

Scheme 3 shows the formula of one of the compounds referred to above. According to Hassel and Seebach in 1979 (Ref. 8), an allylmagnesium compound prepared by metalation of the urea 12 exhibits excellent γ- as well as Z-selectivity upon reaction with carbonyl compounds. It was believed in 1979 that the urea carbonyl group must be protected, by steric shielding in the case of 12, against attack of strong base. To the best of our knowledge, the Hassel-Seebach reagent has not found use in synthetic chemistry. Despite of this, we considered it a good starting point for three reasons:

1. Recent work, notably of Seebach's group (Ref. 9), has clearly shown that simple N-alkyl-urea derivatives are compatible with strong base. Accordingly, we felt that a compound as simple as the N-allylimidazolidone 13 would be a suitable reagent.
2. It was anticipated that a sufficiently hard metal ion would prefer the allylic α-position due to coordination to the urea carbonyl group (14).
3. It was hoped that introduction of a suitable substituent at the starred center of 14 would impart chirality of one definite sense to the allylic side chain.

It might appear that amalgamation of the Hassel-Seebach example with clever stereochemical engineering was a guaranty for success. Unfortunately, this was not the case. There is ample evidence (Ref. 6) that apparently similar allylic metalorganic compounds react in bewilder-ingly different manner. That means, one had to face the prospect that reactions of 14 with carbonyl compounds might give rise to a plethora of regio- and stereoisomers (cf. Scheme 3).

Scheme 3

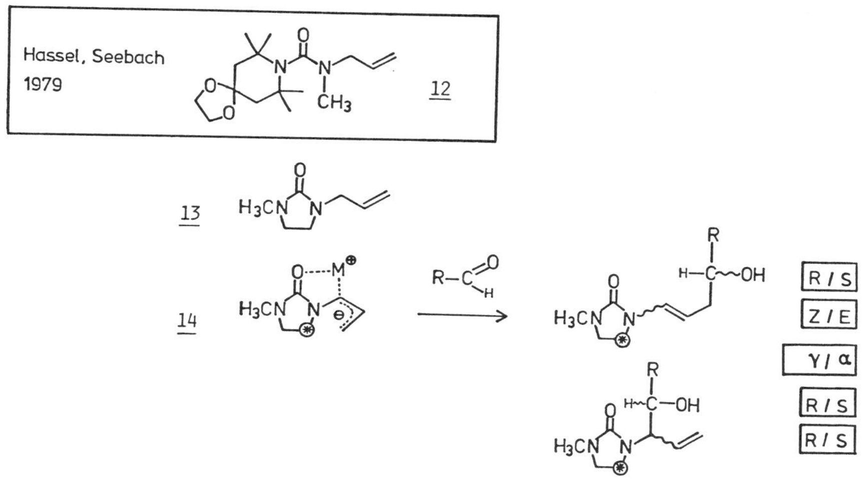

In carrying out the project, the first task was to search for a compound which would satisfy the design criteria outlined above. A promising example was found in the literature (Ref. 10): Dr. Close, of the Abbott Laboratories, had reported in 1950 that fusion of racemic ephedrinium chloride (15) with urea gives rise to an imidazolidone. Close had assigned the trans-configuration to his product, assuming a two-step reaction, Wöhler urea synthesis involving the amino group of 15 followed by internal nucleophilic substitution of the hydroxyl group with inversion of configuration. Today we know that this type of reasoning is dangerous with five-membered ring compounds. Indeed, x-ray crystal structure analysis has established the cis-configuration of 16 (Ref. 11). The N-allylurea derivative 17 was prepared in good yield (NaH, DMF, allyl chloride).

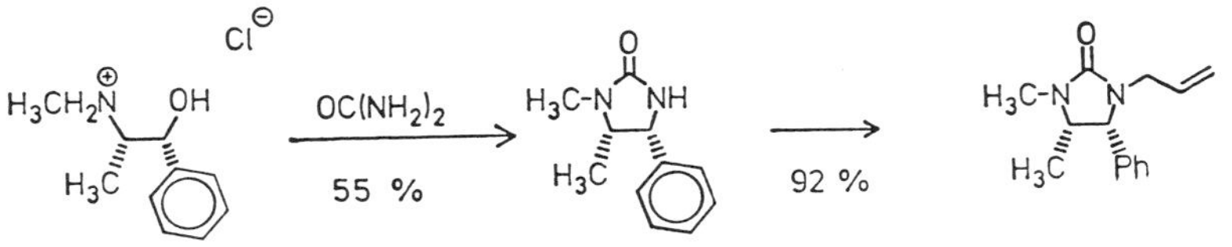

With the N-allylimidazolidone 17 introduced as a promising reagent, it is possible to present a precise outline of the ideas that are sketched on the preceding page.

Scheme 4

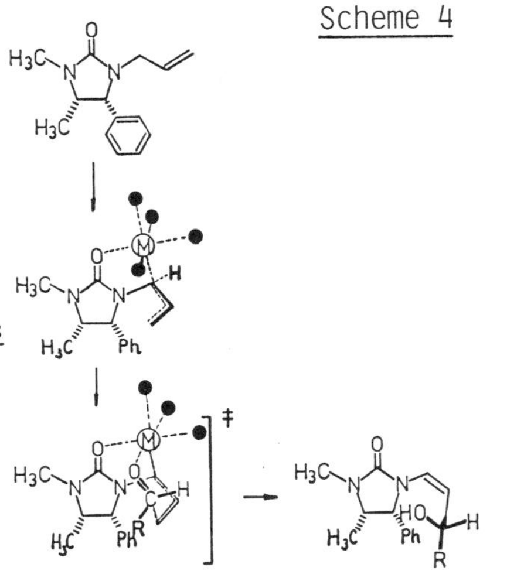

In Scheme 4 the formula of the reagent 17 is thought to be placed in such a way that the plane of the five-membered ring coincides with the plane of the paper and the substituents (Ph, Me) are located behind the paper. Assuming the metal derivative 18 to be a tight ion pair or a covalent structure, transoid disposition of the allylic side chain with respect to the phenyl group is expected to be preferred. Thus, a rigid, concave receptor is created by coordination in conjunction with steric repulsion.

Reactions of 18 with aldehydes are expected to be chelation controlled pericyclic reactions and thus to proceed via a chair-like transition state 19 wherein the larger of the substituents at the aldehyde carbonyl group, R, preferentially adopts an equatorial position in order to minimize interactions with the ligands at M. This model predicts addition products described by the formula 20a to be prevailing (Ref. 12).

Scheme 5

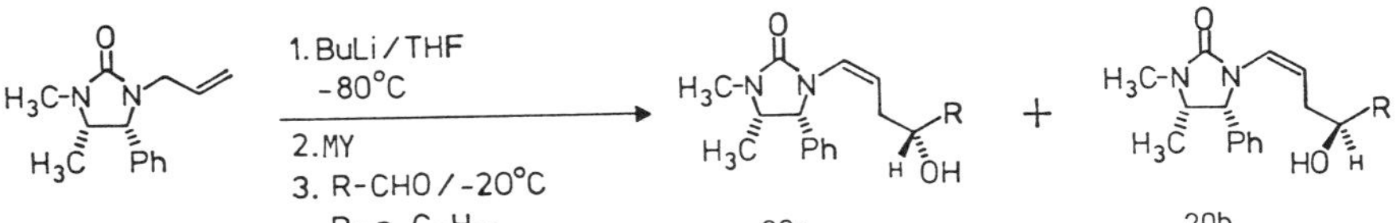

MY	20a		20b	γ/α	Y
MgBr$_2$·Et$_2$O /-20°C	1	:	1	99:1	76%
(iPrO)$_3$TiCl / 0°C	94	:	6	70:30	54%
(NEt$_2$)$_3$TiCl /-20°C	94	:	6	99:1	96%

Scheme 5 presents results obtained by addition of n-nonanal to various metal derivatives of 17. Initially, conditions as given by Hassel and Seebach were applied, involving metalation with butyllithium followed by transmetalation with $MgBr_2 \cdot 2 \ Et_2O$. Reaction of the resultant allylmagnesium compound with n-nonanal proceeded with excellent γ-regioselectivity to give crystalline products, 20a,b; but a discouragingly low level of stereoselectivity was found (Ref. 13). Transmetalation was next carried out with chlorotitanium triisopropoxide. This resulted in good to excellent diastereoselectivity but rather low yield and, a failure again, a very low degree of regioselectivity. Success came, however, upon variation of ligands of the titanium center. Proceeding from isopropoxy to amino ligands was a logical step since the presence of amino ligands at titanium generally enhances thermal stability as well as oxophilicity (Ref. 14).

21, the tris(diethylamino)-titanium derivative of 17, was found to be a generally applicable reagent. Results obtained with this propionaldehyde homoenolate equivalent are summarized in Scheme 6. At the head of this scheme our general procedure is described. Reactions of 21 with aldehydes or ketones generally furnish crystalline products 22a,b. These were usually further

Scheme 6

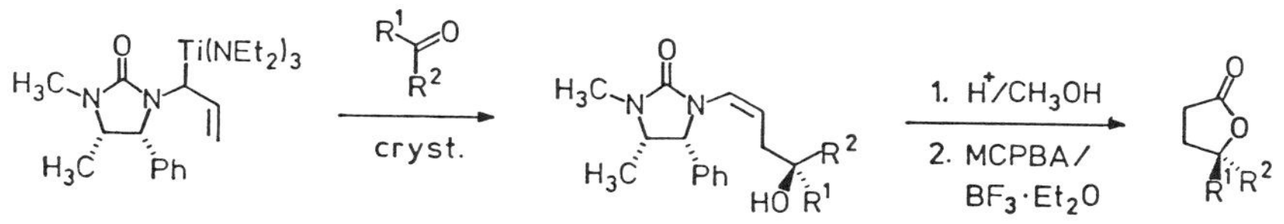

	Y[%]	diastereoselectivity			Y[%]	$[\alpha]_D$	
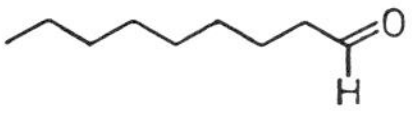 (n-nonanal)	96 / 61	94 : 6 / > 200 : 1			99	−41.1 (+41.1)	(S)
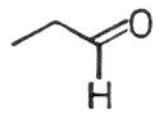 (propionaldehyde)	94 / 63	96 : 4 / > 200 : 1			79	−53.7 (+53.2)	(S)
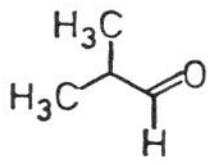 (isobutyraldehyde)	95	96 : 4					
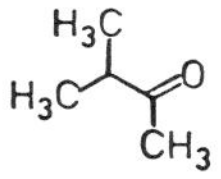 (isopropyl methyl ketone)	93 / 81	98.5 : 1.5 / 150 : 1			97	−10.5 (−10.2)	(R)

processed by methanolysis to yield crude lactol ethers (Ref. 15), oxydation of which according to Grieco (Ref. 16) gives γ-lactones 23; many of these, being natural products, are well characterized with respect to chiroptical data and absolute configuration. Of the table, data within boxes refer to "constitutionally pure" reaction products, others to recrystallized material. Thus, the crude product obtained from the reaction of 21 with nonanal in 96 % yield with diastereoselectivity of 94:6 was twice recrystallized to give a diastereochemically homogeneous compound in 61 % yield. From this, enantiomerically pure (-)(S)-4-octyl-γ-butyrolactone was obtained in quantitative yield. Except for a slightly higher level of stereoselection, 96:4, propionaldehyde and isobutyraldehyde gave results very similar to those obtained with nonanal (Ref.17). Configurational relationships observed for these reactions conform to the mechanistic rationale proposed in Scheme 4 (Ref. 12).

The last entry of the table (Scheme 6), describing the reaction of 21 with isopropyl methyl ketone, shows particularly interesting results: 1. Unexpectedly, the methyl ketone yields a higher degree of stereoselection than the corresponding aldehyde (98.5:1.5 vs. 96:4, respectively). 2. As is reviewed by Scheme 7, isopropyl methyl ketone obviously is an excellent model for the methyl ketone 2 (Scheme 1) that is envisaged to be a key intermediate of the α-bisabolol synthesis. As a discrepancy it is noted, however, that application of 16 as reagent furnishes a γ-lactone with (R)-configuration whereas the (S)−configuration is required for the α-bisabolol series. Fortunately, both enantiomers of ephedrine, precursor of the reagent, are commercially available at nearly equal cost. 3. 4-Isopropyl-4-methyl-γ-butyrolactone, the product from the model reaction, is only 2 steps away (Ref. 18) from an important pheromone discovered by Francke at Hamburg (Ref. 19).

Scheme 7

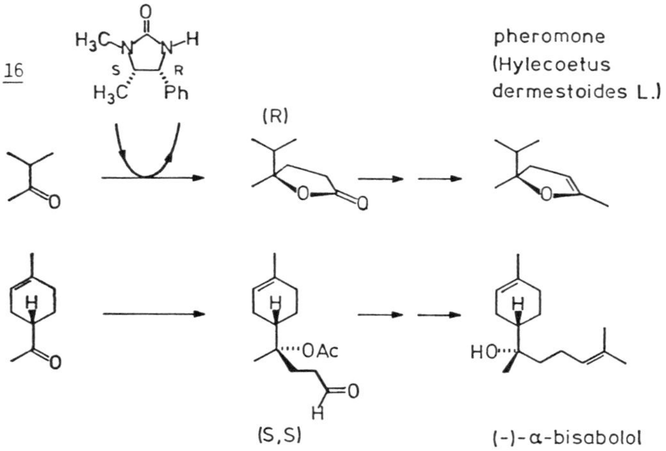

Considering problems associated with the homoaldol reaction as solved, the asymmetric Diels-Alder synthesis required for construction of the 6-membered ring is addressed in the following.

LACTIC ACID ESTERS: PRACTICAL REAGENTS FOR ASYMMETRIC DIELS-ALDER ADDITIONS (T. POLL)

Scheme 8 presents background information pertaining to our work. Following an established tradition, we are using acrylates of chiral alcohols R*-OH as dienophiles for the asymmetric Diels-Alder reaction. In general, Lewis acid catalysis has to be employed in order to achieve high levels of stereoselection (Ref. 20).

Scheme 8

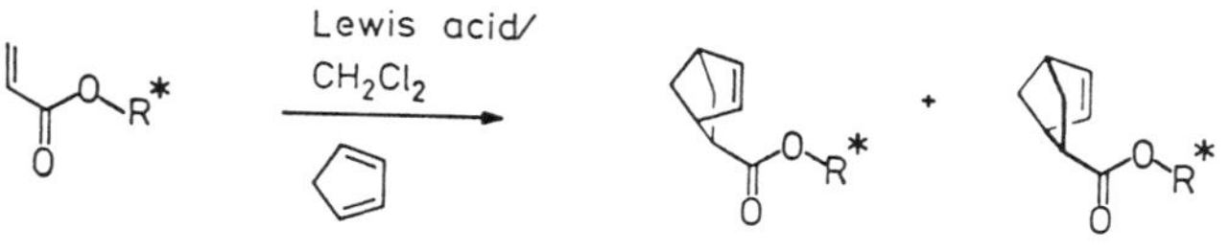

	HO-R*	Lewis acid	temp [°C]	diastereoselectivity		
24						
	R = H	BF$_3$·Et$_2$O	– 70	92	:	8
	R = H	TiCl$_4$	– 20	81	:	19
	R = Ph	TiCl$_4$	– 20	95	:	5
25						
	R = CONHPh	AlCl$_3$	– 45	96	:	4
	R = CONHPh	TiCl$_4$	– 40	98.7	:	1.3
	R = CH$_2$tBu	TiCl$_2$(OiPr)	– 40	99.3	:	0.7

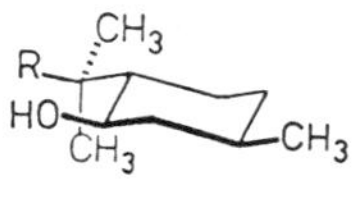

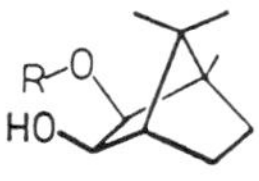

In the early 1960s already, impressive results were achieved by Walborsky and by Sauer with the menthyl group (24 R=H) as chiral auxiliary. With cyclopentadiene, traditionally a standard for evaluation of dienophile efficiency, and Lewis acids BF_3 and $TiCl_4$, mainly endo adducts are obtained with diastereoselectivity of 92:8 and 81:19, respectively (Ref. 21). Still more encouraging results were achieved by Corey in 1975 with the acrylate of 8-phenylmenthol (24 R=Ph): diastereoselectivity of 95:5 as compared to 81:19 for the menthyl ester (Ref. 22). In 1980, based on classifying reagents into convex and concave types, the then novel reagents 25 derived from camphor were introduced by this group (Ref. 23). With shielding groups R being benzyl or N-phenylcarbamoyl diastereoselectivity of up to 98.7:1.3 can be achieved (Ref. 24). Finally, two years later Oppolzer and co-workers, using 25, R=t-Bu, were able to traverse the 99:1 mark (Ref. 25).

With respect to configurational relationships of preferred endo adducts a very regular pattern was found for all Lewis acid promoted Diels-Alder additions of acrylates. A transition state model (Ref. 23a) which at least roughly accounts for the data was proposed by Walborsky and was later modified in order to incorporate modern views on conformations of esters. The model is based on two assumptions (cf. Scheme 9):

Scheme 9

bond formation at Si$_{acr.}$ face (preferred)

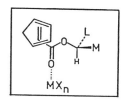
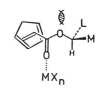

bond formation at Re$_{acr.}$ face

1. The reacting species is a complex wherein the Lewis acid is coordinated to the oxygen atom of the enoate carbonyl group.

2. The conformation of the $COOCR^1R^2R^3$ group, as shown in Scheme 9, is identical to that of the (non-coordinated) ester and is conserved along the reaction path.

Given these assumptions, we are left with four transitions states which have to be ranged according to their energy content. A first decision is based on the customary evaluation of shielding capabilities of the ligands at the carbinyl center (L,M,S). With the largest group at the backside, presumed for Scheme 9), attack of the diene from the front is expected to be preferred. The alternatives left are transition states with anti- or syn-planar conformations of the enoate group. Preference must be given to the former in order to rationalize configurations of preferred products.

The modified Walborsky model given above certainly is a rather crude transition state model. It served well, however, as basis for development of the highly selective bornane derivatives 25 (Scheme 8). As is illustrated by formula A of Scheme 10, the concept was to ensure effective shielding of one face of the enoate moiety by providing a concave environment. However, good results are often due to serendipity. Conspicuously, in all particularly selective reagents of type 25 the shielding group R is connected via an ether or carbonyl bridge to the bornane skeleton. Therefore, proposal B of Scheme 10 was later considered: conformational immobilization as well as shielding due to bidentate coordination of the ester. The validity of this proposal has not yet been substantiated; but it inspired the idea that a segment of the structure B, represented by C of Scheme 10, might exhibit properties similar to B due to structural organization by an appropriate Lewis acid. This proposal was all the more appealing because corresponding reagents, various esters of lactic acid of both (R)- and (S)-configuration, are available at the price of less than ca. 10 US$/kg that was considered desirable for the α-bisabolol synthesis.

Scheme 10

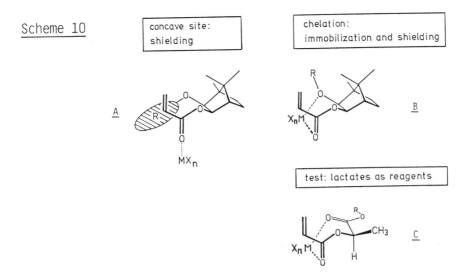

Scheme 11 displays results achieved for reactions of the acrylate of (S)-ethyl lactate (26) with cyclopentadiene (Ref. 26). At -63 °C endo adducts are obtained in ca. 95 % yield. This had been anticipated; but we were quite astonished about the rather high level of stereose- lection found for the diastereoisomers 27a/27b. Thus, under optimized conditions use of 0.7 eq. of TiCl₄ results in stereoselectivity of 93:7; with AlCl₂Et, a somewhat lower degree of stereoselection is found, but with the adduct of opposite configuration being preferred.

Scheme 11

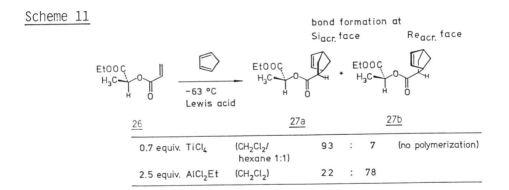

26		27a		27b	
0.7 equiv. TiCl₄	(CH₂Cl₂/ hexane 1:1)	93	:	7	(no polymerization)
2.5 equiv. AlCl₂Et	(CH₂Cl₂)	22	:	78	

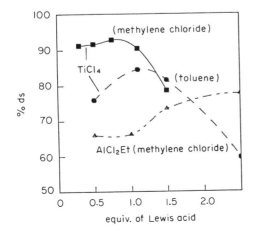

Fig. 1. Plot of diastereoselectivity (Ref. 27) versus ratio of Lewis acid/acrylate 26.

Configurational dichotomy of this kind has not been previously observed for Lewis acid pro- moted Diels-Alder reactions. In the present case, differing effects of Lewis acids (Ref.26) were found for all reaction variables investi- gated. An example is shown in Fig. 1 which de- scribes dependence of diastereoselectivity on the ratio of Lewis acid:acrylate 26. The TiCl₄- promoted reaction, in solvents methylene chlo- ride as well as toluene, displays a distinct maximum of selectivity with ca. 1 eq. of the Lewis acid. In contrast, with AlCl₂Et continous increase of selectivity is observed.

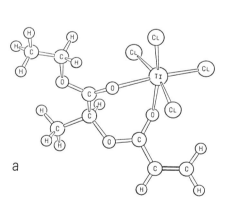

28

Properties shown by the AlCl₂Et-catalyzed reaction are reminiscent of behaviour displayed by menthyl acrylate and similar compounds. Accordingly, the modified Walborsky model is applied which implies for the corresponding transition state, described by formula 28, an anti-planar enoate conformation; attack of the diene then must be assumed to occur at the enoate face opposite to the COOR group in order to rationalize preferential formation of 27b.

On the other hand, data collected for the TiCl₄-catalyzed reaction indicate that a 1:1 acrylate-TiCl₄ chelate complex is the reactant. However, the structure of this complex must be different from that described by formula C (Scheme 10) because this would be expected to yield 27b as major product upon reaction with cyclopentadiene. The structural problem was eventually solved by crystallizing the acrylate-TiCl₄ complex. Its structure, determined by x-ray crystal structure analysis, is described by Fig. 2.

The structure displays the expected octahedral coordination of the titanium atom; but the enoate group is found in the syn-planar rather than the anti-planar conformation. In order to relate the crystal structure to asymmetric synthesis the molecule (Fig. 2a) is displayed in such a way that the plane of the double bond coincides with the plane of the paper. Obviously, one chlorine atom shields the front face of the double bond, though apparently not in a very impressive way. However, a computer-drawn (Ref. 28) space-filling model (Fig. 2b) reveals that chlorine atom as quite heavily intruding into the double bond region. Fig. 2b, however, does not show the facial bias at the double bond. For better assessment Fig. 2c

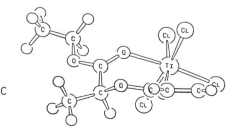

28   29

shows a model which is rotated by 90° around a horizontal axis so that the enoate plane is perpendicularly arranged to the paper. The top face obviously is heavily shielded; there is almost no shielding at the bottom face.

The structure of the reactant in solution may not be identical to that found in the crystal. However, assuming identity allows to rationalize our observations coherently (Ref. 29). At the left, the corresponding transition state model 29 is shown in juxtaposition with 28, that of the AlCl₂Et-catalyzed reaction already discussed. For both models back-face shielding of the enoate group is proposed, but one displays an anti-planar the other a syn-planar conformation of the enoate group; thus, a conformational "two-dimensional" dichotomy is seen as causing the configurational dichotomy observed experimentally.

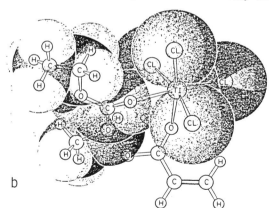

a

b

c

Fig. 2. Crystal structure of the TiCl₄-acrylate 26 complex (Ref. 30).

With a mechanistic rationale at hand that allows prediction of configurational relationships for the asymmetric Diels-Alder addition with lactates as chiral auxiliaries, the stage is set to address our main issue again.

TOTAL SYNTHESIS OF (-)-α-BISABOLOL (A. SOBCZAK, H. RODER)

The strategy outlined earlier (cf. Scheme 1) requires as first step Diels-Alder addition of isoprene to an acrylate. Since a prevailing adduct of (S)-configuration was needed (Scheme 7) the acrylate of (R)-isobutyl lactate (30) was used. As is shown in Scheme 12, stereoselectivities obtained closely parallel those achieved for the reaction of cyclopentadiene with the acrylate of (S)-ethyl lactate (Scheme 11).

Scheme 12

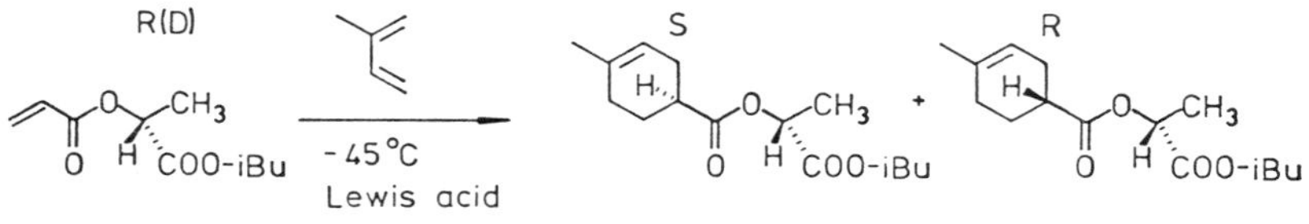

| 0.7 equiv. $TiCl_4$ | $(CH_2Cl_2/$ hexane 1:1) | 94 | : | 6 |
| 3.0 equiv. $AlCl_2Et$ | (CH_2Cl_2) | 19 | : | 81 |

Saponification of the crude product from the $TiCl_4$-promoted reaction gave crystalline acid 4 with 82 %ee (Scheme 13) which corresponds to a 91:9 mixture of (S)-4 and (R)-4, respectively. Treatment (cf. Scheme 12) with LiOH had thus resulted in ca. 3 % racemization. Though it was not necessary for the synthesis of (-)-α-bisabolol (see below) we wanted to prepare enantiomerically pure 4. This was possible via formation of the iodolactone 31 and its recrystallization and reduction (Ref. 31). Reaction of enantiomerically pure (S)-4 with methyllithium furnished the ketone (S)-2 in 92 % yield. As side product 5 % of enantiomerically pure (-)-terpineol was isolated. Thus, no racemization occurs in the addition step. Following our plan, ketone 2 was next subjected to the homoaldol reaction using ent-16 as reagent. The crystalline product 32 was obtained with diastereoselectivity of 98.5:1.5 according to high-pressure liquid chromatographic analysis. Recrystallization gave diastereomerically pure 32 in 83 % yield. Repetition of the same sequence of operations with the enantiomerically impure sample

Scheme 13

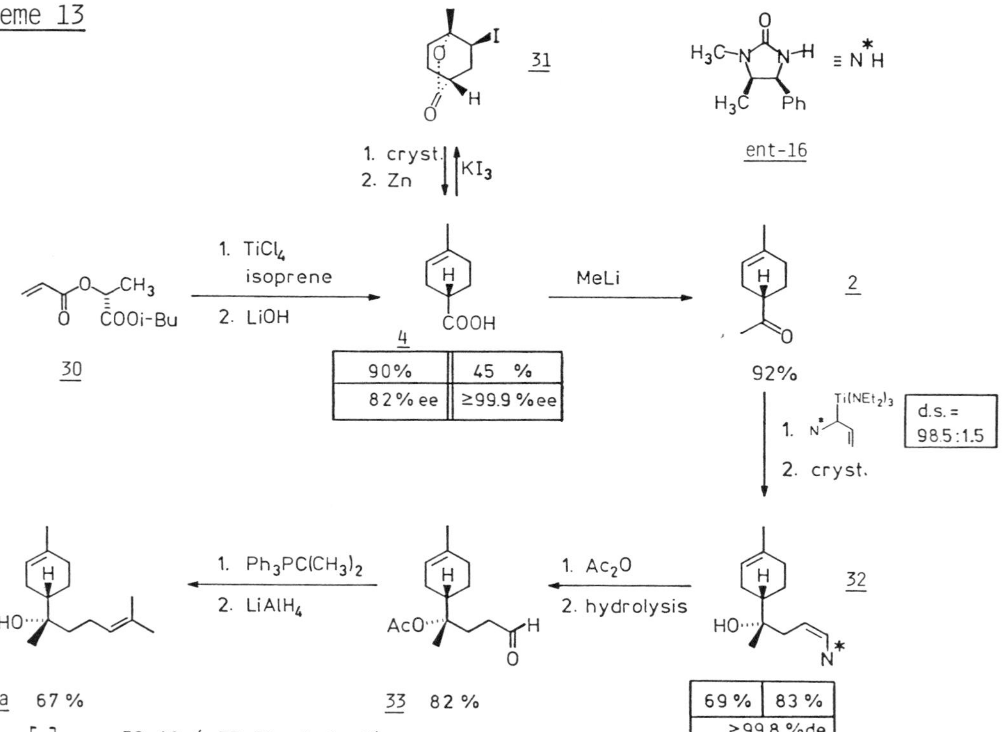

$[\alpha]_D$ = -58.1° (-57.7°, Ref. 3)

of 4 furnished diastereomerically pure 32 in 69 % yield. We regard this result as convincing demonstration of the practical utility of the new chiral auxiliaries.

The urea derivative 32 was acetylated and the product hydrolyzed, using weak aqueous acid in a two-phase system, to give the aldehyde 33 in 82 % yield. Protection of the OH group was necessary because of the subsequent Wittig reaction that completed the side chain. Finally, reductive ester cleavage produced (-)-α-bisabolol (1a) in 67 % yield. The overall yield of diastereomerically pure (Ref. 32) 1a from acrylate 30 was 31 %. This compares favourably with a yield of 16 % to be expected for a route starting from natural (-)-limonene (cf. Scheme 1).

Acknowledgement - It is a pleasure to acknowledge the contributions of my enthusiastic and able collaborators H. Roder, T. Poll and A. Sobczak. The crystal structure analysis mentioned in the text was carried out by J.O. Metter, Institut für Kristallographie der Universität Frankfurt (M); the program SCHAKAL used for the drawings shown in Fig. 2 was provided by Dr. E. Keller, Institut für Kristallographie der Universität Freiburg i. Br. We gratefully acknowledge financial assistance from the Deutsche Forschungs-gemeinschaft, the Fonds der Chemischen Industrie, and the Deutscher Akademischer Austauschdienst.

REFERENCES

1. P.A. Bartlett, Tetrahedron 36, 2 (1980).
2. D.A. Evans, Aldrichim. Acta 15, 23 (1982); S. Masamune, W. Choy, Ibid. 15, 47 (1982).
3. 1a-1c: E. Flaskamp, G. Nonnenmacher, O. Isaac, Z. Naturforsch. 36b, 114 (1981); references cited therein. 1d: E.-J. Brunke, F.-J. Hammerschmidt, Dragoco-Report 37 (1984). Stereo-selective syntheses of racemic 1a: M.A. Schwartz, G.C. Swanson, J. Org. Chem. 44, 953 (1979); T. Iwashita, T. Kusumi, H. Kakisawa, Chem. Lett. 947 (1979). Non-stereoselective partial syntheses of optically active 1a-1d: A. Kergomard, H. Veschambre, Tetrahedron, 33, 2215 (1977); D. Babin, J.-D. Fourneron, M. Julia, Ibid. 37, 1 (1981).
4. W. Knöll, C. Tamm, Helv. Chim. Acta 58, 1162 (1975).
5. H. Ahlbrecht, Chimia, 31, 391 (1977); references cited therein.
6. D. Hoppe, Angew. Chem. in press.
7. R.W. Hoffmann, Angew. Chem. 94, 569 (1982); Angew. Chem. Int. Ed. Engl. 21, 555 (1982).
8. (a) T. Hassel, D. Seebach, Angew. Chem. 91, 427 (1979); Angew. Chem. Int. Ed. Engl. 18, 399 (1979); (b) A. M. Tischler, M. H. Tischler, Tetrahedron Lett. 3407 (1978); (c) Ref. 6, references cited therein.
9. T. Mukhopadhyay, D. Seebach, Helv. Chim. Acta 65, 385 (1982).
10. W. J. Close, J. Org. Chem. 15, 1131 (1950).
11. H. Roder, G. Helmchen, E.-M. Peters, K. Peters, H.-G. von Schnering, Angew. Chem. in press.
12. The configurational descriptor (R or S) of the hydroxylated center is a function of the ligand R and thus may change for a series of aldehydes RCHO.
13. All product diastereomer analyses were carried out by NMR or high-pressure liquid chromato-graphy.
14. M. T. Reetz, Top. Curr. Chem. 106 (1982).
15. The reagent 16 can be recovered in ca. 95 % yield by extraction with aqueous acid.
16. P. A. Grieco, T. Oguri, Y. Yokohama, Tetrahedron Lett. 419 (1978).
17. For data and references see Ref. 11.
18. K. Mori, T. Ebata, S. Takechi, Tetrahedron 40, 1761 (1984).
19. W. Francke, W. Mackenroth, W. Schröder, A. R. Levinson, Les Colloques de l'INRA 7, 85 (1982).
20. Reviews: Y. Mori, J. Synth. Org. Chem. Jpn. 40, 321 (1982); P. Welzel, Nachr. Chem. Techn. Lab. 979 (1983); L. A. Paquette in J. D. Morrison, Asymmetric Synthesis 3B, 455 (1984); H. Wurzinger, Kontakte (Darmstadt) 3 (1984).
21. J. Sauer, J. Kredel, Tetrahedron Lett. 6359 (1966).
22. E. J. Corey, H. E. Ensley, J. Am. Chem. Soc. 97, 6908 (1975).
23. (a) R. Schmierer, Dissertation, Stuttgart 1980; (b) E. Ade, G. Helmchen, G. Heiligenmann, Tetrahedron Lett. 1137 (1980); (c) G. Helmchen, R. Schmierer, Angew. Chem. 93, 208 (1981); Angew. Chem. Int. Ed. Engl. 20, 205 (1981).
24. T. Poll, A. Sobczak, unpublished results.
25. W. Oppolzer, C. Chapuis, G. M. Dao, D. Reichlin, T. Godel, Tetrahedron Lett. 4781 (1982).
26. T. Poll, G. Helmchen, B. Bauer, Ibid. 2191 (1984).
27. For definition of this term see D. Seebach, R. Naef, Helv. Chim. Acta 64, 2704 (1981).
28. Van der Waals radii according to L. Pauling, Die Natur der Chemischen Bindung, Verlag Chemie, Weinheim 1962.
29. Guided by this rationale, we recently found D-pantolactone to be an extremely effective reagent for the asymmetric Diels-Alder reaction.
30. T. Poll, J. O. Metter, G. Helmchen, Angew. Chem. in press.
31. Optical rotation of enantiomerically pure (S)-4: $[\alpha]_{589}^{20}$ -107.5° (c=4.0, 95 % EtOH).
32. Reliable determination of diastereomeric purity of 1a is possible by ^{1}H-NMR (Ref. 3); determination of enantiomeric purity is currently undertaken in these laboratories.

Some synthetic applications of alcohol dehydrogenases and esterases

J. Bryan Jones

Department of Chemistry, University of Toronto,
Toronto, Ontario, Canada M5S 1A1

Abstract - The enzymes of most general and continuing value
in organic synthesis are those that can operate
stereospecifically on a wide range of structurally varied
substrates. Horse liver alcohol dehydrogenase and pig liver
esterase are such enzymes. The asymmetric synthetic
potentials of these enzymes are illustrated by examples of
enantiomeric distinctions, of discrimination between
enantiotopic groups of substrates possessing prochiral
centres or of meso compounds, of regiospecific
transformations, and of combinations of various
specificities. The value of such enzyme-derived products as
chiral precursors of molecules of current interest is also
shown.

In order for enzymes to be accepted into routine use in synthesis they must
be readily, preferably commercially, available. Also, they should accept a
broad structural range of substrates, while operating stereospecifically in
each individual transformation. Although these two specificity criteria are
somewhat antithetical, they are satisfied by a number of enzymes. The
availability of guidelines permitting prediction of stereospecificity is
also helpful. Knowledge of the mechanism of catalysis is advantageous in
that, for example, it can enable potential inhibitors to be identified.
Above all, the experimental procedures for the preparative reactions should
be easily performed in the organic chemical laboratory and should not
require sophisticated biochemical equipment. Horse liver alcohol
dehydrogenase (EC 1.1.1.1) and pig liver esterase (EC 3.1.1.1) satisfy
most of these criteria and will be used to illustrate the scope of synthetic
applicabilities of versatile enzymes.

Horse Liver Alcohol Dehydrogenase (HLADH)

HLADH is a commercially available, nicotinamide adenine dinucleotide
(NAD/H)-dependent, alcohol dehydrogenase that catalyses $CH(OH) \rightleftharpoons C=O$
oxidoreductions of a broad structural range of substrates of organic
chemical interest. In our work, the high costs of the NAD/H coenzymes are
overcome by using the ethanol coupled-substrate regeneration method (1) for
reductions, and flavin mononucleotide-recycling for oxidations (2), using
very simple experimental procedures in each case. The preparative-scale
reactions are generally performed on 1-2 g of substrate. Scaling up to 20 g
or more is easily achieved if more material is required. Reaction times
vary from a few hours in the most favourable cases to 2-3 weeks for the
slowest substrates. Accurate predictions regarding the stereospecificity of
the HLADH-catalysed oxidoreduction process can be made using the Prelog rule
(3) for simple acyclic substrates, and with a cubic-space model of the
enzyme's active site for more complex acyclic and cyclic substrates (4).
All of the HLADH-catalysed reactions discussed below are in accord with the
predictions of these models. Throughout the paper, all enantiomeric
excesses of chiral products, which were measured directly by NMR or GLC, are
⩾97-99% unless specifically indicated otherwise. The isolated yields are in
the 70-90% range for all reactions.

Exploiting the enantiomeric stereospecificity of HLADH

The enantiomeric specificity of HLADH has been widely exploited for the resolution of racemic ketones and alcohols (5). Resolutions of the racemic bridged bicyclic compounds 1, 2 and 3 exemplify this aspect. Such resolutions are very tedious by traditional methods. In contrast, the HLADH-catalysed reductions and oxidations of these compounds permit highly enantiomerically enriched materials to be isolated in a few hours (6). Optically pure materials can then be obtained by recrystallisation. Furthermore, the reductions of the carbonyl groups of the reactive enantiomers of (+)-1 and (+)-2 are diastereotopically specific for one face of the carbonyl group, giving the endo- and exo-alcohols, (+)-4 and (-)-5 respectively. The formation of the exo-isomer (-)-5 on reduction of (+)-3 represents a real bonus, since this is the thermodynamically less stable epimer and is not easily available by direct chemical reduction. The specificity of the enzymic reduction step is not influenced by the thermodynamic stabilities of the alcohol products (+)-2 and (-)-5. The product-alcohol stereochemistries are determined only by the orientation of the carbonyl group with respect to the direction of hydride donation from the coenzyme in the favoured ES complex preceding the transition state.

HLADH is very happy to accept heteroatoms such as O and S (but not N, which complexes the active site Zn^{2+} atom) in its substrates. The reduction of (±)-6 is an interesting case. The transformation proceeds smoothly but is enantioselective only, with the (+)-enantiomer being reduced only a little faster than the (-). The immediate conclusion in such cases is often that enzymic resolution of the racemates is not feasible under these conditions. However, when an enzyme binds both enantiomers of a racemate to form productive ES complexes, it operates stereospecifically on each individual substrate stereoisomer within the two diasteromeric ES complexes. Thus in the HLADH-catalysed reduction of (±)-6, hydride delivery occurs to the Re-face of the carbonyl group of each enantiomer in its respective ES-complex. The corresponding alcohol products (+)-7 and (+)-8, being diastereomeric, can be separated chromatographically. Subsequent chemical oxidation then provides the individual ketone enantiomers, (-)- and (+)-6 respectively (7).

Exploiting the prochiral stereospecificity of HLADH

While the abilities of enzymes to discriminate between enantiomers is very important for resolution and asymmetric synthesis applications, the maximum yield of material of the desired configuration is 50% since the unwanted enantiomer must be discarded unless its structure lends itself to ready recycling back to the starting racemate. The ability of enzymes to be prochirally stereospecific in their catalyses overcomes this problem because direct asymmetric synthesis of chiral products from symmetrical starting materials becomes possible.

HLADH is a powerful enzyme in this regard. For instance, enantiotopic groups attached to prochiral centres can be distinguished. When operating in its oxidative mode on C-3 substituted pentane-1,5-diols 9 the enzyme is enantiotopically selective for the pro-S hydroxyethyl groups to give the hydroxyaldehydes 10. These cyclise spontaneously in situ to the hemiacetals 11 which are themselves substrates of HLADH and undergo further oxidation to give the (S)-lactones 12 directly (8). The direct formation of chiral lactones from these symmetrical diols represents an exciting asymmetric synthetic opportunity. The reaction is general for a wide range of R-substituents (8) but gives lactones 12 of useful ee levels only for the C-3 methyl, ethyl and propyl compounds.

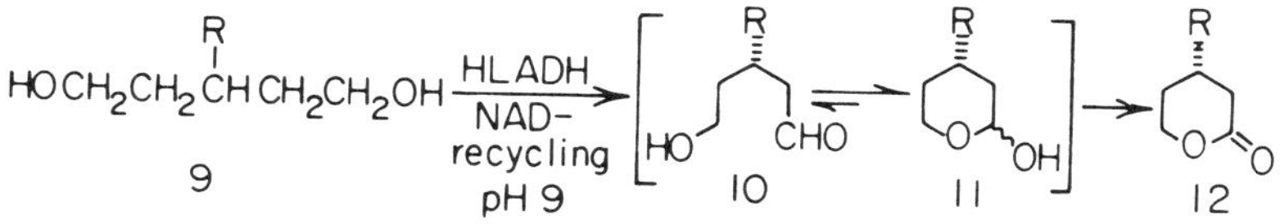

In addition, enantiotopically specific reductions of carbonyl groups can be effected. For HLADH-catalysed reductions of the decalindiones 13 and 14 are specific for the pro-R carbonyl groups, giving the hydroxydecalones 17 and 18 respectively (9,10). Even with the still more symmetrical substrates 15 and 16, the stereospecificity of reduction is retained in the formation of 19 and 20.

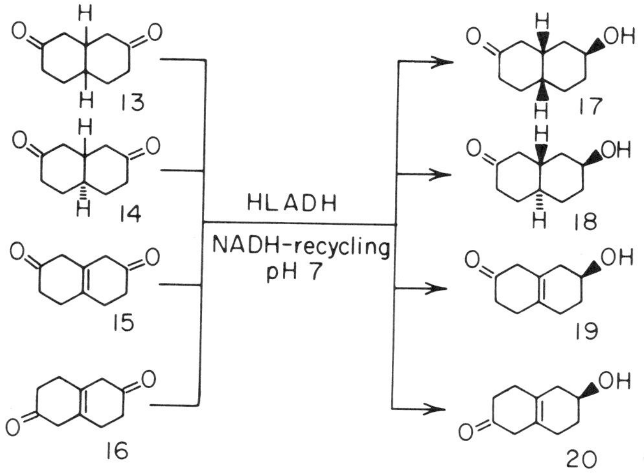

Meso-compounds are another attractive group of symmetrical substrates for which HLADH can catalyse enantiotopically group-specific transformation of asymmetric synthetic value. Meso diols are good substrates, the course of their HLADH-mediated conversions being typified by the oxidation of 21 to 24. The reaction proceeds with pro-S-hydroxymethyl stereospecificity to give the hydroxyaldehyde 22. This is in equilibrium with its hemiacetal isomer **23**, which itself undergoes enzyme-catalysed oxidation to give **24** directly.

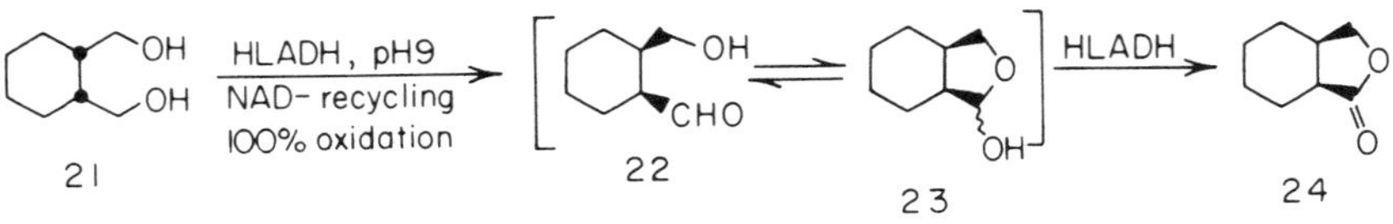

The structural range of meso diols that undergo stereospecific HLADH-mediated oxidation is very broad, as illustrated by the conversions of **25** - **34** to **35** - **44** respectively. The enzyme is remarkable in its ability to discriminate between the enantiotopic hydroxyl groups of such structurally varied acyclic, monocyclic and bridged bicyclic substrates. The oxidations all occur in the same absolute configuration sense. This involves pro-S hydroxyl specific oxidations for all the acyclic and carbocyclic diols and with pro-R enantiotopic selection for the heterocyclic substrates **31** - **33**, X = 0 (11).

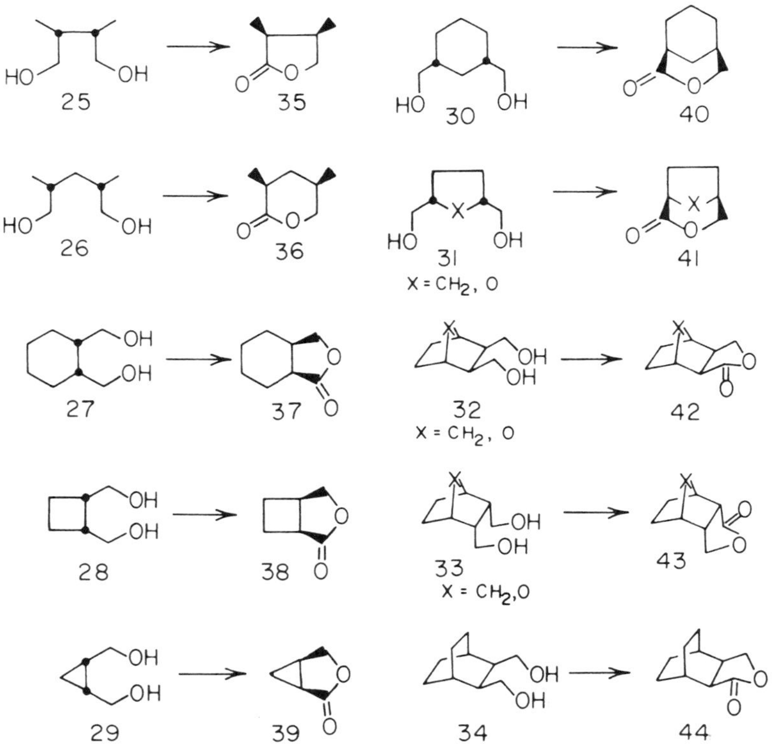

Similar control can be exerted in reductions of meso diketones. HLADH-promoted reduction of trans-decalin-2,6-dione (**45**) is specific for Re-face of the pro-R carbonyl group to give the hydroxyketone **46** (9).

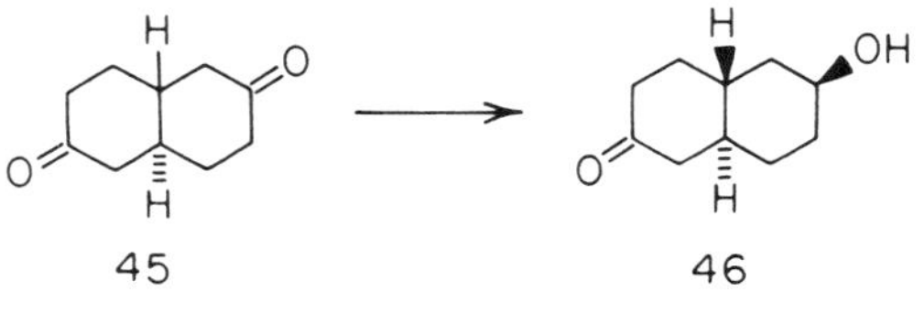

A number of the chiral compounds listed above are valuable precursors of target molecules of current interest such as grandisol (12), (+)-methyl chrysanthemate, pyrethroids, and prostaglandins (11), of macrolides (13), and of (+)-4-twistanone (9,10).

Exploiting combinations of HLADH specificity

The ability of enzymes to combine different specificities in a single step and thus achieve a degree of control presently unattainable in any other way represents a major advantage of the approach.

The reactions described above have already provided some examples of multiple specificity. In the reductions of (±)-1 and (±)-2, HLADH is enantiomerically selective for the (+)-ketone stereoisomers and also diasterotopically specific in its delivery of the hydride equivalent to the Si-face of the (+)-1 and the Re-face of the (+)-2 carbonyl groups respectively. Also, the enzyme is enantiotopically specific in the same step for the pro-R carbonyl group and the Re-face of that carbonyl group for both of 13 and 14. The enantiomeric selectivity combined with regiospecificity manifest in the conversion of 47 and 48 to 49 and 50 respectively is another useful combination. Unless protecting groups are used, discrimination between unhindered primary and secondary alcohol functions within diols such as (±)-47 and (±)-48 is difficult to achieve in single step reactions. However, since only hydroxyl groups that can locate at the oxidoreduction site of HLADH will be oxidized, primary and secondary functions are discriminated on a regional basis determined by enzyme fit. Neither enantiomer of 47 can fit into the active site in a manner that positions the primary alcohol group at the enzyme's oxidation site. Thus oxidation of the hydroxyethyl function is precluded. Furthermore, with the secondary alcohol function at the oxidation site, only the (+)-enantiomer can form a productive ES-complex (4). This combination of regiospecificity and enantiomeric specificity results in only the hydroxy ketone 49 being formed. The regional, as opposed to chemical, basis of the functional group selection is demonstrated further by HLADH-catalysed oxidation of the cyclopentenyl diol (±)-48, for which primary alcohol oxidation, also accompanied by enantiomeric selectivity, becomes favoured. The initial hydroxyaldehyde product 50 undergoes further oxidation via its hemiacetal form 51 in another enantiomerically selective process to give the Fried-Corey lactone (+)-52 of prostaglandin interest (14). The natural prostaglandin precursor (-)-52 can be obtained by chemical oxidation of (-)-51 recovered when (±)-51 is subjected to HLADH-catalysed oxidation (8).

Analogous regiospecific transformations of diketones are illustrated by the reductions of (±)-53 and (±)-54 to (+)-55 and (+)-56 respectively. For these compounds, discriminating between the unprotected carbonyl functions in a chemical reduction would be difficult. In contrast, the enzyme is completely regiospecific for the cyclohexanone function in each case. In addition, the reduction is concurrently enantiomerically specific, and enantiotopically specific for the Re-faces of the carbonyl groups reduced (15).

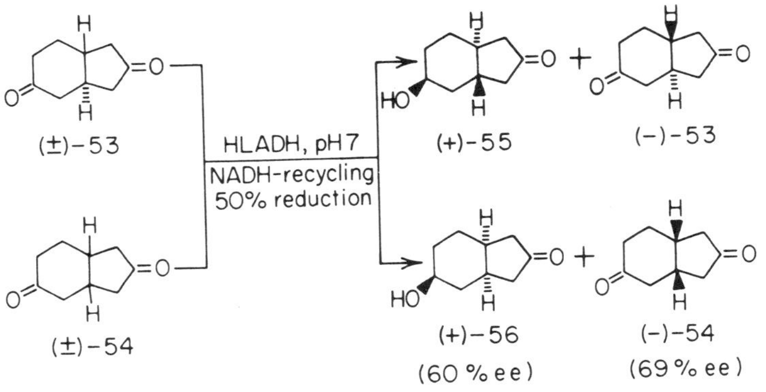

From the above results, it is clear that HLADH is a very versatile enzyme, and its potential for producing valuable chiral synthons is nowhere near exhausted. However, it is by no means the only enzyme of broad asymmetric synthetic value. Another that meets similar criteria is pig liver esterase.

Pig Liver Esterase (PLE)

PLE is a commercially available serine protease that is capable of exerting considerable stereospecific control in its catalysis of the hydrolyses of a broad range of esters. Esterases of this type possess an advantage over oxidoreductases such as HLADH in that they do not require a coenzyme.

While PLE can be used to resolve racemic esters (16), it is its ability to effect stereospecific hydrolyses of symmetrical diester substrates that presents the most exciting asymmetric synthesis opportunities (16,17). For example, hydrolyses of a broad range of C-3 substituted glutarate diesters 57 proceed with complete enantiotopic specificity at pH 7 to give 58. (However, it must be cautioned that ee levels can vary from one batch of PLE to another, and with changes in reaction conditions). These half acid-esters can then be converted to either enantiomers of 59 at will. The C-3 substituent can be varied considerably. Enantiomerically pure products are obtained when R = CH_3, CH_2CH_3, $(CH_2)_2CH_3$, $CH(CH_3)_2$, cyclohexyl, phenyl, $CH_2C_6H_5$ and $(CH_2)_2C_6H_5$. Optically pure piperidones 60 are then readily prepared from the (3R)- or (3S)-lactones.

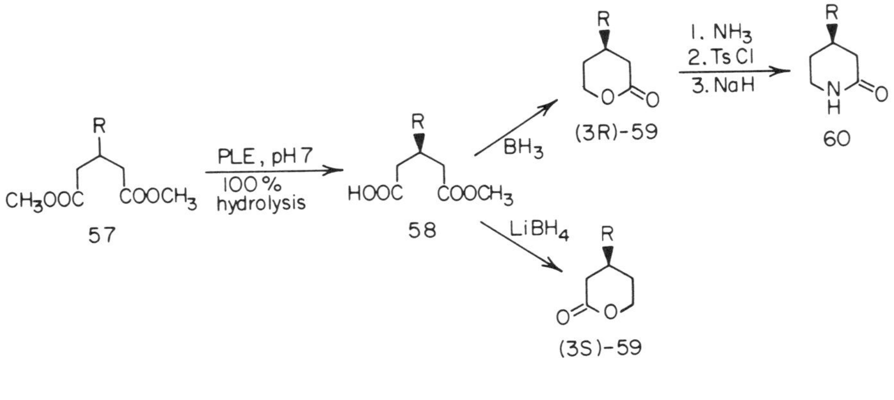

Enantiotopically selective hydrolyses can also be effected on meso diesters (18). Our results for the PLE-catalysed hydrolyses of the monocyclic 1,2-diesters **61 - 64** to the acid-esters **65 - 68**, which are readily convertible to either the (+)- or (-)-lactones **37, 38, 39,** and **69,** show an unprecedented reversal of stereospecificity. The catalyses are R-centre specific for **64,** but S-centre specific for **61** and **62,** with the cyclopentane diester **63** hydrolysis being the switch-over point.

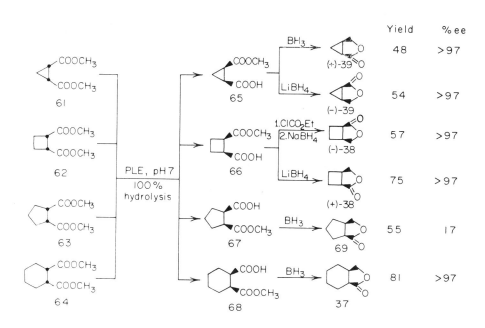

A similar reversal of stereospecificity is observed in hydrolyses of the monocyclic 1,3-diester series **70 - 72** to **73 - 75.** For the carbocyclic substrate **70** the catalysis is enantiotopically selective for the S-centre ester group, while for the heterocyclic substrates, the R-centre group is preferentially hydrolysed. Reversal of enzymic stereospecificity induced by heteroatom replacement of carbon is also, to our knowledge, unprecedented. The acid-esters **73-75** are readily converted into their respective hydroxyesters **76-78.** Lactonization of **76** and **77** proceeds smoothly to give **41,**X=CH$_2$ and O, of opposite absolute configurations to the lactones obtained earlier from HLADH-catalysed oxidations of **31,**X=CH$_2$ and O.

We believe that these reversals of stereospecificity are the consequence of the isozymic composition of PLE (19), and that the different isozymes may have different stereospecificities. We are currently investigating this aspect, which is clearly of great importance with respect to future synthetic exploitation of this enzyme. For HLADH, the fact that the commercial enzyme is also a mixture of isozymes is not serious, since the purchased enzyme does not display the aberrant behaviour of PLE.

The above results are merely illustrative of the exciting opportunities that the use of enzymes now provide, and I am confident that many new applications will soon emerge as the advantages that enzymes offer are increasingly recognized by synthetic chemists. Already, many of the new syntheses that appear combine enzymic and chemical methods, and this trend will undoubtedly strengthen in the near future. Some reviews providing broad perspectives of the synthetic opportunities provided by enzymes in general are listed in reference 20.

Acknowledgement - I am pleased to have this opportunity to thank all the past and present members of my research group for their enthusiasm and creativity. The generous support of the Natural Sciences & Engineering Research Council of Canada, the Atkinson Foundation of Toronto, and Hoffmann-La Roche Inc. (Nutley) is also gratefully acknowledged.

REFERENCES

1. B. Zagalak, P.A. Frey, G.L. Karabatsos, R.H. Abeles, J. Biol. Chem., 241, 3028-3035 (1966).
2. J.B. Jones, K.E. Taylor, Can. J. Chem., 55, 2969-2973 (1976).
3. V. Prelog, Pure Appl. Chem., 9, 119-130 (1964).
4. J.B. Jones, I.J. Jakovac, Can. J. Chem., 60, 19-28 (1982).
5. J.B. Jones, J.F. Beck, Tech. Chem. (NY), 10, 107-401 (1976)
6. A.J. Irwin, J.B. Jones, J. Am. Chem. Soc., 98, 8476-8482 (1976).
7. J. Davies, J.B. Jones, J. Am. Chem. Soc., 101, 5405-5410 (1979).
8. A.J. Irwin, J.B. Jones, J. Am. Chem. Soc., 99, 1625-1630 (1977).
9. D.R. Dodds, J.B. Jones, J. Chem. Soc., Chem. Commun., 1080-1081 (1982).
10. M. Nakazaki, H. Chikamatsu, M. Taniguchi, Chem. Lett., 1761-1764 (1982).
11. G.S.Y. Ng, L.C. Yuan, I.J. Jakovac, J.B. Jones, Tetrahedron, 40, 1235-1243 (1984); A.J. Bridges, P.S. Raman, G.S.Y. Ng, J.B. Jones, J. Am. Chem. Soc., 106, 1461-1467 (1984); C.J. Francis, J.B. Jones, Can. J. Chem., (1984) in press; I.J. Jakovac, H.B. Goodbrand, K.P. Lok, J.B. Jones, J. Am. Chem. Soc. 104, 4659-4665 (1982); I.J. Jakovac, K.P. Lok, J.B. Jones, J. Am. Chem. Soc. (1984), submitted.
12. J.B. Jones, M.A.W. Finch, I.J. Jakovac, Can. J. Chem. 60, 2007-2011 (1982).
13. C.S. Chen, Y. Fujimoto, C.J. Sih, J. Am. Chem. Soc. 103, 3580-3582 (1981); D.B. Collum, J.H. McDonald, W.C. Still, J. Am. Chem. Soc. 102, 2118-2120 (1980); J.W. Cornforth, F.P. Ross, C. Wakselman, J. Chem. Soc. Perkin I, 429-432 (1975).
14. J.J. Partridge, N.K. Chadha, M.R. Uskokovic, J. Am. Chem. Soc. 95, 7171-7172 (1973).
15. A. Krawczyk and J.B. Jones, to be published.
16. C.-S. Chen, Y. Fujimoto, and C.J. Sih, J. Am. Chem. Soc. 103, 3580-3582 (1981); C.-S. Chem, Y. Fujimoto, G. Girdaukas and C.J. Sih, J. Am. Chem. Soc. 104, 7294-7299 (1982); W.K. Wilson, S.B. Baca, Y.J. Barber, T.J. Scallen and C.J. Morrow, J. Org. Chem. 48, 3960-3966 (1983); B. Cambou and A.M. Klibanov, J. Am. Chem. Soc. 106, 2687-2692 (1984).
17. F-C. Huang, L.F.H. Lee, R.S.D. Mittal, P.R. Ravikumar, J.A. Chan, C.J. Sih, E. Caspi and C.R. Eck, J. Am. Chem. Soc. 97, 4144-4145 (1975); M. Ohno, S. Kobayashi, T. Iimori, Y.-F. Wang, and T. Izawa, J. Am. Chem. Soc. 103, 2405-2406 (1981); P. Herold, P. Mohr, and C. Tamm, Helv. Chim. Acta 66, 744-754 (1983); D.W. Brooks and J.T. Palmer, Tetrahedron Lett. 24, 3059-3062 (1983); M. Schneider, N. Engel, and H. Boensmann, Angew. Chem. Int. Ed. 23, 66 (1984).

18. C.-S. Chen, Y. Fujimoto, and C.J. Sih, J. Am. Chem. Soc. 103, 3580-3582 (1981); Y. Ito, T. Shibata, M. Arita, H. Sawai, and M. Ohno, J. Am. Chem. Soc. 103, 6739-6741 (1981); S. Iriuchijima, K. Hasegawa, and G. Tsuchihashi, Agric. Biol. Chem. 46, 1907-1910 (1982); P. Mohr, N.W. Surcevic, C. Tamm, K. Gawronska and J.K. Gawronski, Helv. Chim. Acta 66, 2501-2511 (1983); M. Schneider, N. Engel, P. Honicke, G. Heinemann and H. Gorisch, Angew. Chem. Int. Ed. 23, 67-68 (1984); H.-J. Gais and K.L. Lucas, Angew. Chem. Int. Ed. 23, 142-143 (1984).
19. E. Heymann and W. Junge, Eur. J. Biochem. 95, 509-518 (1979); D. Farb and W.P. Jencks, Arch. Biochem. Biophys. 203, 214-226 (1980), S.E. Hamilton, H.D. Campbell, J. de Jersey, B. Zerner, Biochem. Biophys. Res. Commun. 63, 1146-1150 (1975).
20. J.B. Jones and J.F. Beck, Tech. Chem. (NY) 10, 107-401 (1976); C.J. Suckling and H.C.S. Wood, Chem. Brit. 15, 243-246 (1980); J.B. Jones in Enzymic and Non-Enzymic Catalysis, P. Dunnill, A. Wiseman, and N. Blakeborough (eds.), Ellis Horwood/Wiley, Chichester/New York, 54-83 (1980), in Asymmetric Synthesis, J.D. Morrison (ed.), Academic, New York, Vol. 5, in press (1984); C.H. Wong and G.M. Whitesides, Aldrichimica Acta 16, 27-34 (1984).

Enantioselective synthesis of some carbapenem, aminosugar, and nucleoside antibiotics by enzymatic approach

Masaji Ohno

Faculty of Pharmaceutical Sciences, University of Tokyo,
Tokyo 113, Japan

Abstract- Retrosynthesis was carried out to generate symmetric diesters from the target molecules in the pro-chiral or meso form (symmetrization to simpler substrates with σ-symmetry). The symmetric diesters were subjected to asymmetric hydrolysis with pig liver esterase to create the corresponding chiral half-esters (asymmetrization to intermediates with C_1-symmetry). The chiral half-esters were converted into the target molecules by non-enzymatic procedures. Thus, various types of carbapenems, negamycin, C- and N-nucleosides, carbocyclic nucleosides, and amino-cyclitols of fortimicins were efficiently and enantioselec-tively synthesized. The enzymatic approach to the new chiral synthons developed here has opened up a new avenue for natural product synthesis.

Introduction

Asymmetric synthesis has come to the forefront of modern organic synthesis. The efficient creation of new chiral synthons with desired asymmetric centers and functional groups may be most ideal if it is carried out in a catalytic manner. Our synthetic strategy for obtaining various antibiotics has been designed by the following principle.

(1) Symmetrization: retrosyntheses of most of complex natural products can be designed in such a way to start with symmetric and simplified diesters having σ-symmetry in the prochiral or meso form as schematized in Scheme 1.

(2) Asymmetrization: the symmetric diesters are then subjected to asymmetric hydrolysis with pig liver esterase (PLE) to create the corresponding chiral half-esters. (enzymatic conversion of σ-symmetry-substrates to C_1-symmetry-intermediates)

(3) The chiral half-esters are converted to key intermediates not only for the target molecules but also for the related molecules (structure-activity-relationships) by non-enzymatic procedures including some new methodologies essential for the present approach.

Enantioselective synthesis of (+)-thienamycin, (+)-PS-5, (-)-carpetimycin A, and (-)-asparenomycin C from dimethyl β-aminoglutarate[1-5]

Naturally occurring β-lactam antibiotics which belong to the carbapenem group

Scheme 1

Retrosynthesis of Natural Products
based on Symmetrization–asymmetrization Concept

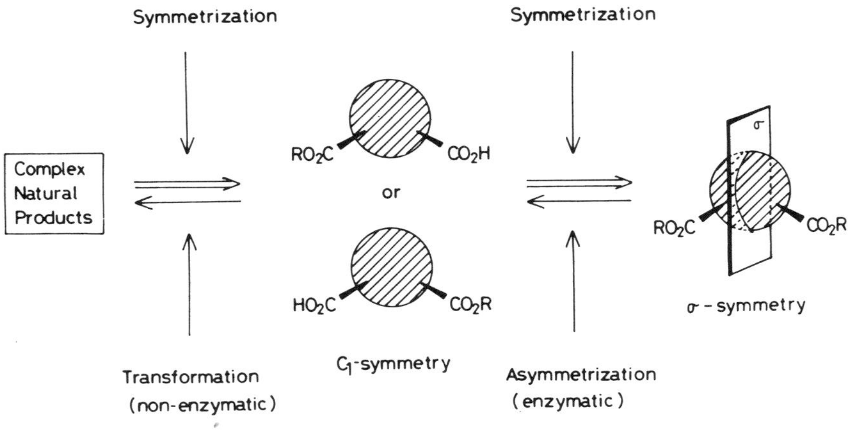

Scheme 2   Naturally Occurring Carbapenem Antibiotics

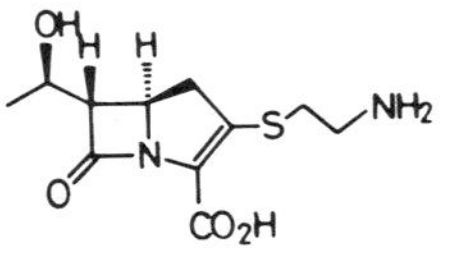

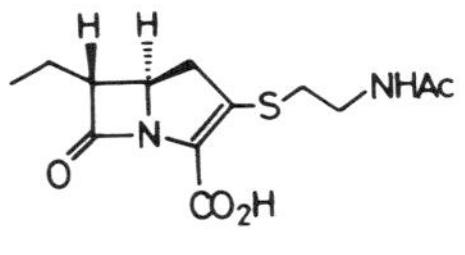

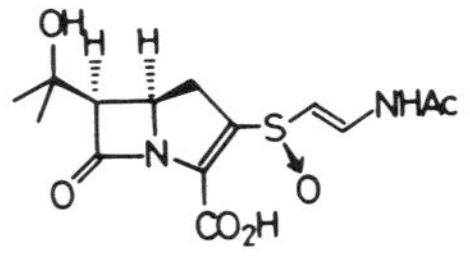

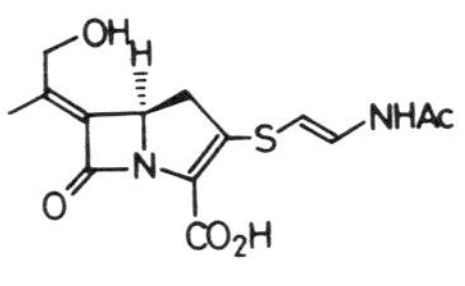

have been the subject of a great deal of synthetic study because of their unique structures and interesting biological activity[6] and can be classified into three groups (trans, cis, and ene types) according to the structural mode of the side chain of the ß-lactam ring as shown in Scheme 2. Synthetically, trans-substituted carbapenem antibiotics (1) have been most extensively studied by a number of research groups.[7] However, most syntheses produce racemic carbapenem derivatives and in the synthesis of optically active compounds multistep non-enzymatic reactions are required to obtain the azetidinone moiety starting from a chiral synthon such as L-aspartic acid.[8] We noticed that the problems of enantioselective synthesis for the carbapenem antibiotics can be extremely simplified by considering symmetric factors within the carbapenem nucleus. Three different approaches to the carbapenem antibiotics were considered according to the symmetrization-asymmetrization concept as shown in Scheme 3, 4, and 5. First of all, 4-[(Methoxycarbonyl)methyl]-2-azetidinone 3 (R=COOMe) was considered to be a key intermediate for the trans- and ene-carbapenem nuclei (1 and 9), and we thought that the chiral half-ester 4 could by easily obtained by the asymmetric hydrolysis of the prochiral symmetric diester 5 as shown in scheme 3 (1 → 2 → 3 → 4 → 5; 9 → 10 → 3 → 4 → 5). For cis-carbapenem 6, trans-δ-lactone 8 was considered to be a key compound, since the hydrolysis of 8 followed by formation of ß-lactam ring will give cis-ß-lactam 7 (6 → 7 → 8 → 4 → 5). Secondly, cis-fused ß-lactam compound 11 was considered to be another key intermediate which will be formed from a chiral half-ester 13. If the symmetric diester 14 is a good substrate for PLE, this route (Scheme 4; 1 or 6 → 11 → 12 → 13 → 14) will be also a very interesting synthesis. Thirdly, symmetrization retrosynthesis generates the meso diester 16 of dihydropyrrole ring system and the chiral half-ester 15 as shown in Scheme 5 (1 → 15 → 16): Dimethyl ß-aminoglutarate 5 was first subjected to asymmetric hydrolysis with α-chymotryspin, but the rate of hydrolysis was extremely slow. We, therefore, tried pig liver esterase (PLE) which had been used first by Sih et al to hydrolyze ß-hydroxy-ß-methyl dimethyl glutarate.[9] PLE hydrolyzed 5 very efficiently to give an optically active half-ester 4(R). The absolute configuration and optical purity of 4(R) were determined by comparison with an authentic sample after conversion to 2-azetidinone 3a with a $Ph_3P-(PyS)_2-CH_3CN$ system. The results showed that PLE cleaved the pro-S methyl ester group of 5 in a low optical yield (about 40%ee). It was also shown that substrate 5 was partly hydrolyzed under the present reaction conditions even in the absence of the enzyme. Therefore, the free amino group was protected by the bynzyloxycarbonyl group (Z) to give 15. Surprisingly, the incubation of 15 with PLE under the same reaction conditions and the hydrogenolysis of the product 16a afforded the (3S)-half-ester 4(S). After cyclization to 2-azetidinone, (S)-4-[(methoxycar-bonyl)methyl]-2-azetidinone (3) was obtained in excellent chemical (82%) and optical (93%) yields. In a typical experiment, the substrate 15 (465mg, 1.5mmol) dissolved in 1.5ml of acetone was added to 0.05M phosphate buffer solution (pH 8.0, 45ml), and pig liver esterase (0.22ml; 300units) was added to the solution at 25^OC, and the mixture was incubated for 7hr. After the reaction, the mixture was acidified with HCl and the solution was extracted

Scheme 3   Synthetic Strategy for trans-, cis-, and ene-Carbapenem
            Antibiotics based on Symmetrization-asymmetrization
            Concept (1)

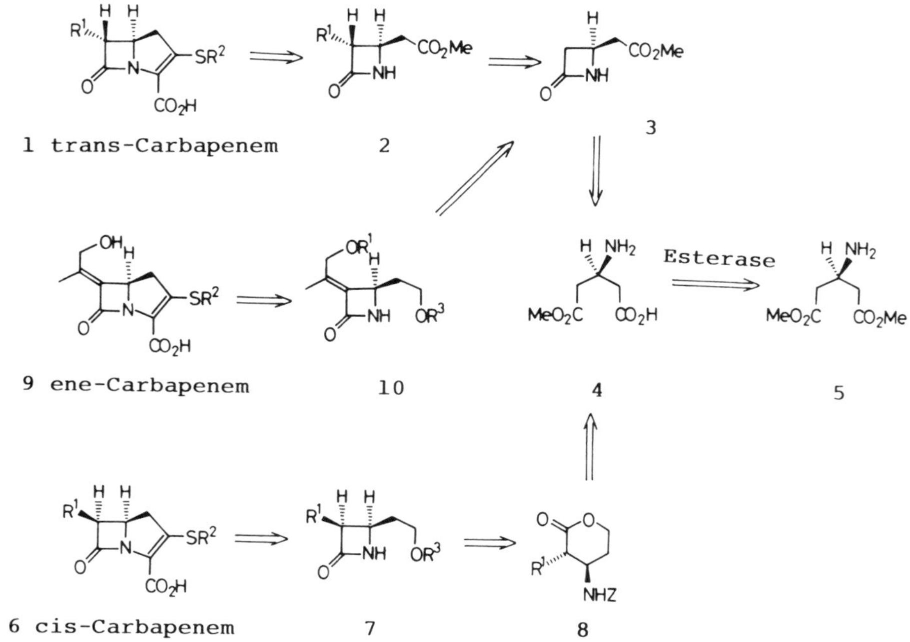

1 trans-Carbapenem   2   3

9 ene-Carbapenem   10   4   5

6 cis-Carbapenem   7   8

Scheme 4   Synthetic Strategy for trans- and cis-Carbapenem
            Antibiotics based on Symmetrization-asymmetrization
            Concept (2)

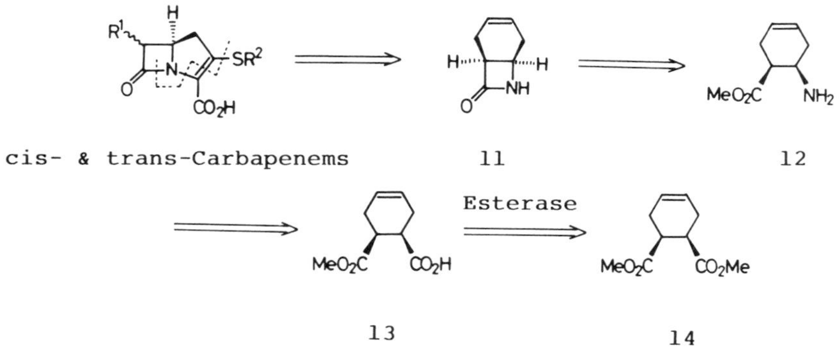

cis- & trans-Carbapenems   11   12

13   14

Scheme 5   Synthetic Strategy for Carbapenem Antibiotics based on
Symmetrization-asymmetrization Concept (3)

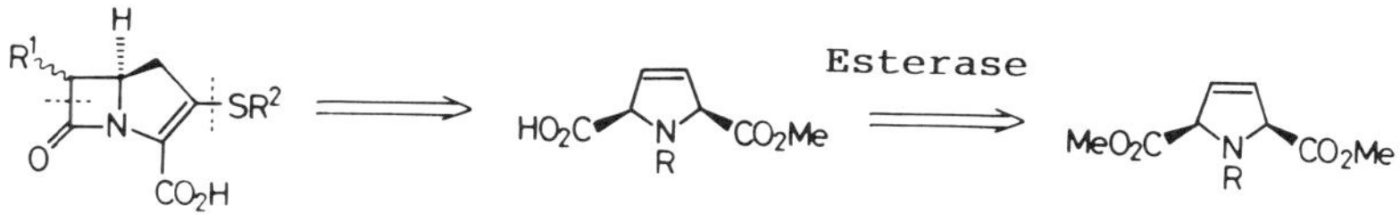

Carbapenem   15   16

Scheme 6

Asymmetric Hydrolysis of Dimethyl β-Amino and β-Z-Aminoglutarates with PLE

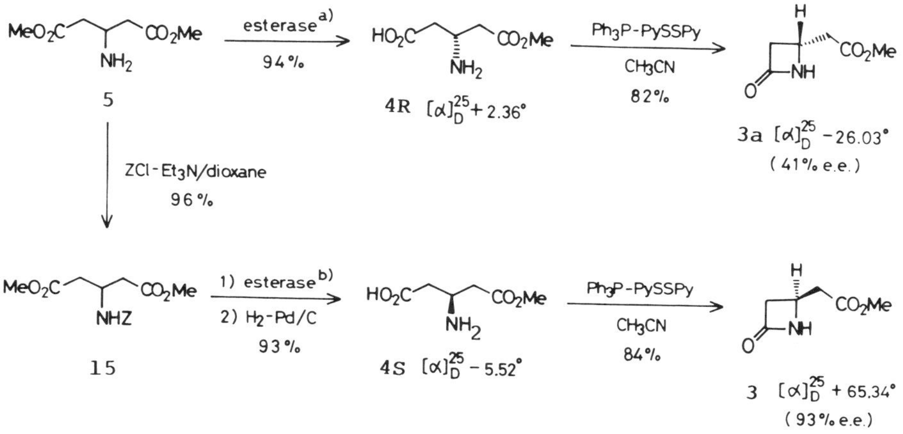

Conditions : 200 units   pig liver esterase /1mmol substrate
a) pH 8.0, 0.05M phosphate buffer , 25°C , 1.5hr
b) pH 8.0, 0.05M phosphate buffer-acetone, 25°C, 7hr

Table 1

Dependence of the Absolute Structures of
Hydrolyzed Products on the N-Protective Groups

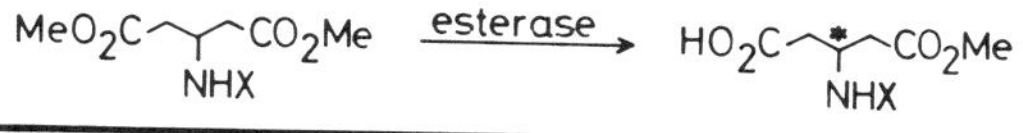

X	Chem. Yield(%)	Opt. Yield(%)	Config.
H	94	41	R
Ac	81	93	R
Bzl	94	50	S
Bz	60	72	S
Z	93	93	S
Boc	93	90	S

with CH_2Cl_2. The organic layer was dried over $NaSO_4$ and concentrated. The
residue was chromatographed on SiO_2 (ether) to afford 410mg (1.39mmol) of
the protected half-ester **16a**. Yield, 93% $[\alpha]_D^{25}+0.69^\circ$ (c 7.5, $CHCl_3$). In
the absence of the enzyme, essentially no hydrolysis of **15** took place. The
results confirm that PLE stereospecifically cleaved the pro-R methyl ester
group of **15**. It is worthy noting that the protection of the amino group at
the prochiral center in **5** with ZCl reversed the absolute configuration of
the product obtained with the same enzyme by selective hydrolysis of one
of the enantiotopic (methoxycarbonyl)methyl groups (Scheme 6). We became
interested in the dependence of the absolute configuration of the hydrolyzed
products on the protective groups. The results are shown in Table 1.
tert-Butyloxycarbonyl-, benzoyl-, and benzylamino derivatives of **5** afforded
the half-ester with the S configuration more selectively, but the acetyl
derivative preferentially afforded the half-ester with R configuration.
These findings are not only synthetically useful but also important in applying a working hypothesis of the topography of the active site of PLE. The
(3S)-half-ester **4(S)** was successfully converted to (+)-thienamycin[3] and
(+)-PS-5,[3] (-)-carpetimycin A,[4] and (-)-asparenomycin C,[5] as outlined in
Scheme 7, 8 and 9. It shold be mentioned here that the cyclization procedure
$Ph_3P-(PyS)_2-CH_3CN$ system, developed here was not only essential for our
synthetic strategy but also presents a novel and versatile method for the
formation of β-lactam compounds from β-amino acids, since known methods were
found unsatisfactory in the yields. For instance, DCC method most commonly
applied to β-lactam formation from β-amino acid gave **3a** only in poor yield
(∿20%) from **4(R)**. The second approach starting with dimethyl cis-1,2-cyclo-
hex-4-enedicarboxylate (**14**) was most beautifully realized to afford the
chiral half-ester **13** converted to the key intermediate **11** through **12** as shown
in Scheme 10. The chiral half-ester **13** will be one of the most versatile
synthons for natural product synthesis. The racemic compound **11** was already
transformed to a <u>cis</u>-carbapenem.[10] The third approach was investigated with
a saturated diester of the dihydropyrrole ring system, and the half-ester
was obtained in reasonable chemical and optical yields as shown in Scheme
11.

<u>Stereocontrolled synthesis of (+)-negamycin from a chiral half-ester **4S**</u>
As shown Scheme 12 of the synthetic strategy for negamycin, a chiral homoal-
lylamine was considered to be a key intermediate for asymmetric induction;
the chiral half-ester **16a** was chosen as the starting material, because it
is now easily available in quantity by the enzymatic hydrolysis of the
prochiral ester **5**. However, the half-ester **16a** has the S-configuration,
while the allylamine must be in the R-configuration. Such enantiomer
conversion has been proved to be another excellent characteristic of the
present approach. The conversion of **4** to the chiral allylamine was easily
carried out as shown in Scheme 13, 14, and 15, and the latter was successfully transformed to (+)-negamycin by using the iodocyclocarbamation newly
developed by us during the present study.[11,12]

Scheme 7

Enantioselective Synthesis of (+)-Thienamycin

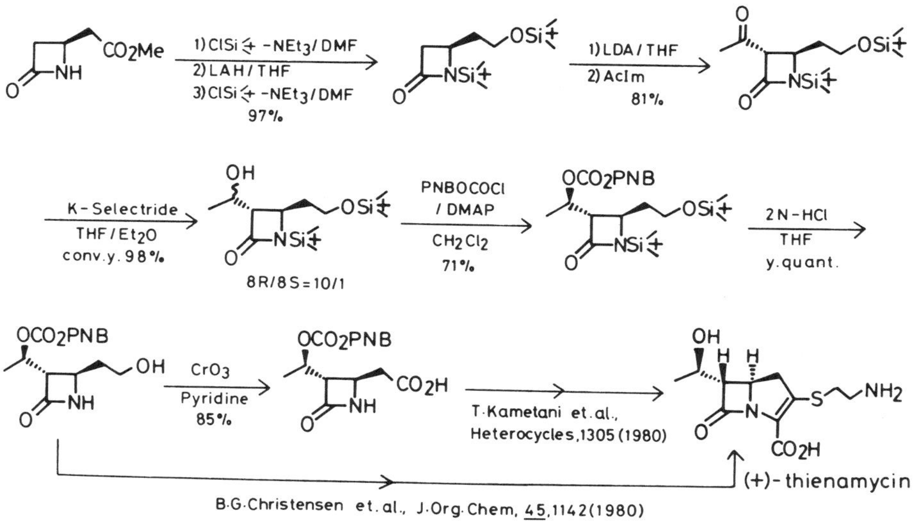

B.G.Christensen et.al., J.Org.Chem, <u>45</u>,1142(1980)

Scheme 8

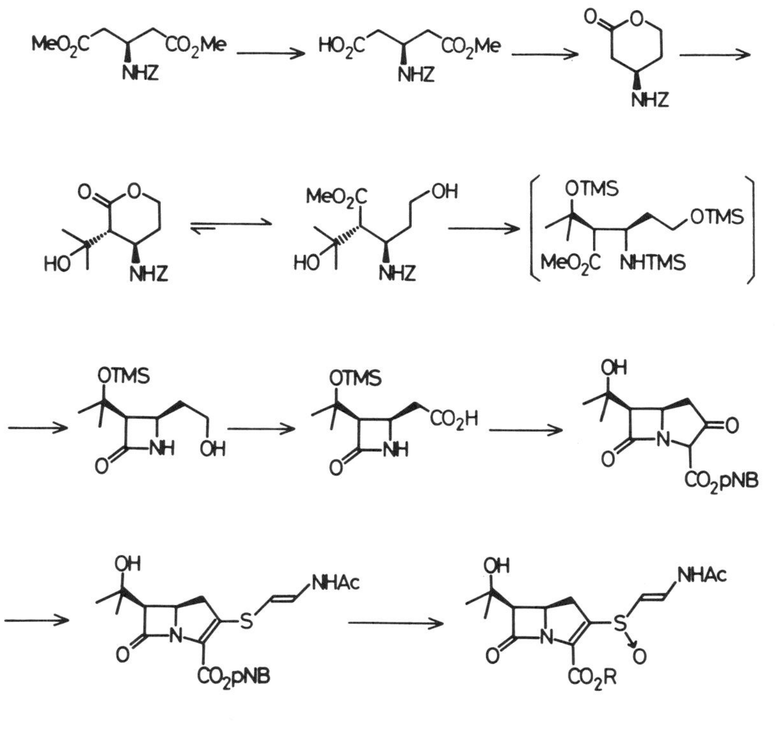

Scheme 9

Highly Stereoselective Introduction of Alkylidene Unit
into Azetidin-2-one by Peterson Olefination

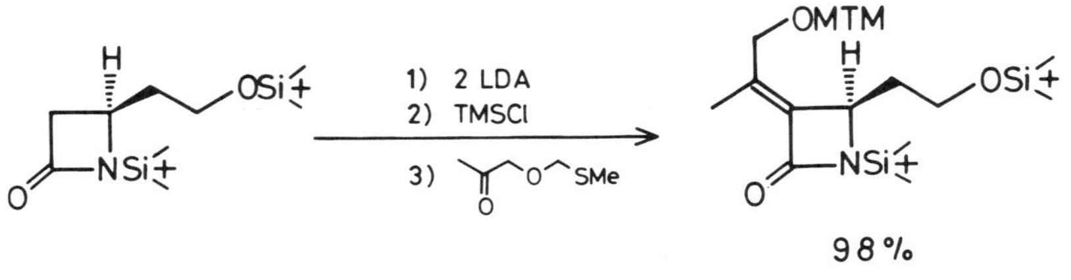

1) 2 LDA
2) TMSCl
3) (acetone with OCH$_2$SMe group)

98%

Scheme 10

14	13 98%	12 68%	11 70%
	96%ee		

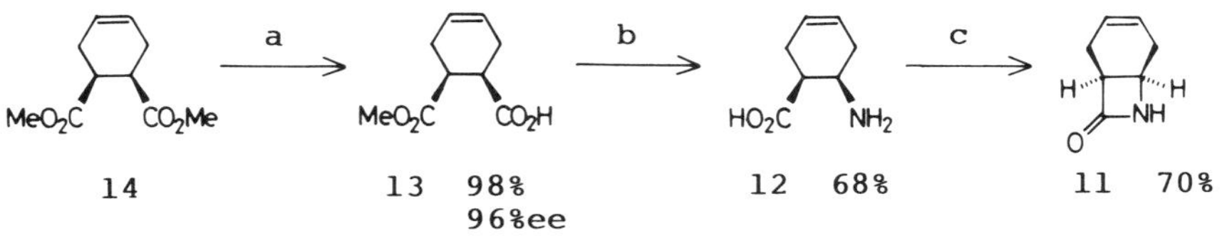

a) PLE/pH8.0 phosphate buffer -10% acetone

b) (1) ClCO$_2$Et-Et$_3$N/acetone   (2) NaN$_3$ (3) C$_6$H$_6$reflux
   (4) BzlOH-TsOH/C$_6$H$_5$reflux   (5) NaOH/acetone   (6) HBr-AcOH

c) Ph$_3$P-(PyS)$_2$-CH$_3$CN

Scheme 11

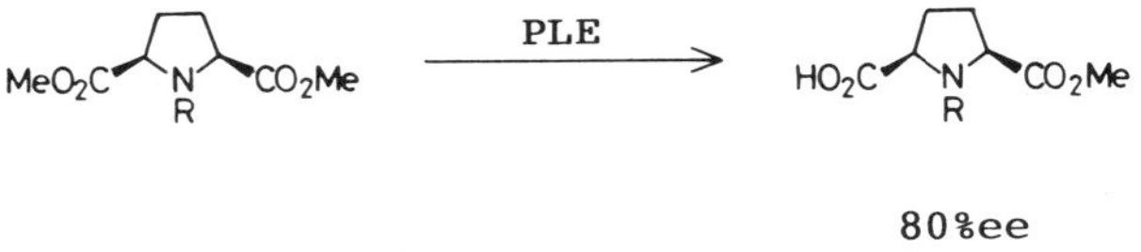

PLE

80%ee

Scheme 12

Synthetic Strategy for Negamycin

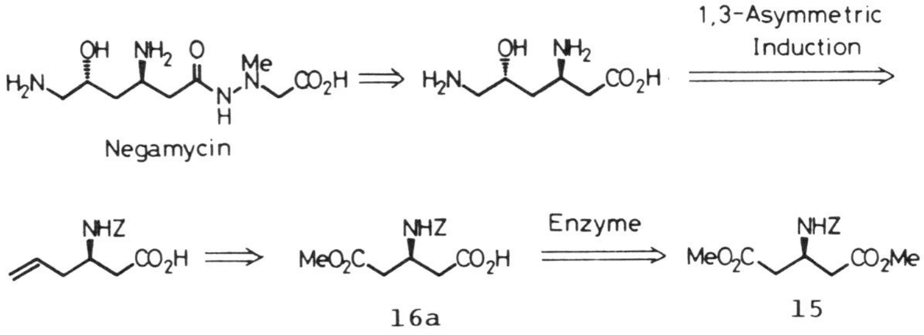

1,3-Asymmetric
Induction

Negamycin

Enzyme

16a   15

Scheme 13

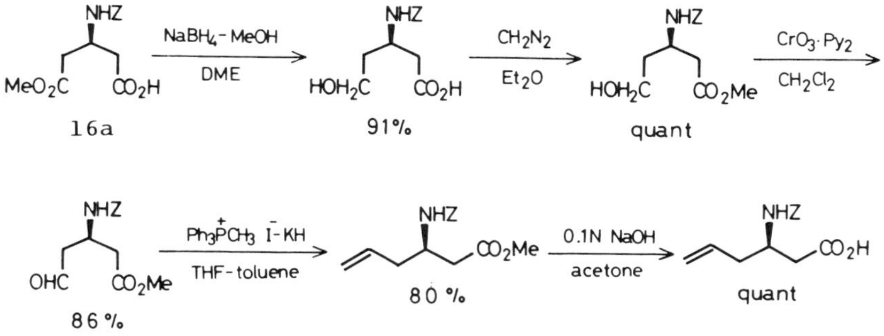

16a
91%
quant

86%
80%
quant

Scheme 14

Preparation of the Versatile Key Intermediate

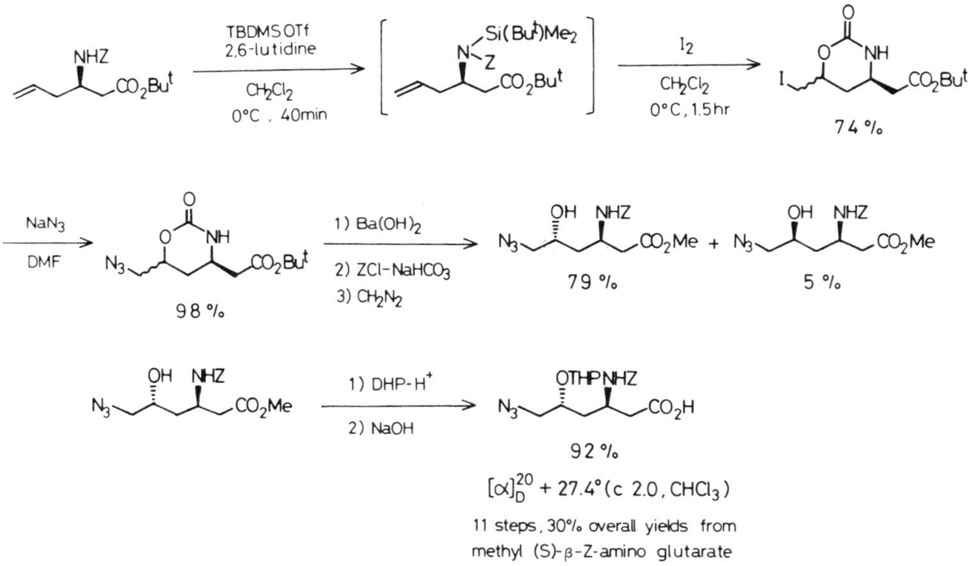

74%

98%
79% + 5%

92%

$[\alpha]_D^{20}$ + 27.4° (c 2.0, CHCl$_3$)

11 steps, 30% overall yields from
methyl (S)-β-Z-amino glutarate

Scheme 15

Synthesis of Negamycin

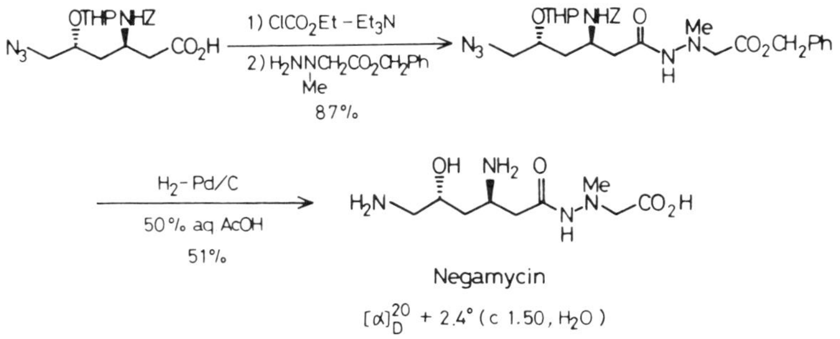

87%

50% aq AcOH
51%

Negamycin

$[\alpha]_D^{20}$ + 2.4° (c 1.50, H$_2$O)

Asymmetric hydrolysis of prochiral dimethyl Z-aminoglutarate 15 with
esterases of microbial origin

Enzymes from mammalian organs such as pig liver esterase are considered to
be expensive and impractical for the large-scale production of any chiral
compounds, and so we have looked for enzymes of microbial origin which
hydrolyze 15 more efficiently and more stereoselectively. Thus, about 500
species of microorganisms, including molds, yeasts, and bacteria were
screened: some of the results are shown in Table 2. All of the tested
strains preferentially hydrolyzed the pro-R group of 15 to yield 16 in
varying degrees of optical purity. It has been shown that Flavobacterium
lutescens IFO 3084 and IFO 3085 hydrolyzed most specifically and efficiently
the pro-R ester of 15 with about 98%ee. The bacteria can be most conveniently used in an immobilized form on carrageenan; the immobilized enzyme was
found to be very stable at room temperature for a period of more than two
years.[13]

Enantioselective synthesis of C- and N-nucleosides, showdomycin,
6-azapseudouridine and cordycepin[14,15,16]

Although the study of the asymmetric synthesis of natural products has been
considerably intensified in recent years, no successful methodology is
available in the nucleoside field. As an extension of our enzymatic approach
to natural product synthesis, sugar moieties of various nucleosides were
considered to be a good target for demonstrating our concept. Two symmetrically constituted diesters, 17a and 18, were selected as the substrates
for the asymmetric hydrolysis by esterases, although such bicyclic and rather
rigid meso compounds had not yet been subjected to an enzyme-mediated
reaction. The substrates, 17a and 18, both easily available from a Diels-
Alder reaction, were subjected to the enzyme reaction, separately (Scheme
16). A preliminary study showed that the rate of hydrolysis by PLE superior
to that by α-chymotrypsin. It was gratifying to find that the unsaturated
diester 17a was more efficiently hydrolyzed than the saturated diester 18.

 In a typical experiment, 17a (3g) in 0.1M phosphate buffer (300ml, pH 8.0)
and acetone (30ml) were incubated with PLE (4140units) at 32°C for 4hr, and
optically active half-ester 20a was thus obtained in 96% chemical yield and
about 80%ee after usual work-up. On the other hand, a symmetric epoxy
diester 19 was also considered to be a good substrate for the synthesis of
another nucleoside, cordycepin, which belongs to the family of N-nucleosides
with a 3-deoxyribose moiety. The substrate 19 was treated with PLE to yield
the expected chiral half-ester 22 in an excellent chemical yield and about
80%ee. The absolute structures and optical purity of the half-esters were
determined by conversion to the known natural products. The half-ester 20a
and 22 described above were transformed to methyl L- and D-ribosides,
(+)-showdomycin, (-)-6-azapseudouridine, and (-)-cordycepin, as summarized
in Scheme 17. It should be mentioned here that a symmetric diester 23 with
the endo configuration was completely inert to PLE and the unsaturated
diester 24 was hydrolyzed with PLE to give 25 in a low optical yield (Scheme
18). These findings also throw significant light on the topography of the

Table 2

Distribution of Esterases in Microorganisms

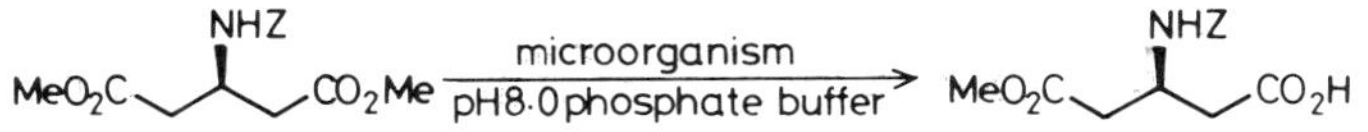
NHZ
MeO2C
CO2Me
microorganism
pH8.0 phosphate buffer
NHZ
MeO2C
CO2H

microorganism	absolute configuration	e.e. (%)
Acromobacter parvuls	S	40
Acromobacter lyticus	S	93
Chromobacterium chocolatum	S	51
Flavobacterium lutescens	S	98
Gluconobacter dioxyacetonicus	S	96
pig liver esterase	S	93

Scheme 16

Asymmetric Hydrolysis of Bicyclic diesters with Pig Liver Esterase

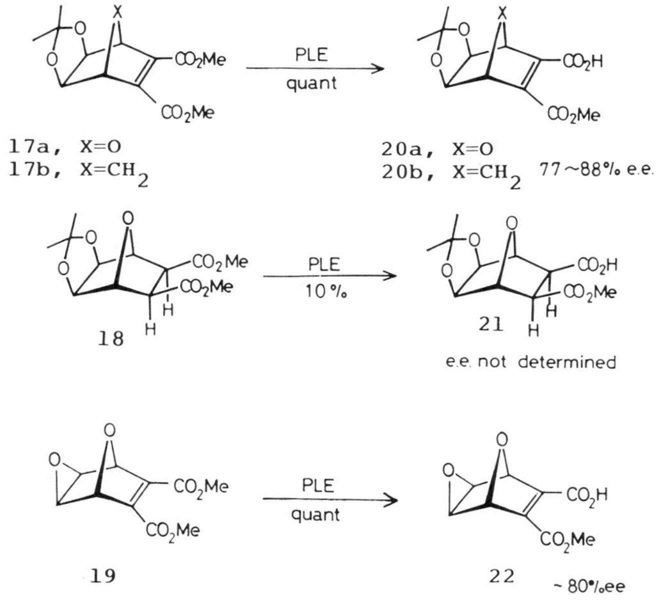
X
O
O
CO2Me
CO2Me
PLE
quant
CO2H
CO2Me
17a, X=O
17b, X=CH2
20a, X=O
20b, X=CH2 77~88% e.e.
CO2Me
CO2Me
H
H
18
PLE
10%
CO2H
CO2Me
21
e.e. not determined
19
PLE
quant
22
~80%ee

active site of PLE, as will be discussed later.

Enantioselective synthesis of carbocyclic nucleosides, (-)-aristeromycin
and (-)-neplanocin A[17]

Since the isolation of aristeromycin from a microorganism in 1968, the
interest in this class of compounds has grown rapidly, and other carbocyclic
analogues of purine and pyrimidine nucleosides have attracted a great deal
of synthetic and biological attention. Furthermore, a new antibiotic,
neplanocin A, was isolated from Actinoplanacea ampllariella in 1981, and
it has been shown that neplanocin A is a novel carbocyclic analogue of
adenosine with a cyclopentene moiety and it exhibits remarkable antitumor
activity against L1210 leukaemia in mice. All of the cyclopentane nucleosides previously synthesized were obtained in racemic forms, and the chiral
carbocyclic moiety seems to be not easily accessible by conventional
synthetic means or by the partial degradation of aristeromycin and
neplanocin. As summarized in Scheme 16, and efficient access to the chiral
cyclopentane derivatives was considered to be the enantioselective generation
of an asymmetric compound **20b** from a meso starting material **17b**. The
half-ester **20b** was obtained in optically active form (80-88%ee) and in
quantitative yield. The optically pure material was most easily and
preferably obtained by recrystallization of a δ-lactone, a key intermediate
to (-)-aristeromycin and (-)-neplanocin A (Scheme 17). By this route,
various carbocyclic nucleosides were synthesized to test the biological
activity.[18,19]

Efficient introduction of chiral centers into cyclohexane ring[20]

The stereocontrolled introduction of polyfunctional groups into the cyclohexane ring has been well developed in recent years, but the original
synthons for such a ring system with the desired absolute configuration are
not readily available from natural sources. We thought, however, that even
such a chiral synthon can be prepared by our enzymatic approach. Thus, as
shown in Scheme 19, the symmetric unsaturated diester **14** was treated with
PLE to afford the chiral half-ester **13** in 98% chemical yield and 96% optical
purity. The absolute structure of **13** was unambiguously verified by the X-ray
analysis of an iodolactone **29**. All the derivatives **26**, **27**, and **28** are very
useful chiral synthons for the natural product synthesis such as fortimicins.

Active site model of pig liver esterase

The asymmetric hydrolysis of a wide variety of symmetrically constituted
diesters with PLE showed quite hopeful and synthetically useful results for
the creation of new chiral half-esters. The structre-stereospecificity
relationships disclosed by the present study already give significant and
useful information regarding the topography of the active site model
(Y-shaped model). Typical examples of the accommodation of prochiral or
meso substrates to the active site are illustrated in Scheme 20. The working

Scheme 17

Chemicoenzymatic Approach to Nucleosides

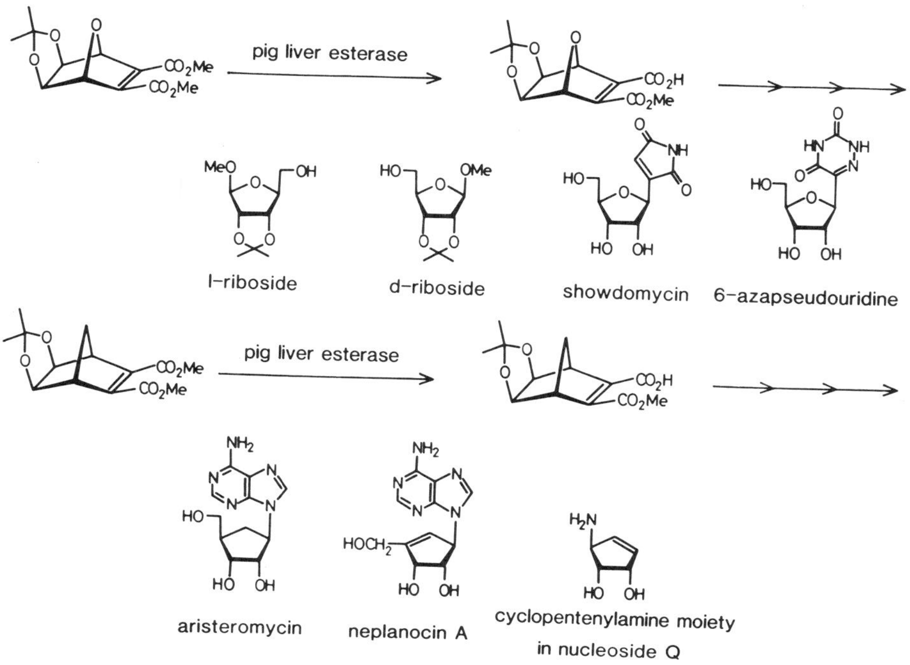
pig liver esterase
CO2Me
CO2Me
CO2H
CO2Me
l-riboside
d-riboside
showdomycin
6-azapseudouridine
pig liver esterase
aristeromycin
neplanocin A
cyclopentenylamine moiety
in nucleoside Q

Scheme 18

Asymmetric Hydrolysis of Bicyclic diesters with Pig Liver Esterase

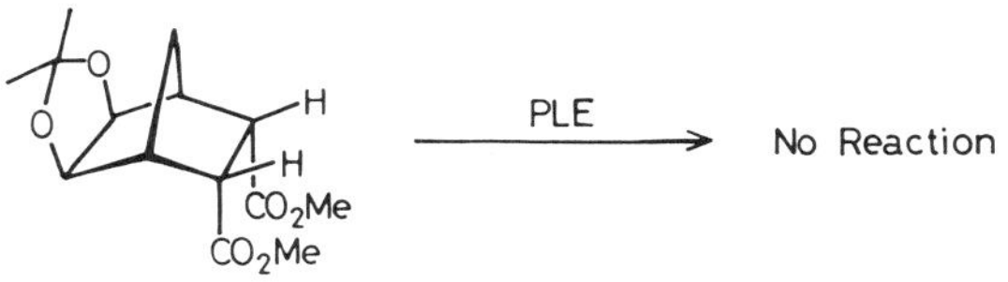
PLE
No Reaction
CO2Me
CO2Me

23

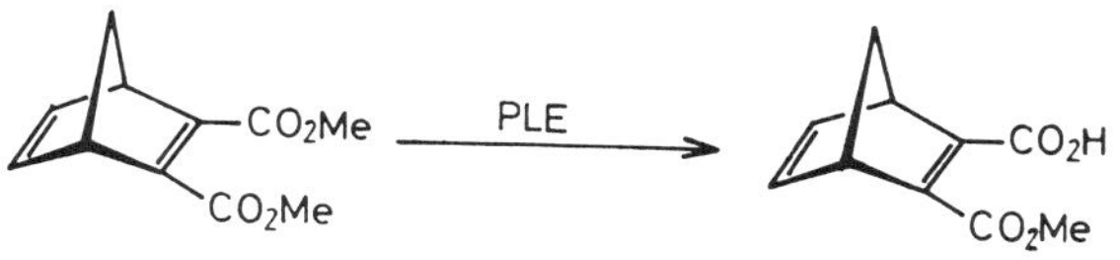
CO2Me
CO2Me
PLE
CO2H
CO2Me

24 25

41 % e.e.

Scheme 19

Absolute Structure of Half-ester Prepared by Asymmetric Hydrolysis with Pig Liver Esterase (PLE)

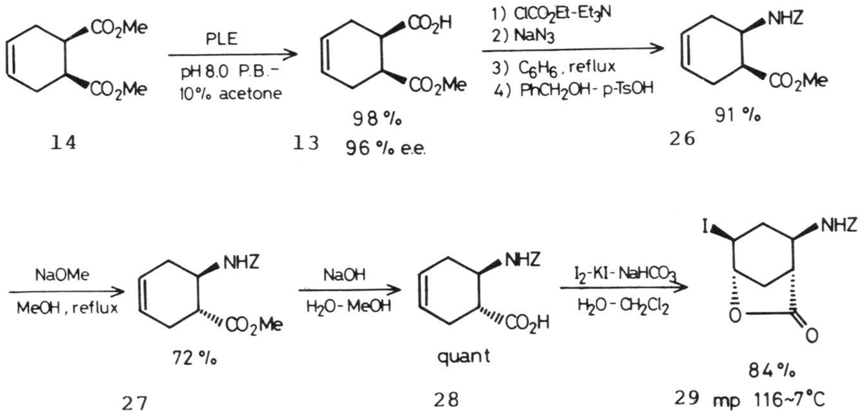

14 13 26 27 28 29

Scheme 20

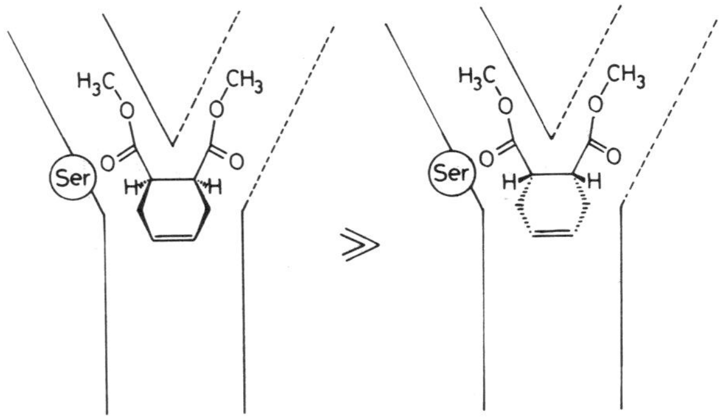

hypothesis of the model is useful not only for the explanation and prediction of the absolute configuration of the major chiral half-ester, but also for the design of good substrates.

Conclusions

The enzymatic approach to the new chiral synthons described here has opened up a new avenue to natural product synthesis. More significantly, we have developed the symmetrization-asymmetrization concept in which a wide variety of diester-substrates possessing σ-symmetry undergo efficient asymmetric hydrolysis with PLE. It is also worthy of note that the optical yields of the chiral half-esters vary in response to even small synthetic change in the substrates and even the absolute configurations of some chiral half-esters may be synthetically controlled.

Acknowledgements- I would like to express my cordial thanks to my collabrators, Associate Professor S. Kobayashi (throughout the course of this work), Drs Y. Ito, Y. Wang, T. Iimori, K. Okano, T. Izawa, M. Arita, H. Nakai, H. Sawai, and T. Murakami, and Messrs T. Shibata, K. Adachi, M. Kurihara, M. Nakada, K. Kamiyama, T. Tsuri, H. Yamashita, Y. Takahashi and T. Isobe, for their great contribution, and Mr. M. Yoshioka for his stimulating discussion.

REFERENCES AND NOTES

1. M. Ohno, S. Kobayashi, T. Iimori, Y-F. Wang, and T. Izawa, J. Am. Chem. Soc., **103**, 2405 (1981).

2. S. Kobayashi, T. Iimori, T. Izawa, and M. Ohno, J. Am. Chem. Soc., **103**, 2406 (1981).

3. K. Okano, T. Izawa, and M. Ohno, Tetrahedron Letters, **24**, 217 (1983).

4. T. Iimori, Y. Takahashi, T. Izawa, S. Kobayashi, and M. Ohno, J. Am. Chem. Soc., **105**, 1659 (1983).

5. K. Okano, Y. Kyotani, H. Ishihama, S. Kobayashi, and M. Ohno, J. Am. Chem. Soc., **105**, 7186 (1983).

6. (a) D. B. R. Johnson, S. M. Schmitt, F. A. Bouffard, and B. G. Christensen, J. Am. Chem. Soc., **100**, 313 (1978). (b) J. J. Tufariello, G. E. Lee, P. A. Senaratne, and M. Al-Nuri, Tetrahedron Letters, 4349 (1979). (c) R. J. Ponsford, and R. Southgate, J. Chem. Soc., Chem. Commun., 845 (1979). (d) T. Kametani, S. Huang, S. Yokohama, Y. Suzuki, and M. Ihara, J. Am. Chem. Soc., **102**, 2060 (1980). (e) L. Cama, and B. G. Christensen, Tetrahedron Letters, **21**, 2013 (1980).

7. An efficient synthesis of dl-thienamycin has recently been reported: D. G. Melillo, I. Shinkai, T. Liu, K. Ryan, and M. Sletzinger, Tetrahedron Letters, **21**, 2783 (1980).

8. T. N. Salzmann, T. W. Ratcliffe, and B. G. Christensen, Tetrahedron Letters, **21**, 1193 (1980).

9.   F. Huang, L. F. H. Lee, R. S. D. Mittal, P. R. Ravikumar, J. A. Chan, C. J. Sih, E. Caspi, and C. R. Eck, J. Am. Chem. Soc., **97**, 4144 (1975).

10. J. H. Bateson, R. I. Hickling, P. M. Roberts, T. C. Smale, and R. Southgate, J. Chem. Soc., Chem. Comm., 1084 (1980).

11. Y-F. Wang, T. Izawa, S. Kobayashi, and M. Ohno, J. Am. Chem. Soc., **104**, 6465 (1982).

12. M. Ohno, S. Kobayashi, T. Izawa, and Y-F. Wang, J. Chem. Soc. Jpn., Chem. & Ind. Chem., **12**, 99 (1983).

13. H. Kotani, Y. Kuze, S. Uchida, T. Miyabe, T. Iimori, K. Okano, S. Kobayashi, and M. Ohno, Agric. Biol. Chem., **47**, 1363 (1983).

14. Y. Ito, T. Shibata, M. Arita, H. Sawai, and M. Ohno, J. Am. Chem. Soc., **103**, 6739 (1981).

15. Y. Ito, M. Arita, K. Adachi, T. Shibata, H. Sawai, and M. Ohno, Nucleic Acid Research, Symposium Series, No.10, 45 (1981).

16. M. Ohno, Y. Ito, M. Arita, T. Shibata, K. Adachi, and H. Sawai, Tetrahedron Symposia in print, **40**, 145 (1984).

17. M. Arita, K. Adachi, Y. Ito, H. Sawai, and M. Ohno, J. Am. Chem. Soc., **105**, 4049 (1983).

18. M. Arita, K. Adachi, Y. Ito, H. Sawai, and M. Ohno, Nucleic Acids Research, Symposium Series, No.11, 13 (1982).

19. M. Arita, K. Adachi, H. Sawai, and M. Ohno, Nucleic Acids Research, Symposium Series, No.12, 125 (1983).

20. S. Kobayashi, K. Kamiyama, T. Iimori, and M. Ohno, Tetrahedron Letters, **25**, 2557 (1984).

Chemical and enzymatic synthesis of genes

J. ENGELS, M. LEINEWEBER AND E. UHLMANN

HOECHST AG, P.O. BOX 80 03 20, 6230 FRANKFURT (MAIN) 80,FRG

Abstract For the chemoenzymatic synthesis of genes two approaches are feasible. The first uses the 5'-phosphorylated deoxyoligonucleotides and combines them to a double strand by hydrogen-bonds which is then joined enzymatically with the help of T4-ligase. Or with longer deoxyoligonucleotides the DNA-duplex can be faster generated by partial overlap thus providing the template for DNA-polymerase to fill in completely the missing nucleotides. Both approaches are demonstrated in the synthesis of the γ-interferon gene. This gene incorporated into E.coli produces biologically active γ-interferon, the polypeptide. Not content with synthesizing a gene we have also put it under the control of a synthetic gene control region in order to optimize its yield. We built in cutting sites for restriction enzymes to allow specific modifications of certain regions within the gene in order to test modified interferons.

> "We shall assume the structure of a *gene* to be that of a huge molecule, capable only of discontinuous change, which consists in a rearrangement of the atoms and leads to an isomeric molecule. The rearrangement may affect only a small region of the *gene* , and a vast number of different rearrangements may be possible."
>
> Erwin Schrödinger, 1944

INTRODUCTION

The tool of recombinant DNA gave a major impetus to chemical DNA synthesis of defined sequence (1). In the course of this development the once underestimated DNA synthesis has opened up unsuspected vistas of total chemoenzymatic gene syntheses.Although the concept of the gene has been the focus of some hundred years of work there seems to be no easy definition to it. As for the practical aspect of this presentation we will concentrate on the classical: One Gene - One Polypeptide description in the reverse order One Polypeptide - One Gene. The major emphasis thus lies in the point that we have an alternative route to enter the world of proteins. We can now make proteins from a master copy like a printer from a press. As organic chemists we sometimes tend to overlook the fact that the number of different polypeptides in nature overrides the number of known organic compounds probably at least by a factor of hundred.

OLIGODEOXYNUCLEOTIDE SYNTHESIS

The secret behind the enormous progress in DNA-synthesis is based on two major developments. First the polymer supported synthesis (2) and secondly the high yield high efficiency of each individual coupling step (3, 4). We start from the suitably protected nucleosides (1a - d) which react with succinic anhydride to the esters (2a - d). These succinic esters can directly be attached to the functionalized polymer support.

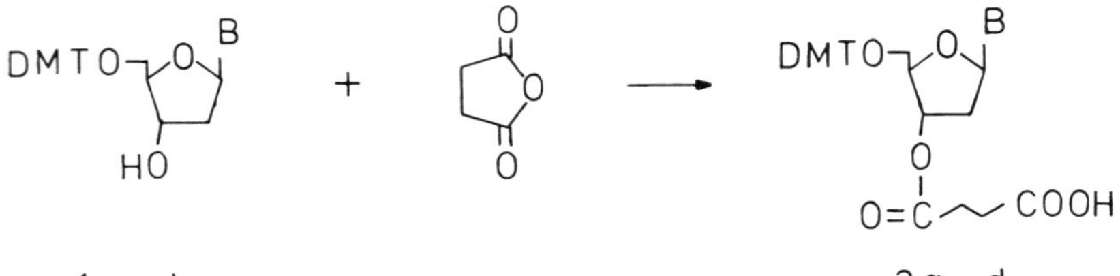

1a – d   2a – d

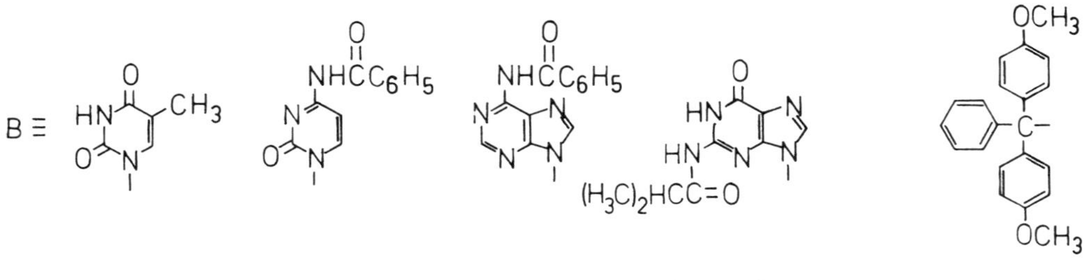

B ≡   a : T   b : bzC   c : bzA   d : ibuG   DMT

As polymer support we initially utilized silica gel but have now even greater success with controlled pore glass beads (CPG) (5). These long chain alkylamine glass beads (CPG, LCAA, Pierce) (6) (3) are coupled with the succinic esters (2a - d) by DCC.

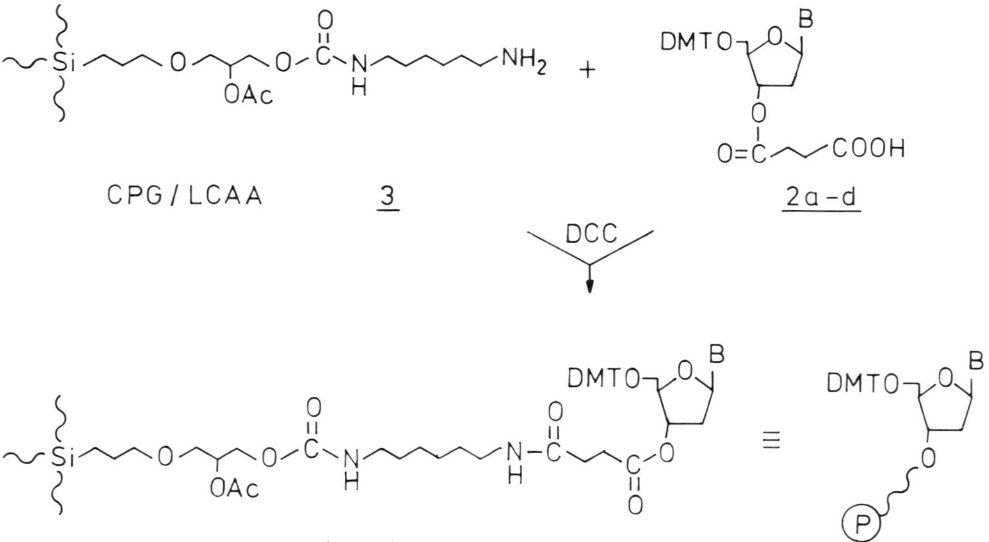

CPG / LCAA   3   2a–d

DCC

4a – d

The synthesis cycle now starts with the deprotection of the 5'-dimethoxy-trityl group with the aid of trichloroacetic acid in methylene chloride (7).

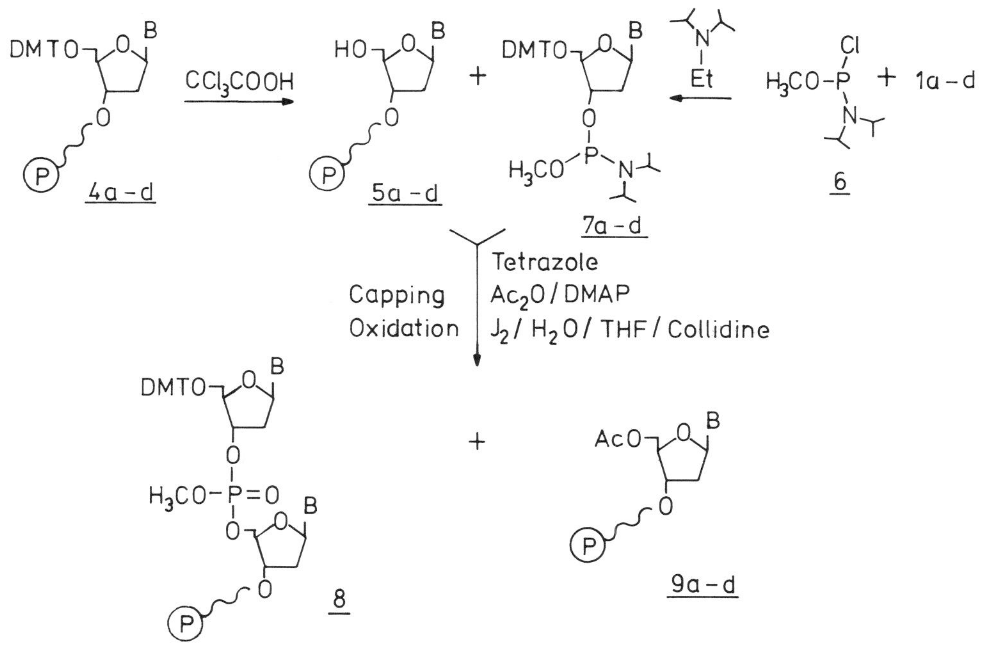

The monomeric building blocks (7a - d) are activated with tetrazole (8) - a mild acid - to furnish as intermediates the phosphite triesters. These can in turn be oxidized with iodine or t-butyl hydroperoxide (9) to the phosphate triesters (8), which are stable products under the reaction conditions. In order to avoid the problem of wrong addition in the next step a capping reaction is included into the cycle. Acetylation with acetic anhydride in the presence of dimethylaminopyridine results in (9a - d). The consecutive addition of the appropriate nucleoside phosphoramidites (7a - d) gives us the deoxyoligonucleotides of defined sequence.

Table 1: Cycle for deoxyoligonucleotide synthesis

Function	Reagent, Solvent	Vol.		Reaction time
1. Detritylation	3 % trichloroacetic acid in dichloroethane	2	ml	1- 2 min
2. wash	dichloroethane	5	ml	
3. wash	methanol	5	ml	
4. wash	dry acetonitrile	5	ml	
5. phosphite addition	0.4 M tetrazole, 0.1 M amidite in dry acetonitrile	0.5 ml		5-10 min
6. wash	acetonitrile			
7. capping	3(acetic anhydride, colli-dine, tetrahydrofurane 1:2:3 (v,v,v) : 1(10 % 4-dimethylamino-pyridine in tetrahydrofu-rane (w/v)	2	ml	2 min
8. wash	tetrahydrofurane			
9. hydrolysis	tetrahydrofurane, water; collidine (2:2:1, v/v/v)	2	ml	1 min

Function	Reagent, Solvent	Vol.	Reaction time
10. oxidation	3 % iodine in tetrahydro-furane, water, collidine 2:2:1, w/v/v/v.	2 ml	2 min

The deprotection consists of a demethylation with e.g. thiophenolate (10), followed by cleavage of the ester and amide bonds with aqueous ammonia. The final crude product is purified by either polyacrylamide gel electrophoresis or by HPLC (Fig. 1) (11).

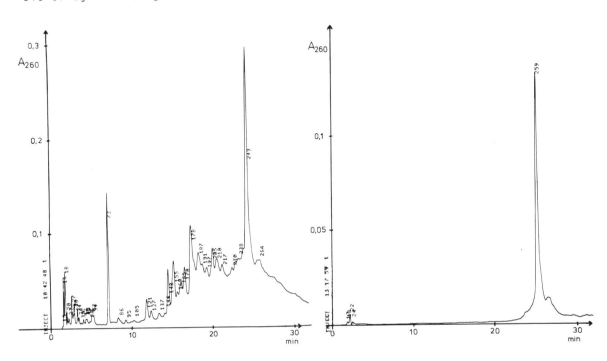

Fig. 1: Reversed phase HPLC (Zorbax C8) of fragment IIIc crude (left) and purified (right) on 15 % polyacrylamide gel. Linear gradient of 0.1 M triethylammoniumacetate and 0.1 M triethylammoniumacetate in 50 % acetonitrile.

Similar results have been obtained with ion exchange HPLC by using Partisil SAX columns and triethylammoniumphosphate-acetonitrile as a buffer system for a gradient elution. The resolution for longer deoxyoligonucleotides - above 30-mers - is somewhat limited within this system. Purification by gel electrophoresis gives always good results. In order to get a sharp peak on reversed phase HPLC one has to rechromatograph the electrophoretically purified nucleotide. For the routine gene synthesis we did not find this to be necessary.

The improvements in the chemical yield of each reaction cycle define the limits of the stepwise oligonucleotide synthesis (Fig. 2). Oligomers in the range of 100-mers are well within the reach of our possibilities now, based on the 98 % repetitive yields obtained.

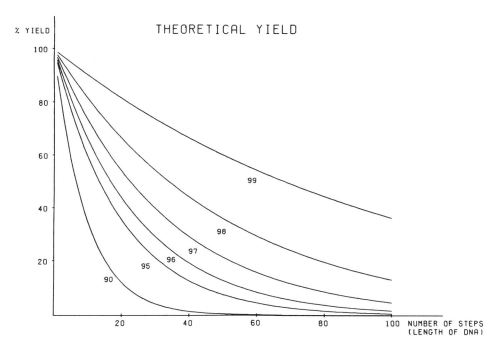

Fig. 2: Graphical representation of the expected length of oligodeoxynucleo-
tides by stepwise addition based on the yields.

After isolating the deoxyoligonucleotides, desalting on a sephadex G 50
column, the monomers are ready for enzymatic reactions.

ENZYMATIC GENE SYNTHESIS

Gene synthesis consists of the correct assembly of DNA-oligomers to the
DNA-duplex. This can be done by two ways. The now classical Khorana approach (12, 13)) uses the enzymes polynucleotide kinase and ligase from
bacteriophage T4. Kinase specifically phosphorylates the deoxyoligonucleotide in the 5'-position on the primary hydroxy function. The phosphorylated
complementary strands are then hybridized through hydrogen bonds to each
other in the aqueous buffer and ligase can seal the nicks. γ-interferon
(14) a polypeptide of 146 amino acids, whose gene we synthesized, will
exemplify the strategy. The gene with the proper start and stop signals as
well as the handles to incorporate it into the plasmid consists of a 450
base pair duplex (Fig. 3).

We constructed the total gene from 34 deoxyoligonucleotides ranging in length from 18 - 33-nucleotides. Based on the reported peptide sequence (15,
16) we arrived at a DNA-sequence that favors the codons occurring predominantly in highly expressed genes in E. coli. The degeneracy of the genetic
code helps to incorporate restriction enzyme cuts at will. This gene
construction thus takes into consideration an easy access to several
portions of the gene as to have an entry later for specific modifications.
Furthermore the total gene was divided into three portions which can be
cloned separately as compared to other recent approaches (17, 18).

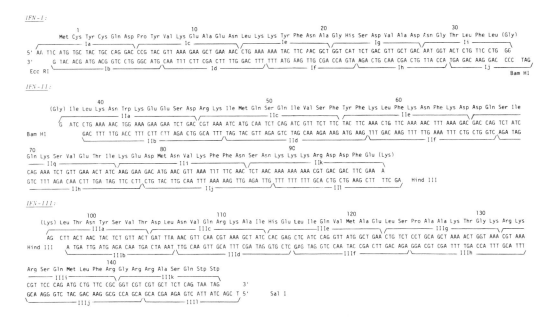

Fig. 3: Nucleotide and amino acid sequence of the synthetic *IFN-γ-gene*. The individual oligonucleotides that were synthesized and ligated to form the duplex are numbered and underlined. The construction of the complete DNA-duplex was carried out in three blocks *IFN-γ I - III* .

This subcloning scheme (Fig. 4) offers a relay or deposit system for the synthesis of large proteins. Furthermore by DNA sequencing techniques it furnishes us with a precise and effective control of what has been synthesized on our long march to a gene.

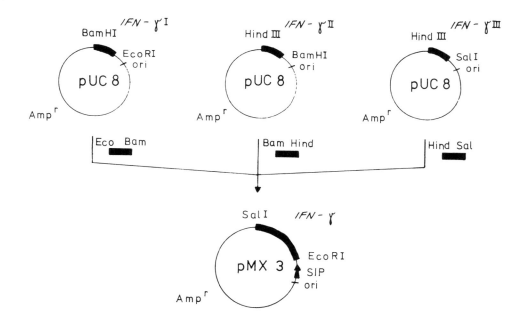

Fig. 4: Construction of the *IFN-γ* recombinant plasmid from the three pUC8 derived plasmids.

The three fragments are defined by the restriction-endonuclease cutting sites Eco RI, Bam HI, Hind III and Sal I. By this subcloning principle we first ligated the fragments *IFN-γI-III* and incorporated them in the plasmids. In E. coli we obtained sizeable quantities of the fragments from restriction enzyme digest (19). The pieces I - III were joined with ligase purified by gel electrophoresis and cloned in E. coli. The DNA-sequence of the resulting *IFN-γ-gene* was established by DNA-sequencing (20).

An alternative approach to generate the DNA-duplex uses the powerful potential of the DNA-polymerase Klenow-fragment (Fig. 5) to synthesize the missing deoxyoligonucleotides (21). The Polymerase is primer dependent and synthesizes in the 5'- 3' direction.

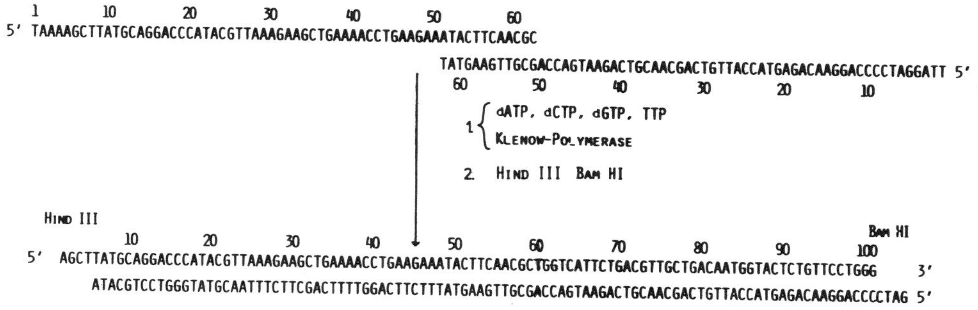

Fig. 5: Construction of a *IFN-γ-I* duplex DNA from two synthetic deoxyoligonucleotides by DNA-polymerase, Klenow-fragment.

The two strands are overlapping in the middle by 10 - 15 base pairs. DNA polymerase in conjunction with the 4 deoxyoligonucleotide triphosphates creates the DNA-duplex. The two 62-mers resulted in the same 100-mer duplex which needed 10-deoxyoligonucleotides by the previous approach. The economy in chemical synthesis using this approach is substantial with an almost 40 % reduction in the amount of chemical synthesis as compared with the before mentioned approach. In order to be sure of the correct fill-in, the DNA has to be sequenced before or after cloning.

SYNTHESIS OF A GENE CONTROL REGION

Within the steps of protein biosynthesis, the unit responsible for the transcription, the RNA-polymerase recognizes a certain region within the DNA the so called promoter (22). In E. coli these sequences (23) are well known. Based on the consensus sequence in certain regions we designed a so called synthetically idealized promoter (SIP))(Fig. 6).

Bam HI　　　　　　　　　　Aha III

GGA TCC TAA ATA AAT TCT TGA CAT TTT TTA AAT AAT TTG GTA TAA TGT GTG

CCT AGG ATT TAT TTA AGA ACT GTA AAA AAT TTA TTA AAC CAT ATT ACA CAC

-35　　　　　　　　-10

Promoter

Aha III　　　Xba I Eco RI

GAA TTG TGA GCG GA TAA CAA TTT CAC TTT AAA GAG GAT CTA GAA TTC ATG

CTT AAC ACT CGC CT ATT GTT AAA GTG AAA TTT CTC CTA GAT CTT AAG TAC

Lac Operator　　　　SD　　　START

Fig. 6: Nucleotide sequence of the synthetic control region (SIP) consisting of the promoter, operator and Shine-Dalgarno (SD) sequence (24).

This unit consits of the promoter, the operator which controls the promoter and a ribosome binding site which regulates translation, the second step in protein biosynthesis. The numbering system begins at the start of transcription. Based on the lactose-system we improved the -35 and -10 region to make the promoter stronger. The operator can be mutated down to render the promoter constitutive as well as by utilizing the essential symmetry axis (26, 25)make it stronger binding to the repressor. We coupled the SIP unit to the γ-IFN-gene in a modified pUC plasmid (26), transformed E. coli with it and expressed the γ-Interferon peptide after induction with IPTG. Purification via precipitation, chromatography and finally immunoaffinity separation (27) gave us the pure γ-Interferon (Fig. 7).

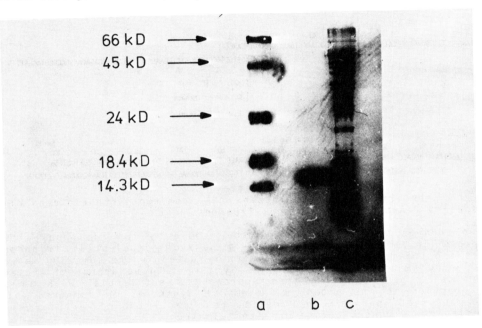

Fig. 7: SDS-polyacrylamide gel electrophoresis of bacterial cells harbouring the IFN-γ plasmid (c), the purified IFN-γ (b) and the protein marker (a).

Now after reaching the first rung of the ladder we of course want to climb higher and want, as medicinal chemist, to look to modify the natural product after having mastered the mechanism of action. The question arises what is the essential portion of the molecule. We therefore clipped the molecule on the amino- and carboxyterminus. This can easily be accomplished by using the build-in restriction enzyme recognition sites (Fig. 8). By this we digested our DNA with Eco RI and Ava II, synthesized the appropriate linker to fill in the missing nucleotides and reinserted the new gene fragment into the host vector.

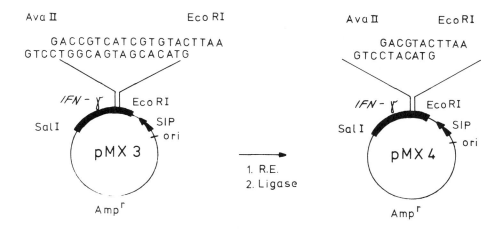

Fig. 8: Schematic representation of the plasmids for the construction of an aminoterminal shortened *IFN-γ-I* segment, using chemically synthesized deoxyoligonucleotides.

The expression of this gene in the above manner resulted in a 3 amino acids shorter γ-interferon peptide (28). The same procedure was adopted for the carboxy terminus. This site specific mutagenesis can create a vast amount of specific base exchanges, deletions or even additions of the underlying polypeptide. We were able to show in biological testing that the amino terminal shortening did not reduce the potency of the γ-interferon, whereas a progressive shortening of the carboxy terminus did. By this approach we hope to get some insight into the locus of activity of γ-interferon.

Acknowledgement: The authors wish to thank Dr. Ulmer for the protein purification, Dr. Rolly for the biological testing and Ms. Kurt and Ms. Löhnert for preparing the manuscript.

REFERENCES

1. S.A. Narang, Tetrahedron 39, 3 - 22 (1983)
2. M.D. Matteucci and M.H. Caruthers, J. Am. Chem. Soc. 103, 3185 - 3191 (1981)
3. S.L. Beaucage and M.H. Caruthers, Tetrahedron Lett. 1859 - 1862 (1982)
4. M.H. Caruthers, S.L. Beaucage, K. Becker, E. Efcavitch, E. F. Fischer, G. Galluppi, R. Goldman, P. de Haseth, F. Martin, M. Matteucci and Y. Stabinsky, Genetic Engineering, Vol. 4, 1 - 17, Plenum Publishing Comp. (1982)
5. G.R. Gough, M.J. Brunden and P.T. Gilham, Tetrahedron Lett. 4177 - 4180 (1981)
6. S.P. Adams, K.S. Kavka, E.J. Wykes, S.B. Holder and G.R. Galluppi, J. Am. Chem. Soc. 105, 661 - 663 (1983)
7. M.J. Gait, S.G. Popov, M. Singh and R.C. Titmas, Nucl. Acids Res. Symp. 7, 243 - 246 (1980)
8. J. L. Fourrey and D. J. Shire, Tetrahedron Lett. 729 - 732 (1981)
9. J. Engels and A. Jäger, Angew. Chem. Int. Ed. Engl. 21, 912 (1982)
10. G.W. Daub and E.E. van Tamelen, J. Am. Chem. Soc. 99, 3526 - (1977)
11. H.G. Gassen and E. Lang, Chemical and Enzymatic Synthesis of Gene Fragments, Verlag Chemie, Weinheim (1982)
12. H.G. Khorana, Science 203, 614 - 625 (1979); H.G. Khorana, K.L. Agarwal, H. Büchi, M.H. Caruthers, N.K. Gupta, K. Kleppe, A. Kumar, E. Oktsuka, U.L. Rajbhandary, J.H. van de Sande, V. Sgaramella, T. Terao, A. Weber and T. Yamada, J. Mol. Biol. 72, 209 - 217 (1972)
13. M.D. Edge, A.R. Greene, G.R. Heathcliffe, P.A. Meacock, W. Schuch, D.B. Scanlon, T.C. Atkinson, C.R. Newton and A. Markham, Nature 292, 756 - 762 (1981)
14. W.E. Stewart, The Interferon System, Springer, New York, (1979)

15. P.W. Gray, D.W. Leung, D. Pennica, E. Yelverton, R. Najarion, C. Simonsen, R. Derynck, P.J. Sherwood, D.M. Wallace, S.L. Berger, A.D. Levinson and D.V. Goedell, Nature 295, 503 - 508 (1982)
16. R. Devos, H. Cheroutre, Y. Taya, W. Degrave, H. Van Heuverswyn and W. Fiers, Nucl. Acids Res. 10, 2487 - 2501 (1982)
17. S. Tanaka, T. Oshima, K. Ohsuye, T. Ono, A. Mizono, A. Ueno, H. Nakazato, M. Tsujmoto, N. Higashi and T. Noguchi, Nucl. Acids Res. 11, 1707 - 1723 (1983)
18. E. Jay, J. Rommens, L. Pommeroy-Cloney, D. Mac Knight, C. Lutze-Wallace, P. Wishart, D. Harrison, W.Y. Lui, V. Asundi, M. Davood and F. Jay, Proc. Natl. Acad. Sci. USA, 81, 2290 - 2294 (1984)
19. T. Maniatis, E.F. Fritsch and J. Sambrook, Molecular Cloning, Cold Spring Harbor (1982)
20. W. Gilbert and A. Maxam, Proc. Natl. Acad. Sci. USA, 74, 560 - (1977)
21. J.J. Rossi, R. Kierzek, T. Huang, P.A. Walker and K. Itakura, J. Biol. Chem. 257, 9226 - 9229 (1982)
22. J.H. Miller and W.S. Reznikoff, The Operon, Cold Spring Harbor (1980)
23. D.K. Hawley and W.R. Mc Clure, Nucl. Acids. Res. 11, 2237 - 2255 (1983)
24. J. Shine and L. Dalgarno, Proc. Natl. Acad. Sci. USA, 71, 1342 - 1346 (1974)
25. J.R. Sadler, H. Sasmor and J.L. Betz, Proc. Natl. Acad. Sci. USA, 80, 6785 - 6789 (1983)
26. A. Simons, D. Tils, B. von Wilcken-Bergmann and B. Müller-Hill, Proc. Natl. Acad. Sci. USA 81, 1624 - 1628 (1984)
27. Celltech, U.K., interferon- -sepharose
28. E. Rinderknecht, E.H. O'Connor and H. Rodriguez, J. Biol. Chem. 259, 6790 - 6797 (1984)

Biohydroxylation of non-activated carbon atoms by the fungus *Beauveria sulfurescens*

by R. FURSTOSS

and A. ARCHELAS, J.D. FOURNERON and B. VIGNE

Laboratoire de Chimie Organique et Bioorganique, Faculté des Sciences de Luminy, Case 901, 70 Route Léon Lachamp, 13288 MARSEILLE CEDEX 9, FRANCE.

Abstract - The biohydroxylation of a number of bridged bicyclic or tricyclic amide-type molecules by the fungus <u>Beauveria</u> <u>sulfurescens</u> (ATCC 7159) has been studied. These hydroxylations lead to regio, stereo and sometimes enantioselective functionalization of non activated carbons. The results have been analyzed in order to get some informations about the topology of the enzyme active site. A model has been proposed which allows to enhance the predictability of these processes.

INTRODUCTION

Although numerous attempts have been made in order to achieve specific functionalization of non activated carbon atoms by using various chemical methods or catalysts, this type of reaction still remains an important challenge in organic chemistry. Because of the ease and the high specificities shown by microbial enzymes which are able to perform such transformations, biohydroxylations constitue a highly powerfull tool for the organic chemist.

Owing to their great commercial importance, steroids have been the most actively studied group of compounds. As shown in Fig. 1, the first demonstration of a microbial hydroxylation of a non activated methylene group, that is the conversion of progesterone to 11-α hydroxyprogesterone by <u>Rhizopus arrhizus</u>, triggered two decades of intensive efforts, which have led to huge scale production of raw material for medicinally important steroids like cortisone . This hydroxylation occurs with complete regio and stereoselectivity (1,2).

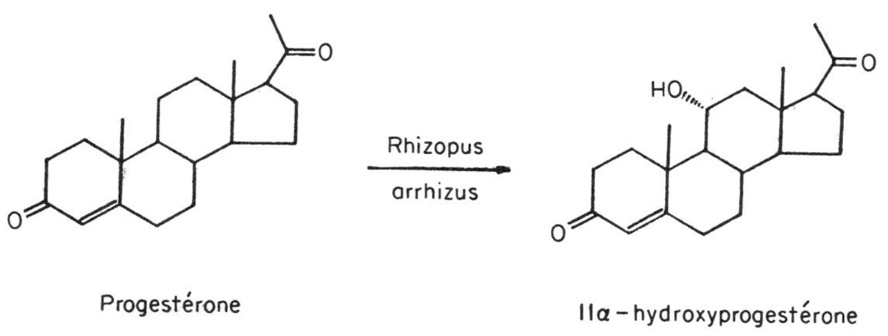

Progestérone   11α – hydroxyprogestérone

Fig. 1

Another very important type of specificity is the prochiral stereospecificity. This is very well illustrated by the specific hydroxylation of the enantiotopic pro-S methyl group

of isobutyric acid by Pseudomononas putida (3), which leads to the very versatile chiral
unit   hydroxyisobutyric acid of S configuration. The preparation of the two enantiomeric
C-4 bromobenzylethers derived from this same chiral building block, enabled further synthesis of optically active R and S muscone, of the natural isomer of vitamine E as well as
of different macrolide antibiotics (Fig. 2)(4).

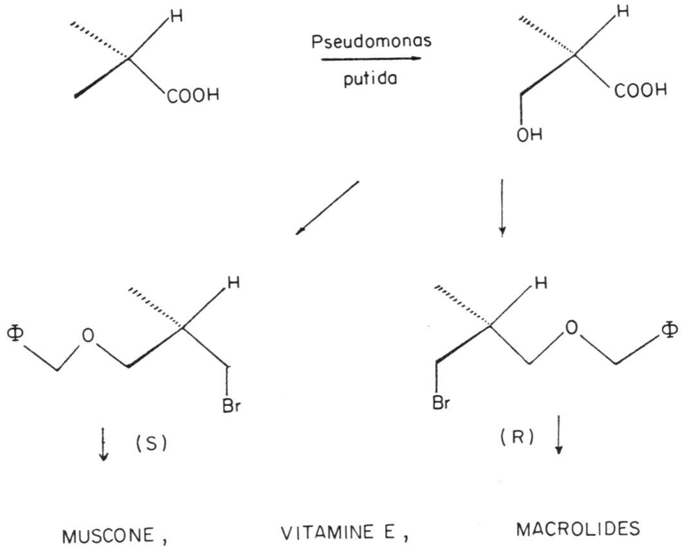

Fig. 2

There is , however, a very important drawback to the use of such methods in organic
chemistry, that is the difficulty to make previsions concerning the specificities of such
hydroxylations on a given substrate.

One way to approach this problem is to study the bioconversion of different structurally
related molecules, which can then be considered as beeing some kind of molecular probes,
and can therefore allow to get informations concerning the influence of different geometric or steric parameters involved in these transformations. With these results in hand,
one can hope to rationalize the metabolic pathways and, as a consequence, try to reach a
high degree of predictability for these enzymatic processes (see for instance ref. 5-8).
Having this point in mind, we decided to study the biohydroxylation of various polycyclic
amides or lactams by the fungus Beauveria sulfurescens (ATCC 7159). This fungus has been
described by a group directed by Fonken to achieve regiospecific hydroxylation of a great

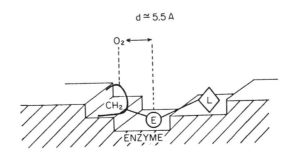

Fig. 3

number of essentially linear or monocyclic amides (9,10). On the basis of the thermodynamically more stable conformations of these compounds, these authors have proposed a model
(11) of the hydroxylating enzyme active site, which implies – first, that the molecule is
fixed to this site by an interaction between the electron rich oxygen atom of the amide
function and the enzyme, – secondly that, as a subsequent step, hydroxylation occurs on a
carbon atom located at a distance of about 5,5 Å of this oxygen atom (Fig. 3)

These considerations led us to raise two type of questions – would it be possible to
exploit these facts in order to achieve the synthesis of hydroxylated building blocks,
which could subsequently be used for synthesis? – would it be possible to get some more
informations about the topology of the active site of the hydroxylating enzyme by studying
conformationally rigid molecules, and especially lactam type molecules? In order to get
some answers to the first question we devised the following strategy . This consists in
fact in three steps (Fig. 4).

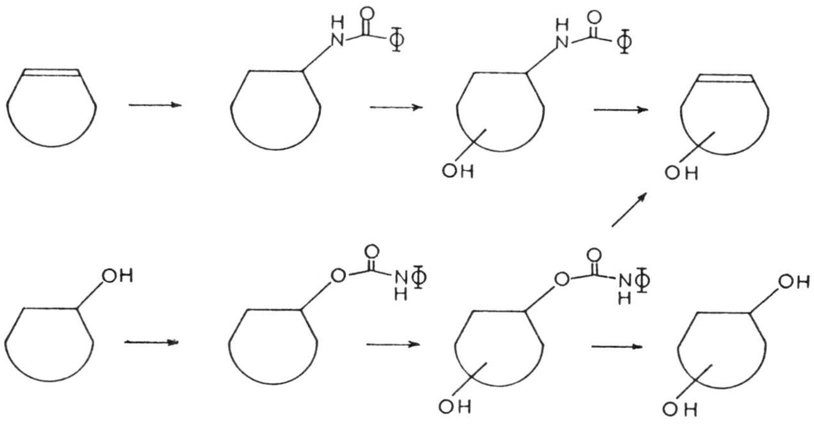

Fig. 4

Starting from compounds owing such functionalities as for instance a double bond or an
hydroxyl group, the first step consists to transform this function into an amide-type
group, which can be an amide or a urethane function. As a second step, which in fact is
the key step, this derivative will possibly be hydroxylated by the fungus, to lead to an
alcohol. The third stage will involve chemical elimination, through one way or another, of
this activating group, in order to recover the starting double bond or alcohol. As can be
seen, everything happens, in the first case, as if one would have hydroxylated the starting material whereas , in the second case, one also can stop at a stage where one forms a
diol, specifically protected on one of the two alcohol functions.

This scheme immediately raises two subsequent questions – what will be the influence of
the stereochemistry of the amide group on the specificity of the hydroxylation, in other
words can one hope to influence the selectivity of the reaction by varying the amide group
stereochemistry ? – will there be a difference between the hydroxylation of an amide and
that of the corresponding urethane ?

RESULTS

The following results do shed some light on these two questions. As can be seen, hydroxylation of the two amide epimers prepared from norbornene do effectively work in a highly
regio and stereospecific way, but neither the regio nor the stereoselectivity of the
hydroxylation process have been altered by the variation of stereochemistry of the amide

group. The same result is observed starting from the epimeric N-methylated amides having a
camphor skeleton. Here again, the reaction is highly regio and stereoselective and no
change is observed dependent on the stereochemistry of the amide. One also must emphasize
that none of the obtained alcohols are optically active, which means the hydroxylation is
not enantioselective in these cases.

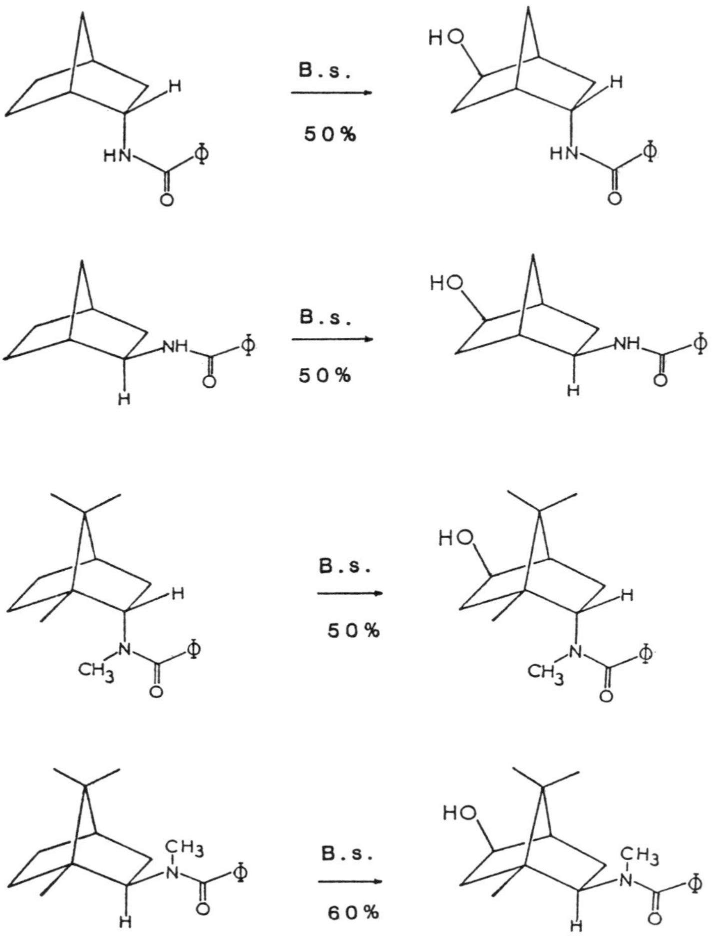

Fig. 5

This is however no longer true for the homologous secondary amides, where we observe a
largely different behaviour between the two epimers. In that case, the racemic exo isomer
still leads to a single racemic alcohol, whereas the endo isomer furnishes three optically
active compounds.

That means that, for this compound, the regioselectivity of the hydroxylation varies from
one enantiomer to another. Thus, the 1R isomer is preferentially hydroxylated at the C-5
position , whereas the 1S isomer is essentially attacked on the C-8 methyl group. Each of
the starting compounds can of course be prepared optically pure starting from the pure
enantiomers of camphor.

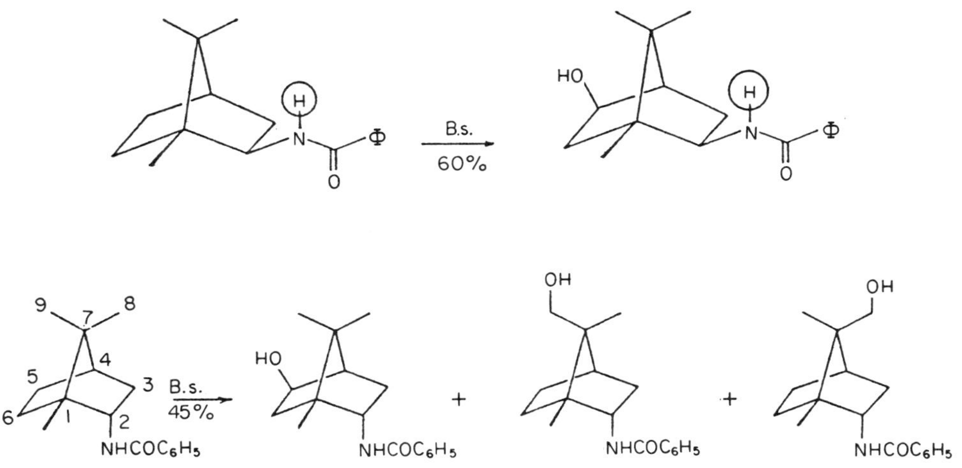

PROPORTIONS

I RS	36	20	44
I R	48	22	30
I S	11	15	74

Fig. 6

No such behaviour has been observed for the different molecules we have prepared starting from α -pinene. These compounds, which differ from one another by the stereochemistry of either the amide or the C-2 methyl group, should also be different as far as their conformations are concerned. In spite of these facts, biohydroxylation of these molecules always occurs at the C-8 methyl group.

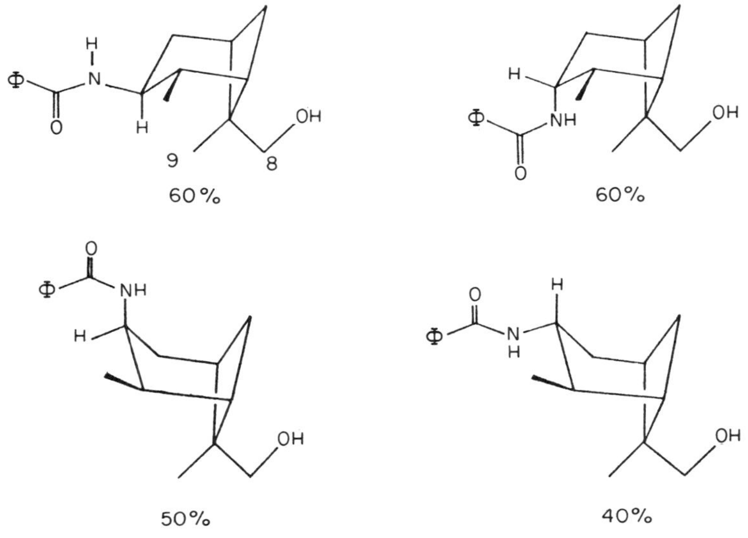

Fig. 7

This is actually a very interesting fact, as far as synthesis is concerned, this methyl group being located on the chemically unattainable side of these molecules. Starting from the hydroxylated compounds ,we are actually studying the third step of the previously proposed scheme, that is to say the elimination of the amide moiety , in order to come back to the starting double bond. Mainly, elimination of the amide on these pinene models should lead to an olefinic alcohol which appears to be a key building block in the synthesis of different terpenes like for instance ∝ -transbergamotene, longipinene, copaene or ylangene. Of course, these terpenes will be easily prepared in their optically active forms, simply by starting with optically active ∝-pinene.

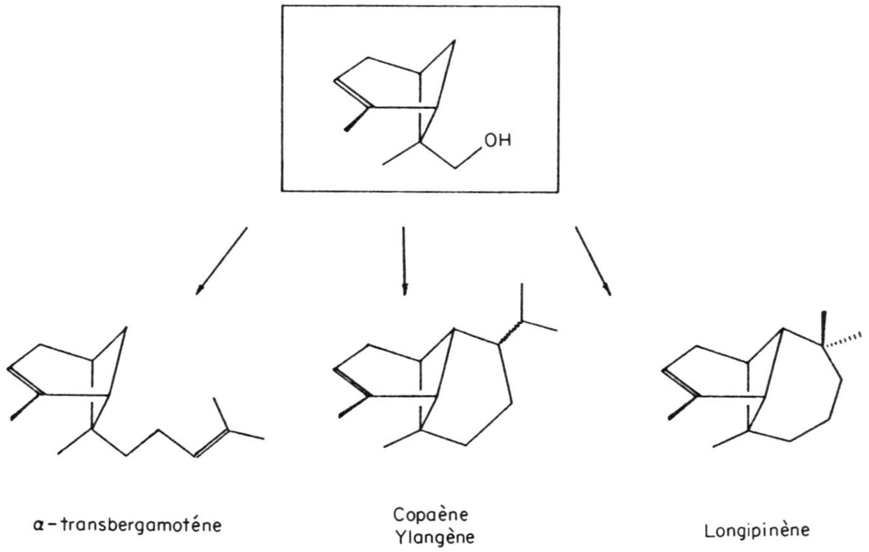

α-transbergamoténe   Copaène   Longipinène
                               Ylangène

Fig. 8

As we mentionned before, it also was of interest to us to check if the hydroxylation of urethanes would work similarly to that of the parent amides. As shown in Fig. 9, one indeed observes hydroxylation on the carbon skeleton with the same regio and stereoselec-

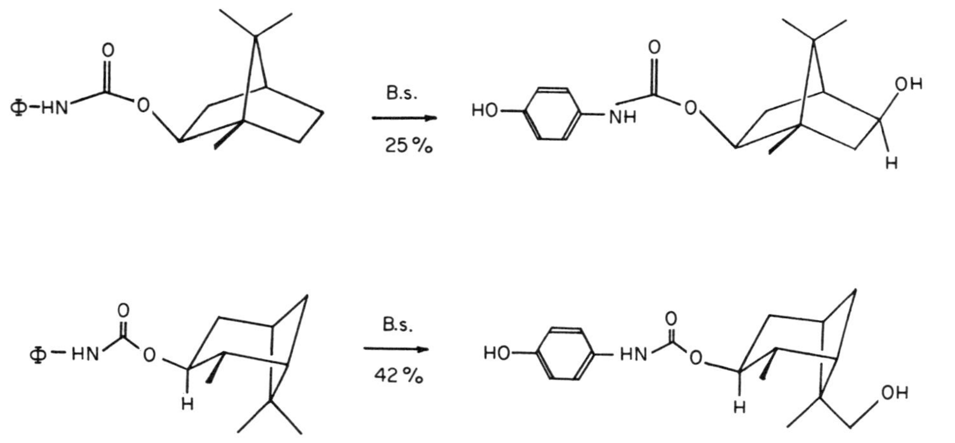

Fig. 9

tivities as those observed for the homologous amides. However, a second hydroxylation occurs at the para position of the aromatic ring. Once again, the enzymatic system achieves a reaction which is quite difficult to realize specifically by chemical ways. Interestingly, the ease with which this aromatic hydroxylation is performed depends upon the lipophilic character of the molecule. For instance, in the case of the cyclohexane derivative , the aromatic hydroxylation is quite slow, and one can analytically follow the formation of the monohydroxylated compound and his subsequent hydroxylation on the aromatic nucleus. If the alkyl group becomes even smaller, as in the case of the cyclopentane derivative, aromatic hydroxylation occurs no more, and one only observes formation of the alkyl hydroxylated compound.

Finally, if the alkyl group becomes too small to be hydroxylated, as in the case of the isopropyl derivative, which is a very efficient pre-emergence herbicide called Propham (12,13), the aromatic hydroxylation again occurs, followed by conjugation with

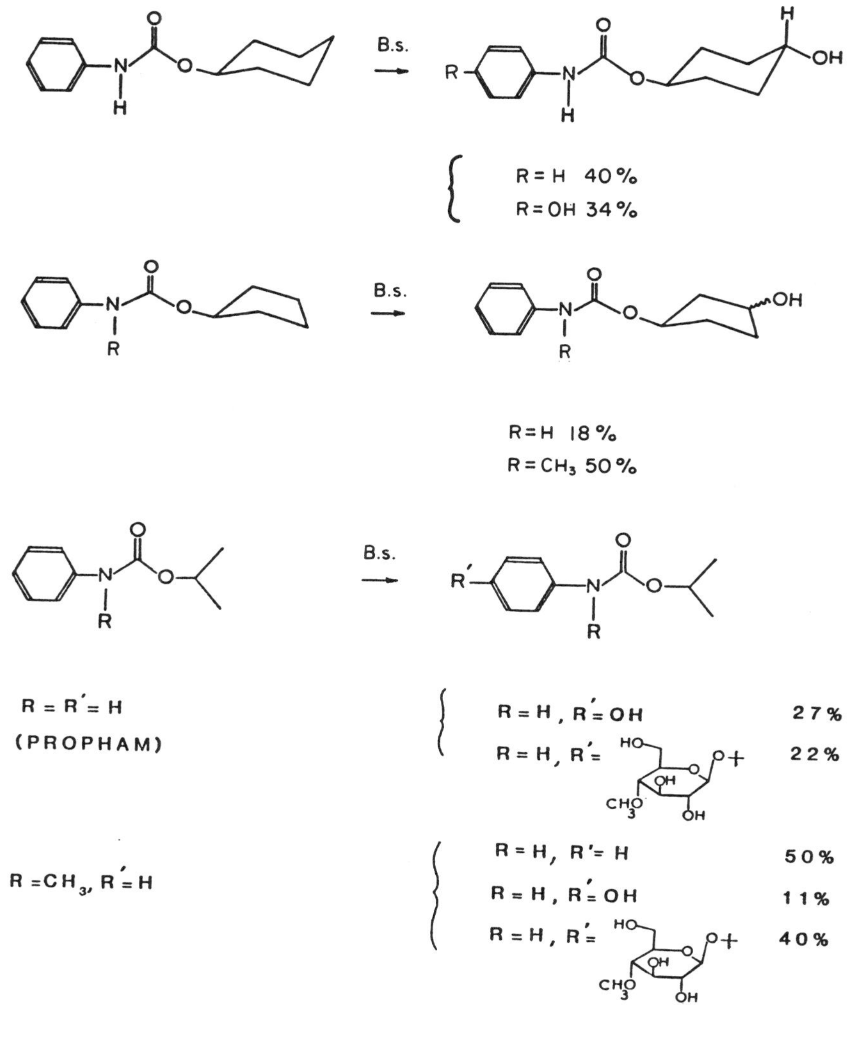

Fig. 10

a carbohydrate moiety. If one starts with the corresponding N-methyl derivative, one also observes demethylation on the nitrogen atom, which is quite a common process in metabolic pathways. These results indicate that, most likely, all these processes are part of the detoxification process of this fungus, designed to transform such xenobiotics into more water soluble compounds (Fig. 10).

The second question which we raised at the beginning of this work was : would it be possible to get some more informations about the topology of the active site of the hydroxylating enzyme by studying conformationally rigid molecules ? The molecules we have described up to now can all be represented as shown on the following scheme.

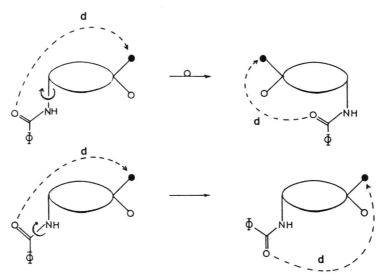

Fig. 11

As one can see, these molecules can therefore exist in quite different conformations, due to rotation around the C-N bond, and also to the existence of the amide as a cisoïd and a transoïd form (14). If one comes back to Fonken's hypothesis concerning the positioning of the amide oxygen atom on the enzymatic active site, it is obviously very difficult to make any prevision concerning the regioselectivity of the hydroxylation on the basis of the proposed 5,5 Å distance between the carbonyl oxygen atom and the hydroxylated carbon atom. Therefore we decided to study some molecules where one, or both, of these degrees of freedom have been suppressed. This is realized for the compounds shown in Fig. 12. In the case of the amides, the only remaining possibility of conformational difference is the cisoïd and transoïd form of the amide function, whereas even this possibility is suppressed in the corresponding lactam type structures.

Also, owing to the previous hypothesis, one could hope that the different localization of the carbonyl function on the carbon skeleton should lead to a different regiospecificity for the hydroxylation. However, this appears not to be the case, since the bioconversion of each of these compounds do lead to a single alcohol, resulting from stereospecific hydroxylation on the same carbon atom (15). This quite surprising result is also observed in the case of the bridged tricyclic amides and lactams (16). Here again, hydroxylation occurs on the same carbon atom, independently of the localization of the carbonyl function on the carbon framework.

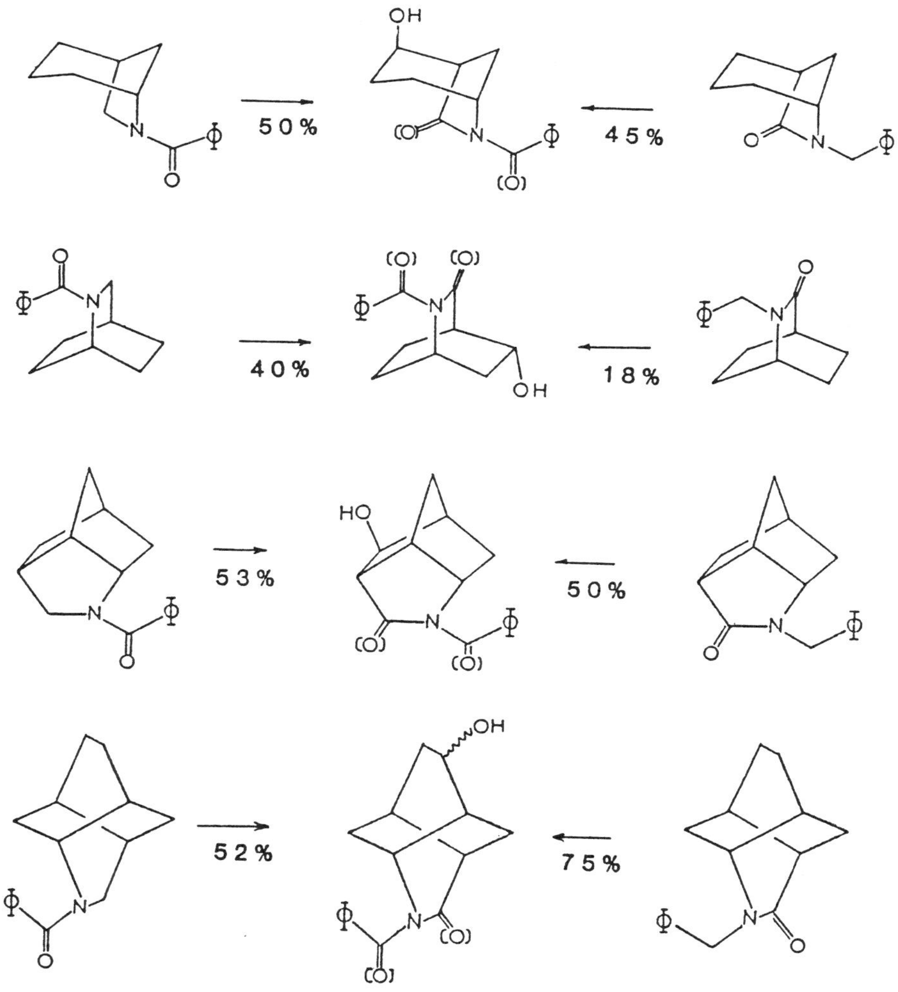

Fig. 12

These observations once again raise the question, whether the oxygen atom does play any role at all in the interaction between the enzyme and the substrate. The answer is still yes : indeed, whereas the alcohols obtained by hydroxylation of the racemic amides are optically inactive, those formed from the racemic lactams are optically active. As the only difference between the two types of models is precisely the localization of the oxygen atom on the carbon skeleton , we must conclude that this atom is somehow involved in the process.

Another interesting feature is the stereoselectivity of these hydroxylations. As seen previously, all the hydroxylations we have observed are highly stereoselective, except the azatwistane one where we observe the formation of two stereoisomeric alcohols. However, as shown in Fig. 13,these two alcohols are diastereoisomers. They do in fact result from hydroxylation of each one of the enantiomers, identically positionned on the enzyme active site, and then hydroxylated from a relatively precise region of the space. This does lead to two alcohols which are in both cases of S configuration, but where the helicities of the molecule are opposite.

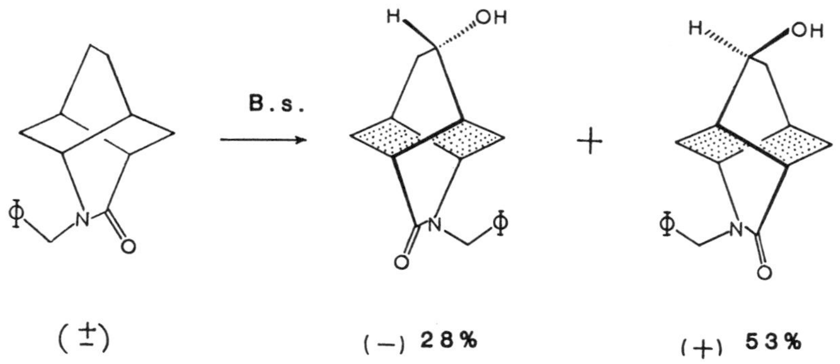

Fig. 13

Interestingly, hydroxylation of N-benzyl, aza-2 ,bicyclo [2,2,1] heptanone-3 does lead to the optically active 5-hydroxylated compound, which appears to be an interesting building block for synthesis of carbocyclic analogs of nucleosides like desoxyaristeromycin (18).

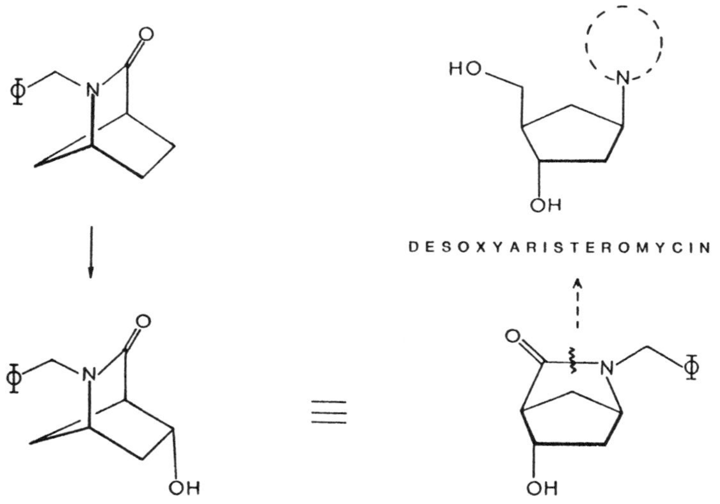

DESOXYARISTEROMYCIN

Fig. 14

DISCUSSION

All the described results do lead to two observations related to the predictability of these hydroxylations. The first one relies to the use of the 5,5 Å distance proposed by Fonken. Owing to the fact that , as we have seen before ,the localization of the oxygen atom does not play a dramatic role as far as the regioselectivity is concerned, we feel that it may be more accurate to use the distance between the nitrogen atom and the hydroxylated carbon as a predictive tool (17). However , this distance does vary from about 2.8 to 5.8 Å in the different models we have studied . Therefore, everything seems to happen as if a two step process would occur, where the first one would be - positioning of the substrate on the enzyme by interaction whith the amide group, - the second step being hydroxylation by way of an oxygen activating site located on a relatively flexible arm (Fig. 15).

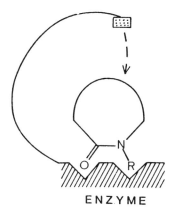

Fig. 15

This model does account for two important features - the very low substrate specificity of the fungus, which is able to hydroxylate numerous structurally different compounds - also, the very high regio and stereoselectivity observed for these hydroxylations, which would occur because of a relatively precise trajectory of the oxygen activating locus.

Besides this notion of distance, there is another factor which appears to be very important as underlined by the formation of opticaly active lactams from racemic starting material. That is the spacial location of the hydroxylating site. In order to localize this region ,we determined the absolute configuration of the obtained optically active alcohols . These are represented on Fig. 16 , each amide being placed in the plane, the carbonyl function towards the left. It thus appears that the hydroxylated carbon atoms effectively are located in the same part of the space, that is on the left handside, generally behind the plane of the amide moiety. It therefore is quite justified to postulate that this corresponds to the spacial localization of the hydroxylating locus.

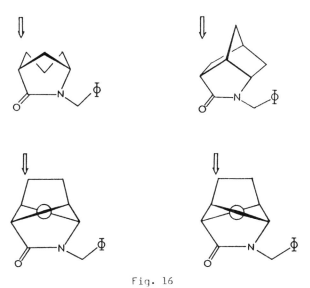

Fig. 16

CONCLUSION

As a conclusion, we would like to emphazise the fact that biohydroxylations may be a very powerfull tool in the hands of the organic chemist who wants to achieve functionalization of non activated carbon atoms. Indeed, these transformations do lead to highly regio and stereoselective processes, and even sometimes to enantioselective ones, allowing the synthesis of optically active compounds from racemic material. In order to get some more insight into these processes, we are carrying on our work by studying different structurally related compounds. Also, we are actually trying to rationalize all the obtained informations by using computerized structure-activity approaches. So we hope to be able to greatly enhance the predictability of these processes, and therefore make these techniques more attractive to the synthetic organic chemist.

REFERENCES

1. D.H. Peterson and H.C. Murray, J. Am. Chem. Soc. 74, 1871 (1952).
2. A. Čapek, O. Hanč and M. Tadra, Microbial Transformations of steroids, Academia, Prague (1966) .
3. Q. Branca and A. Fischli, Helv. Chim. Acta. 60, 925 (1977).
4. A. Fischli, Modern Synthetic methods, 2, Otto Salle Verlag, Frankfurt am Main (1980).
5. V. Prelog, Pure Appl.Chem. 9, 119 (1964).
6. E.R.H. Jones, Pure Appl.Chem. 50, 39 (1978).
7. M. Nakasaki, J. Org. Chem. 45, 4432 (1980).
8. P. Mohr, N. Waespe-Sarcevic, C. Tamm and K. Gawronska, J.K. Gawronski, Helv. Chim. Acta. 66, 2501 (1983).
9. R.A. Johnson, Oxidation in Organic Chemistry, Part C, p.31, Academic Press (1978).
10. G.S. Fonken and R.A. Johnson, Chemical Oxidation with microorganisms, Dekker, New York (1972).
11. R.A. Johnson, M.E. Herr, H.C. Murray and G.S. Fonken, J. Org. Chem. 33, 3217 (1968).
12. C.G. Clark and S.J.L. Wrighy, Soil Biol. Biochem. 2, 217 (1970).
13. D.D. Kaufman, J. Agr. Food. Chem. 15, 582 (1965).
14. S. Pataï, (Ed), Chemistry of Amides, p.7, Wiley Intersciences, New York (1970).
15. R. Furstoss, A. Archelas, B. Waegell, J. Le Petit and L. Deveze, Tetrahedron Letters. 451 (1980).
16. R. Furstoss, A. Archelas, B. Waegell, J. Le Petit and L. Deveze, Ibid. 445 (1981).
17. A. Archelas, R. Furstoss, B. Waegell, J. Le Petit and L. Deveze, Tetrahedron 40, 355 (1984).
18. A. Archelas and C. Morin, Tetrahedron Lett. 1277 (1984).

Carbohydrate-like chiral synthons

Claudio Fuganti

Istituto di Chimica del Politecnico, 20133 Milano, Italy

summary:

As the result of a complex aldehyde condensation-reduction enzymic reaction, baker's yeast fermenting on $\underline{D}$-glucose converts C_6-C_3, α,β-unsaturated aromatic aldehydes (1) into C_6-C_5, $(2\underline{S},3\underline{R})$ 2,3 -methyl diols (4) in $\underline{ca}$. 25% yield. Compound (4) can be viewed as carbohydrate-like chiral synthon. Mechanistic aspects of the enzymic reactions involved in the formation of (4) from (1) and synthetic applications of (4) will be discussed.

A current approach to the synthesis of enantiomerically pure forms of natural products and drugs is based on the use as starting materials of components of the collection of inexpensive, readily available, optically active compounds produced by nature (pool of chirality)[1]. Chemists are, however, not fully satisfied with the present composition of the pool of chirality, since most of the products are accessible in only one enantiomeric form, a circumstance which dictates, when the absolute configuration of the target molecule is opposite to the one of the chosen material, chemical manipulation of the chiral center(s), usually through multistep, low-yield sequences. This is the case of the synthesis of the 3-amino-2,3,6-trideoxy hexoses of the $\underline{L}$ series present in the therapeutically important anthracycline glycosides adriamycin and its 4'-epimer[2] which can be prepared either from inexpensive hexoses of the $\underline{D}$ series, but with a critical inversion of configuration at position 5 at some stage of the sequence, or from the rather rare 6-deoxysugar $\underline{L}$-rhamnose[3,4].

Accordingly, there is considerable interest in finding new chiral products which can be used as starting materials for syntheses of enantiomerically pure forms of elaborated substances, and a particular rich source of chiral products are expected to be the transformations of non-conventional substrates by microorganisms[5]. Microbes are indeed capable of performing a variety of transformations of non-conventional substrates using enzymes either of the primary or of the secondary metabolism, and this property has been recognized by organic chemists since a long time, as we can judge from the impressive number of examples reported on this topic[6].

From a practical point of view, we would expect to be synthetically useful those transformations of non-conventional substrates leading to optically active products performed by microorganisms commercially available at low cost, possessing large quantities of enzymes (usually of the primary metabolism), and showing a wide substrate specificity, while mantaining a precise reaction stereospecificity. Baker's yeast meets most of the above requisites and, accordingly, it has been widely used. However, most of the products obtained by these procedures contain chiral centers of the type R,R'CHR" and/or R,R'CHOH, formed by stereoselective addition of hydrogen across (activated) double bonds and carbonyl groups. The chiral materials obtained in this way contain the same carbon framework of the precursors, and, in most instances, appear also accessible by alternative non-enzymic methods.

In this context, the baker's yeast mediated conversion of aromatic, α,β-unsaturated aldehydes into the set of products indicated in eq. 1 is particularly interesting.

(1)   (2)   (3)   (4)   1

(2)+(3) = 75-70%   (4) = 25-30%

R = H, Me, Br

Indeed, whereas the production of (2) and (3) from (1) falls amongst the known capacities of baker's yeast, the formation of (4), containing two additional carbon atoms respect to the precursor aldehyde and two adjacent chiral centers of the type R,R'CHOH is new and results quite fruitful from a synthetic point of view.[7]

The obtainment of the (2S,3R) diol (4) from the aromatic, α,β-unsaturated aldehyde (1) and fermenting baker's yeast must be regarded as the overall consequence of a complex aldehyde condensation-reduction, involving two distinct chemical operations (eq. 2).
1) Addition of a C_2 unit equivalent of acetaldehyde onto the _si_ face of the carbonyl carbon of the unsaturated aldehyde forms a (R) α-hydroxyketone, in an acyloin-type condensation, and
2) Reduction of the latter intermediate on the _re_ face of the carbonyl gives rise to the diol actually isolated. Indeed, on incubation of α-methylcinnamaldehyde with actively fermenting baker's yeast in the presence of acetaldehyde, at pH 5, (3R) 5-phenyl-4-methyl-pent-4-en-3-ol-2-one was obtained as the sole transformation product.[8] The latter material was converted by yeast at pH 7.5 into the diol (8), quantitatively.

Experiments designed to gain information on the substrate specificity of the sequence of eq. 2 indicated that there is some tolerance by the enzymic system(s) involved as far as the structure of the aromatic aldehydes and the substituents in α-position are concerned.[9] However, α-ethyl and α-_n_-propyl cinnamaldehydes are not converted into the corresponding diols. Furthermore, acetaldehyde is the only aldehyde accepted as second terminus of the reaction, since, when cinnamaldehyde was incubated with yeast in the presence of propional-dehyde or butyraldehyde, (2S,3R) (7) was isolated as the sole transformation product.

The lack of incorporation of α-ethylcinnamaldehyde and propionaldehyde into the type of diols of eq. 2 is likely to be due to the inability of these materials to be accepted as substrates by the condensing enzyme(s) (first part of eq. 2), since the synthetic α-hydroxy-ketones derived from the above aldehydes, as well as those from other carbonyl compounds are stereospecifically reduced by yeast. This has been shown submitting to the yeast transformation the synthetic, racemic α-hydroxyketones (5), (6), (13), (14), (17) and (20), at pH 7.5-6.5. There is, however, a considerable difference in the yield of conversion and in the type of products between the methylketones (5) and (6), and their higher homologues (13) and (14). Whereas (5) and (6) afforded *ca.* 70-80% of a *ca.* 6:4 mixture of (2*S*,3*R*) (7) and (8) and (2*S*,3*S*) (9) and (10), respectively, α-hydroxyketones (13) and (14) gave rise in 20% yield to the *erythro* diols (15) and (16), contaminated by *ca.* 10% of the *threo* isomeric materials. The α-hydroxyketones (17) and (20) behave towards yeast reduction as the sets (5)-(6) and (13)-(14), respectively. Whereas compound (17), a methylketone similar to (5) and (6) afforded in 80% yield a *ca.* 1:1 mixture of diastereoisomeric diols (18) and (19), product (20) afforded 20% of (3*S*,4*R*) (21). The hydroxyketone recovered from the production of (21) was shown to contain an excess of the (4*S*) enantiomer (22). In this context, the α-hydroxyketone (11) affords with fermenting baker's yeast the (2*S*,3*R*) diol (8) and the *threo* isomer (12), enantiomer of (10), in *ca.* 6:4 ratio and 70% overall yield.

(5) R= H
(6) R= Me
(7) R= H
(8) R= Me
(9) R= H
(10) R= Me

(11) (8) (12)

+ THREO ISOMER (10%)

(13) R= Et; R^1 = H
(14) R= H; R^1 = Me
(15) R= Et; R^1 = H
(16) R= H; R^1 = Me

(17) (18) (19)

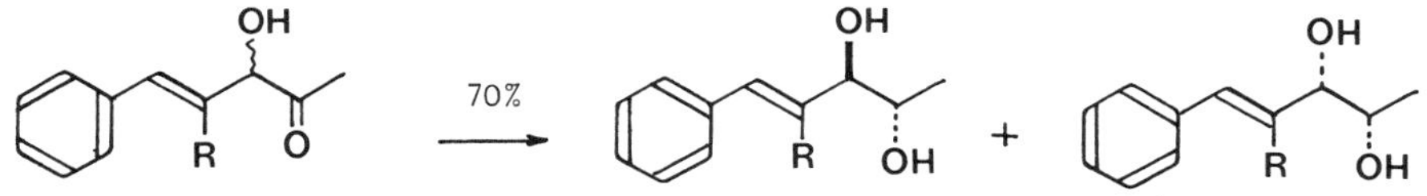

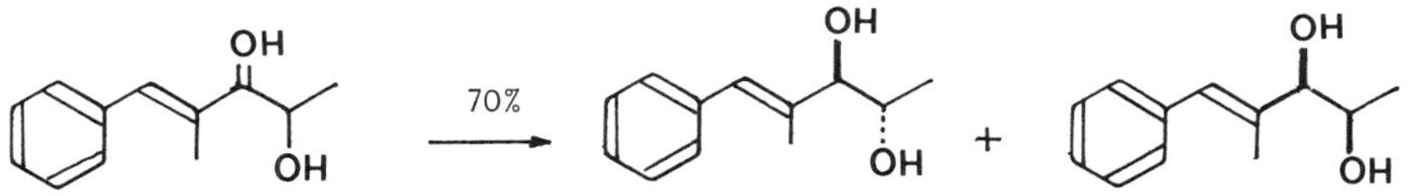

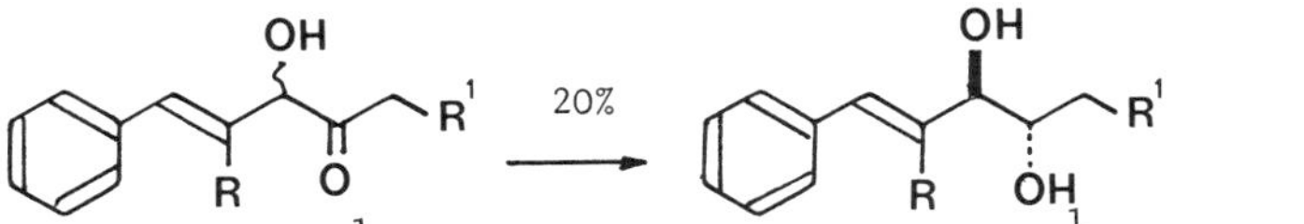

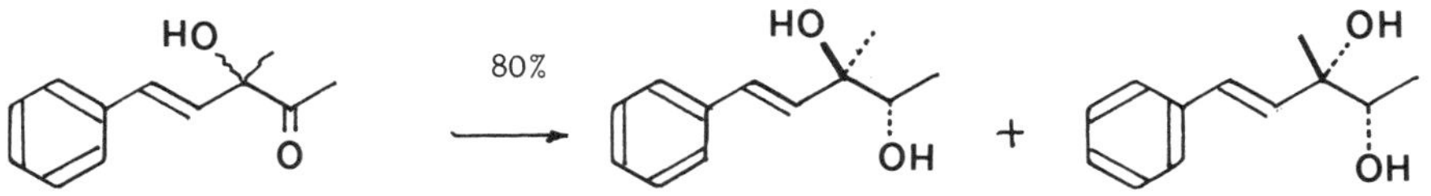

Seen together, the above results point to the following conclusions. 1) For (5) and (6), and probably for (17), hydride addition to the carbonyl occurs on the re face regardless of the configuration of the adjacent chiral center. For (13), (14) and (20), however, only the (R) enantiomer is reduced to a significant extent. 2) Hydride addition onto the si face of the carbonyl takes place in (11) irrespective of the configuration of the adjacent centre. 3) Synthetic hydroxyketones (5) and (6) are better substrates for the yeast enzyme(s) than (13) and (14). The first two are intermediates in the conversion of the aldehydes into the diols (eq. 2), whereas the second two are not formed directly from the corresponding aldehydes in yeast.

The above optically active diols show some relevant structural features which render them quite useful starting materials in the synthesis of enantiomerically pure forms of natural products. 1) The (2S,3R) absolute configuration of the diols (7) and (8) matches that at positions 5 and 4 of 6-deoxysugars of the L series like L-amicetose (23) and L-olivomicose (24), which have been indeed prepared from (7) and (8) in the work designed to establish their absolute configuration.[10] 2) The double bond of the diols can be stereospecifically functionalized, being the stereochemistry of the process dictated by the stereochemistry of the adjacent allylic center. 3) Once the double bond has been saturated it is possible to functionalize the derived products regioselectively at the benzylic position. 4) In saturated products, degradation with O_3 of the benzene ring affords 6-deoxy-hexonic acids, bearing all the chiral centres present in the side chain of the parent compound. D-allomuscarine[11] and the enantiomeric forms of γ-hexanolide[12] have been prepared from (7) taking advantage of the above properties. 5) More, important, from suitably protected forms of the abovementioned diols it is possible to extrude with ozone C_4, C_5 and C_6 carbonyl compounds, bearing, when the isopropylidene derivatives are used, in α and β positions two chiral oxygen substituents embedded in a cyclic acetale framework.

Furthermore, the erythro forms of the latter carbonyl compounds are converted into their α-epimers, with threo stereochemistry, upon basic treatment. Accordingly, starting from the above diols, the chiral carbonyl compounds (25)-(35) have been prepared. The enantiomer of the aldehyde (35) has been recently used as starting material in the synthesis of (+)-

methynolide,[13] whereas product (28) embodies the chiral centers of the side chain of pumiliotoxin.[14]

The value of this type of synthons, which seem particularly useful as starting materials for the synthesis of optically active molecules containing in their frameworks relatively few carbon atoms chiral due to oxygen substitution, is enhanced from the fact that some of them are available in both the enantiomeric forms. Furthermore, taking advantage of the increasing knowledge of methods for stereocontrolled chain elongation through nucleophilic addition onto sp^2 carbon, the number of the derived synthons has been increased considerably. We can thus consider the diols obtained through the above mentioned methods as carbohydrate-like chiral synthons which can be used, in some instances, advantageously as alternative to natural carbohydrates as starting materials for the synthesis of optically active forms of natural products belonging to quite different structural classes.

The most interesting application of products (7) and (8) is in the synthesis of N-trifluoro-acetyl-L-daunosamine (38) and L-vancosamine (49), and of their configurational isomers. Our synthesis of (38) starts off with the C_4 aldehyde (26), obtained from (25) by α-epimerization, which is converted into the phenylsulfenimine (36). Addition onto the latter of diallyl zinc proceeds stereoselectively to afford the C_7-N, carbohydrate-like, non-carbohydrate derived adduct (37), possessing the correct stereochemistry and the required functionalities for a direct conversion into (38), which has been indeed achieved by N-trifluoroacetylation and ozonolysis.[15] N-Trifluoroacetyl-L-acosamine (41), the C-4 epimer of (38), was obtained from the diol (7) in a quite direct manner. Thus, the isopropylidene derivative of (7), upon sequential treatment, in the same pot, with i) O_3 in CH_2Cl_2; ii) 1 mol eq of Ph_3P, and iii) $Ph_3P=CHCO_2Et$, afforded (65% for the three steps) the C_6 chiral ester (39) and ethyl cinnamate. Product (39) adds ammonia stereoselectively to give the C_6-N product (40), subsequently converted into (41) by acid hydrolysis, N-trifluoroacetylation and DIBAH reduction of the intermediate lactone.[16] Similarly, from the C_5 methyl ketone (27), prepared from the diol (8), the phenylsulfenimine (42) was prepared. Addition of diallyl zinc onto the latter afforded (43), with 3,4-erythro stereochemistry, whereas allylmagnesium bromide afforded the C_7-N, 3,4-threo adduct (44). These two materials gave rise, in the usual way, to the L-arabino- and L-ribo-3-C-methyl-3-amino-2,3,6-trideoxysugar derivatives (45) and (46).

From the enantiomeric diols (10) and (12), the enantiomeric ketones (29) and (30) were prepared. The latter materials, via phenylsulfenimines and addition of diallyl zinc afforded the C_7-N adducts (47) and (48), from which L- and D-vancosamine as the N-trifluoroacetyl-derivatives (49) and (50), respectively, were obtained.[17,18]

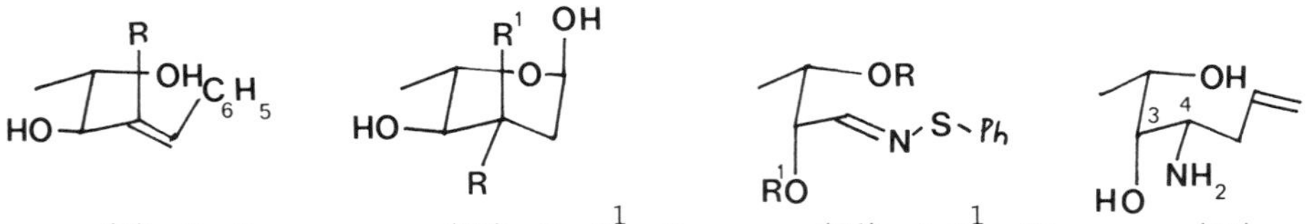

(7) R= H
(8) R= Me

(23) R= R^1= H
(24) R= OH; R^1= Me

(36) R,R^1= CMe_2

(37)

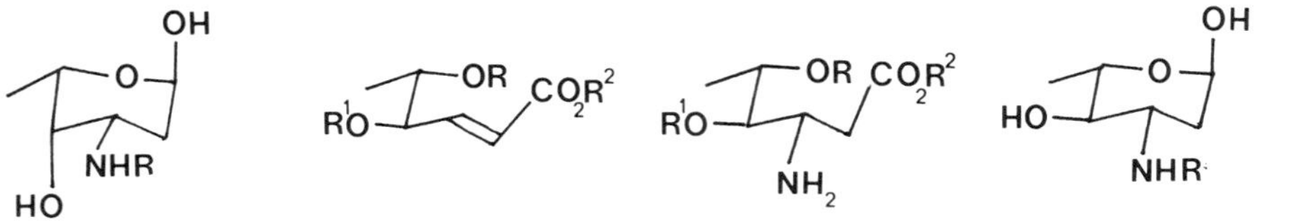

(38) R= $COCF_3$

(39) R,R^1= CMe_2; R^2= Et   (40)

(41) R= $COCF_3$

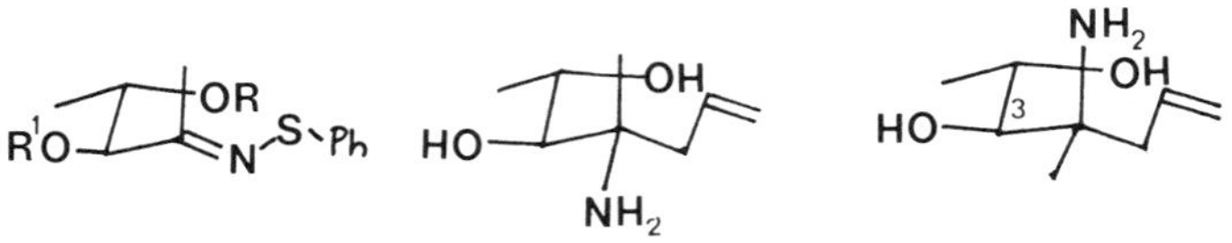

(42) R,R^1= CMe_2

(43)

(44)

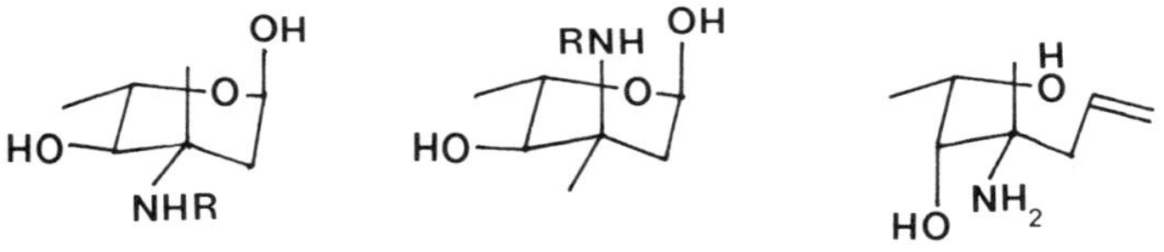

(45) R= $COCF_3$

(46) R= $COCF_3$

(47)

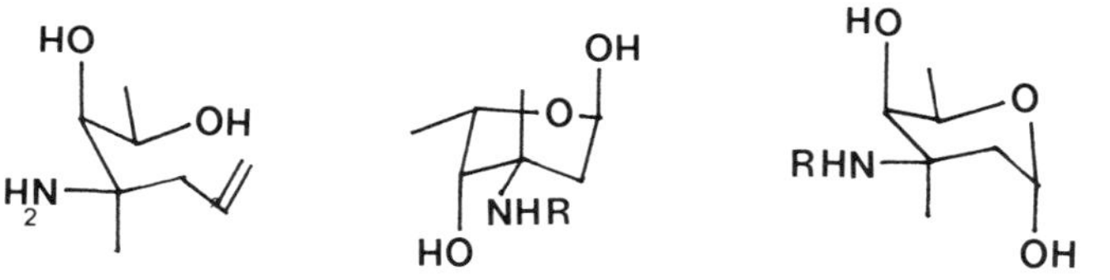

(48)

(49) R= $COCF_3$

(50) R= $COCF_3$

When the aldehyde (25) was reacted with diallyl zinc, the C_7, carbohydrate-like, non-carbohydrate-derived synthon (51) was obtained.[19] The latter material, via unexceptional steps was converted into 2,6-dideoxy-L-ribohexose, the insect pherormones (+)-exo-brevicomin (52) and (R) 5-hexanolide (53), the 6-octanolide (54), the enantiomer of the material used in the synthesis of the side chain of trichoverrins B[20] and into the C_{10} (2R) aldehyde (55), one of the chiral synthons required in convergent syntheses of natural LTB$_4$.[21]

Finally, from product (51) through suitable manipulation of the protecting groups, the epoxyalcohols (56) and (57) have been prepared. The latter materials have been converted into the 4-amino-2,4,6-trideoxy hexoses of the L-series (58) and (59), respectively, which are isomeric of L-daunosamine and L-ristosamine.

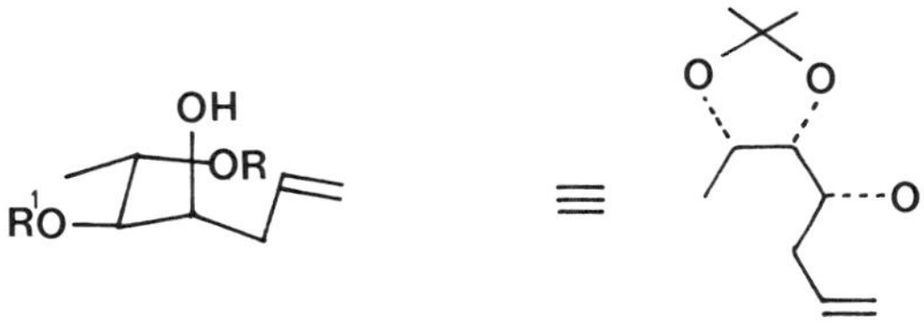

(51) R,R^1= CMe$_2$

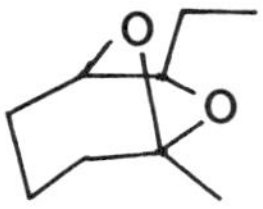

(52)

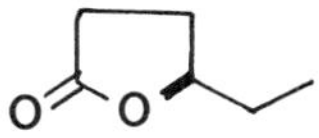

(53)

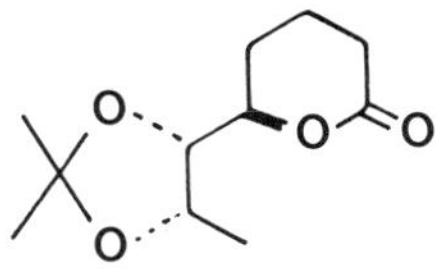

(54)

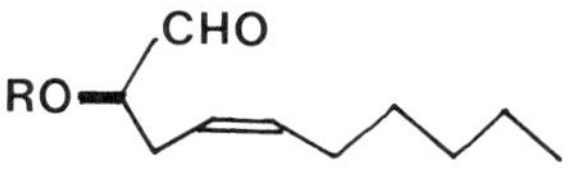

(55) R= SiPh$_2$Bu^t

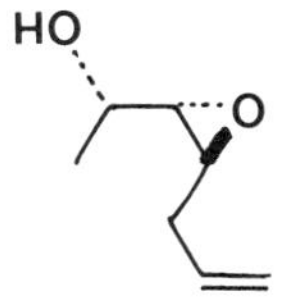

(56)

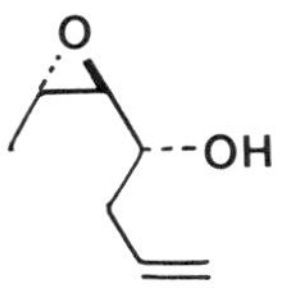

(57)

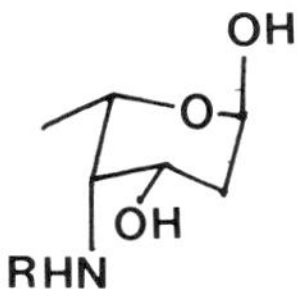

(58) R= COCF$_3$

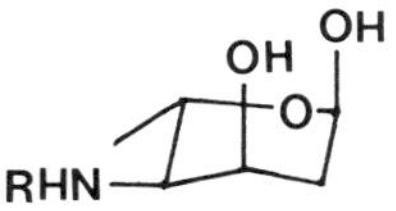

(59) R= COCF$_3$

From the diols (7) and (8) we synthetized also (-)-frontalin,[22] the enantiomeric forms of the insect pheromone 6-acetoxy-5-hexadecanolide in the <u>erythro</u> and <u>threo</u> forms,[23] (3<u>S</u>,4<u>S</u>) 4-methyl-3-heptanol[24] and several C-methyl analogs of (38) and (41).[25,26] However, the most significant example of the relevance to the synthesis of enantiomerically pure forms of natural products of baker's yeast mediated transformations of non-conventional substrates as source of chirality is the synthesis of natural α-tocopherol (vitamin E) from the educts

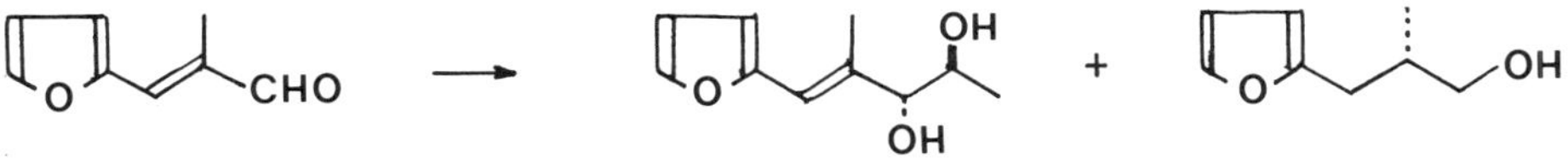

(60)

(61) 15-20%

+

(62) 65-70%

(61) and (62), obtained in baker's yeast from α-methyl-β(2-furyl)acrolein (60). The C_{29} chiral framework of natural α-tocopherol (71) is accessible, according to known procedures, from a chiral C_{14} chromanyl moiety (70) and a Wittig reagent prepared from the terpenoid C_{15} chiral alcohol (66), followed by reduction of the C_{29} unsaturated intermediate. Most of the effort in the field of the synthesis of (71) has been made in the preparation of the two chiral fragments (70) and (66). We used alcohol (62), arising by stereospecific reduction of (60), as masked C_5 chiral terpenoid unit, which, via tosylate, was alkylated to the masked C_{10} chiral terpenoid unit (63). From the latter, the C_{10} chiral alcohol (64) was extruded by ozonolysis and reduction. This material was used to alkylate a second molecule of the above tosyl derivative of alcohol (62) to give the masked C_{15} chiral terpenoid unit (65), eventually transformed into the required C_{15} alcohol (66).

(67) R= Me (27) (63) (64) (68) R= Me (65) (69) (66) (70) R= CH_2Ph (71)

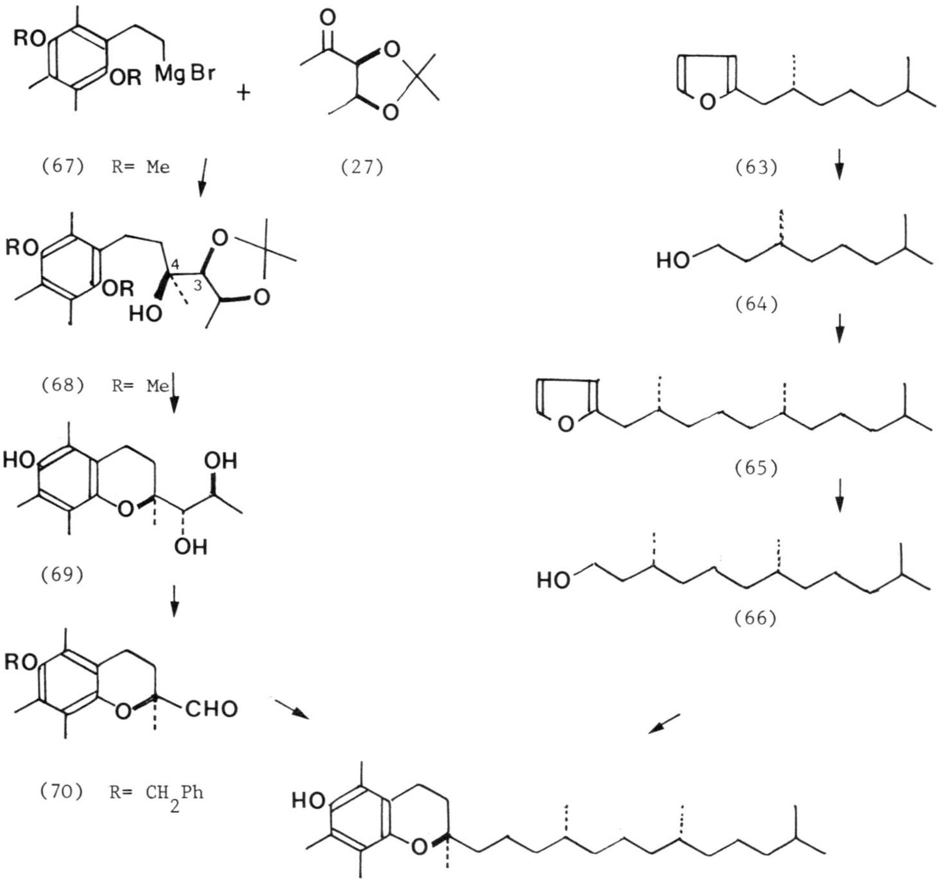

The chromanyl moiety (70) was synthetized via the C_5 chiral ketone (27), extruded from protected (61), to give the 3,4-threo adduct (68), on addition of the Grignard reagent (67). The construction of the chromanyl ring from the adduct (68) followed known procedures. The aldehyde (70) was obtained by oxidation of the chromanyl diol (69). In this last step, the two chiral centers of (27) which assisted the formation of the chiral center at position 4 of (68) are destroyed. The formyl group of (70) is the formyl group originally present in the aldehyde (60). Thus, the C_{29} framework of natural α-tocopherol (71), bearing three chiral centers of two different type (R,R'CHR" and R,R'C(OH)R") becomes accessible from the chiral educts (61) and (62), obtained in a single operation from readily available (60), using commercially available baker's yeast.[27]

In addition to the substances mentioned above, many other chiral products are being prepared from the educts (7) and (8), as examples of the versatility of this type of materials as chiral synthons complementary and/or alternative to the components of the 'pool of chirality' produced by Nature through the normal biosynthetic processes. The general significance of these products is further enhanced from the fact that they are obtained using commercially available baker's yeast, working in tap water, at room temperature and in short reaction times, whereas most of the chiral synthons obtained by microbial transformations of non-conventional substrates require the use of expecially grown cultures, not so easily accessible to synthetic organic chemists as baker's yeast, which might be now considered as a common shelf reagent.

REFERENCES

1 D.Seebach and H.-O.Kalinowski, Nachr.Chem.Techn.,1976,24,415

2 F.Arcamone, G.Franceschi, S.Penco and A.Selva, Tetrahedron Letters,1969,1007; F.Arcamone, S.Penco, A.Vigevani, S.Redaelli, G.Franchi, A.Di Marco, A.M.Casazza, T.Dasdia, F.Tormelli A.Necco and C.Soranzo, J.Med.Chem.,1975,18,703

3 D.Horton and W.Weckerle, Carbohydr.Res.,1976,46,227

4 J.P.Marsh, C.W.Mosher, E.M.Acton and L.Goodman, Chem.Commun.,1967,973

5 A.Fischli in "Modern Synthetic Methods"; R.Scheffold, Ed., Salle and Sauerländer: Frankfurt am Main, 1980; Vol 2, p 269

6 K.Kieslich, Microbial Transformations of non-steroid Cyclic Compounds, G.Thieme: Stüttgard, 1976

7 C.Fuganti and P.Grasselli, Chem.Ind.(London),1977,983

8 G.Bertolli, G.Fronza, C.Fuganti, P.Grasselli, L.Majori and F.Spreafico, Tetrahedron Letters, 1981,965

9 C.Fuganti, P.Grasselli and G.Marinoni, Tetrahedron Letters,1979,1161

10 C.Fuganti and P.Grasselli, J.C.S.Chem.Commun.,1978,299

11 G.Fronza, C.Fuganti and P.Grasselli, Tetrahedron Letters,1978,3941

12 R.Bernardi, C.Fuganti, P.Grasselli and G.Marinoni, Synthesis,1980,50

13 A.Nakano, S.Takimoto, J.Inanaga, T.Katsuki, S.Ouchida, K.Inoue, N.Okukado and M.Yamaguchi, Chem.Lett.,1979,1019

14 M.Uemure, K.Shimada, T.Tokuyama and J.W.Daly, Tetrahedron Letters,1982,4369

15 C.Fuganti, P.Grasselli and G.Pedrocchi-Fantoni, J.Org.Chem.,1983,48,905

16 G.Fronza, C.Fuganti and P.Grasselli, J.C.S.Chem.Commun.,1980,442

17 G.Fronza, C.Fuganti, P.Grasselli and G.Pedrocchi-Fantoni, Tetrahedron Letters,1981,5073

18 G.Fronza, C.Fuganti, P.Grasselli and G.Pedrocchi-Fantoni, J.Carbohydrate Chem.,1983,2,225

19 G.Fronza, C.Fuganti, P.Grasselli, G.Pedrocchi-Fantoni and C.Zirotti, Tetrahedron Letters,
   1982,4143

20 C.Fuganti, P.Grasselli, G.Pedrocchi-Fantoni, S.Servi and C.Zirotti, Tetrahedron Letters,
   1983,3753

21 C.Fuganti, S.Servi and C.Zirotti, Tetrahedron Letters,1983,5285

22 C.Fuganti, P.Grasselli and S.Servi, J.Chem.Soc.Perkin I,1983,241

23 C.Fuganti, P.Grasselli and S.Servi, J.C.S.Chem.Commun.,1982,1285

24 C.Fuganti, P.Grasselli, S.Servi and C.Zirotti, Tetrahedron Letters,1982,4269

25 G.Fronza, C.Fuganti, P.Grasselli, L.Majori, G.Pedrocchi-Fantoni and F.Spreafico, J.Org.Chem.,
   1982,47,3289

26 G.Fronza, C.Fuganti, P.Grasselli, G.Pedrocchi-Fantoni and C.Zirotti, Chem.Lett.,1984,225

27 C.Fuganti and P.Grasselli, J.C.S.Chem.Commun.,1982,205

Synthetic control leading to the synthesis of carbohydrates

Teruaki MUKAIYAMA

Department of Chemistry, Faculty of Science, The University
of Tokyo, Hongo, Bunkyo-ku, Tokyo 113, JAPAN

Abstract - Highly stereoselective reactions developed
based on the concept of "Synthetic Control" are described.
Various sugar derivatives are successfully synthesized by
the application of these reactions.

INTRODUCTION

The concept of "Synthetic Control" is characterized by the utilization of
chelate complexes of common metals for the intra- or inter-molecular
interactions of reacting species leading to highly stereospecific and/or
entropically advantageous reactions. This article summarizes a variety of
highly stereoselective reactions involving asymmetric reactions and useful
methodologies developed for the synthesis of sugar derivatives guided by
the principle of "Synthetic Control".

(I) STEREOSPECIFIC REACTIONS LEADING TO SUGAR DERIVATIVES
Various monosaccharides are generally synthesized starting from the readily
available sugars, such as glucose, galactose, etc. In recent years, much
interest has been focussed on the synthesis of various sugars by the
stereoselective carbon-carbon bond formation between simple organic
molecules. For this purpose, (1,2,5) several new reactions using common
metal salts such as potassium enolate, cadmium salt, zinc enolate, boron
enolate etc. were developed. Next, (3,4) a novel aldol reaction and
allylation of carbonyl compounds using divalent tin species were explored
and various highly stereoselective reactions in the acyclic system were
exploited. Convenient synthesis of a variety of monosaccharides has been
achieved by use of these reactions.

1. Asymmetric aldol reaction by the use of optically active imines

The potassium enolate prepared from atrolactic acid derivative ($\underline{1}$) and
potassium diisopropylamide undergoes aldol type reaction to give $\underline{2}$ in a
highly stereoselective manner. Further acid treatment gives threo-amino
acid ($\underline{3}$) in good yield. 2-Acetamido-2-deoxy-D-arabinose ($\underline{4}$) and 2-
acetamido-2-deoxy-D-ribose ($\underline{5}$) are prepared from (R)- and (S)-atrolactic
acid derivatives respectively by the above reaction.

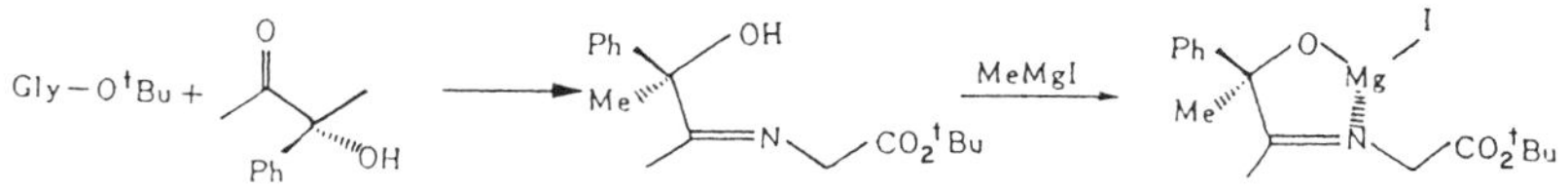

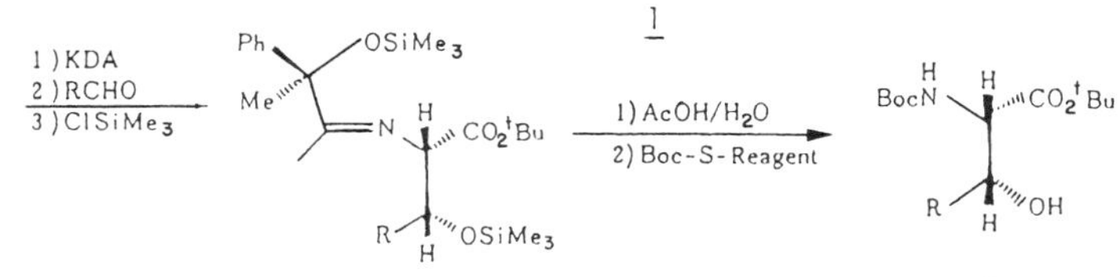

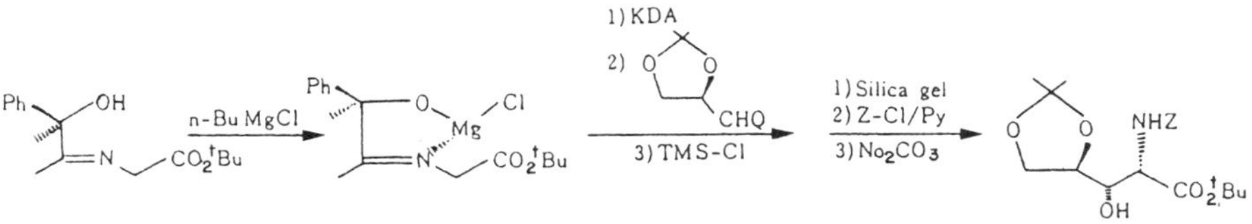

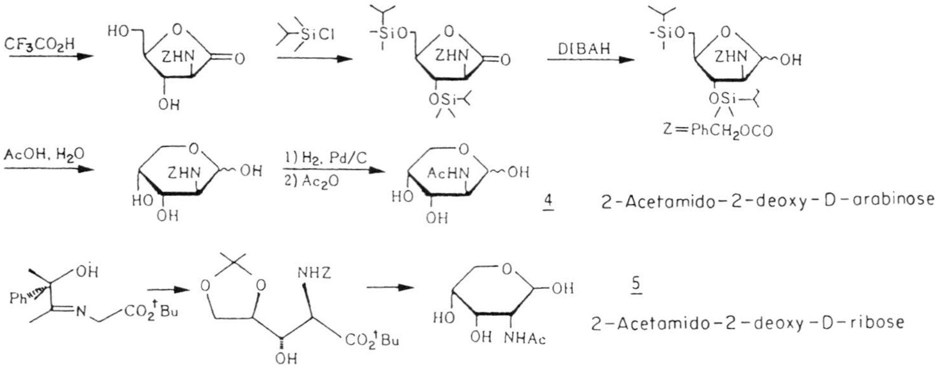

4 2-Acetamido-2-deoxy-D-arabinose

5 2-Acetamido-2-deoxy-D-ribose

2. Stereoselective additions to 2,3-O-isopropylidene-D- or L-glyceraldehyde

The cadmium salt of the 2-allyloxybenzimidazole derivative (7) reacts with various aldehydes to afford adducts (8) with high regio- and stereo-selectivities. The adducts (8)are subsequently transformed into trans-vinyloxiranes (9).

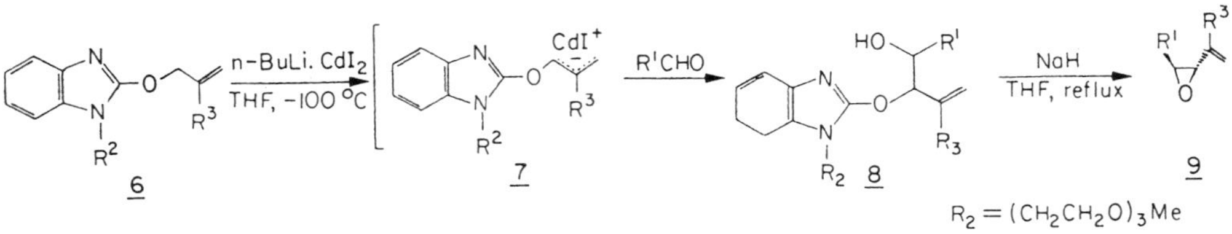

D- and L-Ribose are synthesized starting from 2,3-O-isopropylidene-D- and L-glyceraldehyde, respectively, by the application of this reaction.

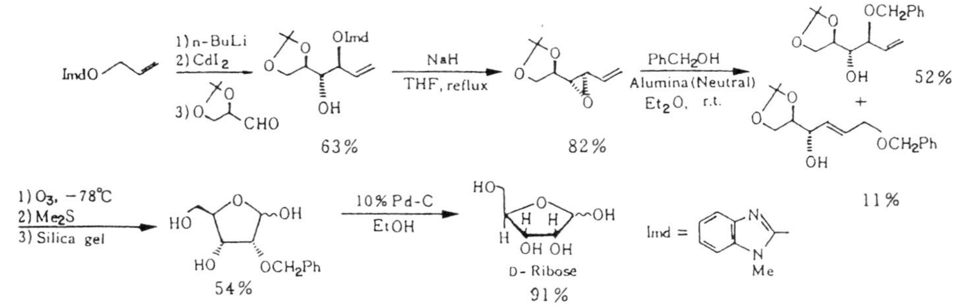

Convenient methods for the preparation of cis-vinyloxiranes (10) or homoallylic alcohols (11) are also explored by employing Et_3Al or R_3B in place of CdI_2 in the above reaction.

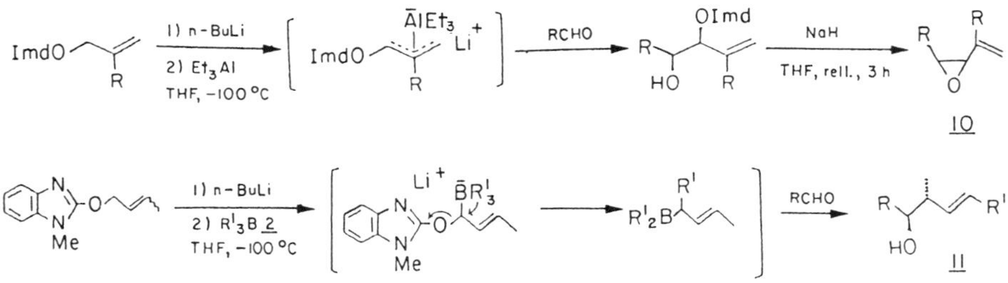

Zinc enolate (13) prepared from acetylene ether (12), pyridin-1-oxide, mercuric chloride, and zinc adds to aldehydes resulting in the formation of α-chloro-β-hydroxy ester (14) in good yield. Subsequent treatment with base gives epoxyester, which in turn is converted to 2-amino-2-deoxy-D-ribose stereoselectively in good yields.

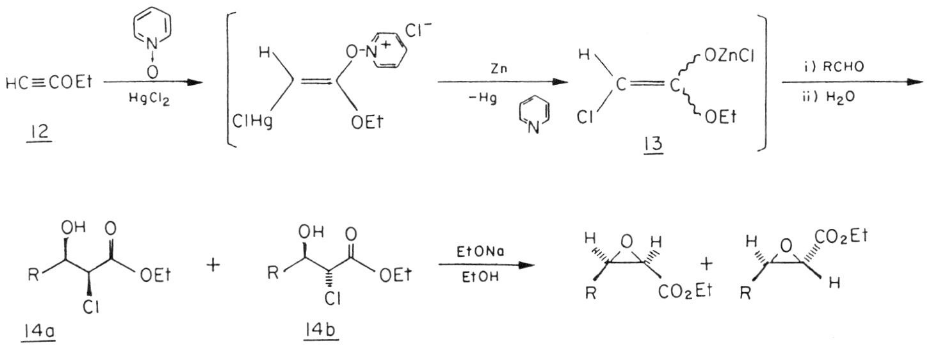

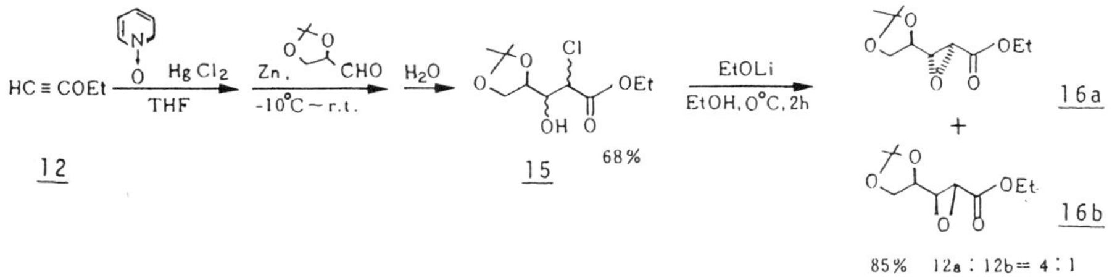

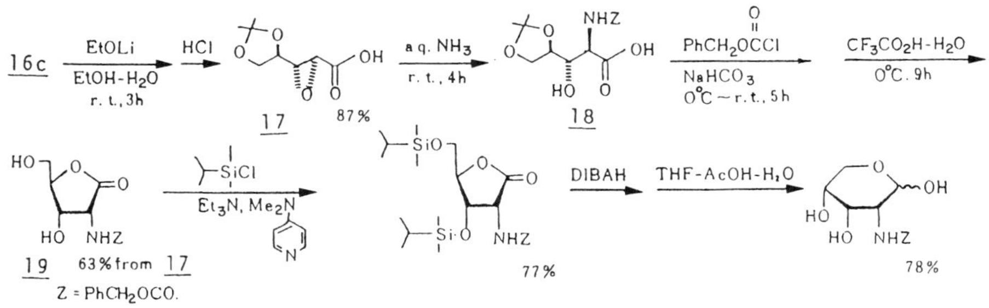

Replacement of pyridin-1-oxide by borinic acid in the above reaction gives vinyloxyborane intermediate (18) which further reacts with aldehydes to give the aldol products (19 and 20). 2-Deoxy-D-ribose is prepared efficiently by application of this reaction.

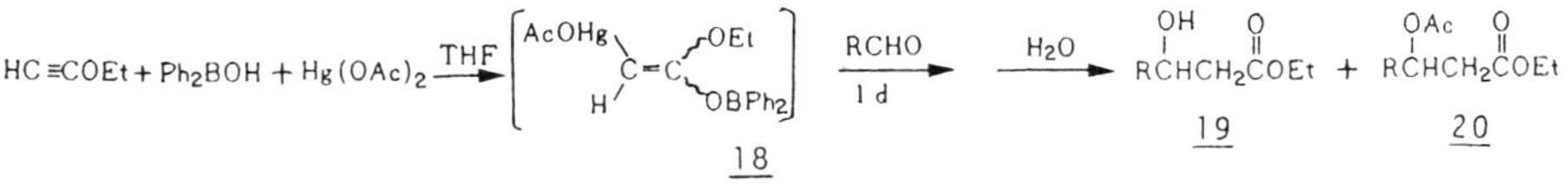

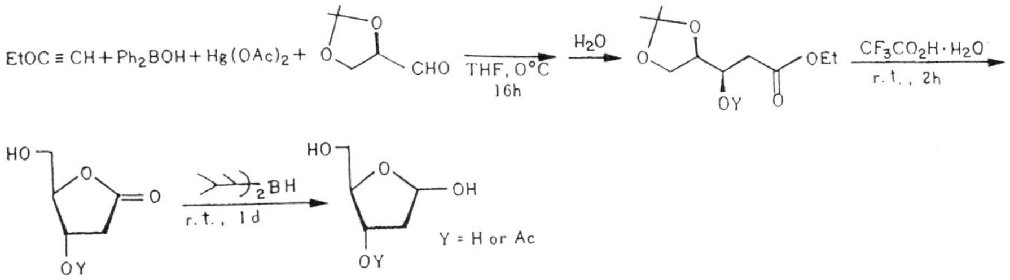

Stereocontrolled addition of 2-furyl anion to 2,3-O-isopropylidene-D-glyceraldehyde can be achieved by the addition of some metal salts. In the presence of zinc iodide, 2-furyllithium undergoes stereoselective addition to afford the anti adduct (22a) with high selectivity. This adduct can be transformed into D-ribulose through standard sequences.

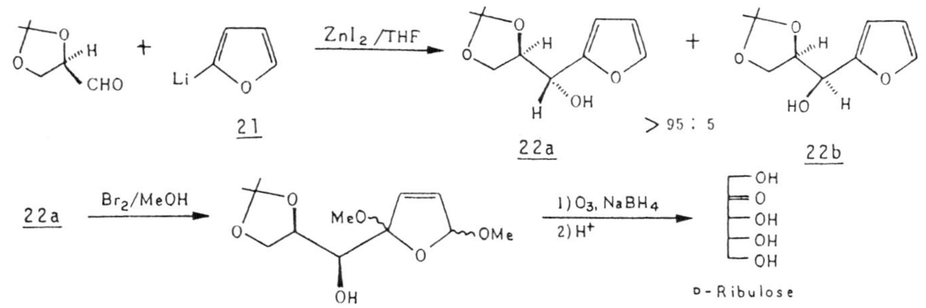

3. New and useful synthetic reactions by the use of stannous fluoride or metallic tin

Allyltin difluoroiodide, formed in situ by the oxidative addition of SnF$_2$ to allyl iodide, is found to react with carbonyl compounds to give homoallylic alcohols in excellent yields under mild reaction conditions.

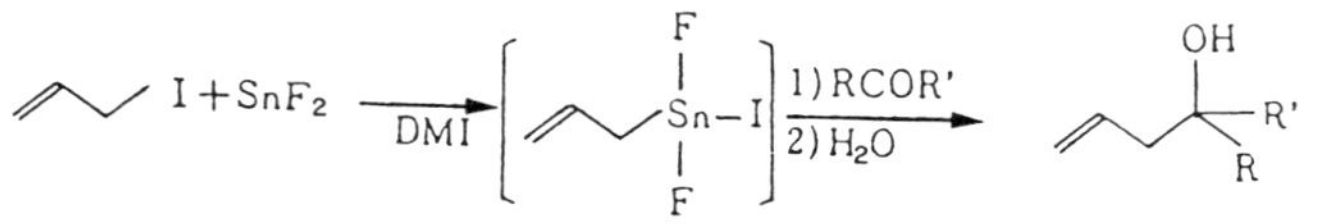

This reaction is applied to the synthesis of 2-deoxy-D-ribose.

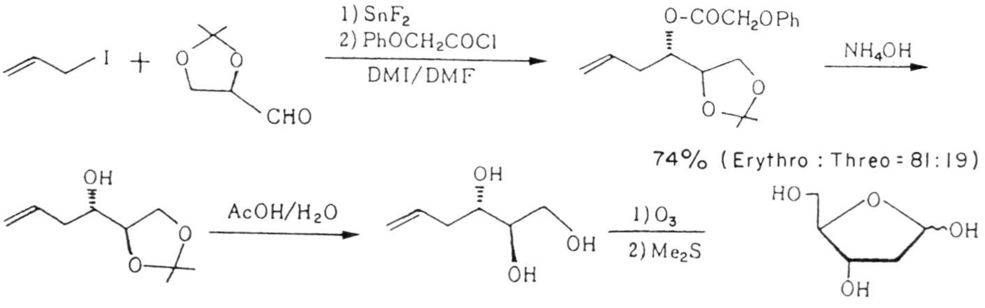

2,2,2-Tribromoethanol derivatives are also obtained by the reaction of carbon tetrabromide with aldehydes in the presence of SnF$_2$. 2,3-Di-O-acetyl-D-erythronolactone (23) is synthesized as shown below.

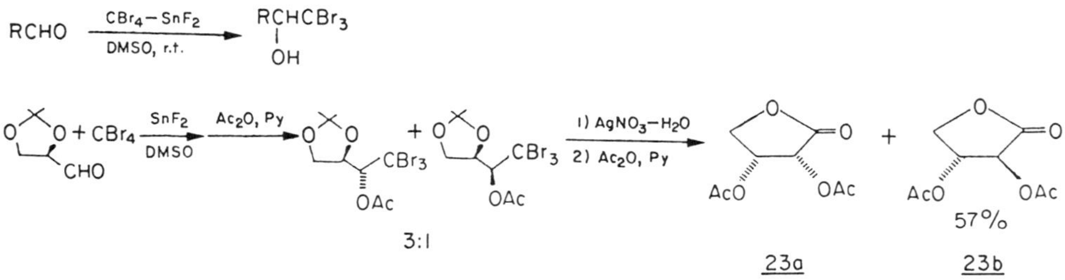

Metallic tin, Sn(0), is more effectively employed; for example, allylation reaction proceeds even with allyl bromide, and α-haloesters react with carbonyl compounds in the presence of Sn(0) to give β-hydroxyesters in good yields.

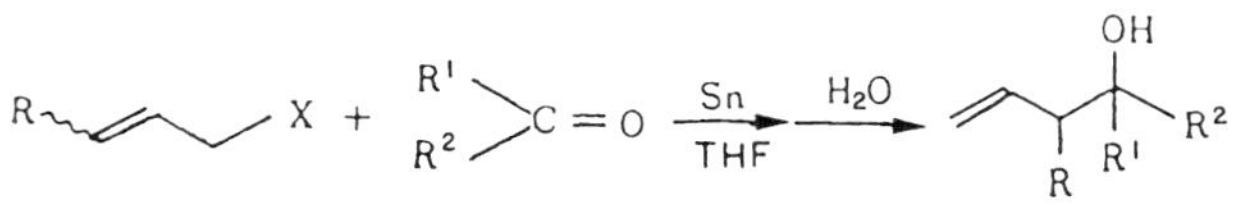

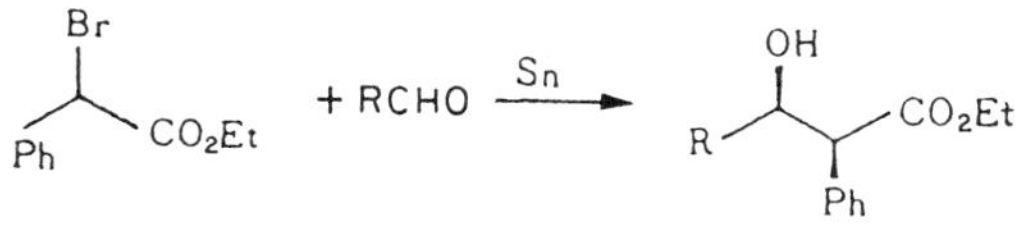

4. Stereoselective aldol reaction by the use of stannous enolates

Aldol reaction is one of the most fundamental and useful synthetic method in organic synthesis and enol borate is known to be the most efficient intermediate in view of its mild reaction conditions and stereoselectivity. Recently, stannous enolate has been found to have prominent features; stannous enolates generated from α-bromoketone and Sn(0) react with aldehydes in a highly regio and stereoselective manner. Successful method for the generation of stannous enolate is newly developed; the stannous enolates generated from ketones and stannous triflate display high reactivities as well as high stereoselectivities, and the crossed-aldols between two different ketones is also realizable in good yields.

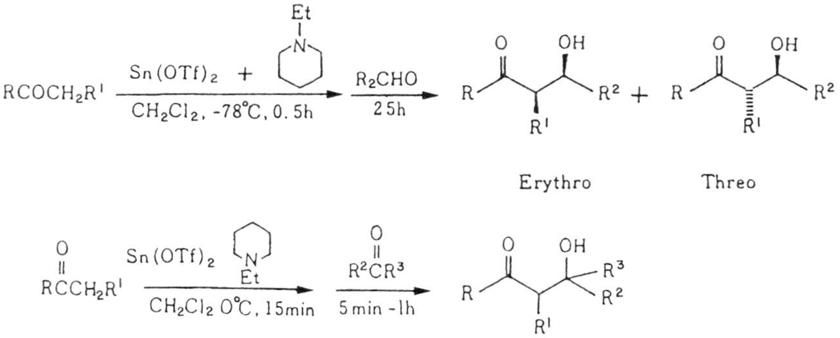

Various cis-α,β-epoxycarbonyl compounds are stereoselectively prepared via α-bromo-β-hydroxycarbonyl compounds by application of this Sn(II) mediated aldol reaction to α-bromocarbonyl compounds.

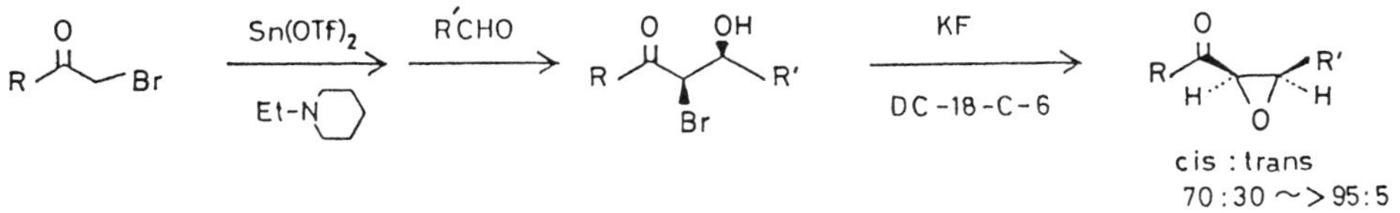

A facile synthesis of the branched chain sugar, 2-C-methyl-D,L-lyxofuranoside (24) has been achieved by using the tin(II) enolate of 1,3-dihydroxy-2-propanone derivative and methyl pyruvate.

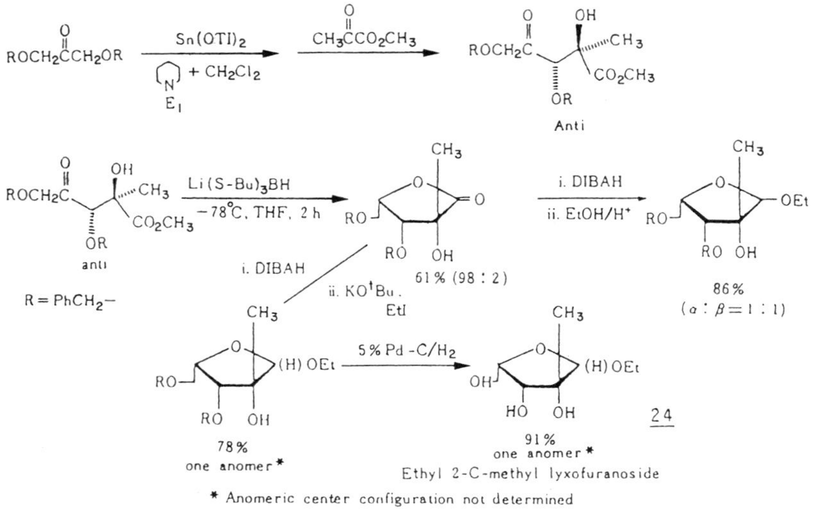

* Anomeric center configuration not determined

Although several asymmetric aldol reactions have been reported recently, chiral auxiliary groups are usually attached to the ketone equivalent molecules in these reactions. No example existed so far in aldol type reaction where two achiral carbonyl compounds are used for constructing a chiral molecule with the aid of a ligand. Based on the considerations that divalent tin having vacant d orbitals is capable of accepting a bidentate ligand coupled with the fact that chiral diamines derived from (S)-proline are efficient ligands in certain asymmetric reactions, enantioselective aldol reaction via divalent tin enolates with chiral diamines was explored. A highly enantioselective cross aldol reaction between aromatic ketones or 3-acetylthiazolidine-2-thione and various aldehydes has been achieved, in which chiral diamines derived from (S)-proline worked very effectively as ligands. This is a first example for the formation of cross aldol in high optical purity starting from two achiral carbonyl compounds by employing chiral diamines as chelating agents.

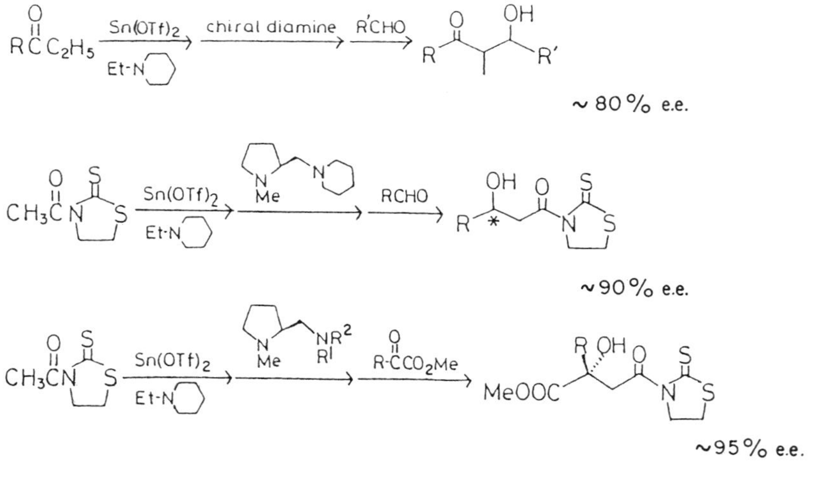

These compounds derived from 3-acetylthiazolidine-2-thione are very versatile chiral materials, capable of being transformed into various synthetic intermediates as demonstrated before. Furthermore, in the stannous enolate mediated aldol-type reactions of 3-(2-benzyloxyacetyl)thiazolidine-2-thione, stereochemical course of the reaction is dramatically altered by the addition of TMEDA as a ligand. High asymmetric induction is also achieved by the addition of chiral diamine derived from (S)-proline.

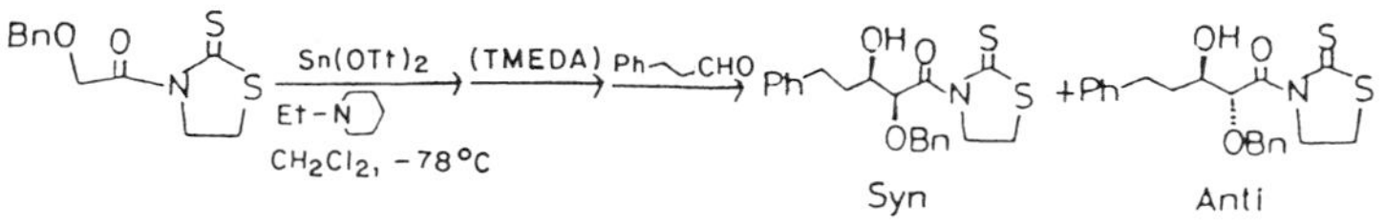

additive	yield(%)	syn:anti
——	62	75 : 25
TMEDA	70	14 : 86
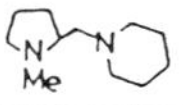 (structure)	81	13 : 87 94% e.e.

Thus the characteristic features of tin(II) enolates enable the stereoselective synthesis of aldol products even from two different ketones. Combination of Sn(OTf)$_2$ and N-ethylpiperidine provides an easy approach to tin(II) enolates, whereas tin(IV) enolates have been prepared through relatively laborious multi step procedures.   Enantioselective aldol reaction effected by chiral diamines also enhances the utility of tin(II) enolates in organic synthesis.

5. New 4-carbon building block for sugar synthesis

Several sugar derivatives are successfully synthesized by employing the newly devised stereoselective carbon-carbon bond forming reactions starting from 2,3-O-isopropylidene-D- or L-glyceraldehyde as mentioned before. With these results in mind, a new and potentially useful 4-carbon building block for the L-sugars, 4-O-benzyl-2,3-O-isopropylidene-L-threose (25), is readily prepared conveniently starting from L-tartaric acid.   L-2-Deoxygalactose(26), 3-amino-2,3-dideoxy-L-xylo-hexose (27) and L-diginose (28) are synthesized conveniently from the aldehyde (25).

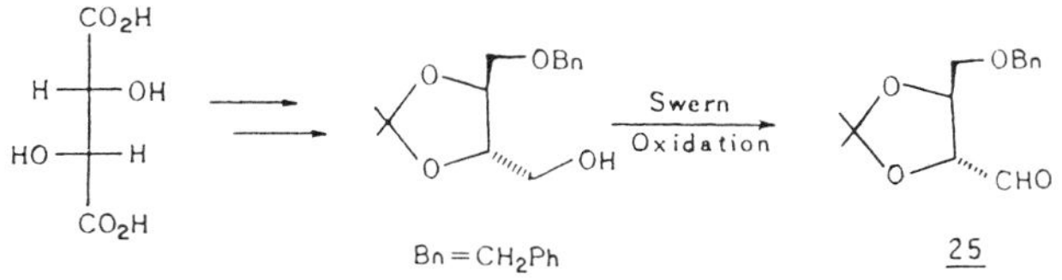

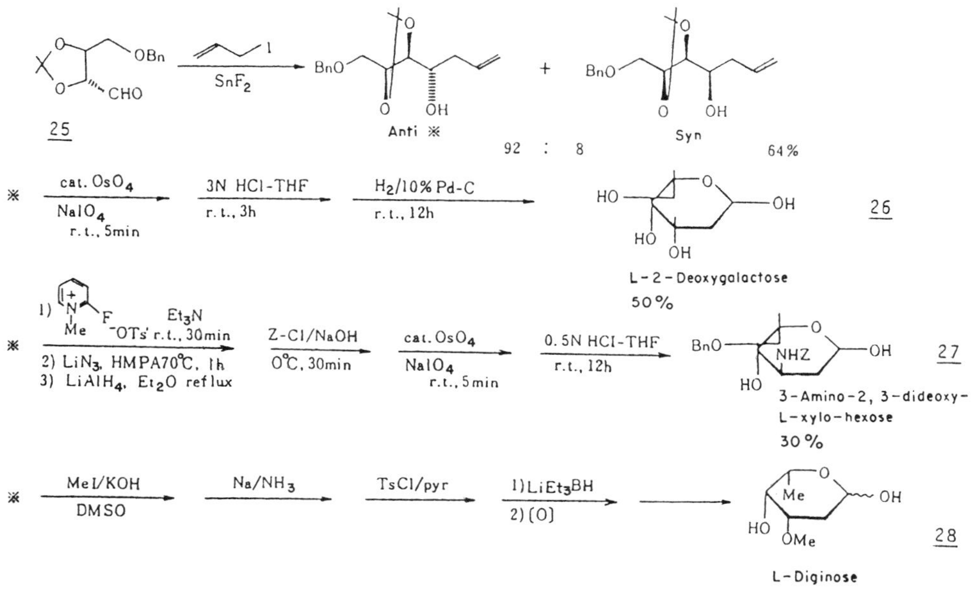

Further, convenient syntheses of the two fragments of Polyoxin J, deoxypolyoxin C and 5-O-carbamoylpolyoxamic acid, have been achieved in a highly stereoselective manner starting from the chiral building block, 4-O-benzyl-2,3-O-isopropylidene-L-threose, which lead to a formal total synthesis of Polyoxin J.

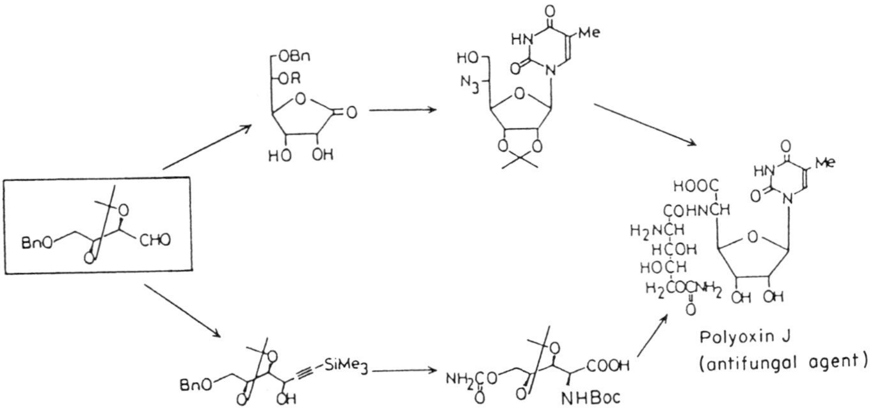

Polyoxin J
(antifungal agent)

(II) Stereoselective Glycosylation Reactions

Besides stereoselective synthesis of various monosaccharides,
stereoselective reaction for the preparation of glycosides is an important
problem in the synthetic field of carbohydrate chemistry. However, the
classical methods, which require the assistance of heavy metal salts or
drastic reaction conditions, are still mostly employed in the synthesis of
such compounds. Taking these disadvantages into consideration, new
glycosylation reactions, which proceed under mild reaction conditions with
high selectivity, are exploited.

1. The glycosyl fluoride method and the trityl perchlorate method

To achieve higher stereoselectivity, the combination of various sugar
derivatives and activating reagents are examined. As a result, α-glucosides
are prepared with high stereoselectivity by the reaction of β-glucosyl
fluoride ($\underline{29}$) with various hydroxy compounds in the presence of $SnCl_2$ and
$AgClO_4$ or $SnCl_2$ and trityl perchlorate.

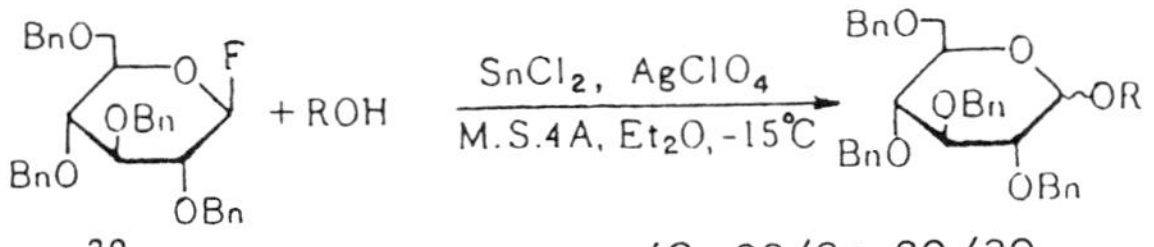

$\alpha/\beta = 92/8 \sim 80/20$

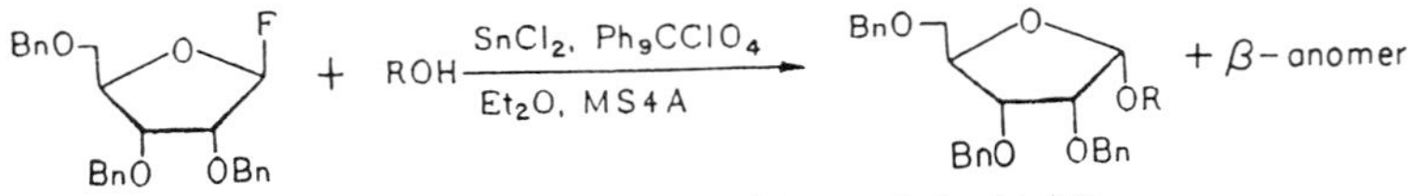

$\alpha/\beta = 88/12 \sim 81/19$

Similarly, α-glycosides are stereoselectively obtained by the reaciton of
1-O-acyl sugars with alcohols in the presence of trityl perchlorate.

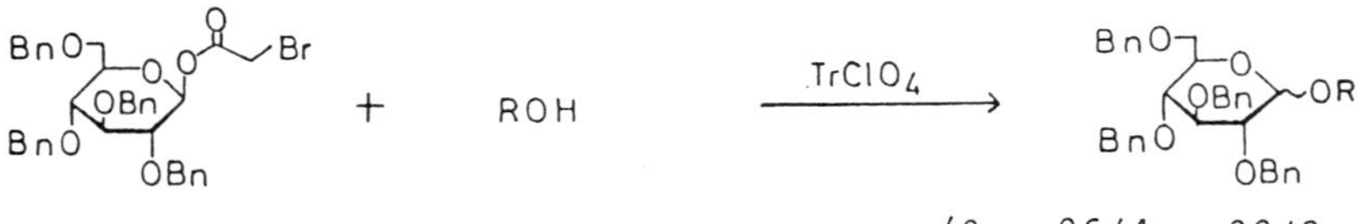

$\alpha/\beta = 96/4 \sim 92/8$

$\alpha/\beta = 79/21 \sim 70/30$

2. New approaches for the synthesis of glucosides. An alkoxylation of silyl enol ethers.

α-Alkoxyketones and α-ketoacetals are prepared in good yields by the reaction of silyl enol ethers and alkyl hypochlorites catalyzed by $(Ph_3P)_4Pd$.

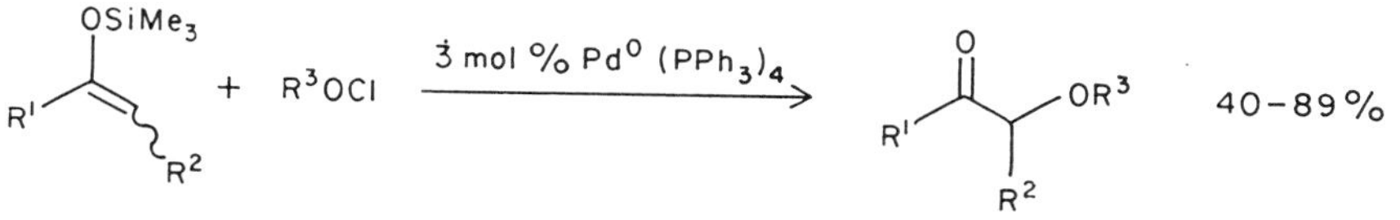

$$40-89\%$$

The reaction is applied to the synthesis of α-glucoside (30) as shown in the following equation.

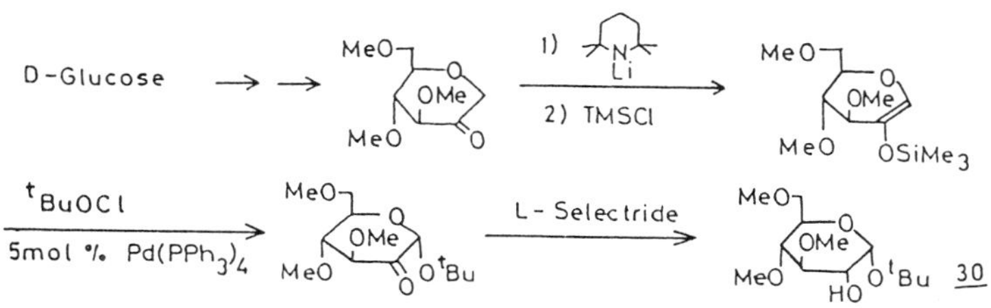

3. A cyclization of hydroxy enol ethers

A stereoselective cyclization of (Z)-(2R, 3R, 4R)-6-cyclohexyloxy-1,3,4-tribenzyloxy-5-hexen-2-ol (32) promoted by $Hg(OCOCF_3)_2$ or PhSeCl is successfully achieved. 2-Deoxy-α-hexopyranoside derivative (32) is obtained almost exclusively by the treatment of 25 with $Hg(OCOCF_3)_2$ followed by reductive work up, while a predominant formation of the β-anomer (33) is performed by the reaction of 31 with PhSeCl, and the successive deselenylation.

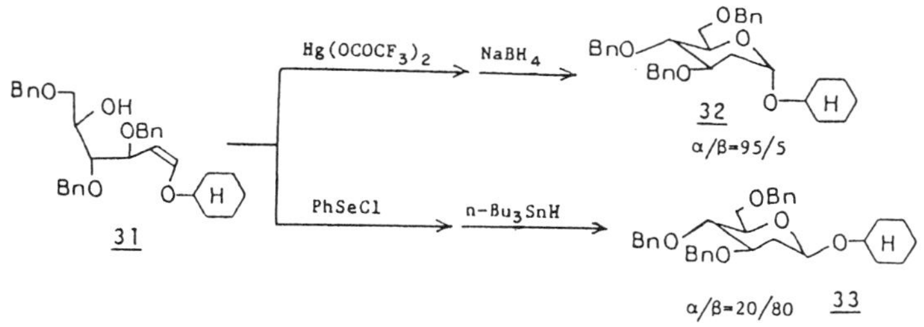

N-Iodosuccinimide also promotes the cyclization to give the α-glucoside derivative, and the disaccharide derivative (35) is synthesized with high stereoselectivity starting from hydroxy vinylether (34).

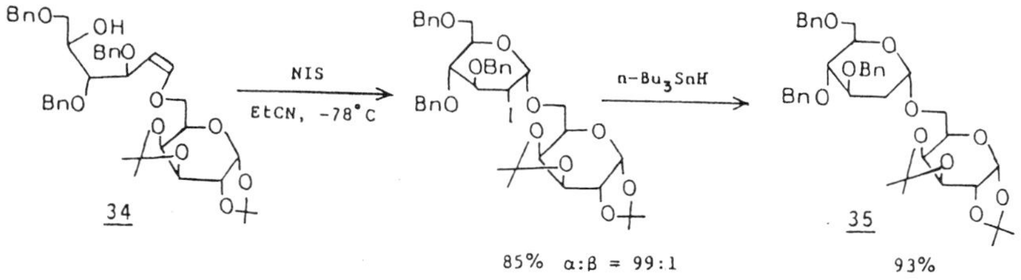

NIS = N-iodosuccinimide

Routes to carbocyclic compounds of pharmaceutical importance from carbohydrates

Steven Chew, Robert J. Ferrier, Petpiboon Prasit, and Peter C. Tyler

Department of Chemistry, Victoria University of Wellington, Wellington, New Zealand

Abstract - A simple method for converting 6-deoxyhex-5-enopyranose derivatives into deoxyinososes offers access to a wide range of functionalised cyclohexane derivatives. From appropriate unsaturated disaccharides glycosylated inosose products are obtainable, and these afford access to compounds of the aminoglycoside antibiotic series. The same rearrangement reaction has been used to obtain tetracyclic compounds with the carbon skeleton of anthracycline aglycones following reaction of a cyclic carbohydrate enone with an o-quinodimethide generated by use of ultrasonics. Functionalised hexahydroanthracenes can be made by a similar approach. Photochemical ring closure of a nona-3,8-dienulose, which was made from a readily available unsaturated hexose derivative, gave a bicyclo[3.2.0]heptane product from which routes to the prostaglandins and carbacyclin have been developed. Efforts to extend the application of the photochemical procedure will be noted.

Although the past decade has seen very considerable advances in the use of sugars as synthetic precursors of optically pure non-carbohydrate compounds (Ref.1) and the consequent rapprochement of carbohydrate chemistry and "orthodox" organic chemistry, developments have mainly resulted in the production of acyclic polyfunctional compounds and saturated heterocyclic compounds - notably, of course, 5- and 6-membered cyclic compounds with oxygen as the hetero atom.

To a lesser extent sugars have been used to prepare carbocyclic compounds, and it is in this area that our interests over recent years have centred. In particular we have noted that few examples have been reported of the conversion of carbohydrate derivatives into functionalised cyclopentanes - despite the abundance of important naturally occurring compounds of this class - and we have reviewed the procedures which have been applied (Ref.2). Methods for obtaining cyclohexanes, which usually involve the generation of carbanionic activity at C-6 of aldohexose derivatives, have been known since 1948 when 6-deoxy-6-nitro-D-glucose was converted by alkali treatment into deoxynitroinositols (Ref.3). More recently, a hexos-5-ulose 6-phosphate, under similar conditions, has been shown to give isomeric inosose monophosphates in a reaction which mimics the cyclisation which occurs during the biosynthesis of inositols (Ref.4). In the course of our work we have encountered new ways of making cyclopentane and cyclohexane derivatives from sugars and have sought to use them in the development of synthetic routes to compounds of pharmaceutical importance. With the prostaglandins, complete routes to versatile synthetic intermediates have been developed, but with the anthracyclines and aminoglycoside antibiotics progress to date has allowed identification of new synthetic routes without the completion of synthesis of specific compounds of pharmaceutical interest.

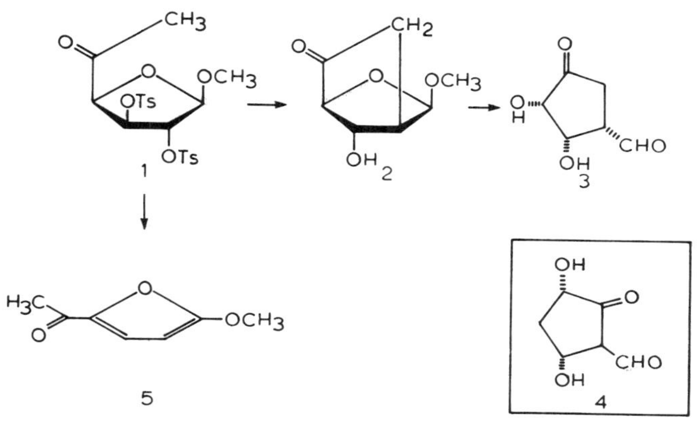

First attempts in our laboratory to prepare a cyclopentane aimed to utilise the potential nucleophile at C-6 of the methyl ketone 1 to displace the leaving group at C-2 and, on treatment with base, the ketone gave some of the bicyclic cyclopentanone 2 by way, no doubt, of the 2,3-D-*ribo*-epoxide (Ref.5). On acid hydrolysis the aldehyde 3 would have been obtainable as a useful synthetic starting material structurally related to the aldehyde 4 which possesses appropriate functionality and stereochemistry to serve as a means of access to the prostaglandins. The product 2, however, was formed by a ring closure reaction which is disfavoured by stereo-electronic factors being similar to a 5-*Endo*-Trig process (Ref.6), and with a β-elimination reaction open to it, it is not surprising that under all the basic conditions tried compound 1 and related ketones gave mainly the furan 5.

In a second attempt at cyclopentane synthesis the methoxymercuration product 6 of the alkene 8 was tested to see whether potential nucleophilic activity at C-6 could be used to give the bicyclic compound 7, but neither 6 nor acyclic 6-mercuriated analogues could be induced to ring close in the desired manner (Ref.7). When, however, the alkene 8 was heated in aqueous acetone with mercury(II) chloride to cause hydroxymercuration at C-5, C-6 and thereby give the hemiacetal 9 in whose carbonyl form C-6 would be more strongly nucleophilic, a high yield of a crystalline product was obtained which was the cyclohexanone 11 (Ref.8). Rather, therefore, than acting at C-2 the nucleophilic C-6 had taken part in an intramolecular aldol reaction with the liberated aldehydic centre at C-1 of the onal 10. The stereoselectivity of the ring closure was very marked with the isolated yield of the axial alcohol 11 being 83%, and in subsequent work others have also shown that this reaction occurs with good (Ref.9), but not complete (Ref.10), selectivity.

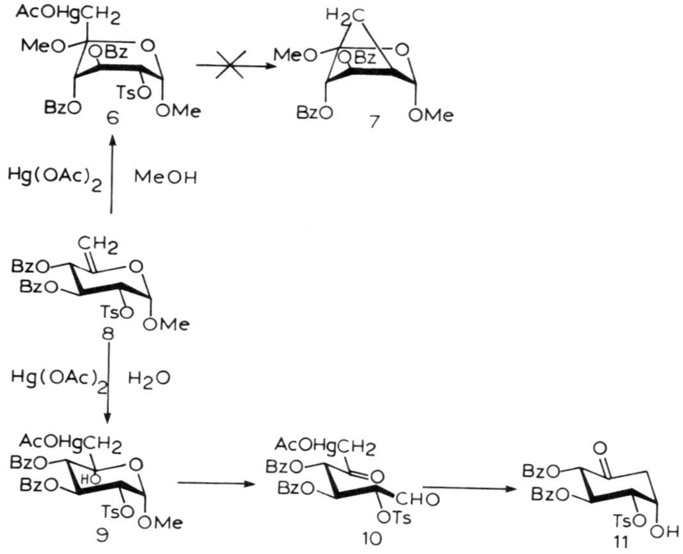

The deoxyinososes, e.g. 11, which are obtainable by this reaction have clear potential as precursors of the aminoinositol components of aminoglycoside antibiotics e.g. kanamycin A(17) (Ref.9-11), and since such compounds contain α-D-glucopyranosyl units or, more commonly, aminodeoxy derivatives thereof bonded to the aminocyclitol units (Ref.12) it is possible to use synthetic strategies which involve glycosylations of appropriate aminocyclitol derivatives or application of the above reaction to α-linked unsaturated disaccharide derivatives followed by α-glucosylation. Such an unsaturated compound has been made by a complex route from methyl hexa-O-acetyl-6-deoxy-6-iodo-β-D-maltopyranoside (Ref.9a), but it was rationalised that octa-O-acetyl-β-D-maltose ought to undergo brominating at C-5 with bromine radicals generated photochemically from N-bromosuccinimide to give compound 12 from which the alkene 13 would be derivable by use of zinc-acetic acid. In practice the alkene (13) was obtainable by this direct approach in 12% yield from maltose octa-acetate following a chromatographic separation, and on heating in aqueous acetone with mercury(II) acetate it gave the derivative 14 in high yield (Ref.13). Amination of this product (Ref.9a) gives access to the pseudodisaccharide class of aminoglycosides (Ref.12) and treatment with 1,5-anhydro-2,3,4,6-tetra-O-benzoyl-D-*arabino*-hex-1-enitol (15) in the presence of catalytic boron trifluoride (Ref.14) gave the α,α-linked 'pseudotrisaccharide' 16 in 68% yield. Appropriate acetate migration in the hydroxyketone 14 (Ref.9a) and α-glycosylation of the isomeric hexa-acetate would give products with the more common cyclitol substitution pattern and appropriate aminations would lead to a range of known aminoglycoside antibiotics; these latter reactions await trial and their successful completion would encompass a new synthetic route to these substances.

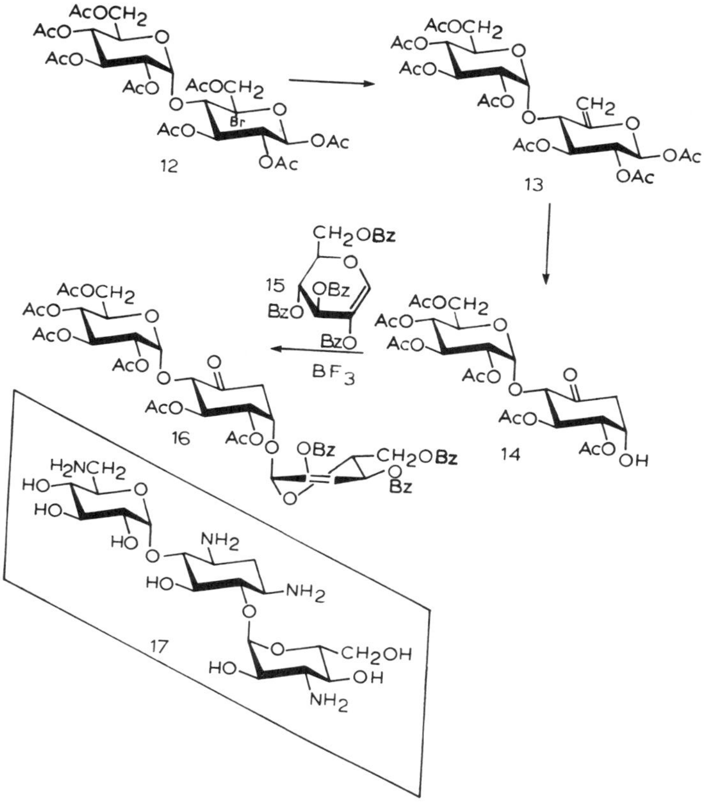

The deoxyinosose synthesis described above has also allowed access to tricyclic and tetra-cyclic compounds of the hydroanthracene and naphthacene series. In particular, a route to the anthracyclinones has been sought and developments to date (Ref.15) have produced compound 28 which has the carbon skeleton and A-ring functionality and stereochemistry of 4-deoxy-β-rhodomycinone 29 (Ref.16). It was found possible to trap o-quinodimethane (18) (Ref.17), generated from α,α-dibromo-o-xylene using zinc under ultrasonic irradiation (Ref.18) with the

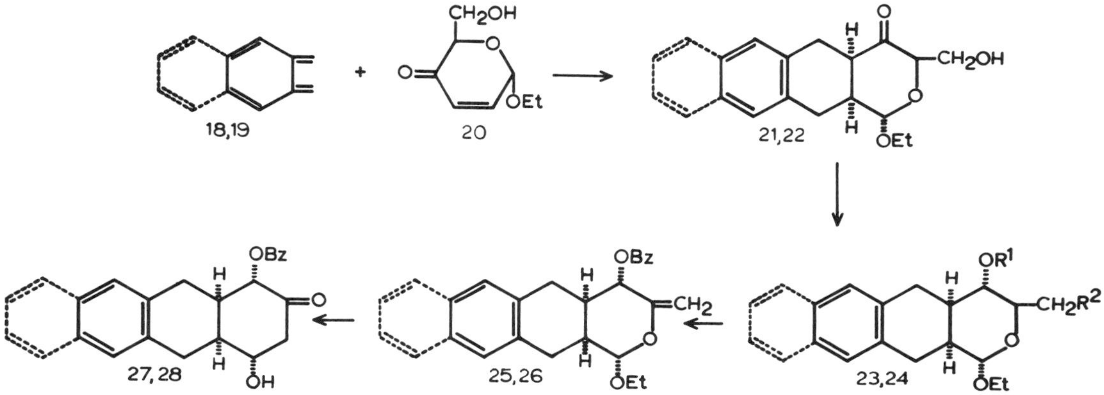

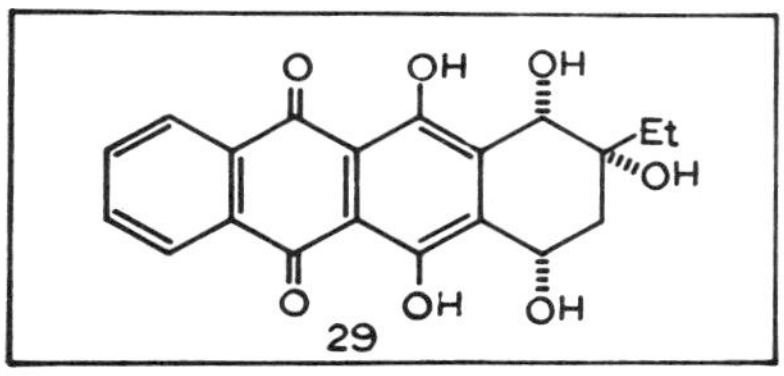

The dotted rings imply that the lower-numbered compounds did not have these rings whereas the higher-numbered compounds did have them.

D-glucose-based enone 20 to give the tricyclic ketone 21 in 70% yield and this, on reduction, afforded in 65% yield the diol 23 (R^1 = H, R^2 = OH) from whose ^{1}H n.m.r. spectrum it was established that the cycloaddition had occurred from the β-face of the enone as had the reduction of the carbonyl group of compound 21. From the diol 23 (R^1 = H, R^2 = OH) the alkene benzoate 25 was obtained by way of the iodide 23 (R^1 = Bz, R^2 = I) and it was converted efficiently into the hexahydroanthracene ketone 27 by the mercury-catalysed process. Parallel work with the o-quinodimethane 19 derived from 2,3-bis(bromomethyl)naphthalene gave an analogous set of tetracyclic products 22, 24 (R^1 = H, R^2 = OH), 24 (R^1 = Bz, R^2 = I), 26 and 28, the yield in the cycloaddition step in this case, however, being about 25%. Current work is concerned with the production of oxygenated derivatives of ketone 28 and with tertiary alcohols derived from the ketonic function.

Bernet and Vasella's reports (Ref.19) that 6-bromo-6-deoxyhexopyranosyl compounds can be readily converted into *aldehydo*-5,6-dideoxyhex-5-enoses by treatment in ethanol with zinc and hence, by 1,3-dipolar cycloaddition processes, into 2-aza-3-oxabicyclo[3.3.0]octanes via nitrone intermediates, describe the first efficient method for obtaining the cyclopentane ring system from sugars. The reaction conditions are attractively mild and when applied to the 6-iodide 30, from which the alkene 8 was derived, afforded first the enal 31 and then the

isoxazolidine 32 in 73% overall yield. Reduction with Raney nickel resulted in cleavage of the N-O bond and spontaneous intramolecular displacement of p-toluenesulphonic acid to give the epimine 33. This was then converted into the iodolactone 34 by treatment with m-chloroperbenzoic acid to give the deaminated alkene and then the side chain was extended by way of the tosyl ester and the derived nitrile. Hydrolysis and iodolactonisation gave the iodide 34 which was converted into the epoxylactone 35 by reduction with tributyltin hydride, deacylation and dehydration in DMF by use of sodium hydride followed by p-toluenesulphonylimidazole (Ref.20). Enantiomerically pure 35, more usually represented as 35A, is a well established synthetic precursor of the prostaglandins, e.g. prostaglandin $F_{2\alpha}$ 36 (Ref.21), and therefore a more general route to these compounds has been defined than was used in Stork's pioneering, specific synthesis of 36 from D-*glycero*-D-*gulo*-heptono-γ-lactone (Ref.22).

A further prostaglandin precursor, related to the epoxylactone 35, is the epoxycyclobutanone derivative 40 (Ref.23) which also became available from the iodide 30 and the enal 31. Chain extension of the latter gave the dienone 37 which, on irradiation at 350 nm, underwent an intramolecular 2+2 cycloaddition reaction to afford the cyclobutane ketone 38 in 86% yield. From this the epoxyacetal 39 was made by Baeyer-Villiger oxidation, hydrolysis of the resultant acetate ester, oxidation to the cyclobutanone, acetalation and de-esterification with potassium carbonate in methanol-dichloromethane (Ref.24). The required product 40 was then produced by silylation, highly selective ring opening of the epoxide ring with lithium aluminium hydride, benzoylation, desilylation, tosylation and treatment of the tosyl-benzoate with base (Ref.25).   Representation 40A reveals the relationship of the product to prostaglandin $F_{2\alpha}$ 36.

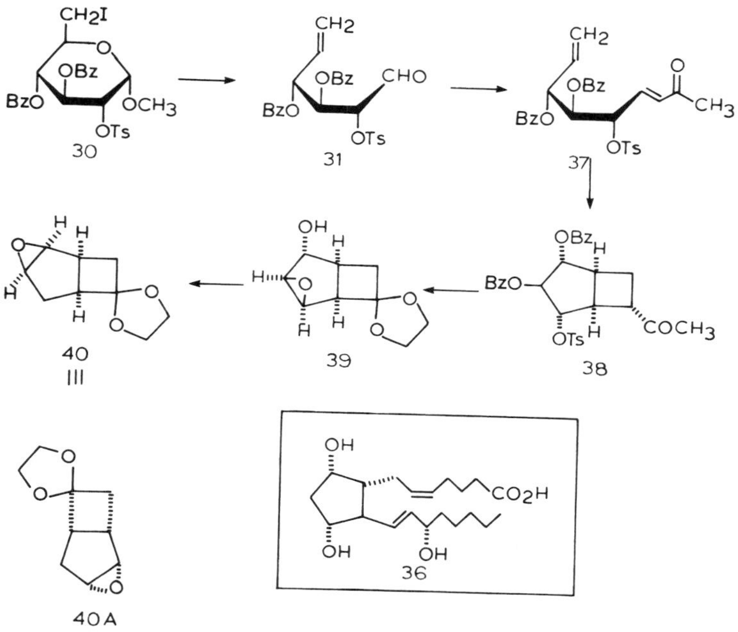

Carbacyclin 46 is an artificial, stabilised modification of the natural prostacyclin 45 and exhibits the potent inhibition of blood platelet aggregation shown by the natural product. It displays no obvious disadvantageous physiological effects and, being no longer a vinyl ether, is resistant to the ready hydrolysis exhibited by 45 (Ref.26).   The cyclopentanone 44, which is a synthetic precursor (Ref.27), has now been made from the epoxyacetal 40.

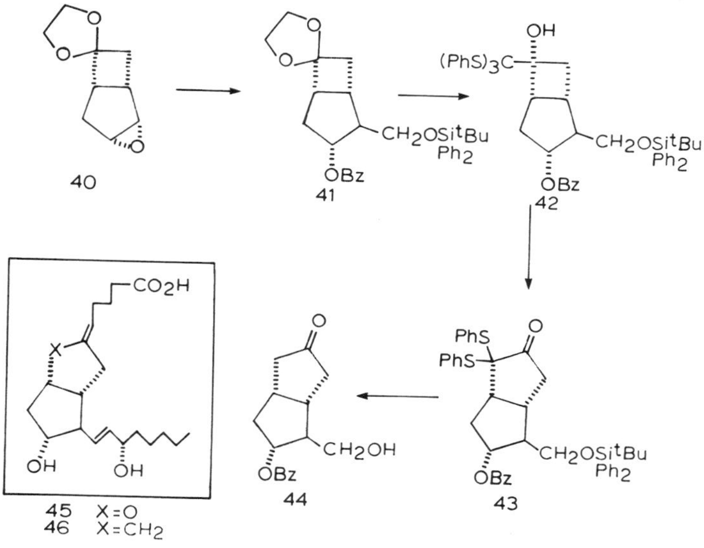

This can be ring-opened with good selectivity with the dithianyl anion (Ref.23a), and then was converted into the silylated benzoate 41 the cyclobutanone analogue of which was treated with tris(phenylthio)methyl anion to give the adduct 42.   Reaction with a mercury(II) salt caused a carbonium ion rearrangement process to occur and gave exclusively the ketone 43 which afforded the target 44 following Raney nickel reduction and hydrolysis of the silyl ether (Ref.25).

Although the above mentioned routes to the prostaglandin intermediates 35 and 40 have the advantage over alternative procedures of affording the compounds in the required enantio-meric forms, they are lengthier than is desirable, and an effort has therefore been made to abbreviate the approach to the cyclobutanone derivative 40 which follows the route 31 → 37 → 38 → 39 → 40.   Abbreviations in the synthesis were sought in three ways: (i) the removal of O-2 from the initial sugar derivative (i.e. OTs in 38), (ii) differentiated substitution at O-3 and O-4 (the benzoyloxy groups in 38) which would permit stereospecific epoxide formation, and (iii) the replacement of the conjugated enone in 37 by a 1,1-disubstituted alkene (i.e. a ketene derivative) to permit the direct formation of a cyclobutanone deriva-tive following the intramolecular 2+2 cycloaddition process.   A series of compounds 51 which met all of these requirements were therefore prepared with the hope that a suitable precursor for 52 could be found.   These could very readily give access to 40.

R= CH₃OCH₂-

Treated with hydrogen bromide in acetic acid followed by methanol, calcium D-gluconate readily gives 2,6-dibromo-2,6-dideoxy-D-mannono-1,4-lactone 47 (Ref.28) which was converted into the differently substituted 48 by selective benzoylation followed by reaction with dimethoxymethane and phosphorus pentaoxide.   Compound 49 was then produced efficiently by use of a zinc-silver couple in refluxing aqueous ethanol, reduction with di-isobutylaluminium hydride gave the free sugar which was silylated at O-4 by way of the N,N-diphenylimidazoli-dine derivative to give the aldehyde 50 and hence the four alkenes 51 (X=Y=Cl; X=Y=Br; X,Y=S(CH₂)₃S; X=CN, Y=NEt₂) (Ref.29).

Neither photochemical nor chemical methods were successful in permitting the intramolecular cyclisation of the four compounds 51, and the methyl ketone 51 (X = COCH₃, Y = H), the analogue of the dienone 37, also failed to undergo the desired reaction, preferring instead to react by deconjugative isomerisation.   We conclude that the efficient ring closure exemplified by 37 → 38 is not a reaction of wide applicability for access to the required ketone 53.

REFERENCES

1. (a) S. Hanessian, Acc. Chem. Res., 12, 159-165 (1979); (b) B. Fraser-Reid and R.C. Anderson, Prog. Chem. Org. Nat. Prods., 39, 1-61 (1980); (c) A. Vasella, Modern Synthetic Methods, 2, 172-267 (1980); (d) Specialist Periodical Reports, Carbohydrate Chemistry, Royal Society of Chemistry, London (1976-); (e) S. Hanessian, Total Synthesis of Natural Products The "Chiron Approach", Pergamon Press, Oxford (1983).
2. R.J. Ferrier and P. Prasit, Pure Appl. Chem., 55, 565-576 (1983).
3. J.M. Grosheintz and H.O.L. Fischer, J. Am. Chem. Soc., 70, 1479-1484 (1948).

4. D.E. Kiely and W.R. Sherman, J. Am. Chem. Soc., 97, 6810-6814 (1975).
5. R.J. Ferrier and V.K. Srivastava, Carbohydr. Res., 59, 333-341 (1977).
6. J.E. Baldwin and L.I. Kruse, J. Chem. Soc., Chem. Commun., 233-235 (1977).
7. R.J. Ferrier and P. Prasit, Carbohydr. Res., 82, 263-272 (1980).
8. R.J. Ferrier, J. Chem. Soc., Perkin Trans. 1, 1455-1458 (1979).
9. (a) F. Sugawara and H. Kuzuhara, Agric. Biol. Chem., 45, 301-304 (1981); (b) N. Sakairi
    and H. Kuzuhara, Tetrahedron Lett., 23, 5327-5330 (1982); (c) I. Pelyvás,
    F. Sztaricskai, and R. Bognár, J. Chem. Soc., Chem. Commun., 104-105 (1984).
10. D. Semeria, M. Philippe, J.-M. Delaumeny, A.-M. Sepulchre, and S.D. Géro, Synthesis,
    710-713 (1983).
11. I. Pintér, J. Kovács, A. Messmer, G. Tóth, and S.D. Géro, Carbohydr. Res., 116, 156-161
    (1983).
12. S. Umezawa, Adv. Carbohydr. Chem. Biochem., 30, 111-182 (1974); Pure Appl. Chem., 50,
    1453-1476 (1978).
13. R. Blattner, R.J. Ferrier, and P. Prasit, J. Chem. Soc., Chem. Commun., 944-945 (1980).
14. R.J. Ferrier, N. Prasad, and G.H. Sankey, J. Chem. Soc. (c), 587-591 (1969).
15. S. Chew and R.J. Ferrier, J. Chem. Soc., Chem. Commun., (1984), in press.
16. J.R. Brown, Prog. Med. Chem., 15, 126-164 (1978).
17. R.L. Funk and K.P.C. Vollhardt, Chem. Soc. Rev., 9, 41-61 (1980).
18. B.H. Han and P. Boudjouk, J. Org. Chem., 47, 751-752 (1982).
19. B. Bernet and A. Vasella, Helv. Chim. Acta, 62, 1990-2016, 2400-2410, 2411-2431 (1979).
20. R.J. Ferrier, P. Prasit, and G.J. Gainsford, J. Chem. Soc., Perkin Trans. 1, 1629-1634
    (1983).
21. E.J. Corey, K.C. Nicolaou, and D.J. Beams, Tetrahedron Lett., 2439-2440 (1974);
    J.J. Partridge, N.K. Chadha, and M.R. Uskoković, J. Am. Chem. Soc., 95, 7171-7172
    (1973).
22. G. Stork, T. Takahashi, I. Kawamoto, and T. Suzuki, J. Am. Chem. Soc., 100, 8272-8273
    (1978).
23. (a) R.J. Cave, C.C. Howard, G. Klinkert, R.F. Newton, D.P. Reynolds, A.H. Wadsworth, and
    S.M. Roberts, J. Chem. Soc., Perkin Trans. 1, 2954-2958 (1979); (b) R.F. Newton and
    S.M. Roberts, Tetrahedron, 36, 2163-2196 (1980).
24. R.J. Ferrier, P. Prasit, G.J. Gainsford, and Y. Le Page, J. Chem. Soc., Perkin Trans. 1,
    1635-1640 (1983); R.J. Ferrier, P. Prasit, and P.C. Tyler, ibid., 1641-1644 (1983).
25. R.J. Ferrier, P.C. Tyler, and G.J. Gainsford, J. Chem. Soc., Perkin Trans. 1, (1984),
    in press.
26. W. Bartmann and G. Beck, Angew. Chem., Int. Ed. Engl., 21, 751-764 (1982).
27. W. Skuballa and H. Vorbrüggen, Angew. Chem., Int. Ed. Engl., 20, 1046-1048 (1981).
28. K. Bock, I. Lundt, and C. Pedersen, Carbohydr. Res., 68, 313-319 (1979).
29. R.J. Ferrier and P.C. Tyler, Carbohydr. Res., (1984), in press.

The kinetic anomeric effect in asymmetric synthesis: a new approach to aminophosphonic acids

Rolf Huber, Andreas Knierzinger, Ewa Krawczyk, Jean-Pierre Obrecht and Andrea Vasella

Organisch-Chemisches Institut der Universität Zürich, Winterthurerstrasse 190, CH-8057 Zürich, Switzerland

Abstract - The 1,3-dipolar cycloaddition of N-glycosylnitrones leading to 5,5-disubstituted isoxazolidines is a highly diastereoselective reaction. The diastereoselectivity is explained on the basis of a kinetic anomeric effect (=lone pair-polar bond interaction in the transition state). C-monosubstituted N-glycosylnitrones undergo (E Z) equilibration and 3-substituted isoxazolidines are obtained with a lower d.e. Thus the C-phosphonoylnitrone 17 added to ethylene with a d.e. of only 50%. The hypothesis of a stereoelectronic control by the kinetic anomeric effect predicts that nucleophiles should add to configurationally stable nitrones with a high diastereoselectivity. The reaction of lithium dialkylphosphites with the crystalline (Z)-nitrones 27 and 37 gave the N-glycosyl,N-hydroxy-1-aminophosphonates 30 (90%, d.e.=90%), 32 (92%, d.e.=83%), and 38 (90%, d.e.=93%), respectively, which were transformed into (+)-phosphoserine and (+)-(S)-phosphovaline, respectively. (+)-(S)-Phosphoalanine was obtained via the non-crystallizing nitrone(s) 3 and/or 4 (41%, d.e.=91%). Tris-(trimethylsilyl)phosphite reacted with the C-phenylnitrone 50 to give the (-)-(S)-N-hydroxyaminophosphonic acid 52 (84%, d.e.=61%) under catalysis by zinc chloride. Proton-catalysis gave the (+)-(R)-N-hydroxyaminophosphonic acid 52 (83%, d.e.=94%). Similar results were obtained with the C-arylnitrone 55. Additions of tris-(trimethylsilyl) phosphite to the nitrone 27 gave under either condition the (S)-N-hydroxyphosphoserine ether 59 (ca. 80%, d.e.=56% and 79%, resp.). Proton-catalyzed addition to 37, however, gave the (R)-N-hydroxyphosphovaline (51%, d.e.=ca. 90%). Oxidation of the N-glycosyl,N-hydroxyaminophosphonate 30 gave the diastereomeric C-alkyl, C-phosphonoylnitrones 61 and 62 (80-90%, 3:1). Methyl- and phenyllithium were added to the (E)-diastereomer 61 to give the hydroxylamines 63 (61%, d.e.=58%) and 64 (ca. 25%), respectively.

INTRODUCTION

Easy availability, high diastereomeric excess (d.e.) and easy removal of the chiral auxiliary are the main condition carbohydrate derivatives should fulfill in order to be useful for asymmetric synthesis (Ref. 1,2). Among the many derivatives satisfying the first condition are the N-glycosyl nitrones (e.g. 3/4 in scheme 1) (Ref. 3-7, 21). In these derivatives, the nitrone function is directly attached to the anomeric centre and thus close to the chiral carbohydrate moiety. This circumstance favours asymmetric induction and allows the influence of the ring oxygen on the reactivity of the nitrones to be studied. Finally, in the reaction product, the carbohydrate moiety may be removed hydrolytically by a glycoside cleavage.

We would like to present two aspects of N-glycosylnitrones, both related to asymmetric induction and concerned with a rationalization thereof, namely 1,3-dipolar cycloadditions of nitrones and the addition of some phosphorous nucleophiles to them, with the purpose to arrive at a general synthesis of α-aminophosphonic acids. Phosphonic acid analogues of α-aminocarboxylic acids have received considerable attention (Ref. 8-13) since the first one was prepared in 1972 (Ref. 14), but none of the methods for their synthesis

(Ref. 14-22) has been claimed to be general.

SCHEME 1.

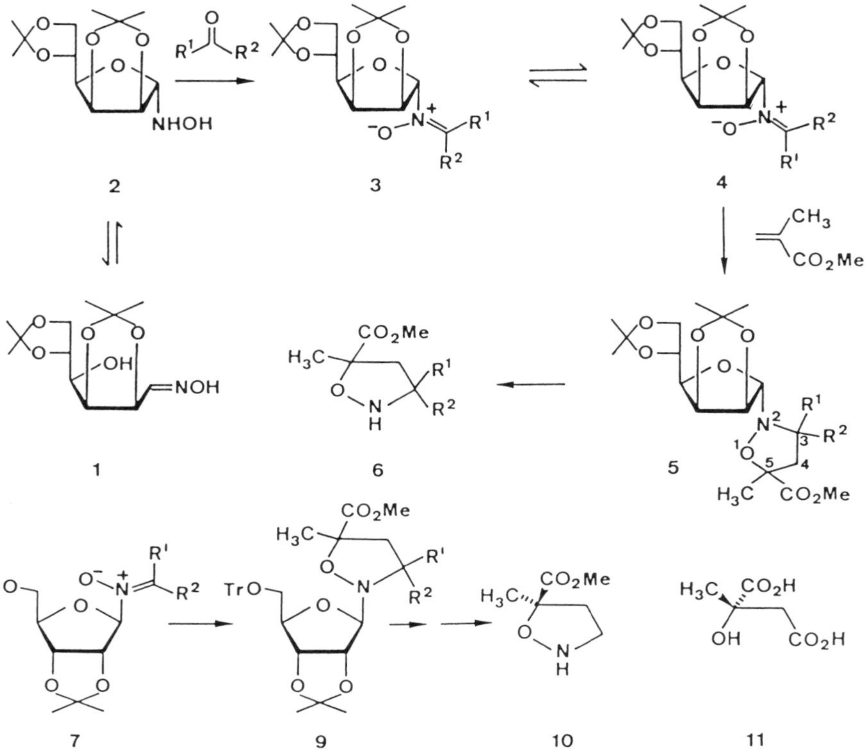

Tr = −CPh₃

a   R¹ = R² = H
b   R' = Me, R² = H
c   R' = R² = Me

1,3 DIPOLAR CYCLOADDITIONS

N-Glycosylnitrones are available from the reaction of aldehydes or ketones with N-glycosylhydroxylamines, which in solution are present in equilibrium with the corresponding hydroxy-oximes, well known and easily available carbohydrate derivatives. The nitrones do not have to be isolated; they react in situ with appropriate dipolarophiles to give N-glycosylisoxazolidines. This sequence of reactions is illustrated in scheme 1 for the oxime of diisopropylidene-D-mannose 1, which is obtained from commercially available mannose in two high yielding steps (Ref. 4). The addition of the nitrones 3/4 a-c derived from 1 and formaldehyde, acetaldehyde or acetone, respectively, to methyl methacrylate leads to the N-glycosylisoxazolidines 5 a-c. In this process a new center of chirality is formed at C(5) and, for 5b, also at C(3). The d.e. related to the formation of the center at C(5) was 79, 87, and ⩾95%, for 5 a-c, respectively. The d.e. related to C(3) in 5b was only 38,5%. This was interpreted by a highly diastereoselective cycloaddition which in the case of the nitrones 3b ⇌ 4b does not strongly depend on the nitrone geometry. Acid hydrolysis then gave the optically active N-unsubstituted (+)-isoxazolidines 6 a-c. A parallel behaviour was observed for the ribose derived nitrones 7 a-c, which gave the N-glycosylisoxazolidines 9 a-c with similar, but inverted diastereoselectivities. The (R)-configuration of the (−)-isoxazolidine 10 obtained from 9a was deduced by a correlation with (S)-citramalic acid (11).

The diastereoselectivity of the cycloadditions may be rationalized on the basis of a stereoelectronic effect in the transition state. This effect is based on the conjugative interaction between the newly formed, doubly occupied sp^3-orbital on the nitrogen atom and the σ*-orbital of the C(1),O bond of the glycosyl moiety (scheme 2). A coplanar arrangement of these orbitals should lead to a lowering of the activation energy and to an increased electrophilicity of the nitrones. The (exo-)anomeric effect which is present in the product may already become effective in the transition state and may stabilize both the product of the cycloaddition ("classic" anomeric effect) and the transition state leading to it ("kinetic" anomeric effect). The increased electrophilicity of the nitrone may be expected on the basis of the lowering of the LUMO, due to an interaction of the LUMO of the nitrone π-system and the σ*$_{C,O}$-orbital. This orbital interaction is possible in both the "O-endo" and "O-exo" conformation of the nitrones (scheme 2).

SCHEME 2.

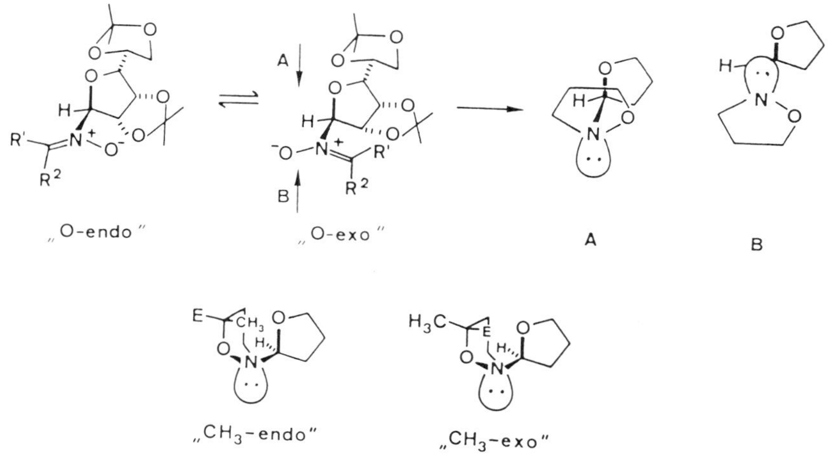

„O-endo"   „O-exo"   A   B

„CH$_3$-endo"   „CH$_3$-exo"

Since the d.e. rose with increasing substitution of the nitrones, the sterically less congested "O-endo" conformation seems to react preferentially. Two more factors determine the asymmetric induction: the direction of approach of the dipolarophile and the endo- vs. exo-orientation of its substituents (scheme 2). The approach from the side of the C,O bond leading to an antiperiplanar arrangement of lone pair and polar bond (as in A) seems to be stereoelectronically - albeit not sterically - preferable to an attack leading to a coplanar arrangement (as in B). The exo/endo-orientation of the methacrylate methyl group is difficult to predict, but a preferred endo-orientation of the methyl group has been demonstrated for such nitrones, which do not possess an exceedingly bulky N-substituent (Ref. 23). Endo-orientation of the methyl group and attack from the side of the C(1),O bond onto the nitrones in an "O-endo" conformation leads to the observed induced sense of chirality.

The cycloaddition of the nitrones 3b/4b proceeds with a high asymmetric induction at C(5) but with a low one at C(3). Can this d.e. be improved so as to get to aminocarboxylic and to aminophosphonic acids? We first sought to answer this question using nitrones derived from 1 and glyoxalates (scheme 3).

SCHEME 3.

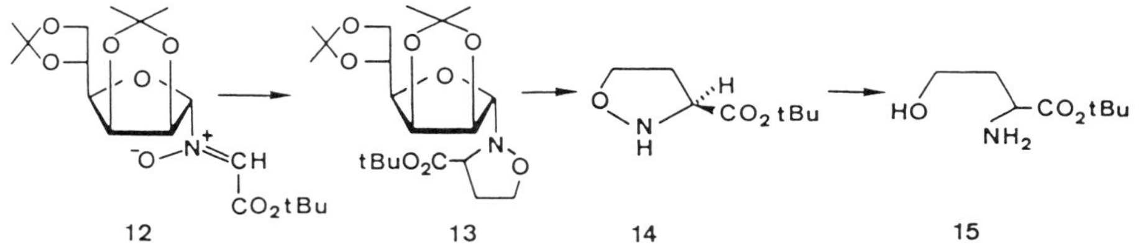

12   13   14   15

The t-butyl nitrone 12 proved to be most convenient. It reacted with ethyl-

ene in over 90% yield and with a d.e. of 54% to give the spontaneously
crystallizing 13 as the main product together with its diastereoisomer. The
glycosylisoxazolidine 13 was hydrolyzed to the 5-oxaproline ester 14. The
(L)-configuration of 14 was proven by correlation (via the homoserine
derivative 15) with aspartic acid (Ref. 5,7).

For an analogous preparation of the α-aminophosphonic acid 20 (scheme 4) we
needed the C-phosphonoylnitrone 17. Its formation in the presence of ethylene
by in situ oxidation of the hydroxylamine 16 (Ref. 21) gave the isoxazolidines 18 (81%) with a d.e. of 50%. The products 18 were separated after
partial deisopropylidenation to 19. The major isomer again possessed the
L-configuration, showing the very similar behaviour of C-alkoxycarbonyl- and
C-phosphonoyl-N-glycosylnitrones.

SCHEME 4.

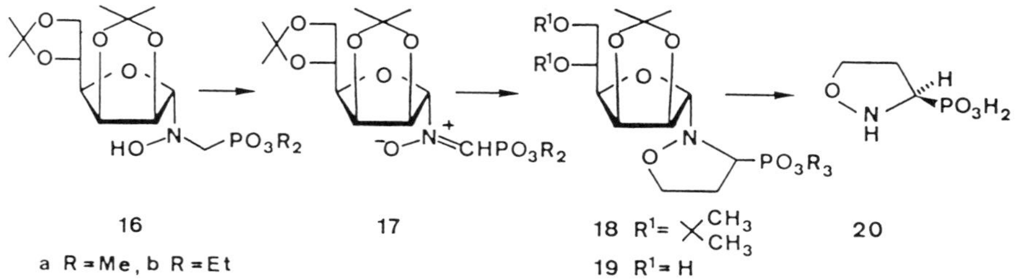

| 16 | 17 | 18 R¹= $\times\begin{matrix}CH_3\\CH_3\end{matrix}$ | 20 |
| a R=Me, b R=Et | | 19 R¹= H | |

The rationalization of the asymmetric induction requires the consideration of
three of the factors discussed above: the (E,Z)-configuration of the nitrone,
the ("O-endo" vs. "O-exo") conformation of the nitrone and the direction of
approach of ethylene ("syn" vs. "anti" to the C(1),O bond). A "syn" approach
of ethylene to the nitrone in an"O-endo" conformation requires the (E)-
nitrone to participate preferentially in a cycloaddition leading mainly to an
L-configurated product (scheme 5). The facile interconversion of (E) and (Z)-
C-alkoxycarbonylnitrones has been demonstrated (Ref. 24,25) and it has been
argued that (E)-nitrones are more reactive (Ref. 26). The relatively low d.e.
of 50% shows how much it would be preferable to use configurationally homogeneous and defined nitrones. For this reason and as a test of the stereoelectronic explanation of the diastereoselectivity based on the "kinetic" anomeric effect, we turned to nucleophilic additions to N-glycosylnitrones. We
chose phosphorous nucleophiles (see e.g. Ref. 27,28) with the aim of finding
a new asymmetric synthesis of α-aminophosphonic acids.

SCHEME 5.

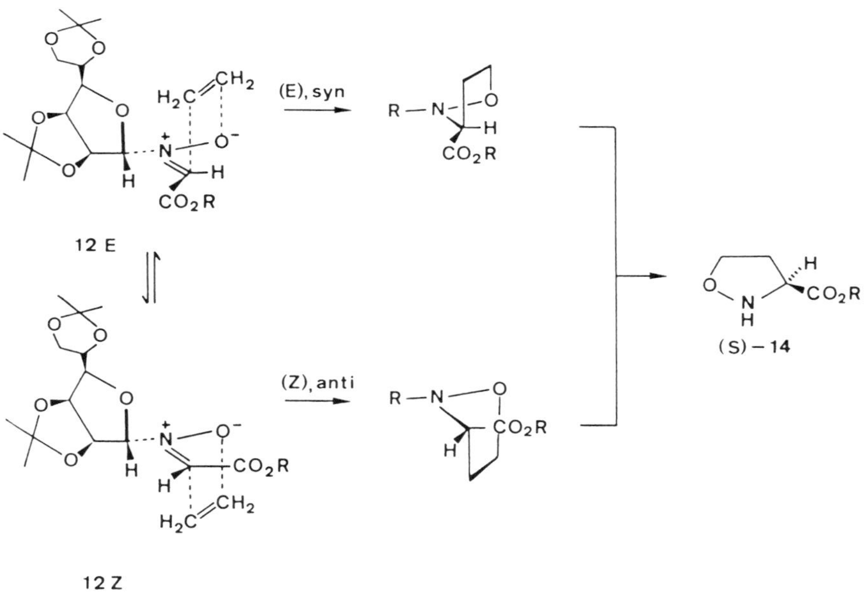

NUCLEOPHILIC ADDITIONS

The addition of a nucleophile to N-glycosylnitrones yields N,N-disubstituted
hydroxylamines. In the course of the addition, a doubly occupied, non-bond-
ing sp^3-orbital at nitrogen is formed, which may be coplanar with the
C(1),O bond and interact with its 6*-orbital. Obviously, there is an analogy
between a LUMO-controlled 1,3-dipolar cycloaddition of a (N-glycosyl)-nitrone
and the addition of a nucleophile to it. We expect that the "kinetic"
anomeric effect will again lead to a selective lowering of the transition
states which have the right geometry. It is conceivable that this effect will
be stronger in this case, due to a conjugative interaction between the lone
pairs at nitrogen and oxygen of the developing hydroxylamino group (resp. its
anion), leading to a higher energy of the nonbonding orbital at nitrogen.

SCHEME 6.

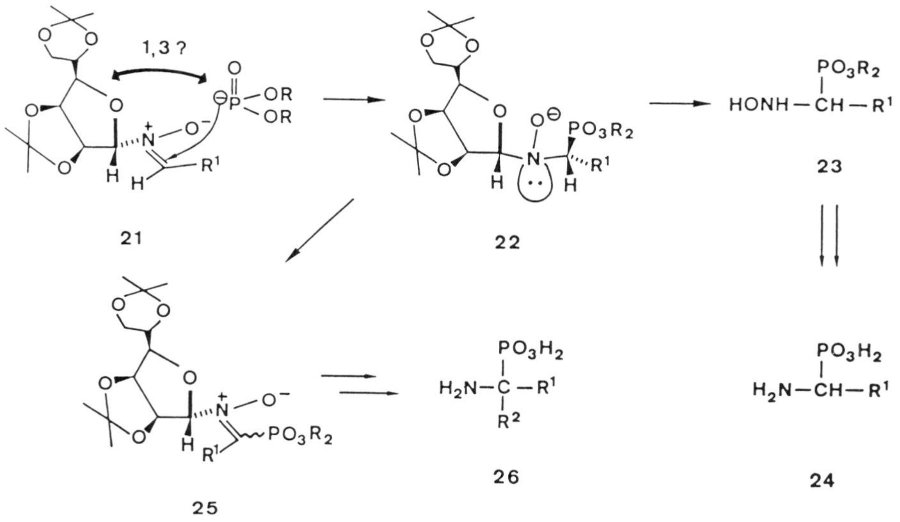

An asymmetric synthesis of α-aminophosphonic acids then requires the preparation of configurationally homogeneous nitrones and the addition of phosphorous nucleophiles which react at a low enough temperature to avoid (E ⇌ Z)-
isomerization of the nitrone. The course of the asymmetric induction is
difficult to predict since in the case of charged nucleophiles, chelation
and charge-dipole interactions may affect both the conformation of the
nitrone and the approach of the nucleophile. The addition of e.g. a dialkylphosphite salt gives N-glycosyl-N-hydroxyphosphonates 22. Glycoside cleavage
will lead to N-hydroxyphosphonates 23, which may be transformed into aminophosphonic acids 24 by hydrogenolysis and hydrolysis. Scheme 6 also indicates
that a reoxydation of the primary addition products 22 forms C-phosphonoylnitrones 25, which may react either with 1,3-dipolarophiles or with nucleophiles, leading to the aminophosphonic acids of type 26.

Aminophosphonic acids from C-alkylnitrones
The first crystalline nitrone we obtained was 27 (scheme 7)(Ref. 29). It is
easy to prepare, stable and its (Z)-configuration was proven by X-ray
analysis. Preliminary experiments with potassium dimethylphosphite showed
that addition occurred between -60 and -20° in methylene chloride to give a
diastereomeric mixture of 28 (ca. 60%) contaminated with varying amounts of
29 (Ref. 28). The reaction with potassium diethylphosphite gave cleanly the
spontaneously crystallizing addition products (30) in a very high yield. The
d.e. was found by ^{31}P-NMR spectroscopy to be 77%. The diastereoselectivity
depends on the counterion and on the dialkylphosphite used, as indicated in
scheme 7. The best d.e. (90%) was obtained using lithium diethylphosphite.
Recrystallization of the product thus obtained gave in a yield of 66% the
pure major isomer, which was hydrolyzed to the protected (+)-N-hydroxyphos-

phoserine 33 (95%). Hydrogenolysis then gave the diethylester of (+)-phosphoserine (35). (+)-Phosphoserine itself was easier to obtain from the dibenzylphosphite adduct 32 (92%, d.e.=83% by ^{31}P-NMR and HPLC). Three recrystallizations afforded the major isomer in 63% yield and a d.e. of 98,7% (HPLC). It was again hydrolyzed to the crystalline (+)34 (85%). Hydrogenolysis of 34 gave (+)-amino-2-hydroxyethylphosphonic acid 36 (phosphoserine) in 77% yield. The minor diastereoisomer of 32 was isolated from the mother liquor and transformed into (-)34. (The absolute configuration of (+)-phosphoserine has very recently been determined to be (S) (X-ray analysis of the MTPA-amide).)

SCHEME 7.

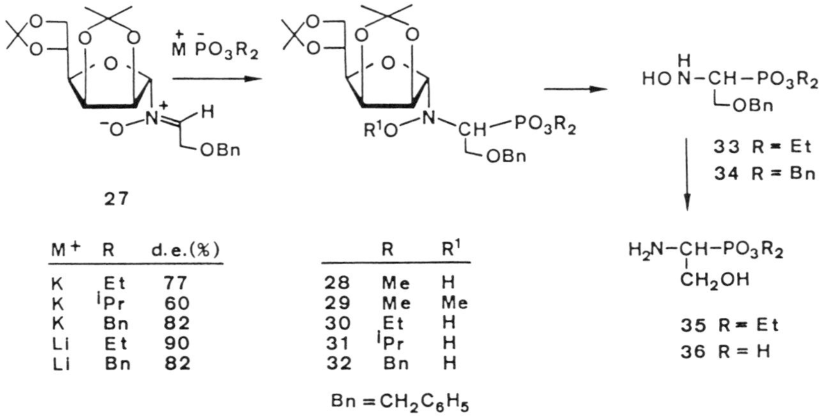
$\overset{+}{M}\ \overset{-}{P}O_3R_2$
27
HON-CH-PO3R2
OBn
33 R = Et
34 R = Bn
H2N-CH-PO3R2
CH2OH
35 R = Et
36 R = H

M+	R	d.e.(%)
K	Et	77
K	iPr	60
K	Bn	82
Li	Et	90
Li	Bn	82

	R	R1
28	Me	H
29	Me	Me
30	Et	H
31	iPr	H
32	Bn	H

$Bn = CH_2C_6H_5$

The next aminophosphonic acid we prepared was (-)-(S)-phosphovaline 41 (scheme 8)(Ref. 30, 31). It is a known compound and its preparation will tell if the method described leads to L- or D-aminophosphonic acids preferentially.

SCHEME 8.

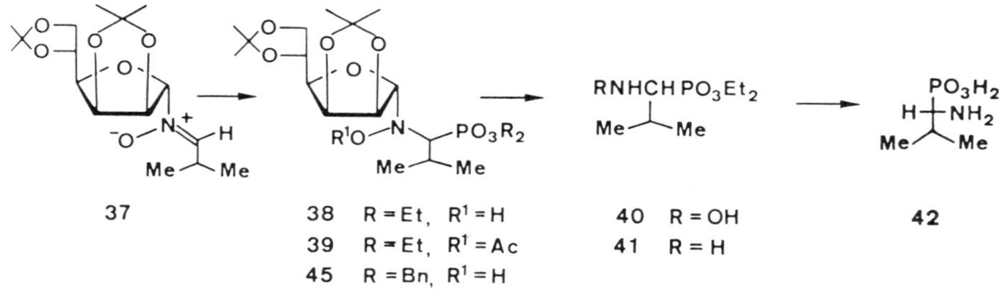
37
38 R = Et, R1 = H
39 R = Et, R1 = Ac
45 R = Bn, R1 = H
RNHCH PO3Et2
40 R = OH
41 R = H
42

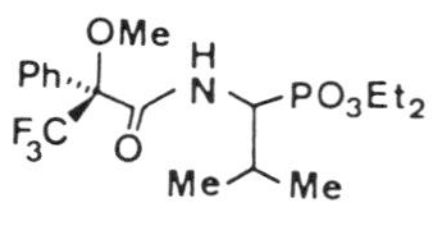

43

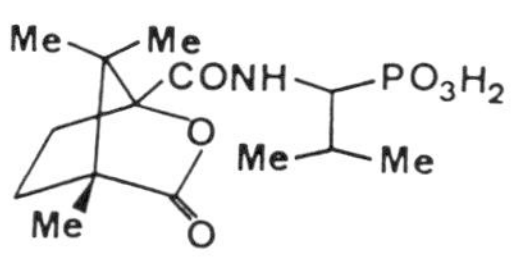

44

Condensation of the mannose oxime 1 with isobutyraldehyde gave the crystalline nitrone 37 in 85% yield. Although 37 did not react with potassium diethylphosphite, it added lithium diethylphosphite. The crude addition product showed only two peaks in the ^{31}P-NMR-spectrum, one corresponding to excess diethylphosphite and the other to the phosphonate. A crude chromatography, whereby separation of potential diastereoisomers was avoided, gave the spontaneously crystallizing product 38 in 63% yield. After recrystallization, a 60% yield of one diastereoisomer was obtained and characterized as the acetate 39. Hydrolysis of 38 gave the somewhat unstable N-hydroxyamino-phosphonate 40, which was transformed by hydrogenolysis and hydrolysis into the (+)-phosphovaline 42 (85% from 38). It showed an $[\alpha]_D^{25} = -0.8^\circ$ (c=1.99, 1N NaOH) corresponding to the (S)-configuration (Ref. 31). When we first obtained (-)-42, we compared the optical rotation of our sample to a literature value of -10° (Ref. 30) and, therefore, carefully checked the enantiomeric purity of our phosphovaline preparation. Racemic diethylphos-phovalinate 41 (Ref. 18) was transformed into the diastereomers 43 according to Mosher (Ref. 32). The diastereomers were easily distinguished by ^{31}P-, ^{13}C- and ^{1}H-NMR-spectroscopy and separated by HPLC. We then prepared the amide 43 from crude and from recrystallized 38, respectively. The d.e. of the "Mosher-amide" prepared from crude 38 was 93% and we accepted this value as indicating the diastereoselectivity of the addition reaction to 37. The d.e. of the amide prepared from recrystallized 38 was 99%. The acid hydro-lysis of 41 to 42 was shown to proceed with a slight loss of enantiomeric purity, since the camphanic amide 44 (Ref. 33) prepared from 41 via 42 (d.e.=99%) showed a d.e. of only 92%. Evidently, the nucleophilic addition of lithium diethylphosphite to the nitrone 37 proceeded with a high degree of diastereoselectivity. The addition of dibenzylphosphite also gave a spontaneously crystallizing product 45. Its ^{31}P-NMR-spectrum also showed a single peak in the phosphonic acid region. But since the hydrolysis of 45 was difficult and led to partial decomposition, we have not explored its chemistry any further.

In both the preparations reported, the nitrones were easy to crystallize. Would we still get a good chemical yield and d.e., if the nitrones were difficult to prepare and to obtain in a pure form? An answer to this question was sought in the preparation of phosphoalanine 46 (Ref. 34)(scheme 9).

SCHEME 9.

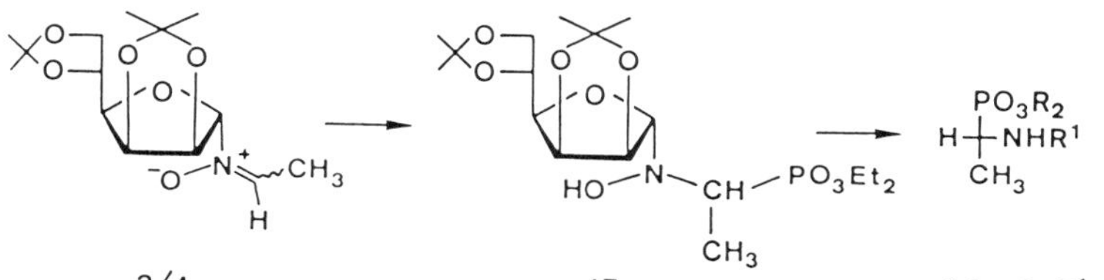

Reaction of the oxime 1 with excess acetaldehyde gave a crude product (3/4), which was freed of excess acetaldehyde and then treated between -70 and -20° with lithium diethylphosphite. Aqueous workup and a filtration over silicagel gave 41% of the addition product 47, which was transformed into the diethyl-phosphoalanine 48 and further into (+)-(S)-phosphoalanine 46. The diastereo-selectivity of the addition was deduced from the optical purity of 46 (90-95%)(no purification of intermediates) and from the d.e. (91.5%) in which the MTPA-amide 49 (Ref. 32) was prepared from the phosphonate 48. Since the yield in which the nitrone(s) 3 and/or 4 were formed is not known, it is difficult to comment upon the yield and diastereoselectivity of the addition reaction. But the experiment shows that reasonable yields of a desired aminophosphonic acid may also be obtained if the necessary nitrone cannot be easily isolated and purified.

Aminophosphonic acids from C-arylnitrones
Nitrones and especially C-arylnitrones usually crystallize well and so did the C-aryl-N-glycosyl nitrones 50 and 55 (scheme 10). The reaction of 55 with lithium diethylphosphite did not, however, lead to the desired addition compound, but several products were formed in varying amounts. So we turned to the nucleophilic tris-(trimethylsilyl)phosphite 51 (Ref. 20,35). It

reacted with the nitrone 50 in refluxing benzene and in the presence of catalytic amounts of zinc chloride (A in scheme 10) to give, after aqueous workup, directly the (-)-N-hydroxyphosphophenylglycine 52 in a yield of 84%. Hydrogenolysis led to the well known (-)-(S)-phosphophenylglycine 53. An enantiomeric purity of only 61% was found for (-)54, prepared from 53 by chromatography on a DNBPG-column according to Pirkle (36). When the nitrone 50 was treated with tris-(trimethylsilyl)phosphite in benzene/ methylenechloride in the presence of catalytic amounts of HClO$_4$ between -50° and room temperature (B in scheme 10), we obtained again an 83% yield of N-hydroxyphosphophenylglycine. To our surprise, however, the product obtained under proton catalysis was dextrorotatory. The diastereoselctivity was found to be 94%, as inferred from the enantiomeric purity of (+)54.

Although, from the results of the 1,3-dipolar addition, we had to expect the formation of the enantiomeric, (R)-aminophosphonic acids by using the ana- logous, ribose derived nitrones, it appeared much more convenient to use one precursor and two sets of reaction conditions to obtain either enantiomer. Only a few experiments have been run so far to see if this dichotomy, which one would like to trace back to chelation effects, is general. The 4-tert.- butylphenylnitrone 55 reacted in a similar way as did 50, to give the N- hydroxyaminophosphonic acids 56 in yields of 80-90%. The enantiomeric purity of (-)-(S)-56, obtained in the zinc chloride-catalyzed reaction, was only 35%. This may be taken to indicate a strong dependence of the diastereo- selectivity upon the (not optimized) reaction conditions. The enantiomeric purity of (+)-(R)-56, obtained in the proton-catalyzed reaction is at least 90% (non base-separated peaks in the chromatography (Ref. 36) of 58).

SCHEME 10.

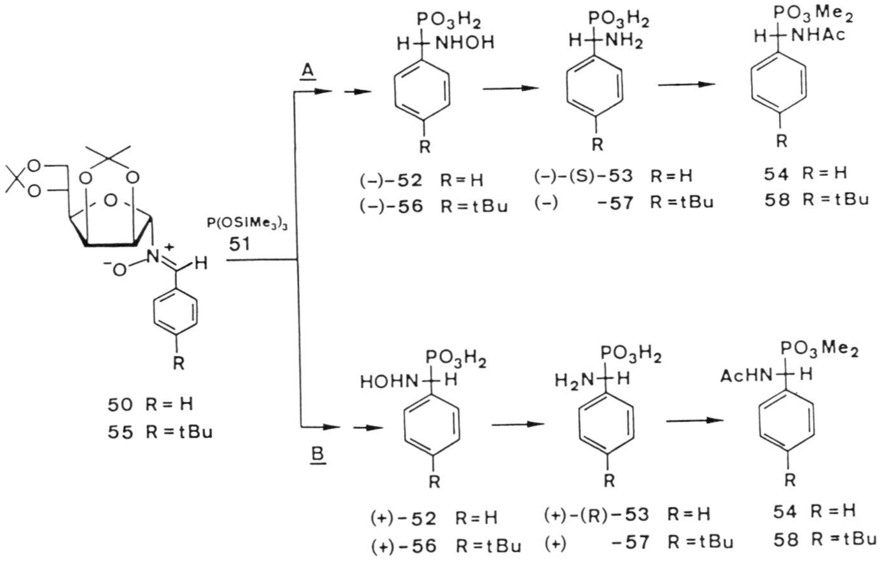

With regard to the "chelation-conjecture", it appeared interesting to look at the tris-(trimethylsilyl)phosphite addition to the nitrone 27 under proton- and zinc chloride-catalysis, respectively. In this case, both conditions gave the (+)-N-hydroxyaminophosphonic acid 59 in a yield of 75-85% (scheme 11). This acid was transformed into (+)-phosphoserine 36, the same (S)-enantiomer, which had already been obtained from the reaction of lithium dibenzyl- phosphite with 27.

The enantiomeric purity of these preparations has not yet been determined, but it appears to be lower for the product from the proton-catalyzed re- action (optical purity=56%) than for the one from the zinc chloride-catalyzed reaction (optical purity=79%). in contradistinction to this result, the proton-catalyzed addition of tris-(trimethylsilyl)phosphite to 37 gave the

(R)-N-hydroxyphosphovaline 60 (51%, unoptimized), which was transformed into
the (-)-(R)-phosphovaline 42, the (L)-enantiomer of the phosphovaline
obtained from the addition of lithium diethylphosphite. We do not yet know
the result of the zinc chloride-catalyzed addition.

SCHEME 11.

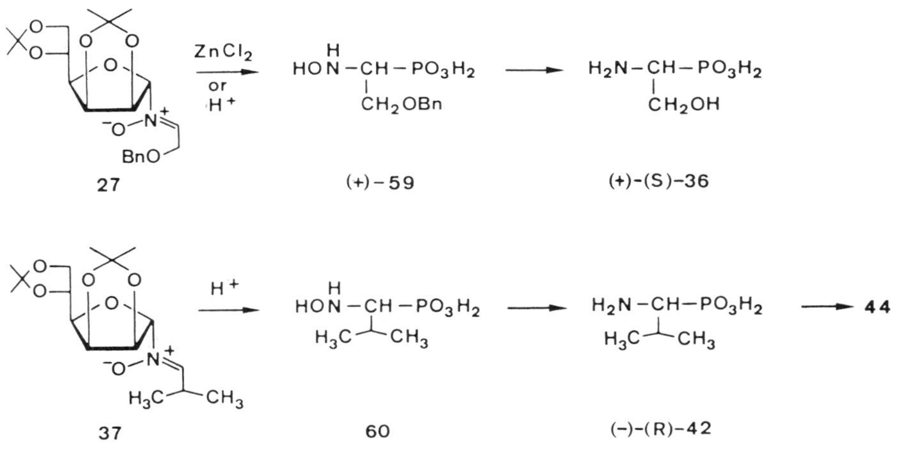

Aminophosphonic acids from C-alkyl,C-phosphonoyl-nitrones.
Both enantiomers of a desired α-aminophosphonic acid might also be obtained
with the help of one enantiomer of a chiral auxiliary by inversion of the
order in which the alkyl- and phosphonoyl moieties are assembled: i.e. by
either alkylation of a C-phosphonoylnitrone or by phosphonoylation of a
C-alkylnitrone. The first one of these possibilities was investigated by
studying the alkylation of the C-alkyl,C-phosphonoylnitrone 61 (scheme 12).

SCHEME 12.

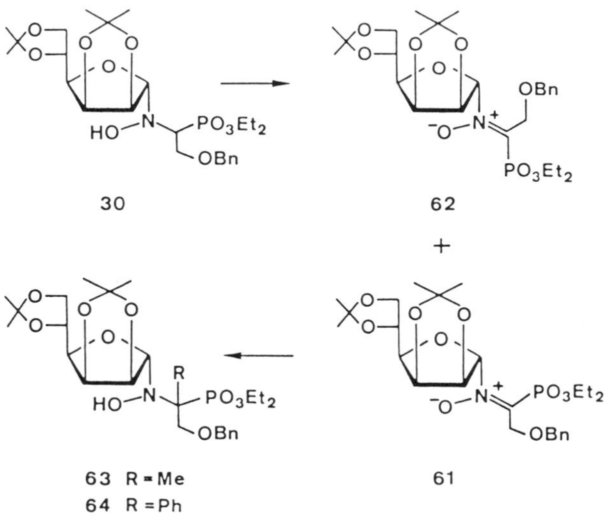

This nitrone was obtained in a yield of about 60% together with its diastereoisomer 62 (ratio 3:1) by oxidation of the hydroxylamine 30 with silver
carbonate in toluene (Ref. 37,38) followed by chromatography. (The assignment of the (E)-configuration to 61 is based on ^{1}H-NMR-arguments). Addition
of methyllithium gave the diastereomeric hydroxylamines 63 in 61% yield and
a d.e. of 58% (HPLC). So far, we have obtained the major isomer in pure form.

Preliminary experiments have also shown that the addition of phenyllithium gives the diastereomeric hydroxylamines 64, so far in rather low yields (d.e.=70%). Further experiments will have to show the viability of the addition of alkylgroups to the C-phosphonoylnitrone 17 and of the 1,3-dipolar addition of its alkylated analogue 61.

In conclusion, we have shown that the hypothesis of the kinetic anomeric effect (see Ref. 39 and 40, particularly chapters 3, 4 and 6), which was advanced as the basic factor governing the asymmetric induction in 1,3-dipolar cycloadditions of N-glycosylnitrones correctly predicted a high asymmetric induction in the nucleophilic addition of phosphorous nucleophiles to these nitrones. It is clear, that the outcome of these asymmetric reactions strongly depends upon several other factors which remain to be investigated in detail. From a preparative point of view, some approaches to enantiomerically pure or enriched α-aminophosphonic acids and to α-N-hydroxyaminophosphonic acids from nitrones have been described, which hopefully will prove to be of general applicability.

Acknowledgement - We thank the Swiss National Science Foundation, Sandoz A.G. Basle and Givaudan A.G. Dübendorf for generous support. We thank Dres. J. Bieri and R. Prewo in our Institute and Dr. J. Daly and Mr. P. Schönholzer, Zentrale Forschungseinheiten, F. Hoffmann-La Roche & Co. A.G., Basle, for X-ray analysis.

REFERENCES

1. T.D. Inch, Adv. Carbohydr. Chem. Biochem. 27, 191 (1972).
2. J.D. Morrison, ed., Asymmetric Synthesis Vol. 2, Academic Press, New York (1983).
3. A. Vasella, Helv. Chim. Acta 60, 426 (1977).
4. A. Vasella, Helv. Chim. Acta 60, 1273 (1977).
5. A. Vasella and R. Voeffray, J. Chem. Soc. Chem. Comm. 97 (1981).
6. A. Vasella and R. Voeffray, Helv. Chim. Acta 65, 1134 (1982).
7. A. Vasella, R. Voeffray, J. Pless and R. Huguenin, Helv. Chim. Acta 66, 1241 (1983).
8. S. Neale, Chem. Biol. Interactions 2, 349 (1970).
9. J.G. Allen, F.R. Atherton, M.J. Hall, C.M. Hassall, S.W. Holmes, R.W. Lambert, W.J. Lloyd, L.J. Nisbet and P.S. Ringrose, Nature 272, 56 (1978).
10. F.R. Atherton, M.J. Hall, C.M. Hassall, S.W. Holmes, R.W. Lambert, W.J. Lloyd and P.S. Ringrose, Antimicrob. Agents Chemother. 18, 895 (1980).
11. T. Kamiza, K. Hemmi, H. Takeno and M. Hashimoto, Tetrahedron Lett. 95 (1980).
12. M.M. Slaughter and R.F. Miller, Science 211, 182 (1981).
13. T. Kametani, K. Kisagawa, M. Hiigari, K. Wakisaka, S. Haga, H. Sugi, K. Tanigawa, Y. Susuki, K. Fukawa, O. Irino, O. Saito & S. Yamabe, Heterocycles 16, 1205 (1981).
14. W.F. Gilmore and H.A. McBride, J. Am. Chem. Soc. 94, 4361 (1972).
15. J. Kowalik, P. Mastalerz, M. Soroka & J. Zoń, Tetrahedron Lett. 3965 (1977).
16. Y.P. Belov, V.A. Davankov and S.V. Rogozhin, Izv. Akad. Nauk SSSR, Ser. Khim 1546 (1977).
17. A. Kotynski and W.J. Stec, J. Chem. Res.(S) 41 (1978).
18. Z.M. Kudzin and W.J. Stec, Synthesis 469 (1978).
19. J.W. Huber & W.F. Gilmore, Tetrahedron Lett. 3049 (1979).
20. J. Zoń, Polish J. Chem. 55, 643 (1981).
21. A. Vasella and R. Voeffray, Helv. Chim. Acta 65, 1953 (1982).
22. J.M. Villanueva, N. Collignon, A. Guy and Ph. Savignac, Tetrahedron 39, 1299 (1983).
23. P.M. Wovkulich, F. Barcelos, A.D. Batcho, J.F. Sereno, E.G. Baggiolini, B.M. Hennessy and M.R. Uskokovič, Tetrahedron 40, 2283 (1984).
24. Y. Inouye, J. Hara and H. Kalisawa, Chem. Lett 1407 (1980).
25. J. Bjørgo, D.R. Boyd and D.C. Neill, J.Chem.Soc.Chem.Comm. 478 (1974).
26. G. Bianchi, C. DeMicheli and R. Gandolfi, J.Chem.Soc.Perkin I 1518 (1976)
27. J.M.J. Tronchet, D. Schwarzenbach, E. Winter-Mihaly, Ch. Diamantides, U. Likič, G. Galland-Barrera, Ch. Jorand, K. Deen Pallie, J. Ojha-

Poncet, J. Rupp, Y. Moret and M. Geoffroy, Helv.Chim. Acta 65, 1404 (1982).

28. P. Milliet and X. Lusinchi, Tetrahedron 35, 43 (1979).
29. R. Badoud and A. Vasella, unpublished results
30. T. Głowiak & W. Sawka-Dobrowolska, Tetrahedron Lett. 3965 (1977).
31. P. Kafarski, B. Lejczak and J. Szewczyk, Can. J. Chem. 61, 2425 (1983).
32. J.A. Dale, D.L. Dull and H.S. Mosher, J. Org. Chem. 34, 2543 (1969).
33. H. Gerlach, Helv. Chim. Acta 61, 2773 (1978).
34. R.F. Atherton, M.J. Hall, C.H. Hassall, K.W. Lambert and P.S. Ringrose, Antimicrob. Agents Chemother. 15, 677 (1979).
35. T. Hata and M. Sekine, J. Am. Chem. Soc. 96, 7363 (1974).
36. W.H. Pirkle, J.M. Finn, B.C. Hamper, J. Schreiner and J.R. Pribish in E.L. Elie and S. Otsuka, ed. Asymmetric Reactions and Processes in Chemistry, ACS Symposium Series No. 185, Am. Chem. Soc. (1982).
37. M. Fétizon and M. Golfier, Compt. rend. Acad. Sci. ser. C 267, 900 (1968).
38. J.A. Maassen and Th.J. deBoer, Recueil 90, 373 (1971).
39. M. Petrzilka, D. Felix, and A. Eschenmoser, Helv. Chim.Acta 56, 2950 (1973).
40. P. Deslongchamps, Stereoelectronic Effects in Organic Chemistry, Pergamon Press (1983).

Carbohydrates *vs.* non-carbohydrates in organic synthesis

Stephen Hanessian

Department of Chemistry, Université de Montréal
Montréal, Québec, CANADA, H3C 3V1

Summary - Strategies for the total synthesis of enantiomerically
pure natural products based on the asymmetric synthesis and chiron
approaches are discussed. The utility of the four major classes of
optically active, chiral starting materials (chiral templates) is
discussed and analyzed by retrosynthetic reasoning.

INTRODUCTION

Target-oriented total synthesis, particularly in the area of natural products, has always
carried with it an aura of challenge and sophistication (1). Reaching the pantheon of
accomplishment as many have done, is due in major part to considerations of design and
strategy, which after fine-tuning, results in a final blueprint ready for execution. Such
is the saga of the long and sometimes arduous journey towards the synthesis of natural
products.

However, emphasis in this area of endeavor has, in fact, shifted in recent years. Judgement
is made on such issues as genius of conception, elegance of design, innovations in
methodology and efficiency of the overall process. The additional requirement to produce
optically pure material in one enantiomeric form or another has created yet another hurdle
to overcome (2). The synthetic organic chemist has taken on these challenges with creative
flair and exemplary execution. The phenomenal growth in novel synthetic methodology,
particularly as it relates to asymmetric synthesis (3), may be attributed in major part to
natural product chemistry. In fact, we can venture to say that the synthesis of a given
target may be the vehicle for the study of certain organic transformations. Conversely, it
may not be presumptious to say that some syntheses are initiated and often completed mainly
as a result of the discovery of an important bond-forming process. However, with regard to
overall strategy, the germination, evolution and conception of an idea which relates the
final target structure to the starting materials, are sometimes never revealed. In other
words, what mental or visual structural and functional associations do we make when we
confront the intricacies of a target structure? Are we not tempted and often driven by a
fixed notion or a unique transform which we wish to be the highlight of the synthesis?
Sharing this knowledge with the community of other practioners of the art may add yet
another dimension to the written account that invariably ensues such achievements.

APPROACHES TO THE SYNTHESIS OF ENANTIOMERICALLY PURE COMPOUNDS

In terms of overall strategy, two principal approaches have been adopted by synthetic
organic chemists, when considering enantiomerically pure targets. In the first approach, a
considerable portion of the stereochemical and functional group information present in the
target is derived from the starting materials to be used. We have recently referred to this
as the 'chiron' approach (4). One can thus select from a variety of naturally occurring
compounds ('chiral templates') (4) such as amino acids (5), hydroxy acids (2,6), sugars
(carbohydrates) (7-9) and terpenes, which are transformed into optically active, chiral
synthons ('chirons') and used as intermediates in the synthesis. Other types of optically
pure starting materials can also be obtained by techniques of resolution, by microbiological
or enzymatic transformations (10), and other methods that allow the establishment of one or
more asymmetric centers.

A second strategy utilizes optically active compounds in a different way (Figure 1). Thus,
asymmetric synthesis (3) is utilized in various bond-forming reactions to carbon for
example, by using reagents that have inherent chirality, or possess chiral auxiliaries,
either affixed to them or as co-reactants. Asymmetric induction to varying degrees may
result, and in turn lead to progressively more functionalized segments of a given target.

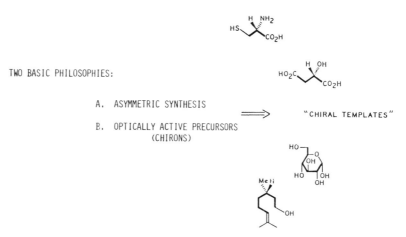

TWO BASIC PHILOSOPHIES:

    A. ASYMMETRIC SYNTHESIS

    B. OPTICALLY ACTIVE PRECURSORS
        (CHIRONS)

"CHIRAL TEMPLATES"

Figure 1.   Approaches and strategies in the total synthesis
of enantiomerically pure natural products

While both approaches rely on the utilization of optically active materials, the 'chiron'
approach sees them as an integral part of the carbon skeleton of the target, whereas in
asymmetric synthesis, the chiral auxiliaries or external catalysts (complexing agents, etc.)
are merely used as stereochemical 'sensors', that provide the desired bias in the systematic
construction of segments of the target.

The two approaches are philosophically and conceptually different. It has been said that a
painter occupies space, while a sculptor displaces it. Since the synthetic chemist is an
astute strategist, as well as an artist and a molecular architect in his own right, one
could consider the analogy which relates synthesis using chiral starting materials to
painting, and asymmetric synthesis to sculpting. In less abstract terms, the two approaches
can be schematically illustrated in Figure 2, where an asymmetric target is assembled using
prefabricated units (chiron approach), or from chiral, 'left' and 'right'-handed basic
components either in racemic or optically active form after resolution for example.

RACEMIC   TARGET   CHIRON   CHIRAL
MIXTURE   TEMPLATES

         ASYMMETRIC   SYNTHESIS WITH OPTICALLY ACTIVE
         SYNTHESIS   (CHIRAL) SYNTHONS
                                         (CHIRONS)

Figure 2.   Synthetic approaches to optically pure targets

Although widely used, neither strategy can be considered to be conceptually new at the
present time. However, much innovation has been brought forth within each domain. This is
particularly the case in asymmetric synthesis, where new developments and additions to the
arsenal of enantio- and diastereo-selective methods of bond formation are announced on
almost a weekly basis. Remarkable advances in the aldol reaction (11) with impressive
diastereoselection, and the phenomenal impact of the Sharpless epoxidation reaction (12),
are but two examples of the organic chemist's ingeneous efforts in the quest to mimick
Nature in stereochemical virtuosity. When faced with the decision of assembling the carbon
framework of a given target, the synthetic chemist normally does so with a preconceived
plan, where critical transforms follow a logical and often predictable sequence. In
published format, the elegance in design and the precision of execution are often
accompanied by an eloquence in narrative (13).

The objective of this article is to compare and contrast the two main approaches to total
synthesis of enantiomerically pure or enriched compounds. In addition, as the title
indicates, we will examine the role of carbohydrates in this area. The format which follows
consists of briefly discussing the functional, topological and structural relationships

between targets and precursors, and to illustrate specific syntheses where 'chirons' derived from carbohydrates and other optically active compounds have been used. Other approaches to the same or related targets, with emphasis on asymmetric synthesis, will also be discussed. The reader will hopefully develop an appreciation for the virtues and shortcomings of each approach, as well as for the underlying strategies.

ASYMMETRIC SYNTHESIS VERSUS THE 'CHIRON' APPROACH

It is difficult to lay down guidelines for a choice of a synthetic approach to a given target. The nature, relative disposition, and number of asymmetric centers, together with the type of carbon skeleton must be regarded as focal points. Considering the combination of these features, one may be inclined to choose an appropriate chiral precursor, such as an amino acid, hydroxy acid, etc. These will provide a framework with a predetermined pattern of substitution which may be subjected to manipulation in order to converge with the desired functionality and sense of chirality in the target. On the other hand, a stepwise construction of the carbon skeleton using asymmetric synthesis as the focal issue may be the tactic of choice. It may therefore be useful to express some views regarding the two approaches and to tabulate some of the advantages and limitations.

A. Asymmetric synthesis.

        Advantages: - generally the method of choice for the generation of one or two asymmetric centers in a single reaction.
                           - practical and efficient at times
                           - intellectually stimulating

        Limitations: - in general, does not provide multiple asymmetric centers per reaction (except for some Diels-Alder reactions) (14)
                           - optimum relationship between desired chain length and the number of asymmetric centers may be difficult to attain
                           - optical purity is dependent on the substrate and the reagent
                           - not applicable for certain classes of compounds.
                           - unpredictable stereoselection at times.

B. 'Chiron' Approach.

        Advantages: - access to polyfunctional carbon chains or rings; can provide substantial segments of the target
                           - absolute configuration and optical purity virtually in hand
                           - high regio- and stereoselectivity
                           - predictive power
                           - aesthetically appealing

        Limitations: - necessity to manipulate and protect functional groups, thus lengthening the sequences at times
                           - inaccessibility or high cost of desired chiral starting materials in some cases
                         - not applicable for certain classes of compounds

In general, asymmetric synthesis could be considered to be the more practical approach when one or two asymmetric centers are to be generated per reaction. It could be the method of choice for targets or segments thereof, where relatively few asymmetric centers are present. Although the prospects of achieving virtually complete asymmetric induction in successive steps may present problems, ingeneous means have been found to surmount them and a number of challenging and complex targets containing several asymmetric carbons atoms have been synthesized (11,15).

When one or two asymmetric centers are involved, 'chirons' derived from amino acids, hydroxy acids and some terpenes may also prove to be practical and efficient provided that their carbon framework can be adapted to chain-elongation or shortening according to the demands of the target structure. The use of carbohydrates as chiral precursors in such cases would entail a greater number of steps caused by the necessity to eliminate unwanted asymmetric centers. In expressing such views, it is not intended to lessen the importance of a given published route relative to another. With the passage of time, better and more efficient routes to the same target invariably ensue, and while they may be overpowering in some aspects, they do not lessen the original impact of previous contributions. In spite of its apparent length, the 'chiron' approach has produced optically pure materials that have been used as stereochemical references in many instances. Moreover, considering the complexity of the targets, it is also an acknowledged fact that the 'chiron' approach has had a remarkably high rate of success in the list of completed syntheses (4). Figures 3-15 illustrate selected examples of the synthesis of relatively simple and complex optically active targets.

The synthesis of the enantiomeric R- and S-sulcatols from R- and S-glutamic acids (16) and carbohydrate precursors (17) illustrates the utility of the amino acid in the synthesis of such a target (Figure 3). It is preferable to the carbohydrate approach because it provides

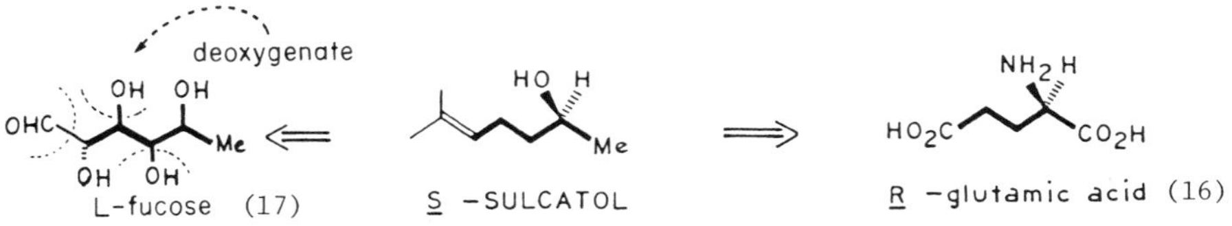

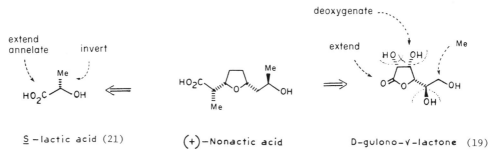

Figure 3. Synthetic precursors to S-sulcatol.

the desired sense of chirality and a significant portion of the carbon skeleton.   The alternative syntheses from carbohydrates achieve the same objective but are lengthy because of the necessity to eliminate functionality.

The synthesis of recifeolide using R-propylene oxide (from R-lactic acid) as a chiral precursor (18) illustrates a logical and practical strategy which takes full advantage of functionality and chirality, without further adjustments except for elaboration of the carbon framework (Figure 4).

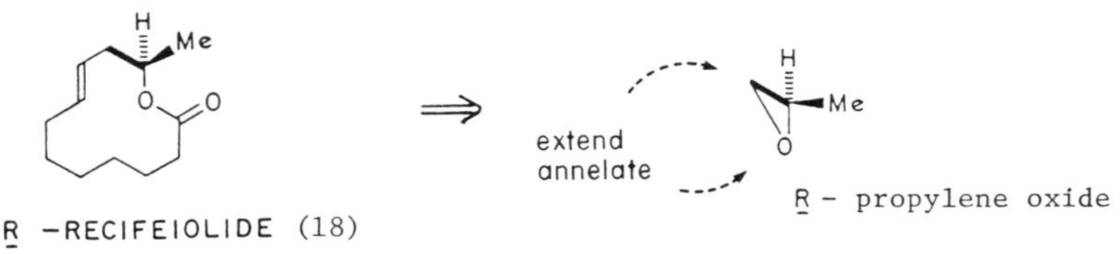

Figure 4. The R-propylene oxide route to R-recifeiolide.

Segments of the nonactic acids have been synthesized from carbohydrate precursors (19,20) as well as from S-lactic acid (21).   The Ireland (19) and Fraser-Reid (20) approaches take advantage of the ring form of carbohydrate precursors which provide good overlap of the carbon framework of the target, including the right-hand side-chain (Figure 5).   A number of functional group adjustments are required, in order to achieve maximum convergence with the desired level of deoxygenation in the ring as well as on the side-chains.   The carbohydrate derived 'chirons' serve a further useful purpose in that they allow the successful application of the Claisen rearrangement for the elaboration of the left-hand side-chain.   The S-lactic acid approach (21) has its merits in providing a three-carbon segment with the intended absolute stereochemistry.   The antithesis of the 'chiron' approach is exemplified in the utilization of D-ribonolactone to construct a five carbon segment of racemic nonactic acid derivative (22).   Indeed, the carbohydrate was stripped of most of its hydroxyl groups, hence the asymmetric centers, by extensive dehydration.   The resulting butenolide was hydrogenated and eventually converted into a pentanetriol which was further elaborated.   The carbohydrate was therefore used as a source of the five-carbon dihydroxy acid.

Figure 5. Synthetic precursors to (+)-nonactic acid.

The synthesis of the pheromone (+)-disparlure from R,R,tartaric acid (23) demonstrates an ideal example of utilization of this useful chiral template in synthesis (Figure 6). However, it should be appreciated that the diacid is symmetrical in terms of the nature of functional groups, and the two ends have to be differentiated in order to elaborate the remaining framework of the target. Methodology has been developed to that end (2) which extends the synthetic utility of this four-carbon precursor in synthesis (24). The advent of the Sharpless epoxidation reaction has made targets such as disparlure much more readily accessible (25).

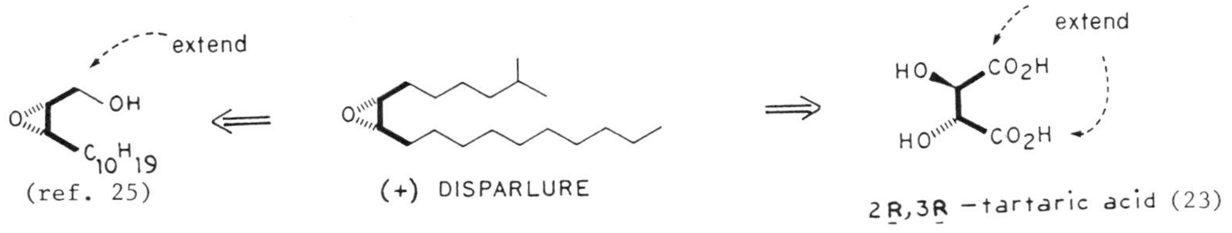

Figure 6. Synthetic precursors to (+)-disparlure.

A combination of microbiological reduction and enolate chemistry (26), resolution of a racemic precursor (27) as well as starting with D-glucose (28) or an optically active butenolide (29) have been used to produce the pheromone 3S, 4S-4-methyl-3-heptanol (Figure 7). As in the case of the sulcatols, it is evident that the carbohydrate approach is not only longer, but less practical since it involves extensive deoxygenation and an extension at C-2 of D-glucose. Here, another 'chiron' derived from an amino or hydroxy acid, or resorting to Sharpless or aldol methodology would seem more appropriate.

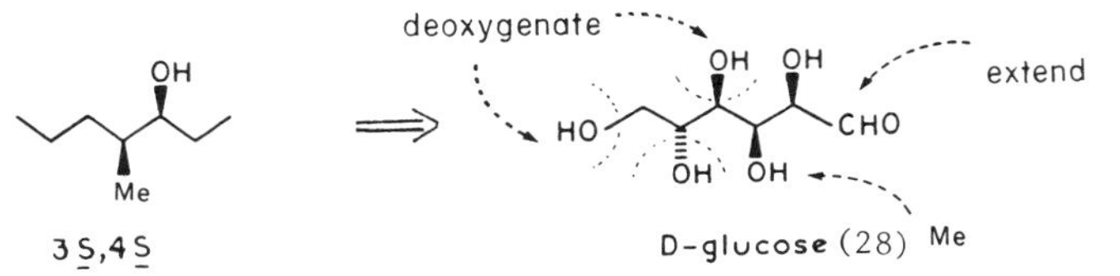

Figure 7. Access to 3S,4S-4-methyl-3-heptanol from D-glucose.

The synthesis of (-)-shikimic acid from carbohydrate precursors (30,31) and by Diels-Alder methodology (32) represents two extremes in conception of design (Figure 8). Thus, Bestmann and Kranz (30) utilized the D-arabinose framework to provide a five-carbon segment of the carbocyclic target, comprising all-three asymmetric centers. An ingeneous intramolecular insertion of a one-carbon unit and an extension by Wittig-based methodology completed the synthesis. The Fleet (31) approach utilizes the inherent disposition of hydroxyl groups in D-mannose and relies on a phosphonate-type ring closure. Masamune and coworkers (32) utilized an asymmetric Diels-Alder reaction relying on chiral dienophiles with remarkable success. Here, a single reaction generates three asymmetric centers with excellent regio- and stereocontrol.

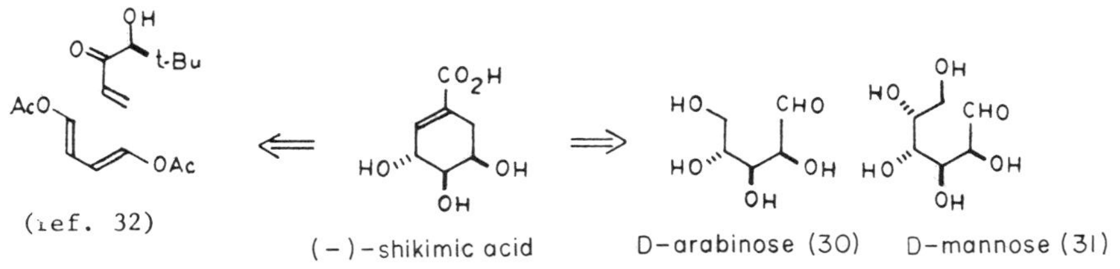

Figure 8. Carbohydrate-based and asymmetric Diels-Alder routes to (-)-shikimic acid.

One of several examples of the utilization of carbohydrates in the total synthesis of heterocyclic natural products is that of anisomycin (Figure 9). Thus C-3/C-4 of D-glucose provides the desired threo orientation of the diol unit in addition to a five-carbon framework, needed for further elaboration (33). Another synthesis starts with D-ribose (34). R,R-Tartaric acid has also served its purpose well in providing the desired sense of chirality on a four-carbon framework (35,36). Note once again that the two carboxyl groups in the diacid have to be distinguished from one another in order to achieve the required level of convergence with the target.

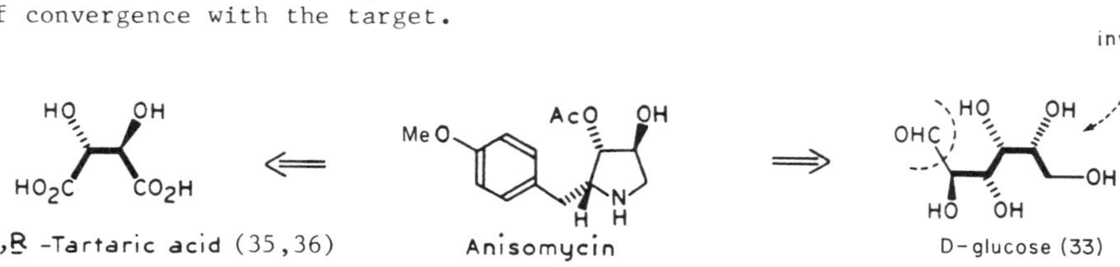

Figure 9. Synthetic precursors to anisomysin.

As a class of natural products, the macrolide antibiotics have captured the interest and aroused the imagination of synthetic chemists for close to three decades (37,38). The advent of novel synthetic methodology such as the aldol reaction (11) as well as the development of ingeneous synthetic strategies have allowed the synthetic conquest of these awesome targets. Carbohydrates have played a pivotal role as chiral templates in the elaboration of the complex structures of a number of macrolide antibiotics (4) as shown initially in the case of erythronolide A (39) (Figure 10). The carbohydrate template

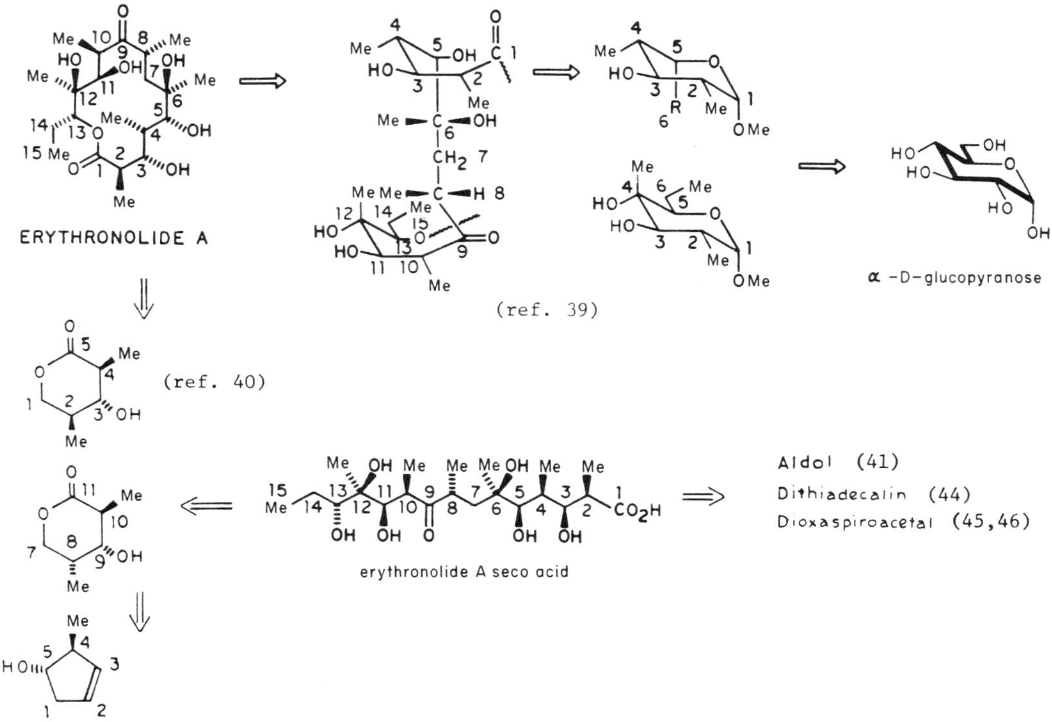

Figure 10. Unravelling hidden symmetry in erythronolide A.

approach afforded two six-carbon segments with appropriate functionality, which were chain-extended and assembled into the seco acid structure. This strategy showed how conformational bias could be used to systematically modify the functional groups in methyl α-D-glucopyranoside. Then, when the template effect had served its useful purpose, acyclic structures were unravelled and the synthetic operations were pursued. This operational amenity is not shared by other optically active synthetic precursors. Stork and coworkers (40) have shown an ingeneous unravelling of symmetry elements by relating the C-1—C-5 and C-7—C-11 segments of dihydro erythronolide A seco acid to a common optically active precursor (Figure 10). Thus 1S,2S-(+)-2-methyl-3-cyclopenten-1-ol was used as a template to introduce sequential functionality and give two lactones. Note that only the C-methyl group which corresponds to C-4 and C-10 in the target was maintained from the precursor. Highly diastereoselective methods of aldol condensation (11), particularly utilizing innovative techniques of asymmetric induction through double kinetic resolution, and/or the influence of chiral auxiliaries have provided much intellectual stimulation in the area of macrolide synthesis (41). Other elegant strategies for the construction of the complex polyfunctional carbon framework of erythronolide A have utilized ingeneous concepts of conformational bias and anomeric stereoselection as exemplified by the Woodward dithiadecalin (44), and the Deslongchamps (45) and Ireland spiroacetal (46) approaches. Impressive syntheses of erythronolides A (42) and B (43), erythromycin A (44), by these and other strategies have been achieved. Carbohydrates have also been useful in the total synthesis of carbomycin B (47,48) and other macrolides (49,50). Figure 11 illustrates the Nicolaou-Tatsuta (47,48) strategy for the synthesis of carbonolide B starting from D-glucose.

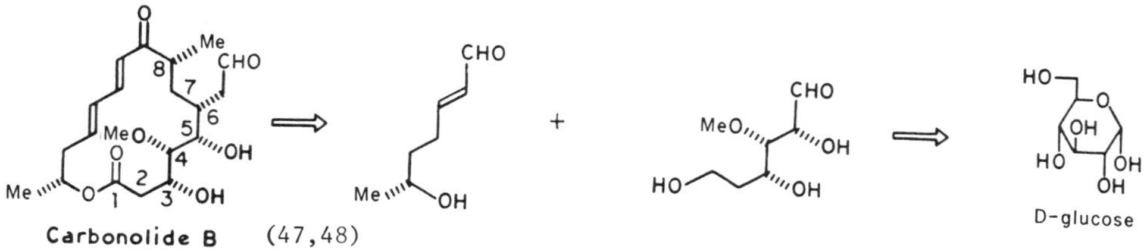

Carbonolide B   (47,48)

Figure 11. Synthesis of carbonolide from D-glucose.

UTILIZATION OF 'CHIRONS' DERIVED FROM CARBOHYDRATES AND NON-CARBOHYDRATES IN ORGANIC SYNTHESIS

We have previously elaborated on the characteristics of the four major classes of optically active starting materials in terms of their carbon framework, sense of chirality, asymmetric centers, sequential functionality and other inherent features (4). These are among the important considerations when choosing one or more of these molecules as synthetic precursors to targets. But prior to such a decision, one must scrutinize the target molecule and locate elements of apparent or hidden symmetry related to the chiral precursor in question (discovery). Does one "see" the disguised carbon framework of an amino acid or a carbohydrate for example in a segment of the target skeleton? Once a "visual dialogue" is established and the potential precursors are located, can one elaborate practical and efficient synthetic routes to such segments with assurances for regiochemical convergence and stereochemical integrity? These issues have been discussed at length elsewhere (4) and it will suffice to enumerate a few guidelines that may assist the reader in the 'discovery' process.

The 'chiron' approach depends primarily on a convergence of the general shape and topology of the carbon framework of the target, the interelationship of its functional groups and their sense of chirality, with corresponding features in the chiral precursor. Allowance is made of course for possible adjustments of functionality through chemical manipulation. For example, amino acids, hydroxy acids and terpenes can be used as templates for segments of targets in which α-amino acid, α-hydroxy acid, diol units, etc. or appropriately transformed variants of these are required for superposition. They can prove to be ideal starting materials for relatively short carbon chains with one or two asymmetric centers, or for carbocyclic or heterocyclic compounds. Figure 12 shows the structures of some non-carbohydrate precursors which were used for the synthesis of a variety of complex natural products (51-55).

In general, sugars offer a greater variety of inherent stereochemical and functional features, including variable chain-length, compared to other chiral precursors. In addition, they offer the option to manipulate cyclic or acyclic forms. Stereoelectronic effects, topology, conformational bias, and stereochemical versatility combine to give the synthetic chemist a good deal of predictive power (4). As a consequence, carbohydrates have been much more versatile as precursors to a larger variety of structures compared to others precursors (4).

Since the 'discovery' process could be somewhat difficult in the case of carbohydrate precursors, it has been useful to adopt the so-called "rule-of-five" (4) which applies to six-membered (pyranose-type) derivatives. The hemiacetal carbon atom and the ring oxygen which is located five bonds away, are made to coincide with and sp, sp^2 (or sp^3) carbon and a hetero atom in the target structure respectively. Thus, two reference points are located by juxtaposition of the carbon framework of the precursor (cyclic or acyclic form) and the target. Substituents 'in-between' these two reference points can then be introduced by systematic manipulation of the hydroxyl groups. There is an added advantage in that carbohydrates can be manipulated as cyclic forms, thus taking full advantage of conformational bias, and proceeding to acyclic forms after the template effect has served its useful purpose. Figure 13 illustrates how this guideline is applicable to the β-lactam antibiotic thienamycin.

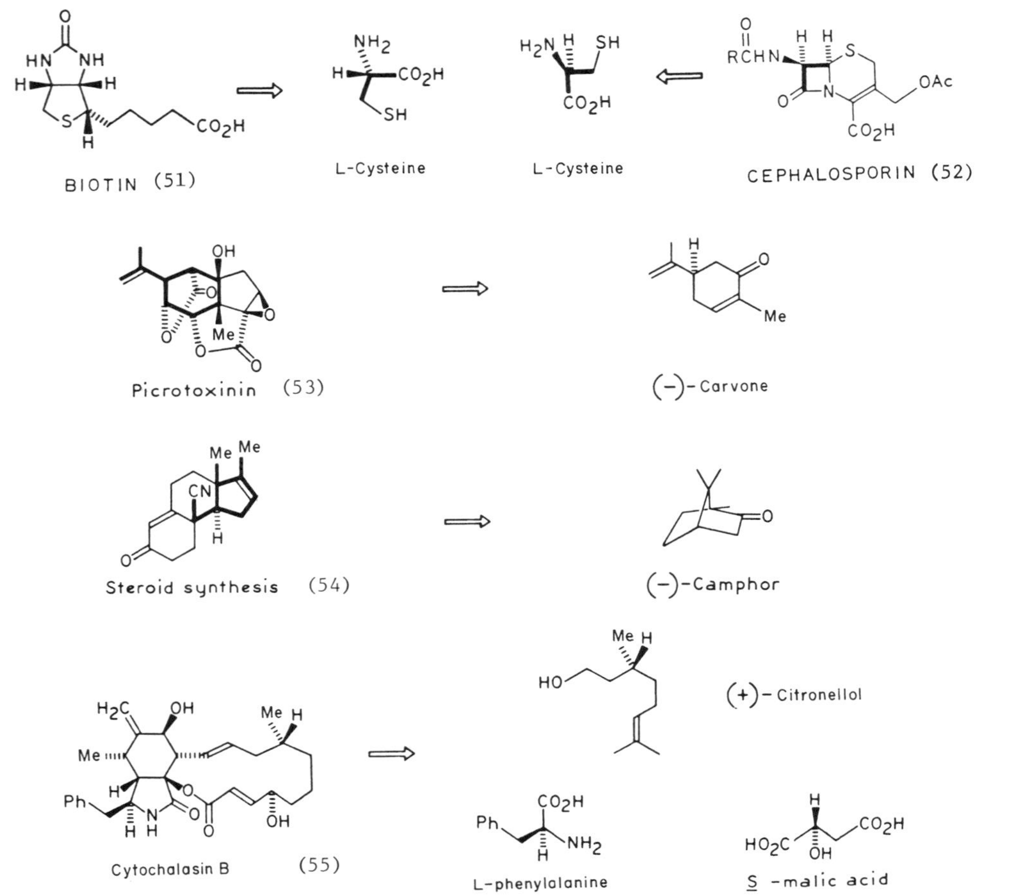

Figure 12. The use of amino acids hydroxy acids and terpenes as chiral templates
in natural product synthesis.

In the case of thienamycin, previous syntheses from L-aspartic acid (56) and L-threonine
(57) have each given access to one of the three asymmetric centers. The synthesis of a

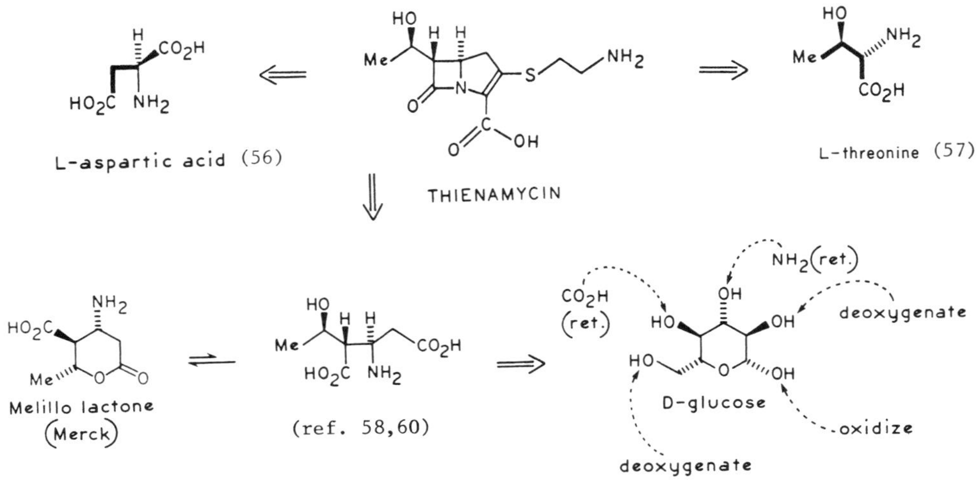

Figure 13. Decoding the functional and stereochemical information in
thienamycin.

critical intermediate (Melillo lactone) (56) from 2-deoxy-D-arabino-hexose (available from
D-glucose or commercially) which already harbors one asymmetric center convergent with the

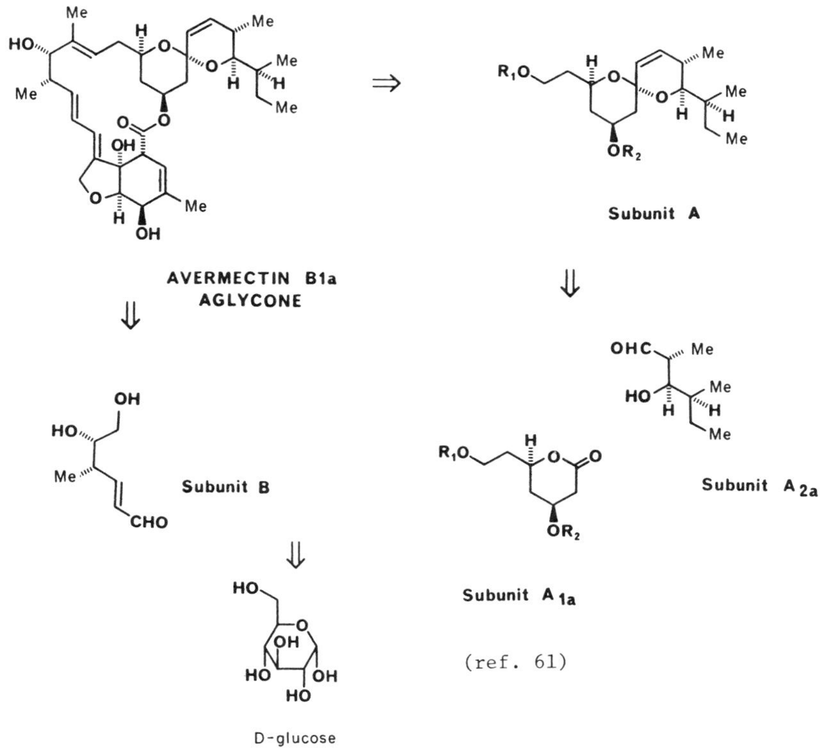

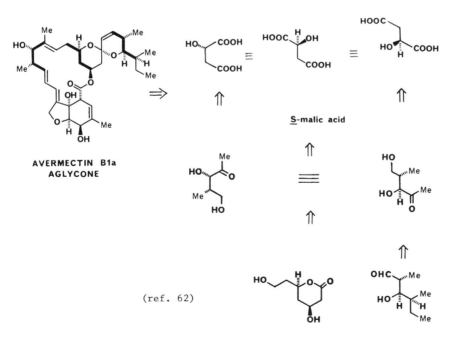

Figure 14.   Retrosynthetic analysis relating segments of Avermectin B$_{1a}$
             aglycone to D-glucose.

target (C-8), illustrates how a ring form could be manipulated to introduce all the desired

Figure 15.   Discovering S-malic acid as a four-carbon chiral template for
             segments to Avermectin B$_{1a}$ aglycone.

functional groups (58). Other approaches from carbohydrates have also been described (59).
Approaches to the synthesis of the complex anthelmintic macrolactone, avermectin B_{1a} (60)
and its analogs from D-glucose have been recently reported (61,62). Retrosynthetic analysis
reveals the possibility to assemble the upper portion of the molecule from D-glucose
(Figure 14) (61).

Interestingly, the molecule presents other avenues of exploitation by different precursors.
Thus, the readily available S-malic acid can be regarded as a valuable 4-carbon template
upon which to build the entire upper segment of the target (Figure 15). Indeed, the entire
upper half of avermectin B_{1a} has been assembled from this hydroxy acid (63).

COMPUTER ASSISTED PRECURSOR RECOGNITION (CAPR)

As already discussed, the selection of potential precursors to a target depends on a variety
of parameters. We have developed a computer program that selects appropriate synthetic
precursors among amino acids, hydroxy acids, carbohydrates and selected terpenes for a variety
of target structures (64). The program will decipher the functional and stereochemical
information (including R,S and E,Z configurations) present in the target structure, and
relate it to corresponding parameters in appropriate precursors. A priority scale (up to
100%) is generated based on favorable or unfavorable matching of end-groups (ex. CO_2H,
CH_2OH, Me, etc.), internally-located groups, and chain-lengths or ring forms. Allowance is
made for reasonable chemical transformations, while maintaining stereochemical integrity.
When a favorable match is found between a segment of the target structure and a given
precursor, it is registered and given a priority by the program. In general, the program
will provide precursors for the longest span of carbon atoms with minimum perturbation of
adjacent or vicinal chiral centers. For the moment, precursors with a maximum length of
eight carbon atoms (including carbohydrates) and a minimum length of three carbon atoms
(amino acids, hydroxy acids, etc.) are considered by the program. The target structure is
entered graphically on the screen and those segments of the carbon skeleton for which a
synthetic precursor is derived are indicated by directing the cursor to each carbon atom or
by asking for the best precursors with a designated length of carbon atoms (three and
over). Alternatively, one may ask for the best precursors (example 〉 90% priority). The
corresponding precursors will appear on the screen with their priority scores. The program
will reproduce the precursor in a perspective drawing that provides the best overlap with
the corresponding segment in the target and it will show all asymmetric centers with α- or
β- bonded (hash-line and solid wedges) functional groups that coincide with that portion of
the carbon framework which overlaps with the target appears in bold lines. Figures 16 and
17 show the output for acivicin, an antitumor antibiotic (65), and for the upper segment
of avermectin B_{1a} aglycone (63). Note how D-mannosamine provides excellent convergence for

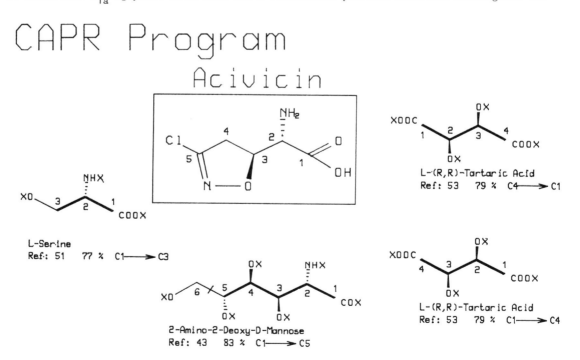

Figure 16. Selected synthetic precursors to acivicin as generated by the CAPR program.

the C-1–C-5 segment of acivicin (COOH taken as C-1) with the correct regio- and
stereochemistry at C-2/C-3 hence its high rating. Adjustment of oxidation and reduction
states at C-1 and C-4 respectively and loss of C-6 are necessary to give the desired

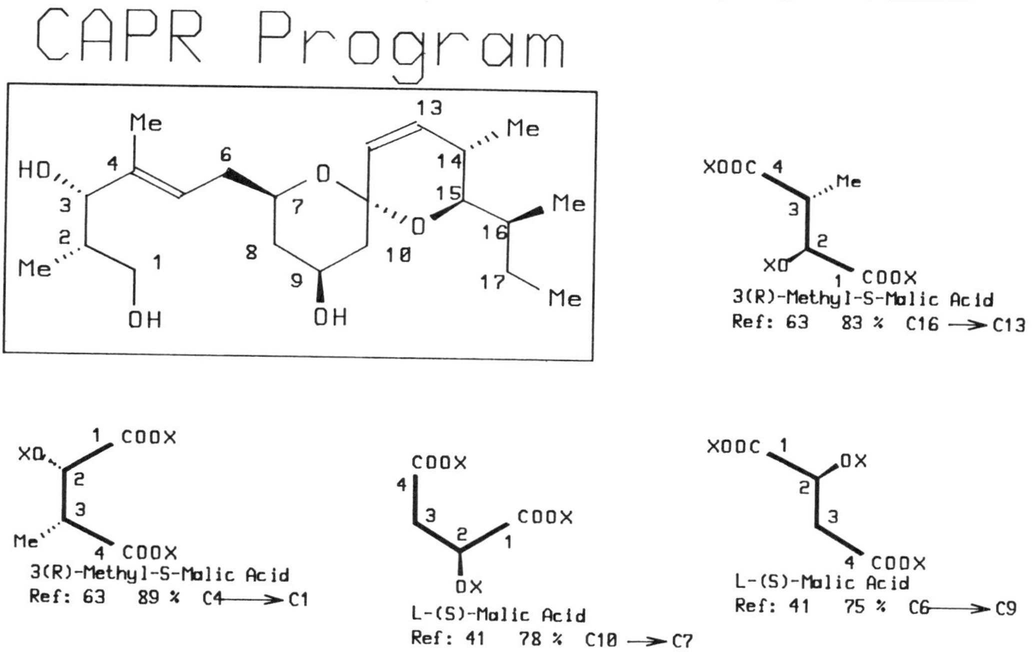

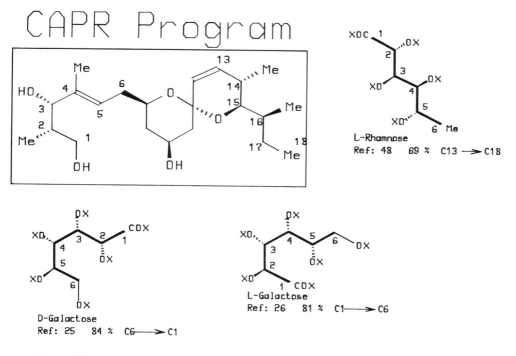

Figure 17a. Three and four-carbon synthetic precursors to the upper segment of
Avermectin B$_{1a}$ aglycone (63).

five-carbon framework incorporating functionality and stereochemistry. Note the cross-line
bisecting C-5–C-6 indicating the site of cleavage. Among the three-carbon precursors,
L-cysteine and L-serine represent excellent templates for further elaboration. However,

Figure 17b. Selected six-carbon synthetic precursors to the upper segment of
Avermectin B$_{1a}$ aglycone (63).

provisions have to be made to lengthen the carbon chain and secure the hydroxyl functionality as well as the correct sense of chirality at the β-carbon atom. In the case of avermectin B_{1a} aglycone (Figure 17a), the program finds 2S, 3R-3-methyl malic acid, which can be easily obtained from S-malic acid (66) as a versatile starting material for two segments of the target encompassing four-carbon atoms and two asymmetric centers. Note that the priority score for the right-hand segment which corresponds to C-16—C-13 is slightly less than that for the left-hand segment (C-4—C-1). The reason being that the precursor must be subjected to a larger number of transformations to overlap with the C-16—C-13 segment compared to the left-hand counterpart.

Figure 17b shows some of the six-carbon precursors suggested by the program. Note that in each case the precursor is depicted with the contours and shape that overlap with the corresponding segment in the target framework. Finally, a bibliographic compilation of key references will lead the user to useful information regarding the utilization of the appropriate precursor in synthesis or a related chemical transformation. The program (Pascal) operates on a VAX-780 computer and a Tektronix 4112A graphics terminal with a 4662 plotter.

CONCLUSION

One of the refreshing aspects of organic synthesis is the individuality of various approaches to a target. There are never enough ways of bond breaking and reconstruction and therein lies a source for innovation. It is clear that when adopting the 'chiron' approach, certain elements of logic and practicality have to be considered. While there are no general rules to be followed, experience has shown that in spite of its relative length at times, the sugar-based strategy has made its mark in the synthesis of polyfunctional targets containing three asymmetric centers or more, particularly if they are contiguous. One can therefore manipulate a cyclic form of a hexopyranose for example, setting up a maximum of five contiguous asymmetric centers (including the anomeric center). At the end of such an operation, and once the molecule is unfolded to an acyclic equivalent for example, a significant amount of functionality and chirality can be in hand. Approaches utilizing other chiral precursors in which one or two asymmetric centers are available for manipulation may be competitive, but the overall outcome may be less predictable with regard to regio and stereocontrol of additional asymmetric centers when the chain is lengthened. Such precursors are usually ideally suited for targets or segments containing less than three contiguous asymmetric centers. As previously mentioned, in general, the sugar-based strategy provides a longer span of the intended carbon skeleton compared to other precursors. This feature loses its appeal if the strategy calls for extensive deoxygenation or chain-shortening. In such cases, other 'chirons', particularly derived from amino acids, hydroxy acids and certain terpenes would be more practical to use.

"If there is one way better than another, it is the way of Nature"

Aristotle

ACKNOWLEDGMENTS

We thank the National Scientific and Engineering Council of Canada, le Ministère de l'éducation du Québec and Merck Frosst (Canada) for generous financial assistance. Special thanks are also due to the following coworkers for their efforts in developing our CAPR graphics program, L. Forest, S. Léger, F. Major et J.-C. Bolduc.

REFERENCES

1. See for example, "The Total Synthesis of Natural Products", Vol. 1-4, J. ApSimon, ed., Wiley-Interscience, New York, N.Y., 1973-81; J.S. Bindra and R. Bindra, "Creativity in Organic Synthesis", Academic Press, Inc., New York, N.Y., 1975.
2. See for example, D. Seebach and E. Hungerbühler, in "Modern Synthetic Methods", R. Scheffold, ed., Otto Salle Verlag, Frakfurt am Main, Germany,p. 91-171 (1980).
3. "Asymmetric Synthesis", J.D. Morrison, ed., Vol. 1-4, 1983-84; see also J. ApSimon and R.P. Seguin, Tetrahedron, 35, 2797-2842 (1979); and references cited therein.
4. S. Hanessian, "Total Synthesis of Natural Products: The 'Chiron' Approach". J.E. Baldwin, ed., Pergamon Press, Oxford, England, 1983.
5. K. Drauz, A. Kleeman and J. Martens, Angew. Chem., Int. Ed. Engl., 21, 584-608 (1982).
6. See for example, K. Mori in "The Total Synthesis of Natural Products", J. ApSimon,ed., Wiley-Interscience, New York, N.Y. Vol. 4, 1981, p. 1
7. B. Fraser-Reid and R.C. Anderson, Fortschr. Chem. Org. Naturst., 39, 1-61 (1980); B. Fraser-Reid, Acc. Chem. Res., 8, 192 (1975).
8. A. Vasella, in "Modern Synthetic Methods", R. Scheffold, ed., Otto Salle Verlag, Frankfurt am Main, Germany, 1980, p. 173-267.
9. S. Hanessian, Acc. Chem. Res., 12, 159-165 (1979).

10. J.B. Jones in "Applications of Biochemical Systems in Organic Chemistry" Part 1, J.B. Jones, C.J. Sih and D. Perlman, eds., Wiley-Interscience, New York, N.Y., 1976, p. 107. A. Fischli, in "Modern Synthetic Methods", R. Scheffold, ed., Otto Salle Verlag, Frankfurt am Main, Germany, 1980, p. 269-350.

11. For recent reviews, see D.A. Evans, J.V. Nelson and T.R. Taber, Topics in Stereochemistry, 13, 1-115 (1982); R.W. Hoffmann, Angew. Chem., Int. Ed. Engl., 21, 555-566 (1982); S. Masamune and W. Choy, Aldrichimica Acta, 15, 47-63 (1982); T. Mukaiyama, Org. React., 28, 203-331 (1982); C.H. Heathcock, Science, 214, 395-400 (1981).

12. T. Katsuki and K.B. Sharpless, J. Am. Chem. Soc., 102, 5976-5978 (1980).

13. See for example R.B. Woodward in "Perspectives in Organic Chemistry", A.R. Todd, ed., Interscience, 1956, p. 155.

14. See for example D.A. Evans, K.T. Chapman and J. Bisaha, J. Am. Chem. Soc., 106, 4261-4263 (1984). W. Choy, L.A. Reed and S. Masamune, J. Org. Chem., 48, 1139-1141 (1983); W. Oppolzer, C. Chapuis, G.M. Dao, D. Reichlin and T. Godel, Tetrahedron Lett., 23, 4781-4784 (1982) and references cited therein.

15. See for example, D.A. Evans, Aldrichimia Acta, 15, 23-32 (1982); S. Masamune and W. Choy, Aldrichimica Acta, 15, 47-63 (1982); Y. Kishi, Aldrichimica Acta, 13, 23-30 (1980).

16. K. Mori, Tetrahedron, 31, 3011-3012 (1975).

17. H.R. Schuler and K. Slessor, Can. J. Chem., 55, 3280-3287 (1977).

18. K. Utimoto, K. Uchida, M. Yamaga and Nozaki, Tetrahedron Lett., 3641-3645 (1977).

19. R.E. Ireland and J.-P. Vevert, J. Org. Chem., 45, 4259-4260 (1980).

20. K.M. Sun and B. Fraser-Reid, Can. J. Chem., 58, 2732-2735 (1980).

21. U. Schmidt, J. Gombos, E. Haslinger and H. Zak, Chem. Ber., 109, 2628-2644 (1976).

22. A.G.M. Barrett and H.G. Sheth, J. Org. Chem., 48, 5017-5022 (1983).

23. K. Mori, T. Takigawa and M. Matsui, Tetrahedron Lett., 3953-3956 (1976).

24. K.C. Nicolaou, D.P. Papahatjis, D.A. Claremon and R.E. Dolle, J. Am. Chem. Soc., 103, 6967-6969 (1981).

25. B.E. Rossiter, T. Katsuki and K.B. Sharpless, J. Am. Chem. Soc., 103, 464-465 (1981).

26. G. Fräter, Helv. Chim. Acta, 62, 2829-2832 (1979).

27. K. Mori, and H. Iwasawa, Tetrahedron, 36, 2290-2213 (1980).

28. J.-R. Pougny and P. Sinay, J. Chem. Res., 5, 1-2 (1982).

29. J.P. Vigneron, R.Meric and M. Dhaenens, Tetrahedron Lett., 21, 2057-2060 (1980).

30. H.J. Bestmann and H.A. Heid, Angew. Chem., Ind. Ed. Engl., 10, 336-337 (1971).

31. G.W.J. Fleet and T.K.M. Shing, J.C.S. Chem. Comm., 849-850 (1983).

32. S. Masamune, L.A. Reed, J.T. Davis and W. Choy, J. Org. Chem., 48, 4441-4443 (1983).

33. J.P.H. Verheyden, A.C. Richardson, R.S. Bhat, B.D. Grant, W.L. Fitch and J.G. Moffatt, Pure Appl. Chem., 51, 1363-1383 (1978).

34. J.G. Buchanam, K.A. MaLean, H. Paulsen and R.W. Wightman, J.C.S. Chem. Comm., 486-488 (1983).

35. C.M. Wong, J. Buccini and J. TeRaa, Can. J. Chem., 46, 3091-3094 (1968).

36. I. Flener and K. Schenker, Helv. Chim. Acta, 53, 754-763 (1970).

37. S. Masamune, G. Bates and J.W. Corcoran, Angew. Chem., Ind. Ed. Engl., 16, 585-607 (1977).

38. K.C. Nicolaou, Tetrahedron, 33, 683-710 (1977); T.G. Back, Tetrahedron, 33, 3041-3059 (1977).

39. S. Hanessian and G. Rancourt, Pure Appl. Chem., 49, 1201-1214 (1977); Can. J. Chem., 55, 1111-1113 (1977); S. Hanessian, G. Rancourt and Y. Guindon, Can. J. Chem., 56, 1843-1846 (1978).

40. G. Stork, I. Paterson and F.K.C. Lee, J. Am. Chem. Soc., 104, 4686-4688 (1982).

41. See for example, S. Masamune, L.D.-L. Lu, W.P. Jackson, T. Kaiho and T.Toyoda, J. Am. Chem. Soc., 104, 5524-5526 (1982); C. H. Heathcock, Science 214, 395-400 (1981).

42. E.J. Corey, P.B. Hopkins, S. Kim, S. Yoo, K.P. Nambiar and J.R. Falck, J. Am. Chem. Soc., 101, 7131-7134 (1979).

43. E.J. Corey, S. Kim, S. Yoo, K.C. Nicolaou, L.S. Melvin, D.J. Brunelle, J.R. Falck, E.J. Trybulski, R. Lett, and P. Sheldrake, J. Am. Chem. Soc., 100, 4620-4623 (1978).

44. R.B. Woodward and coworkers, J. Am. Chem. Soc., 103, 3210, 3213, 3215 (1981).

45. P. Deslongchamps, "Stereoelectronic Effects in Organic Chemistry", J.E. Baldwin, ed., Pergamon Press, Oxford, England, 1983, p. 328.

46. R.E. Ireland, J. Daub, G.S. Mandel and N.S. Mandel, J. Org. Chem., 48, 1312-1325 (1983).

47. K.C. Nicolaou, M.R. Pavia and S.P. Seitz, J. Am. Chem. Soc., 103, 1224-1226 (1981).

48. K. Tatsuta, A. Tanaka, K. Fujimoto, M. Kinoshita and S. Umezawa, J. Am. Chem. Soc., 99, 5826-5827 (1977); K. Tatsuta, Y. Amemiya, S. Maniwa and M. Kinoshita, Tetrahedron Lett., 21, 2837-2840 (1980); K. Tatsuta, T. Yamaguechi and M. Kinoshita, Bull. Chem. Soc. Japan, 51, 3035-3038 (1978).

49. K. Tatsuta, Y. Amemiya, Y. Kanemura, H. Takahashi and M. Kinoshita, Tetrahedron Lett., 23, 3375-3378 (1982).

50. K. Tatsuka, A. Nakagawa, S. Maniwa and M. Kinoshita, Tetrahedron Lett., 21, 1479-1482 (1980).

51. P. Confalone, G. Pizzolato, E.G. Baggiolini, D. Lollar and M.R. Uskokovic, J. Am. Chem. Soc., 99, 7020 (1977).

52. K. Hensler, J. Gosteli, P. Naegli, W. Oppolzer, R. Ramage, R. Ranganathan and H. Vorbrüggen, J. Am. Chem. Soc., 88, 852-853 (1966).
53. E.J. Corey and H.L. Pearce, J. Am. Chem. Soc., 101, 5841-5843 (1979).
54. R.V. Stevens and F.C.A. Gaeta, J. Am. Chem. Soc., 99, 6105-6106 (1977).
55. G. Stork, Y. Nakahara, Y. Nakahara, and W.J. Greenlee, J. Am. Chem. Soc., 100, 7775-7779 (1978).
56. T.N. Salzmann, R.W. Ratcliffe, B.G. Christensen and F.A. Bouffard, J. Am. Chem. Soc.. 102, 6161-6163 (1980); see also D.G. Melillo, T. Liu, K. Ryan, M. Sletzinger and I. Shinkai, Tetrahedron Lett., 22, 913-916 (1981).
57. See for example, M. Shiozaki, N. Ishida, T. Hiraoka, and H. Yanazisawa, Tetrahedron Lett., 22, 5205-5208 (1981).
58. S. Hanessian, G. Rancourt, D. Desilets and R. Fortin, Can. J. Chem., 60, 2292-2294 (1982).
59. P. Durette, Carbohydr. Res., 100, C27-C30 (1982); N. Ikota, D. Yushino, and K. Koga, Chem. Pharm. Bull., 30, 1929-1931 (1982); see also A. Knierzniger and A. Vasella, J.C.S. Chem. Comm., 9-11 (1984).
60. G. Albers-Schonberg, B.H. Arison, J.C. Chabala, A.W. Douglas, P. Eskola, M.H. Fisher, A. Lusi, H. Mrozik, J.L. Smith and R.L. Tolman, J. Am. Chem. Soc., 103, 4216-4221 (1981).
61. S. Hanessian, A. Ugolini and M. Therien, J. Org. Chem., 48, 4427-4430 (1983).
62. R. Baker, R. Herbert, P.E. House, O.T. Jones, W. Fracke and W. Reith, J. Chem. Soc., Chem. Comm., 52-53 (1980).
63. S. Hanessian, A. Ugolini and D. Dubé, unpublished results.
64. S. Hanessian, F. Major and S. Léger, Proceedings 1st Cyprus Conferences on New Methods in Drug Research, Limassol, Cyprus, April 14, 1983, A. Makriyannis, ed., J.A. Prous, S.A., Barcelona, Spain, (in press).
65. D.G. Martin, D.J. Duchamp and C.G. Chidester, Tetrahedron Lett., 2549-2552 (1973).
66. D. Seebach and D. Wasmuth, Helv. Chim. Acta, 63, 197-200 (1980).

Carbohydrates in functionally-substituted vinyl carbanion chemistry

Richard R. Schmidt

Fakultät Chemie, Universität Konstanz, D-7750 Konstanz, Germany

Abstract - Direct C-lithiation of functionally substituted acrylates yields short routes for the synthesis of butenolides tetronates, α-hydroxy butenolides, and α-hydroxy tetronates. These structural units are moieties of many important carbohydrates and related natural products. Approaches to diastereoselective syntheses of ascorbic acids, 3-deoxy-D-*manno*-2-octulosonate (KDO), 3-deoxy-D-*glycero*-D-*gulo*-2-nonulosonate, a precursor for the synthesis of N-acetyl-neuraminic acid (NANA), and of the "top half" of the macrolide antibiotic chlorothricolide are outlined (Ref. 1).

INTRODUCTION

Functionally substituted vinyllithium compounds are of great interest as reactive intermediates especially if other unprotected groups are also present (Ref. 2-5). Our investigations with ß- and α,ß-disubstituted acrylic acids and derivatives have demonstrated that these compounds are acessible to direct lithiation of the ß- and/or α-position (Schemes 1 and 2) (Ref. 2-4). This means that the generation of stable vinyllithium derivatives is faster than the expected 1,2- or 1,4-addition of the lithiating agent to the acrylate system.

Reactions with electrophiles and especially with electrophilic-nucleophilic compounds (aldehydes, ketones, esters, α,ß-unsaturated carbonyl compounds, cumulenes, etc.) afford short syntheses for butenolides, tetronates, α-hydroxy-butenolides, α-hydrotetronates, cyclopentenones, maleic acid derivatives, oxalacetates etc. (Scheme 1) (Ref. 3, 4).

Scheme 1

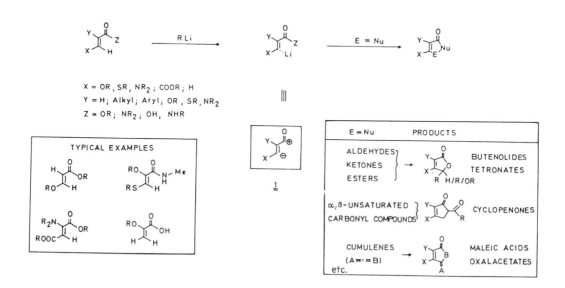

X = OR, SR, NR$_2$; COOR; H
Y = H; Alkyl; Aryl; OR, SR, NR$_2$
Z = OR; NR$_2$; OH, NHR

A wide variety of substituents are compatible with this lithiation: alkyl, aryl, electron-donating, and electron-withdrawing substituents in α- or β-position conducting to interesting regio- and stereoselectivities (Ref. 3, 4). The accessibility to a twofold lithiation leads to a preparative versatility, which for structural reasons is not obtained in aryl and hetaryl systems. The regioselectivity is mainly determined by the β-substituents on one side and the carboxylic functionality on the other side (Ref. 6, 7). In addition, some acrylic acid derivatives could be lithiated selectively in either the α- or β-position by use of kinetic or thermodynamic control (Ref. 6-8). Thus from acrylates in stepwise reaction the valuable synthons 1 (Scheme 1) and 2 (Scheme 2) are generated, which are structural moieties of many natural products.

Scheme 2

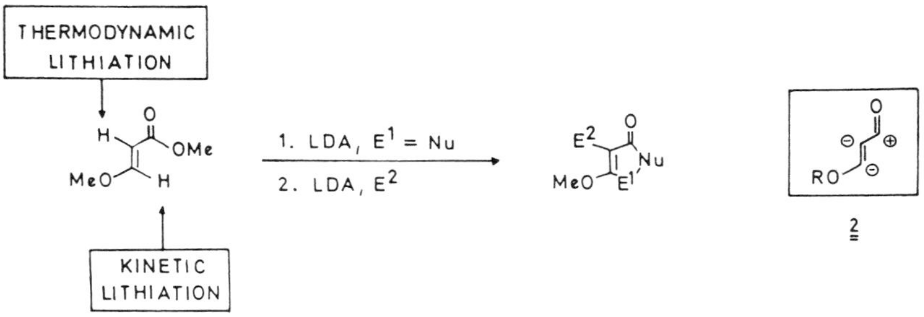

The ease of butenolide and tetronate formation from β-functionally substituted acrylates and carbonyl compounds was used quite successfully (Ref. 3, 4). However, the observation that these compounds are accessible to a subsequent α-lithiation was especially rewarding (Ref. 9). This methodology seemed to be of use for a short synthesis of the antibiotically active aspartetronins and gregatings, for which tetronate (furan-2-one) structures were proposed. Reaction of the β-lithiated acrylate species with (E,E)-3,5-octadien-2-one gave the required α-unsubstituted tetronate. α-Lithiation with LDA and reaction with acetic anhydride led to the expected α-substituted tetronate in an overall two-step reaction (Ref. 9). However, by these investigations it was demonstrated, that the natural gregatin B is isomeric to the compound obtained having a furan-3-one structure (Ref. 9, 10).

β-Functional X-substituents of acrylate synthons 1 and 2 should be accessible to an ensuing nucleophilic or electrophilic displacement reaction generating the new synthons 3 - 6 (Scheme 3). This way an even greater versatility of

Scheme 3

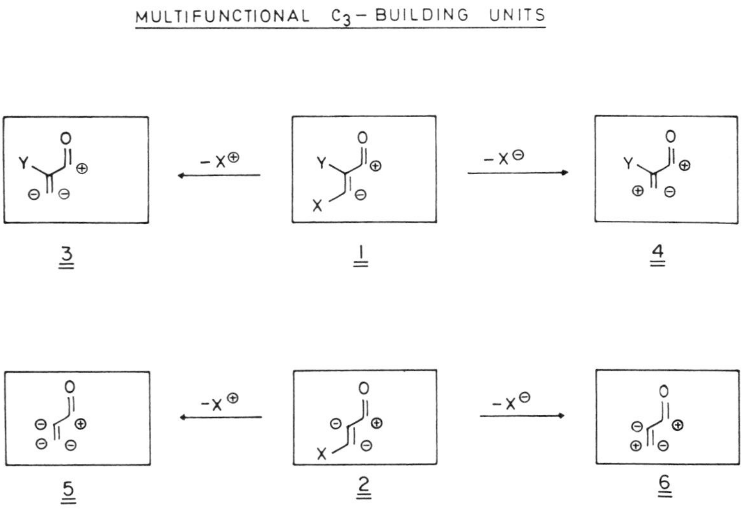

MULTIFUNCTIONAL C_3- BUILDING UNITS

functionally substituted acrylates as synthetic equivalents of multifunction-al C_3-building blocks is obtained. Combination with stereoselectivity by re-actions with chiral electrophiles or by chirality transfer with chiral func-tional substituents in different positions (Scheme 4) is a means for establi-

Scheme 4

DIASTEREOSELECTIVITY WITH CHIRAL ELECTROPHILES:

DIASTEREOSELECTIVITY WITH LITHIATED CHIRAL ACRYLATES:

(X,Y,Z CONTAINING CHIRAL GROUPS)

shing the basis for a great variety of natural product syntheses. Approach-es to diastereoselective syntheses of ascorbic acids (Ref. 4, 11), 3-deoxy-D-*manno*-2-octulosonate (KDO) (Ref. 11, 12), 3-deoxy-D-*glycero*-D-*gulo*-2-non-ulosonate (Ref. 13), a precursor for the synthesis of N-acetyl-neuraminic acid (NANA), and a diastereoselective synthesis for the "top half" of the macrolide antibiotic chlorothricolide (Ref. 14) will be discussed.

SYNTHESIS OF ASCORBIC ACIDS

A straight forward retrosynthetic scheme for the formation of L-ascorbic acid (vitamin C) is shown in Scheme 5, which implicates a *threo*-specific reaction

Scheme 5

L-ASCORBIC ACID: RETROSYNTHETIC SCHEME

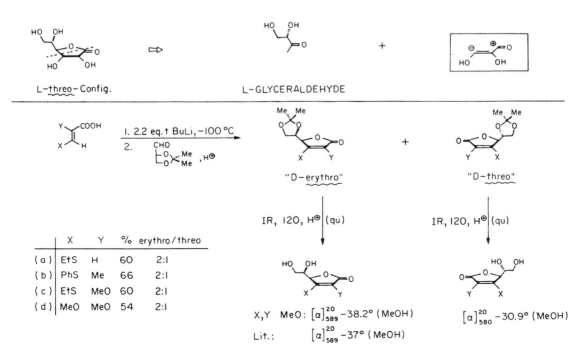

L-threo-Config.   L-GLYCERALDEHYDE

	X	Y	%	erythro/threo
(a)	EtS	H	60	2:1
(b)	PhS	Me	66	2:1
(c)	EtS	MeO	60	2:1
(d)	MeO	MeO	54	2:1

"D-erythro"   "D-threo"

IR, 120, H⊕ (qu)   IR, 120, H⊕ (qu)

X,Y MeO: $[\alpha]_{589}^{20}$ −38.2° (MeOH) $[\alpha]_{580}^{20}$ −30.9° (MeOH)

Lit.: $[\alpha]_{589}^{20}$ −37° (MeOH)

between L-glyceraldehyde and a synthon derived from <u>1</u> (see Scheme 1). Application of this approach to 2,3-dimethoxy acrylic aci<u>d</u> as synthetic equivalent of <u>1</u> and 2,3-O-isopropylidene-<u>D</u>-glyceraldehyde as chiral electrophile led to a convenient diastereoselective formation of 2,3-di-O-methyl-<u>D</u>-araboascorbate and -<u>D</u>-ascorbate in a two-step procedure (54 %, *erythro : threo* = 2:1). The structural assignment was possible by comparison of physical properties with literature values (Ref. 15).

SYNTHESIS OF KDO

3-Deoxy-2-glyculosonates biochemically generated from phosphoenol pyruvate and phosphorylated sugars are an interesting class of natural products (Scheme 6). 3-Deoxy-<u>D</u>-*arabino*-2-heptulosonate is intermediate in the shikima-

Scheme 6

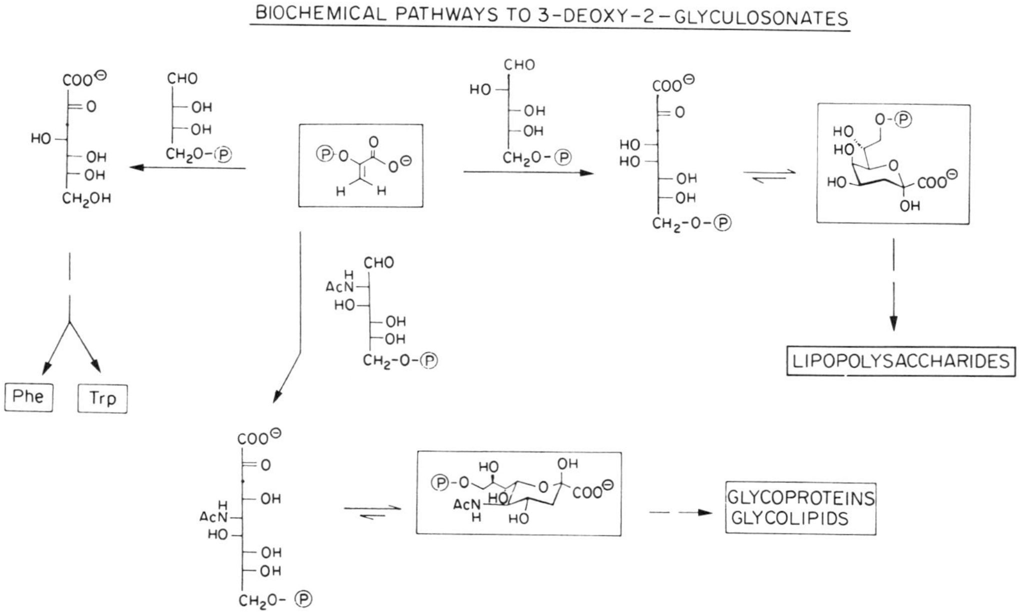

BIOCHEMICAL PATHWAYS TO 3-DEOXY-2-GLYCULOSONATES

te pathway finally leading to the amino acids phenylalanine (Phe) and tryptophan (Trp). 3-Deoxy-<u>D</u>-*manno*-2-octulosonate (KDO) is an integral part of the lipopolysaccharides of gram-negative bacteria (Ref. 16). 5-Acetylamino-2,5-dideoxy-<u>D</u>-*glycero*-<u>D</u>-*galacto*-2-nonulosonate is more known under its trivial name N-acetylneuraminic acid (NANA). It is an integral part of many glycoproteins and glycolipids.

The observed *erythro*-selectivity in the above described synthesis of ascorbic acids makes this reaction a good model for a short and efficient KDO synthesis. Earlier syntheses of KDO from <u>D</u>-mannose derivatives by Wittig-type chain elongation required additional steps for the introduction or liberation of the 2-keto group (Ref. 16-18). Syntheses from <u>D</u>-arabinose and derivatives led to disadvantageous diastereoisomer formation and separation (Ref. 19). The interest in the chemistry of KDO was reason to develop - based on the biochemical pathway - a synthesis with C-lithiated functionally substituted acrylates as pyruvate equivalent C$_3$-building unit and <u>D</u>-arabinose as C$_5$-unit.

The retrosynthetic Scheme 7 outlines the prospected KDO-synthesis via an *a*-hydroxybutenolide intermediate. However, a prerequisite is the generation of ß-lithiated species from *a*-alkoxyacrylates (X=H) or equivalents (X=SR, etc.), which react with <u>O</u>-protected <u>D</u>-arabinose *erythro*-diastereoselectively to the <u>D</u>-*manno*-isomer.

The results shown in Scheme 8 demonstrate, that *a*-methoxyacrylate is directly lithiated with tert.-butyllithium at the unsubstituted vinylic C-atom (Ref. 11, 13, 20). With 2,3:4,5-di-O-isopropylidene-<u>D</u>-arabinose as electrophile mo-

Scheme 7

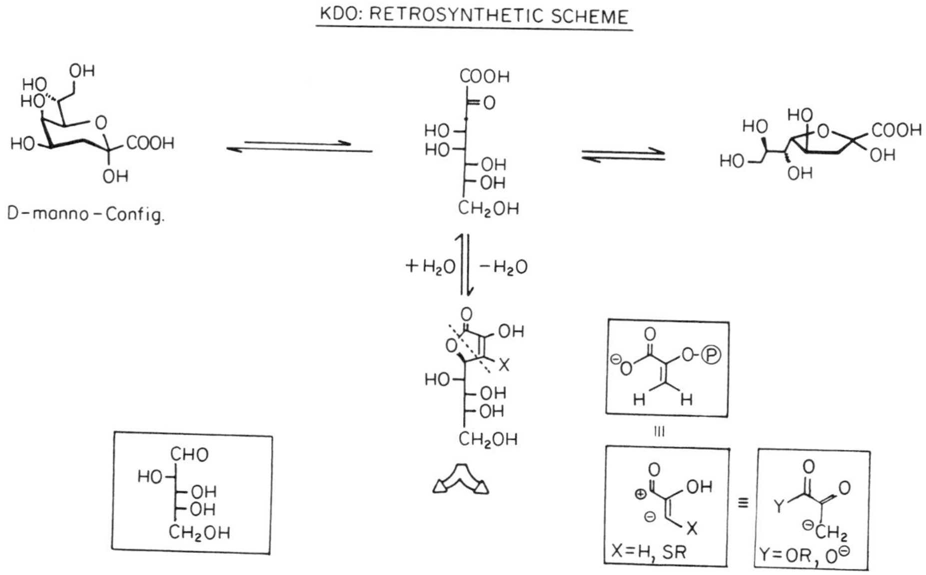

KDO: RETROSYNTHETIC SCHEME

Scheme 8

F.M. Unger, at. (1981)

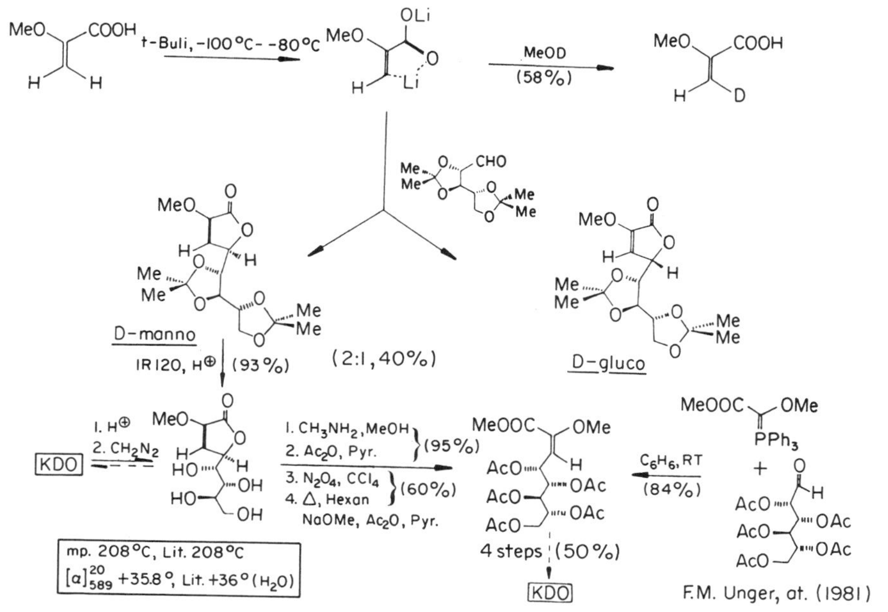

dest yields and low *erythro*-diastereoselectivity (40 %, 2:1) was obtained. Deisopropylidenation of the D-*manno*-isomer with acidic ion exchange resin was nearly quantitative; demethylation, however, was accompanied by byproduct formation, an observation made already with structurally similar compounds. Therefore this quite short route to KDO was not efficient. Structural proof for the deisopropylidenated intermediate, synthesized already by methylation of KDO (Ref. 21), was further obtained by transforming it in a four step sequence into a penta-O-acetyl derivative synthesized by Unger and his group (Ref. 16).

More convenient and comparable with the successful Collins KDO synthesis (Ref. 17) is a route starting from ß-alkylthio acrylates and O-protected D-arabinose. A 2,3-dideoxy-D-*manno*-octonate derivative is obtained in two steps, which is identical to a compound reached after five steps in the Collins synthesis. Additional three steps are required for the introduction of the 2-keto group, for the deprotection, and the final transformation to KDO.

Results with various lithiated acrylates clearly show, that the diastereoselectivities obtained depend on functional groups in α- and ß-position of the acrylate system (Ref. 11). The amount of the D-*manno*-isomer formed is increased by bulkier groups at the α-C-atom. In addition, the reactivity of α-functionally substituted ß-lithiated acrylates is increased by the presence of ß-alkylthio of ß-arylthio substituents.

The experience obtained thus far was used to design a specific acrylate system shown in Scheme 9. The α-benzyloxy group was chosen for the introduction

Scheme 9

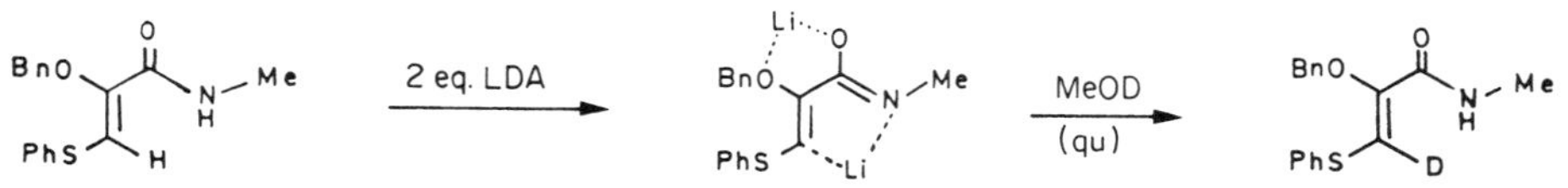

and ease of liberation of the 2-keto group; in addition, this bulky α-substituent should help to enhance the *erythro*-diastereoselectivity. The ß-phenylthio group was selected to increase the overall reactivity of the ß-lithiated species. The N-methyl amide group chosen was suggested to increase the diastereofacial selection due to the rigidity of the overall system. Deuteration experiments with this easily obtainable acrylamide derivative demonstrated the ease of formation of the dilithiated species shown in Scheme 9.

Reaction of this dilithiated species with 2,3:4,5-di-O-isopropylidene-D-arabinose gave the desired D-*manno*-2-octulosonate derivative (depending on the solvent system used) in high yield and in excellent *erythro*-diastereoselectivity (Scheme 10; THF/HMPT: 85 %, *erythro : threo* ·15:1). Clean crystallisation of

Scheme 10

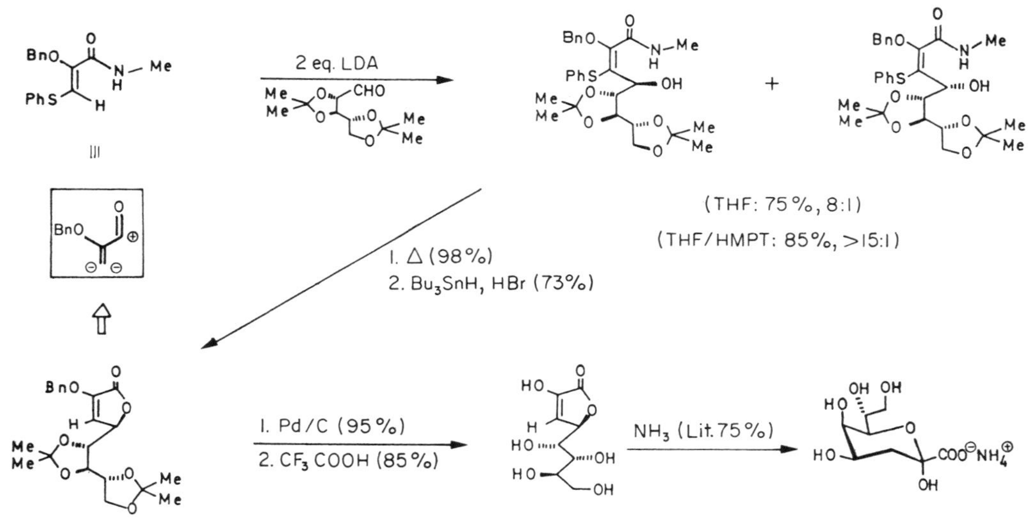

the *erythro*-compound avoided any chromatographic separation of the D-*manno*- and D-*gluco* isomer. The transformation of the D-*manno*-isomer into KDO was carried out in five convenient, high-yielding steps (Ref. 12).

The diastereofacial selection can be explained using the Felkin model (Scheme 11) (Ref. 22). Steric and stereoelectronic effects favor *Re*-attack of the

Scheme 11

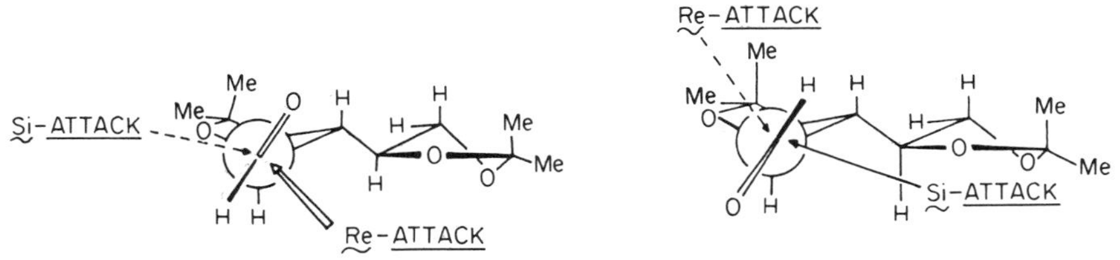

Re-ATTACK → *erythro* – resp. *manno* – PRODUCT

Si-ATTACK → *threo* – resp. *gluco* – PRODUCT

vinyllithium derivative in the (M)-conformer* over *Si*-attack. For the (P)-conformer* *Re*-attack and due to steric interaction with the 4,5-O-isopropylidene group especially *Si*-attack are less favored. This way the preferential D-*manno*-isomer formation can be rationalized.

SYNTHESIS OF 3-DEOXY-D-*glycero*-D-*gulo*-2-NONULOSONATE

Reaction of 2,3-O-isopropylidene-D-glyceraldehyde with the dilithiated acrylamide species of Scheme 11 led again to preferred *erythro*-diastereoselectivity (Scheme 12; overall yield 62 %, *erythro* : *threo* = 4:1). The preference for

Scheme 12

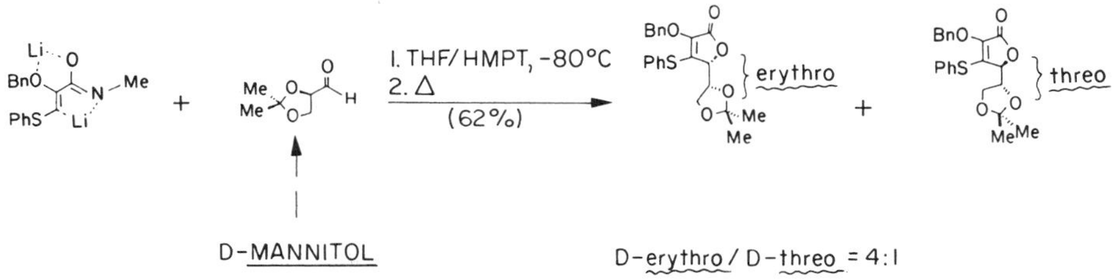

D-MANNITOL

D-*erythro* / D-*threo* = 4:1

Re-ATTACK ⟶ *threo* – PRODUCT

Si-ATTACK ⟶ *erythro* – PRODUCT

the *erythro*-isomer can also be explained by inspection of the preferred trajectories in the M- and P-conformers, which are entirely different from those in the KDO-synthesis (see Scheme 11). *Si*-attack in the P-conformer giving *erythro*-product is clearly favored. However, due to lower steric hindrance *Re*-attack in the M-conformer giving the *threo*-product is less disfavored as in the KDO-synthesis.

---

* M- and P-conformer: CIP-notation referring to the C_1-C_2-bond in the carbohydrate molecule.

A retrosynthetic analysis of NANA applying an analogous methodology leads to a functionally substituted vinyllithium species which requires *threo*-diastereoselective reaction with a D-mannosamine derivative (Scheme 13). However,

Scheme 13

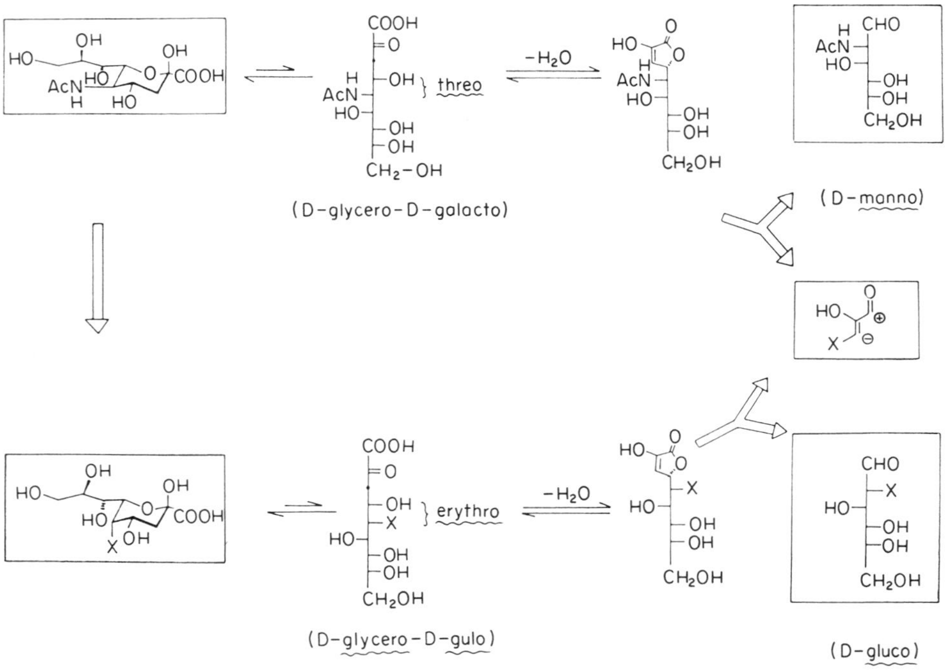
threo
$-H_2O$
(D-glycero-D-galacto)
(D-manno)
erythro
$-H_2O$
(D-glycero-D-gulo)
(D-gluco)

the observed *erythro*-preference in these reactions and the difficulty in arriving at an appropriate D-mannosamine derivative was reason to modify the synthetic plan as shown in Scheme 13. This pathway includes the introduction of the required N-acetylamino group by an S_N2-substitution with an appropriate N-nucleophile. This way an easily available D-glucose derivative could be applied, which again requires for the steric control an *erythro*-selective reaction.

Attempts to verify this synthetic plan are shown in Scheme 14. 2-O-Benzyl-3,4:5,6-di-O-isopropylidene-D-glucose easily obtained from D-glucose (Ref.

Scheme 14

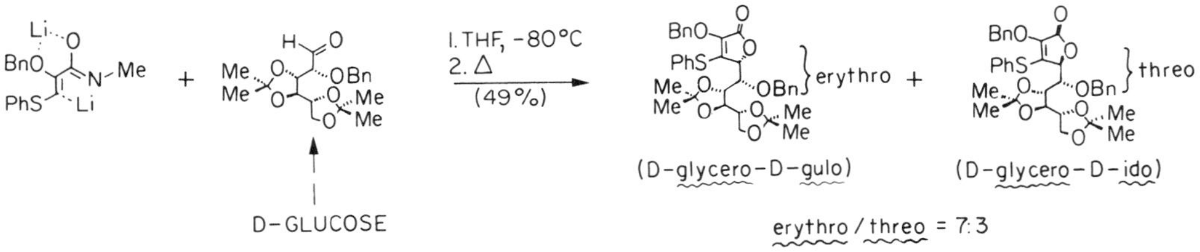
1. THF, -80°C
2. Δ
(49%)
erythro
threo
D-GLUCOSE
(D-glycero-D-gulo)
(D-glycero-D-ido)
erythro / threo = 7:3

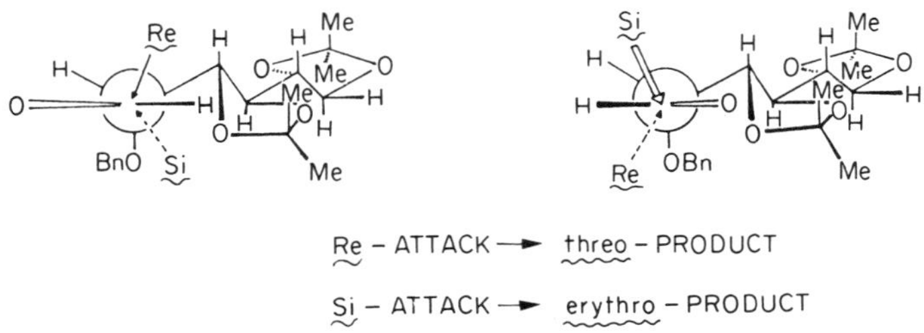
Re - ATTACK → threo - PRODUCT
Si - ATTACK → erythro - PRODUCT

13) was used as electrophile in the reaction with the dilithiated acrylamide species. As expected, mainly the *erythro*-product having D-*glycero*-D-*gulo*-configuration was formed (overall yield 49 %, *erythro* : *threo* = 7:3). The Felkin model (Ref. 22) again helps in understanding the preferred product formation. However, further experiments are needed to increase the efficiency of this reaction.

Removal of the phenylthio group from the D-*glycero*-D-*gulo*-isomer (Scheme 15) was carried out as demonstrated in the KDO-synthesis (see Scheme 10). It led

Scheme 15

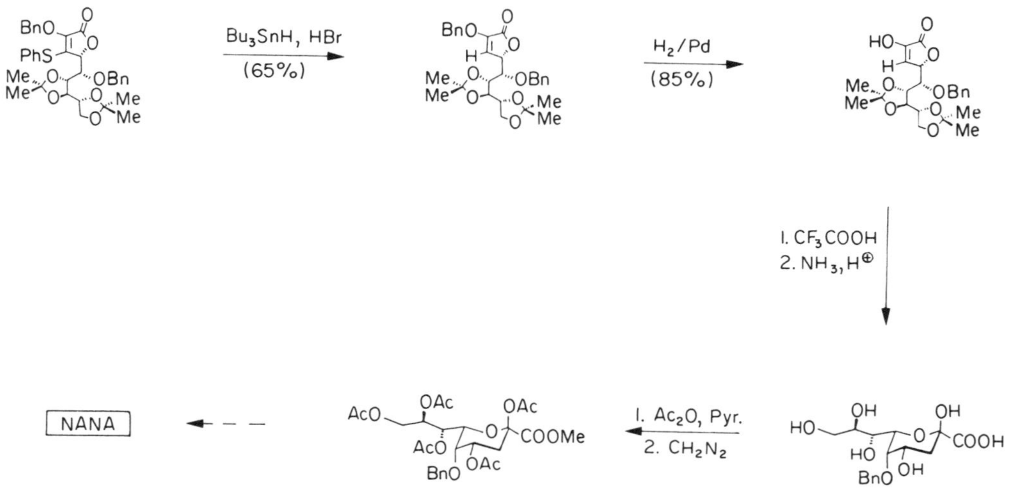

to a di-O-benzyl protected compound, which for synthetic reasons needed regio-selective debenzylation. This selectivity was readily obtained by hydrogeno-lysis in the presence of palladium. Due to the higher reactivity of the eno-late system debenzylation at the 2-position took place at a much faster rate compared with debenzylation at the 5-position. Deisopropylidenation with tri-fluoroacetic acid and ammonia treatment gave the desired 5-O-benzyl-D-*glycero*-D-*gulo*-2-nonulosonate. For structural assignment this compound was per-O-acetylated and transformed into the methylester; the pyranose derivative obtained should be also useful for a convenient transformation into NANA.

INVESTIGATIONS FOR SYNTHESIZING CHLOROTHRICOLIDE

The macrolide antibiotic chlorothricin (Scheme 18) was isolated from *Strepto-myces antibioticus*. This antibiotic is active against gram-positive bacteria because of inhibition of pyruvate carboxylase and maleate decarboxylase. The aglycon chlorothricolide contains part of the original activity of chlorothri-cin (Ref. 23). Several investigations towards the synthesis of partial struc-tures of this complex molecule have been reported (Ref. 24-26). The retrosyn-thetic analysis of the aglycon chlorothricolide (Scheme 16) leads to two sub-structures ("top half" and "bottom half", Ref. 24). The top half contains an *a*-hydroxy-tetronate moiety, which should be directly obtainable from a func-tionally substituted ß-lithiated acrylate species and a cyclohexenone deriva-tive as electrophile (Ref. 14).

The complex macrolide antibiotics kijanimicin and tetracarcin have aglycon structures quite similar to chlorothricin (Ref. 27). Therefore the outlined strategy to the synthesis of chlorothricolide should be especially rewarding because the corresponding aglycon structures kijanolide and tetronolide could be obtained via a similar route.

Scheme 16

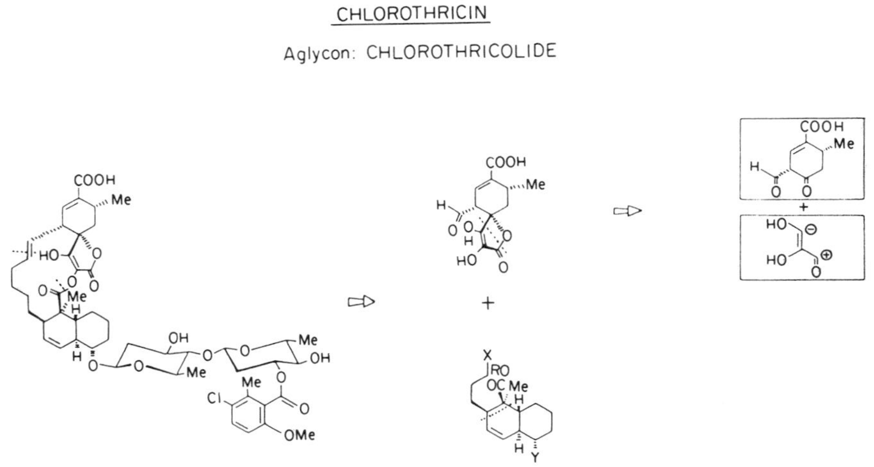

CHLOROTHRICIN

Aglycon: CHLOROTHRICOLIDE

To prove the feasibility of this synthetic plan, the cyclohexenone derivative shown in Scheme 17 was required. For a regioselective reaction of the ketone carbonyl group the aldehydic carbonyl group had to be protected. The cis-geometry required for the methyl and formyl groups can be generated by a Diels-Alder approach. For this aim pent-3-en-2-one was converted in a three step sequence to a butadiene derivative with a protected formyl group and methyl group in 1,4-position. Cycloaddition with methyl propiolate gave the expected cyclohexadiene derivative. However, liberation of the cyclohexenone moiety with tetrabutylammonium fluoride (TBAF) led not unexpectedly to elimination and concomitant loss of the cis relationship.

Scheme 17

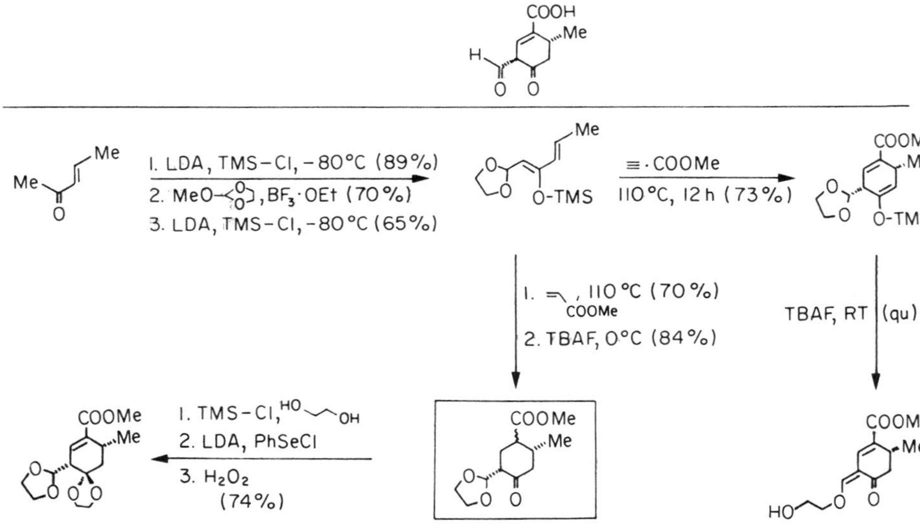

A cycloaddition route with acrylate as dienophile gave the related cyclohexanone derivative. Investigations towards the introduction of the missing CC-double bond afforded the corresponding cyclohexenone derivative quite readily. Therefore this methodology should be also effective at a later stage of the total synthesis of chlorothricolide.

The lower reactivity of ß-lithiated α,ß-functionally substituted acrylates towards cyclohexanone and derivatives (Ref. 11) was reason to use the more reactive ß-lithiated ß-functionally substituted acrylates for the formation of the spiro tetronate moiety (Scheme 18). The reaction of the all-cis substituted cyclohexanone derivative of Scheme 17 with ß-ethylthio dilithio acrylate and ß-methoxy dilithio acrylate gave diastereospecifically the required spirotetronate derivatives. The structure of the methoxy-substituted compound was assigned by X-ray analysis (Ref. 28). The ethylthio-substituted compound was cleanly transformed into the methoxy compound via sulfoxide formation with m-chloroperbenzoic acid (MCPBA) and methoxide treatment. Saponification of the ester group in these compounds led to the corresponding

Scheme 18

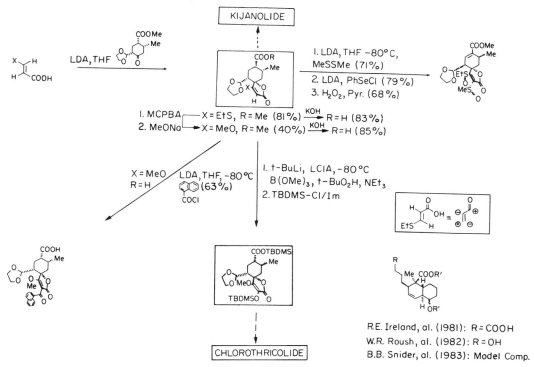

acids. The usefulness of these acids as intermediates in the synthesis of the top half of chlorothricolide was shown by direct C-lithiation with tert.-butyllithium/lithium cyclohexylisopropyl amide (LCIA) at the α-position of the tetronate moiety. The required α-hydroxylation was carried out with the methyl borate tert.-butylhydroperoxide/triethylamine system (Ref. 29). The α-hydroxylated compound obtained which contains the carbon skeleton, the stereochemistry, and the functional groups of the top half of chlorothricolide was characterized as disilylated product.

Syntheses of analogs of the missing dehydrodecaline moiety have been already published (Ref. 24-26). Therefore the construction of the macrolide antibiotic chlorothricolide should now be possible.

The synthetic strategy outlined is applicable to the synthesis of kijanolide and tetronolide as well. The successful α-acylation of the spirotetronate intermediate is a crucial step for this aim. The reactions discussed here demonstrate that ß-methoxy acrylates are synthetic equivalents of multifunctional C_3-building blocks.

Enantioselective syntheses of chiral tetronates may be carried out with the help of chiral substituents at different positions of the acrylate system (see Scheme 4). The ß-position is especially appropriate for the attachment of chiral substituents because nucleophilic addition of chiral alkoxides to propiolate and final removal of this substituent via addition/elimination is well established by previous experiments (Scheme 19). This methodology is exemplified in the reaction of α-tetralone with ß-(2,3:5,6-di-O-isopropylidene-α-D-mannofuranosyloxy)-acrylate readily obtained from the corresponding mannose derivative and propriolate (Ref. 14). The enantiomer ratio finally received was 4:1.

Scheme 19

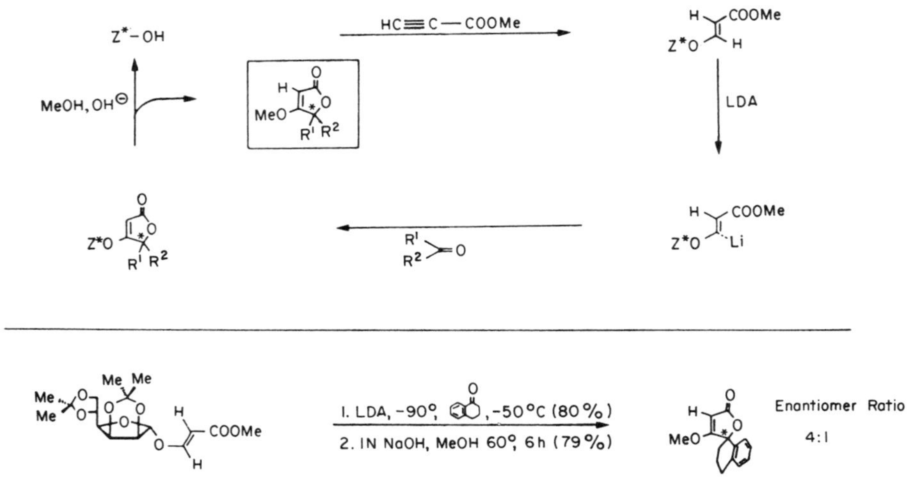

Some preliminary experiments with C-lithiated acrylates having carbohydrates as chiral substituents are summarized in Scheme 20. Electrophiles are a-tetralone and a racemate of the cyclohexanone molecule required for the top half synthesis of chlorothricolide (see Scheme 17 and 18). However, the selectivities obtained suggest further modifications in the chiral substituent. The diastereomeric spirotetronate intermediates obtained could be easily separated by column chromatography this way enabling at least a convenient racemate resolution.

Scheme 20

REACTIONS OF LITHIATED ACRYLATES HAVING CARBOHYDRATES AS CHIRAL SUBSTITUENTS

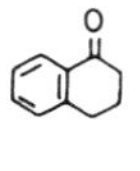 α-tetralone 1.25 eq.	4:1	3:2	3:2
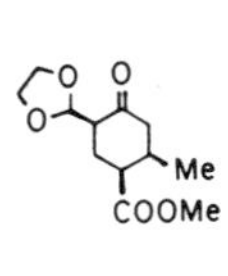 cyclohexanone 1.25 eq.	—	2:1	1:1
3.0 eq	—	—	3:2

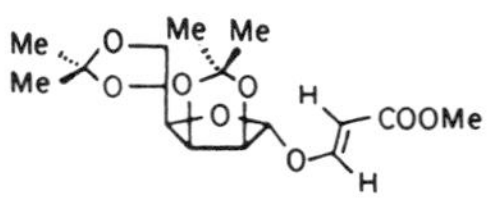

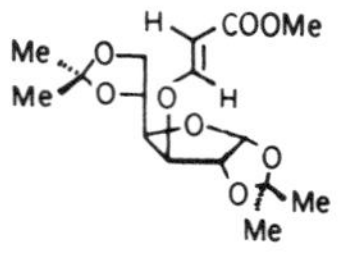

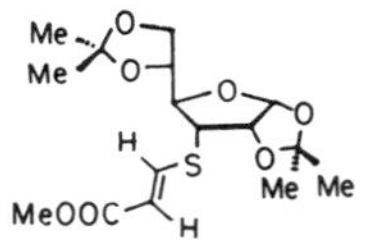

Reaction Conditions: LDA, THF, −90 °C

ACKNOWLEDGEMENTS

Financial support from the Deutsche Forschungsgemeinschaft and the Fonds der Chemischen Industrie is gratefully acknowledged. - I am particularly indebted to my coworkers Rainer Betz, Alfons Enhsen, Rolf Hirsenkorn, Jürgen Kast, Dominique Lafont, Okiko Miyata, Heike Speer, and Jörg Talbiersky.

REFERENCES

1. Vinyl Carbanions, Part 23. - For part 22, see Ref. 20.

2. R.R. Schmidt, Lect.Heterocycl.Chem. 4, 97-114 (1978); Suppl.J.Heterocycl. Chem. 15 (1978), and references cited therein.

3. R.R. Schmidt and R. Hirsenkorn, Tetrahedron 39, 2043-2054 (1983), and references cited therein.

4. R.R. Schmidt, Bull.Soc.Chim.Belg. 92, 825-836 (1983), and references cited therein.

5. H.W. Gschwend and H.R. Rodriguez, Org.Reactions 26, 1-360 (1979), and references cited therein.

6. R.R. Schmidt, J. Talbiersky, and P. Russegger, Tetrahedron Lett. 4273-4276 (1979).

7. R.R. Schmidt and H. Speer, Tetrahedron Lett. 4259-4262 (1981).

8. R.R. Schmidt and J. Talbiersky, Angew.Chem. 89, 891-892 (1977); Angew. Chem.Int.Ed.Engl. 16, 851-852 (1977).

9. O. Miyata and R.R. Schmidt, Tetrahedron Lett. 23, 1793-1796 (1982).

10. N.G. Clemo and G. Pattenden, Tetrahedron Lett. 23, 589-592 (1982).

11. R. Betz, Dissertation, Univ. of Konstanz, 1984.

12. R.R. Schmidt and R. Betz, Angew.Chem. 96, 420-421 (1984); Angew.Chem.Int. Ed.Engl. 23, 430-431 (1984).

13. A. Enhsen, Diplomarbeit, Univ. of Konstanz, 1984.

14. R.R. Schmidt and R. Hirsenkorn, Tetrahedron Lett., accepted for publication; R. Hirsenkorn, Dissertation, to be submitted.

15. E. Hawkins, E. Hirst, and J. Jones, J.Chem.Soc. 246-248 (1939).

16. F.M. Unger, Adv.Carbohydr.Chem.Biochem. 38, 323-388 (1981), and references cited therein.

17. P.M. Collins, W.G. Overend, and T. Shing, J.Chem.Soc.Chem.Commun. 1139-1140 (1981).

18. D. Charon, R.S. Sarfati, D.R. Strobach, and L. Szabo, Eur.J.Biochem. 11, 364-369 (1969); M.B. Perry and D.T. Williams, Methods Carbohydr.Chem. 7, 44-48 (1976).

19. M.A. Ghalambor, E.J. Levine, and E.C. Heath, J.Biol.Chem. 241, 3207-3215 (1966); C. Hershberger, M. Davis, and S.B. Binkley, ibid. 243, 1585-1588 (1968); C. Hershberger and S.B. Binkley, ibid. 243, 1578-1584 (1968); N.K. Kochetkov, B.A. Dimitriev, and L.V. Bachinovsky, Carbohydr.Res. 11, 193-197 (1969).

20. R.R. Schmidt, A. Enhsen, and R. Betz, Synthesis, accepted for publication.

21. D. Charon and L. Szabo, J.Chem.Soc.Perkin I, 1628-1632 (1976).

22. D.A. Evans, J.V. Nelson and T.R. Tabor, Top Stereochem. 13, 1-115 (1982); J. Mulzer, Nachr.Chem.Techn.Lab. 32, 16-18 (1984), and references cited therein.

23. P.W. Schindler and M.C. Scrutton, Eur.J.Biochem. 55, 543-553 (1975).

24. R.E. Ireland and W.J. Thompson, J.Org.Chem. 44, 3041-3052 (1979); Tetrahedron Lett. 4705-4708 (1979); R.E. Ireland, W.J. Thompson, G.H. Srouji, and R. Etter, J.Org.Chem. 46, 4863-4873 (1981).

25. W.R. Roush and S.E. Hall, J.Am.Chem.Soc. 103, 5200-5211 (1981); S.E. Hall and W.R. Roush, J.Org.Chem. 47, 4611-4621 (1982).

26. B.B. Snider and B.W. Burbaum, J.Org.Chem. 48, 4370-4374 (1980).

27. A.K. Mallams, M.W. Puar, R.R. Rossman, A.T. McPhail, and R.D. Macfarlane, J.Am.Chem.Soc. 103, 3940-3943 (1981).

28. We are indebted to O. Scheidsteger and G. Huttner for performing an X-ray analysis of the spirotetronate obtained.

29. V. Jäger and W. Schwab, Angew.Chem. 93, 578-579 (1981); Angew.Chem.Int. Ed.Engl. 20, 603-604 (1981).

The stereospecific synthesis of macro-bicyclic and macro-polycyclic polyethers from carbohydrate precursors

J. Fraser Stoddart, Steven E. Fuller, Simon M. Doughty, Peter C.Y.K. Ning, and M. Kevin Williams

Department of Chemistry, The University, Sheffield S3 7HF, England

David J. Williams, Billy L. Allwood, and Alexandra M.Z. Slawin

Department of Chemistry, Imperial College, London SW7 2AY, England

Howard M. Colquhoun

New Science Group, Imperial Chemical Industries PLC, The Heath, Runcorn WA7 4QE, England

Abstract – A review of progress towards the stereospecific syntheses of the *cis*-dibenzo-3*n*-crown-*n* ethers **1** (*n* = 10) and **2** (*n* = 12), and the *cis-syn-cis* dibenzo-36-crown-12 derivative **3**, starting from meso-tartaric acid is presented. Such compounds have been identified as molecular receptors designed to complex with large organic (*e.g.* [diquat]$^{2+}$) and organometallic (*e.g.* [Rh(cod)(NH$_3$)$_2$]$^+$) cations. The stereospecific synthesis of the [3.3.3]cryptand **33**, which exists as slowly interconverting *in-in ii*-**33** and *out-out oo*-**33** conformational diastereoisomers in solution, is described starting from 2,3-*O*-isopropylidene-D-glycerol D-**34** and 2,3-di-*O*-benzyl-L-glycerol L-**35**. This investigation may be regarded as a model one for the stereospecific synthesis of molecular receptors such as the *syn*-fused bis(benzo-18-crown-6) derivative **4**.

INTRODUCTION

Recent observations [1-5] that dibenzo-3*n*-crown-*n* ethers (DB3*n*C*n*) [6], particularly in the range *n* = 8-11, adopt either U-shaped or V-shaped conformations (Figure 1) in their binding of large cationic guest species such as [Pt(bipy)(NH$_3$)$_2$]$^{2+}$ (bipy ≡ 2,2'-bipyridyl), [diquat]$^{2+}$ (diquat ≡ 6,7-dihydrodipyrido[1,2-*a*:2',1'-*c*]pyrazinediium), [Rh(cod)(NH$_3$)$_2$]$^+$ (cod ≡ 1,5-cyclooctadiene), and [Rh(nbd)(NH$_3$)$_2$]$^+$ (nbd ≡ norbornadiene), have prompted us to design hosts which will bind more efficiently and selectively with these large organic and organometallic guests. Clearly, one of our objectives must be to increase the rigidity of DB3*n*C*n* hosts in such a way that their conformations are preorganised to be either U-shaped or V-shaped and thus hopefully confer upon them some selectivity in the complexation of cationic guest species such as those shown in Figure 1. One way to realise this objective is to introduce bridging units stereospecifically into DB3*n*C*n* constitutions in such a manner as to afford macrobicyclic and macropolycyclic systems, *e.g.* **1** - **4**, where their conformations are dictated to some extent at least by their constitutional and configurational characteristics. With the DB30C10 and DB36C12 derivatives **1** and **2**, the *cis* configurations at the ring junctions between the two macrocyclic rings can be defined [*cf*. 7] by incorporating an erythritol residue during their synthesis.

DB3*n*C*n*	*n*	*j*	*k*
DB18C6	6	1	1
DB21C7	7	2	1
DB24C8	8	2	2
DB27C9	9	3	2
DB30C10	10	3	3
DB33C11	11	4	3
DB36C12	12	4	4

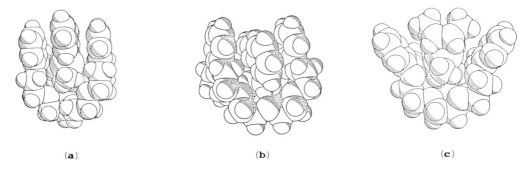

(**a**)   (**b**)   (**c**)

Figure 1. Space-filling representations of the supramolecular structures of
(**a**) [Pt(bipy)(NH$_3$)$_2$.DB30C10]$^{2+}$, (**b**) [diquat.DB30C10]$^{2+}$, and (**c**) [Rh(cod)(NH$_3$)$_2$.DB24C8]$^+$
determined by *X*-ray crystallography.

In the case of the macrotricyclic DB36C12 derivative **3**, two erythritol residues have to be
introduced, preferably by a stereospecific means, such that the two *cis* ring junctions have
relative *syn* stereochemistry. [The thickened bonds in formulae **1** - **3** indicate that erythritol
units have been incorporated at the ring junctions.] Inspection of space-filling molecular
models shows (Figure 2) that, whereas the conformations of hosts **1** and **2** are obliged to assume
a V-shape, host **3** is encouraged to adopt a U-shaped conformation.

The macrobicyclic host **4**, containing two *syn*-fused benzo-18-crown-6 rings offers yet a
further challenge in stereospecific synthesis: that is, to incorporate two glycerol residues
such that the macrobicycle adopts *out-out* (or *in-in*), but not *out-in*, stereochemistries [8].
[The thickened bonds in formula **4** indicate that glycerol units have been incorporated at the
ring junctions.] A V-shaped conformation is indicated for this host on examination of space-
filling molecular models.

We shall now examine the consequent challenges in the stereospecific synthesis and report on
our progress towards solving the problems associated with them.

RESULTS AND DISCUSSION

The stereospecific synthesis of the *cis*-DB30C10 derivative **1** is outlined in Scheme 1. 2,3-*O*-
Isopropylidene-erythritol **5**, obtained by a known sequence of reactions [9] from meso-tartaric
acid, and the ditosylate **6** of the bis-2-hydroxyethyl ether, derived [10] from catechol, were
reacted in tetrahydrofuran with both sodium hydride and lithium hydride present [11] as bases
to afford the benzo-14-crown-4 derivative **7** in 23% yield. The use of the mixed hydrides is
necessary: the lithium ion presumably acts as a template [11-13] in the ring closure step.
Quantitative removal of the isopropylidene protecting group gives a diol **8**, which can be
condensed with the ditosylate **9** of 1,2-bis(5-hydroxy-3-oxa-1-pentyloxy)benzene, derived [14]
from catechol to afford (19%) the desired *cis*-DB30C10 derivative **1** as a crystalline compound.
The literature [10, 14] preparations of the ditosylates **6** and **9** illustrate two methods by
which bismethyleneoxy units can be built into appropriate precursors (*e.g.* catechol).

1   **3**

2   **4**

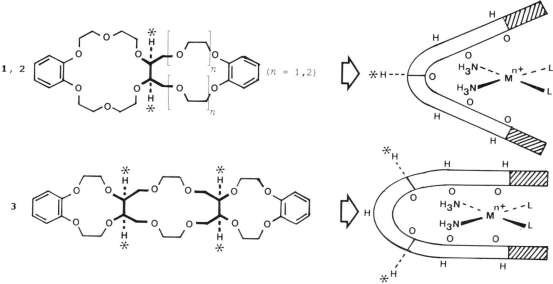

Figure 2. V-Shaped and U-shaped receptor molecules incorporating erythritol.

In Scheme 2, which outlines the stereospecific synthesis of the *cis*-DB36C12 derivative **2**, the first series of steps (**5** → **10**) illustrates yet another approach [15] to the elaboration of polyether chains by the formal insertion of bismethyleneoxy units into hydroxyl groups: thus, conversion of the diol **5** into its diallyl ether was followed by ozonolysis and reductive work-up to afford the extended diol **10** in 61% yield overall. On this occasion, however, reaction of **10** with the ditosylate **6** in tetrahydrofuran afforded (23%) the benzo-20-crown-6 derivative **11** in the presence of sodium hydride alone as base. The other reactions {**11** → **12** (+**9**) → **2**} in Scheme 2 were performed as described previously for those {**7** → **8** (+**9**) → **1**} in Scheme 1 and the *cis*-DB36C12 derivative **2** was isolated as an oil. The stereospecific synthesis of the parent *cis*-36-crown-12 derivative **13** has been reported [7, 13] — starting from 2,3-*O*-isopropylidene-erythritol — and described as a 'jaws-like' host [7]. Preliminary ^{1}H nuclear magnetic resonance spectroscopic investigations have revealed that hosts **1** and **2** form adducts with [Pt(bipy)(NH₃)₂][PF₆]₂ and [Rh(cod)(NH₃)₂][PF₆] in dideuteriodichloromethane solution. Thus, conformations of a V-shape, illustrated diagrammatically in Figure 2 for hosts **1** and **2**, are probably utilised in the binding of [ML₂(NH₃)₂]ⁿ⁺ guests.

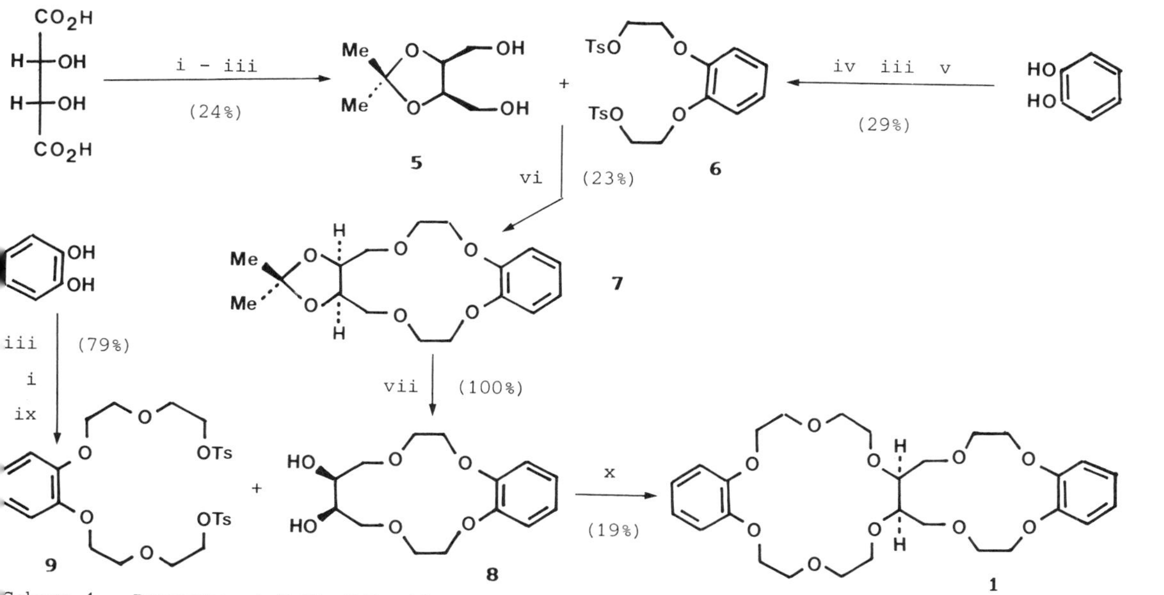

Scheme 1. Reagents: i MeOH, HCl; ii Me₂CO, H₂SO₄; iii LiAlH₄, THF; iv BrCH₂CO₂Et, K₂CO₃, Me₂CO; v TsCl, C₅H₅N; vi LiH, NaH, THF; vii MeOH, H₂O, Resin-H⁺; viii ClCH₂CH₂OCH₂CH₂OTHP, NaOH, n-BuOH; ix TsCl, DMAP, C₅H₅N; x NaH, THF.

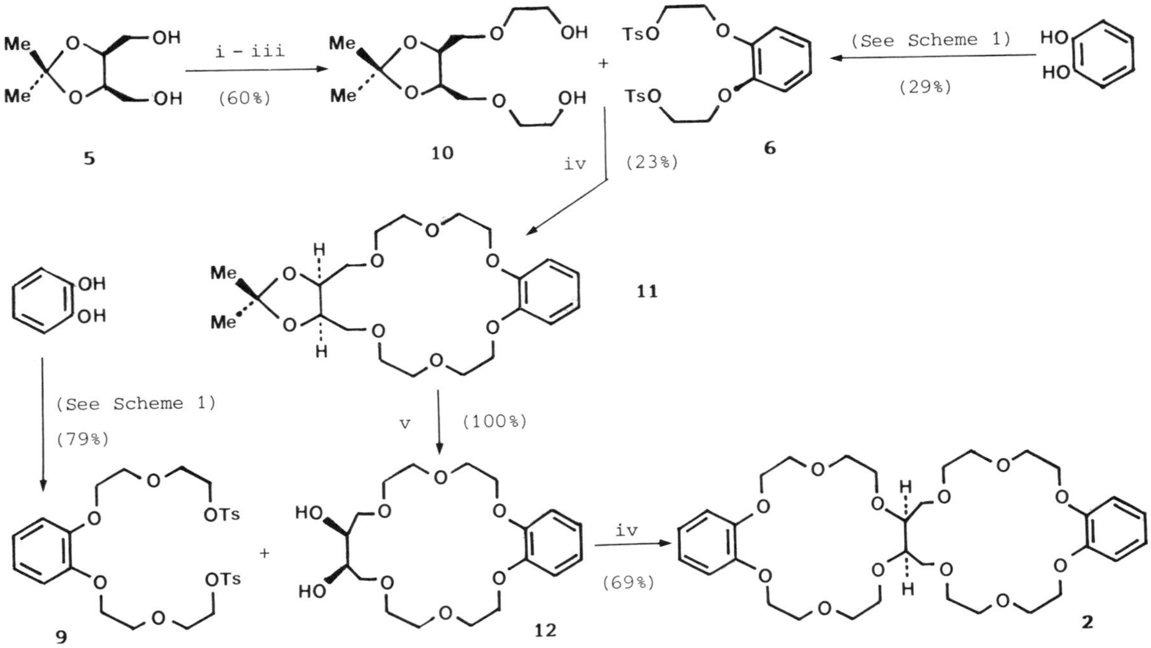

Scheme 2. Reagents: i CH$_2$=CHCH$_2$Br, KOH, PhMe; ii O$_3$, MeOH; iii NaBH$_4$, MeOH; iv NaH, THF; v MeOH, H$_2$O, Resin-H$^+$.

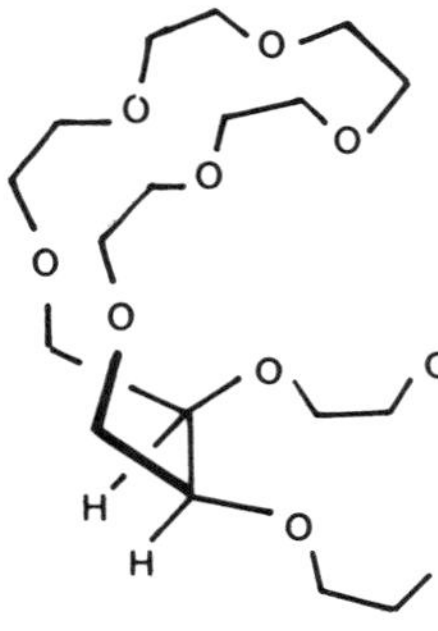

Our first approaches (Schemes 3 and 4) to the synthesis of the U-shaped receptor molecule have involved the attempted synthesis of the *cis-syn-cis*-DB36C12 derivative **3** by routes that are only partially stereospecific in the first instance: subsequently, we hope to modify the second approach (Scheme 4), which has afforded the *cis-anti-cis*-isomer **14**, in addition to **3**, to provide a completely stereospecific synthesis of the desired *cis-syn-cis*-DB36C12 derivative **3**. Here, we describe a partially stereospecific synthesis of **3**.

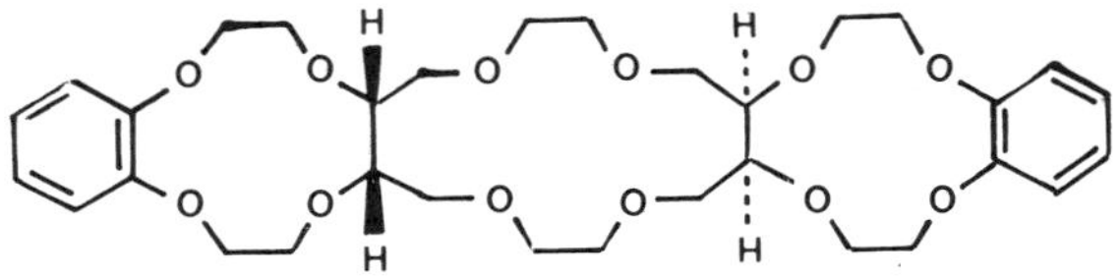

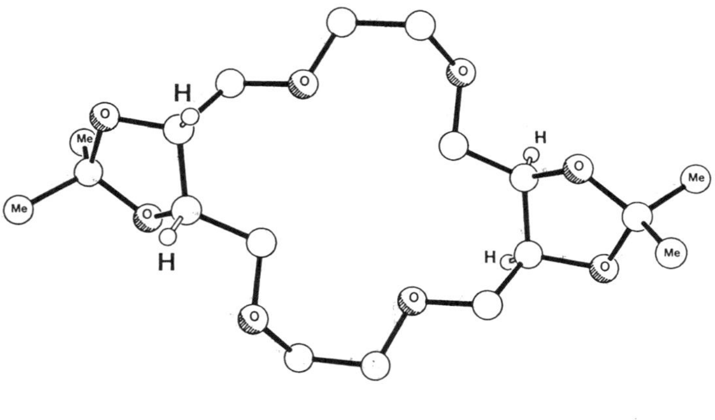

Figure 3. The molecular structure of the *cis-anti-cis*-bis(2,2-dimethyl-1,3-dioxolano)-16-crown- derivative **16** determined by *X*-ray crystallography.

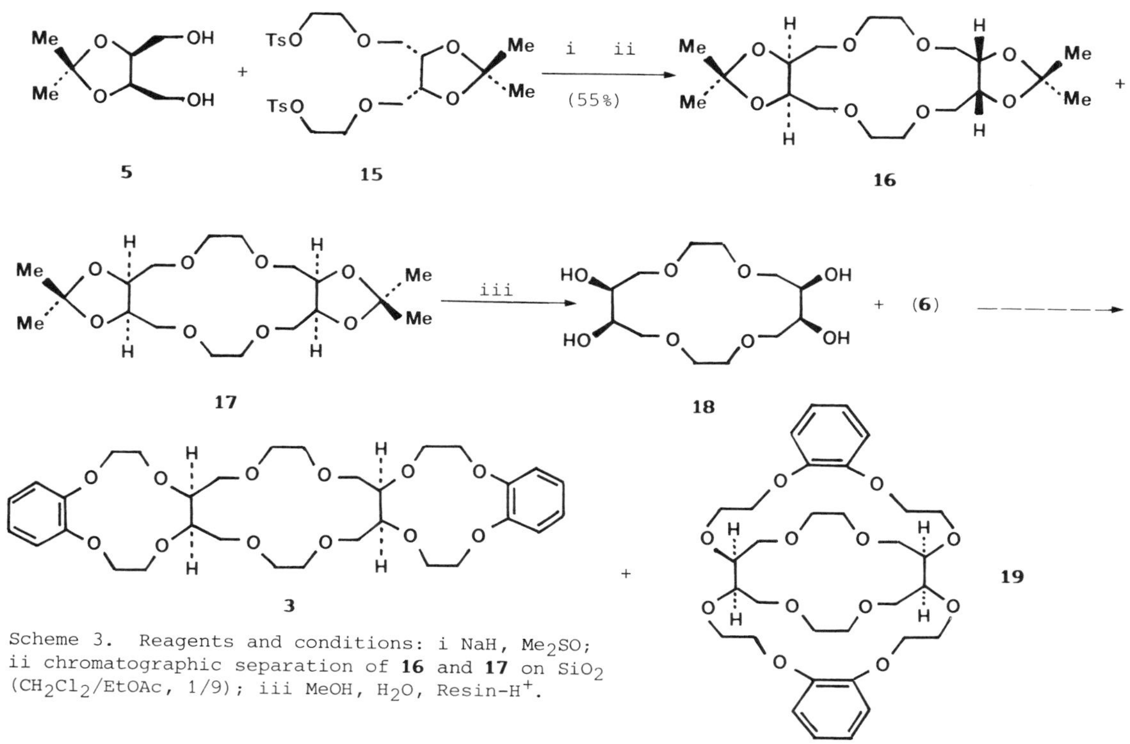

Scheme 3. Reagents and conditions: i NaH, Me₂SO; ii chromatographic separation of **16** and **17** on SiO₂ (CH₂Cl₂/EtOAc, 1/9); iii MeOH, H₂O, Resin-H⁺.

Scheme 3 shows that sodium hydride promoted condensation in dimethylsulphoxide of the diol **5** with the ditosylate **15**, obtained from **10**, affords the bis(2,2-dimethyl-1,3-dioxolano)-16-crown-4 derivative as a mixture of configurational diastereoisomers **16** and **17** in good (55%) yield. The major (2.7 parts) and faster migrating isomer on silica gel (dichloromethane/ethyl acetate, 1/9) crystallised (m.p. 120-122°C) from dichloromethane-pentane as good quality single crystals, suitable for *X*-ray structural analysis: the structure determination demonstrated (Figure 3) that this isomer **16** has the *cis-anti-cis* configuration. Thus, the minor (1.0 part) and slower migrating isomer **17** on silica gel chromatography was assigned the *cis-syn-cis* configuration: acid-catalysed hydrolysis of the isopropylidene protecting groups proceeds quantitatively to give the all-*cis* tetraol **18**. Although the reaction of **18** with the ditosylate **6** has been attempted under various base-solvent conditions (*e.g.* sodium hydride in dimethylsulphoxide at 65°C for 72 hours), neither the *cis-syn-cis*-DB36C12 derivative **3** nor the constitutionally-isomeric bis(benzo-18-crown-6) derivative **19** have been identified as reaction products. On

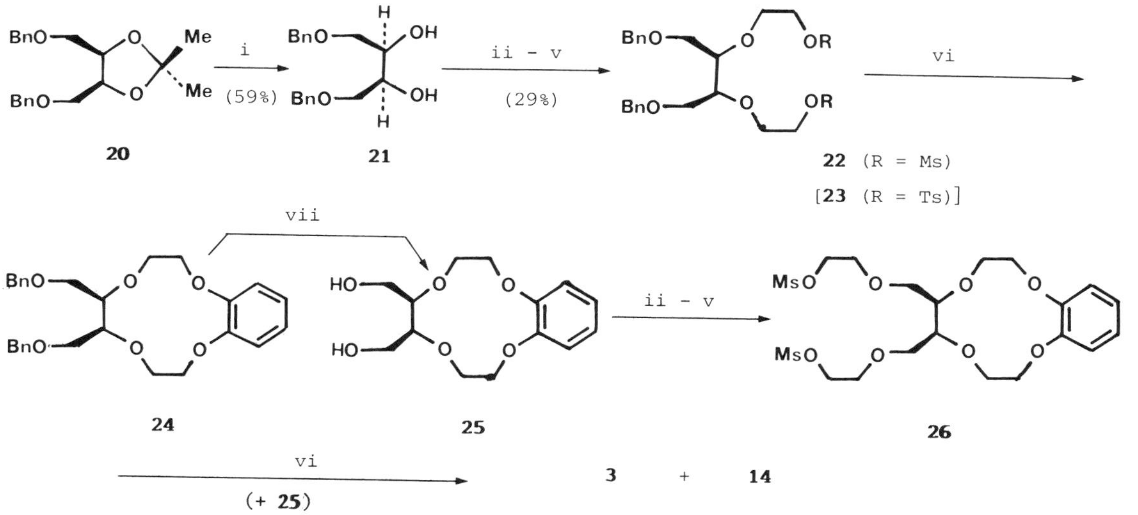

Scheme 4. Reagents: i MeOH, Resin-H⁺; ii CH₂=CHCH₂Br, KOH, PhMe; iii O₃, MeOH; iv NaBH₄, MeOH; v MsCl [or TsCl], C₅H₅N; vi LiH, NaH, THF; vii H₂, Pd/C.

account of the problem raised [17] by the possibility of isomeric products, and the
identifications of their constitutions, if isolated, we decided to search for a solution to
the problem by building on the benzo-12-crown-4 rings before elaborating the central 16-crown-4
ring. 1,4-Di-*O*-benzyl-2,3-*O*-isopropylidene-erythritol **20** was prepared in almost quantitative
yield from the diol **5** using benzyl bromide in toluene with potassium hydroxide as base.
Scheme 4 shows how these benzyl ether functions have been employed to protect the primary
hydroxyl groups of the erythritol whilst building on a benzo-12-crown-4 ring at its secondary
hydroxyl groups. The selective hydrolysis of the *O*-isopropylidene function in **20** was achieved
in refluxing aqueous methanol using a cation exchange resin in the protonated form as catalyst:
since some partial hydrolysis of the benzyl ether functions could not be avoided, the diol **21**
was isolated in somewhat less than quantitative yield. Conversion of **21** into its extended
diol was accomplished by the allylation-ozonolysis-reduction procedure discussed with reference
to **5** → **10** in Scheme 2. Because the crude extended diol proved very difficult to handle by
chromatographic means, it was mesylated directly to give (40%) the extended dimesylate **22**
which could be purified by column chromatography on silica gel without undue loss of material.
[Tosylation of the extended diol afforded only 20% of the pure ditosylate **23** after column
chromatography.] The condensation of the dimesylate **22** with catechol was not straightforward.
The use of lithium and sodium hydrides in admixture is recommended as the base and the choice
of the solvent is also important. Whereas the benzo-12-crown-4 derivative **24** was isolated
from refluxing tetrahydrofuran in only a few percent yield, heating in *N*,*N*-dimethylformamide
at 100°C afforded 57% of pure **24**. Hydrogenolysis of the benzyl ether protecting groups in **24**
gave the diol **25** almost quantitatively and this was subjected to the familiar sequence of
reactions (*i.e.* allylation, ozonolysis, reduction, and mesylation) to prepare the extended
dimesylate **26**. Base-promoted condensation of **25** with **26** in the mixed hydride system in *N*,*N*-
dimethylformamide at 100°C afforded the isomeric DB36C12 derivatives **3** and **14**. Although they
have been separated by column chromatography on silica gel, at the time of writing this account,
they still have to be identified as the *cis-syn-cis* and *cis-anti-cis* isomers **3** and **14**,
respectively. Partly in anticipation of this challenge, we have devised a stereospecific
synthesis (Scheme 5) of the *cis-syn-cis* isomer **3** starting from D-glucose. In its initial
stages, it relies upon the known preparations [18, 19] of 1,3-*O*-ethylidene-L-erythritol (2,4-*O*-
ethylidene-D-erythritol) **28** from 4,6-*O*-ethylidene-D-glucose **27** [20]. Carefully controlled
benzylation provides the monobenzylated derivative **29** of **28** after column chromatography. The
stereospecific synthesis of *cis-syn-cis*-DB36C12 **3** is envisaged to proceed from **29** *via* key
intermediates such as **30**, **31**, and **32**: in the final base-promoted condensation of **32**, it is
anticipated that 'dimerisation' to afford **3** with a central 16-membered ring will be favoured
over intramolecular ring cyclisation to give an 8-membered ring as part of a *cis*-fused benzo-
18-crown-6 derivative.

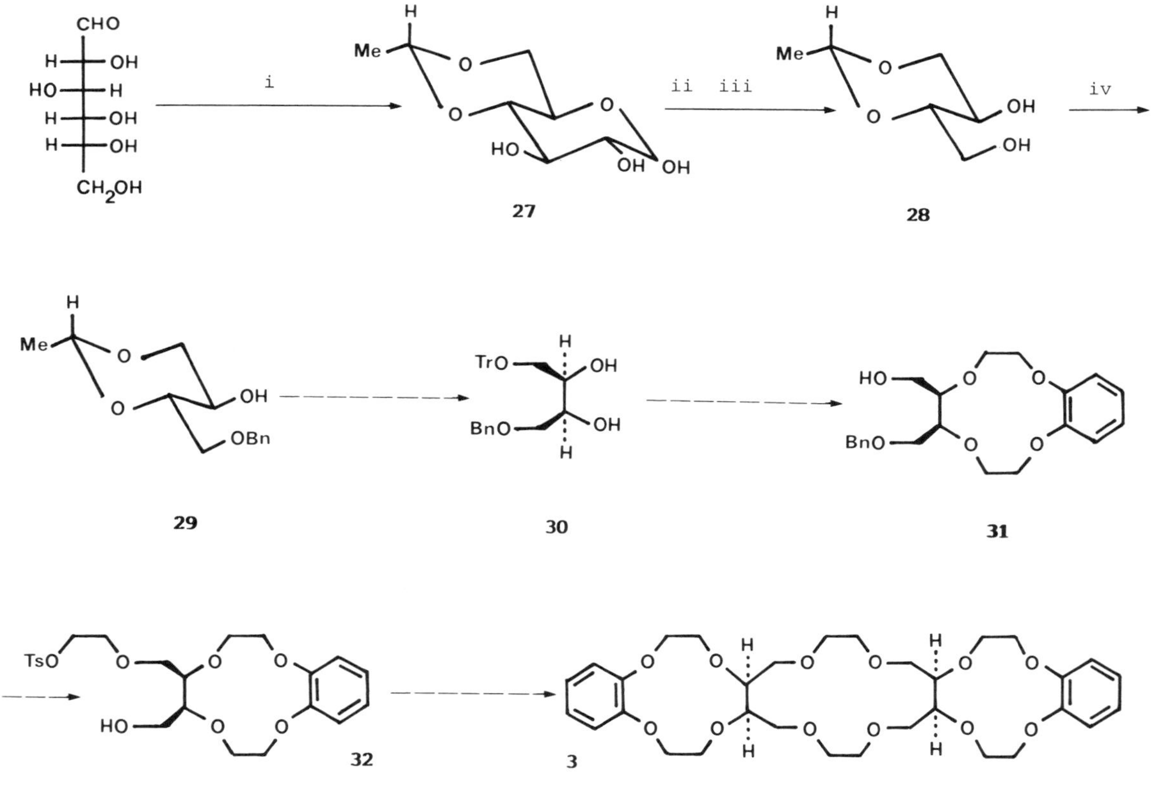

Scheme 5. Reagents: i (MeCHO)$_3$, H$_2$SO$_4$; ii NaIO$_4$, H$_2$O; iii NaBH$_4$, H$_2$O; iv PhCH$_2$Br, NaH, DMF.

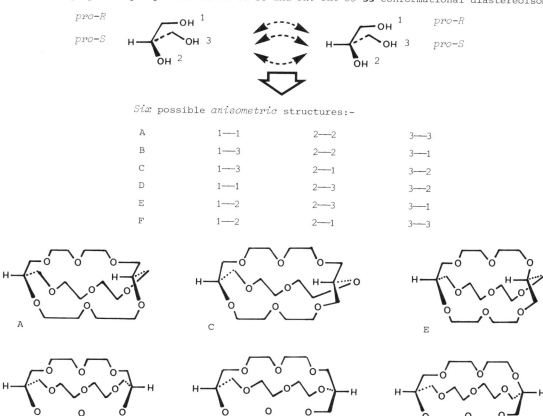

Macrobicyclic hosts such as **4** also provide a synthetic entry into V-shaped receptor molecules. In a stereospecific sense, the challenge offered by the requirement for *syn*-fusion of the two benzo-18-crown-6 rings in **4** is one of bringing together two glycerol units with their prochiral hydroxymethyl groups in one of six different ways. With reference to the model system provided by the [3.3.3]cryptand **33**, the analysis summarised in Figure 4 shows the six possible anisometric structures (A - F) that can be generated by linking two glycerol residues by three -OCH₂CH₂OCH₂CH₂O- chains. Of these structures, only two equate with the desired constitution and only B (*oo*-**33**) with the *out-out* stereochemistry shown for its bridgehead hydrogen atoms corresponds to the desired *syn*-fusion of the polyether chains. The structure (*ii*-**33**) with *in-in* stereochemistry has these two bridgehead atoms oriented inside the cavity of the [3.3.3]cryptand **33** and is related to the *out-out* isomer *oo*-**33** by conformational interconversion. Structures of the type B require linking of the *pro-R* with the *pro-S* hydroxymethyl groups in the two separate glycerol units at the carbon bridgeheads: in a search for a way of preparing *in-in* and *out-out* isomers without having to resort to inconvenient and difficult isomer separations [21, 22], we have devised a stereospecific route where 2,3-*O*-isopropylidene-D-glycerol D-**34** [23] constitutes one bridgehead, and 2,3-di-*O*-benzyl-L-glycerol L-**35** [24], the other bridgehead [25]. Scheme 6 illustrates our approach with reference to the synthesis of the [3.3.3]cryptand **33**, which exists as chromatographically separable *in-in ii*-**33** and *out-out oo*-**33** conformational diastereoisomers.

Figure 4.   The *six* possible anisometric structures generated by linking *two* glycerol residues by *three*   -OCH₂CH₂OCH₂CH₂O- chains.

Base-promoted condensation of D-**34** with an excess (2.3 mol. equiv.) of diethyleneglycol bis-tosylate [26] afforded the extended tosylate D-**36**. Employing the same base (sodium hydride) and solvent (tetrahydrofuran), D-**36** was reacted with L-**35** to give the acyclic polyether DL-**37**, which was deprotected — methanolysis of the acetal function followed by hydrogenolysis of the benzyl ether groups — quantitatively to give the meso-tetraol **38**. Tritylation by a procedure [27], which employs N,N-dimethyl-4-aminopyridine as a catalyst, proceeded regioselectively at the primary hydroxyl groups in **38** to give the ditrityl ether **39** in good yield. Base-promoted condensation of **39** with diethyleneglycol bistosylate [26] afforded the crude 18-crown-6 derivative **40**, which was most easily purified by a stepwise sequence of precipitations that rely upon consecutive complexations of **40** with ammonia borane [28] and acetonitrile [29].

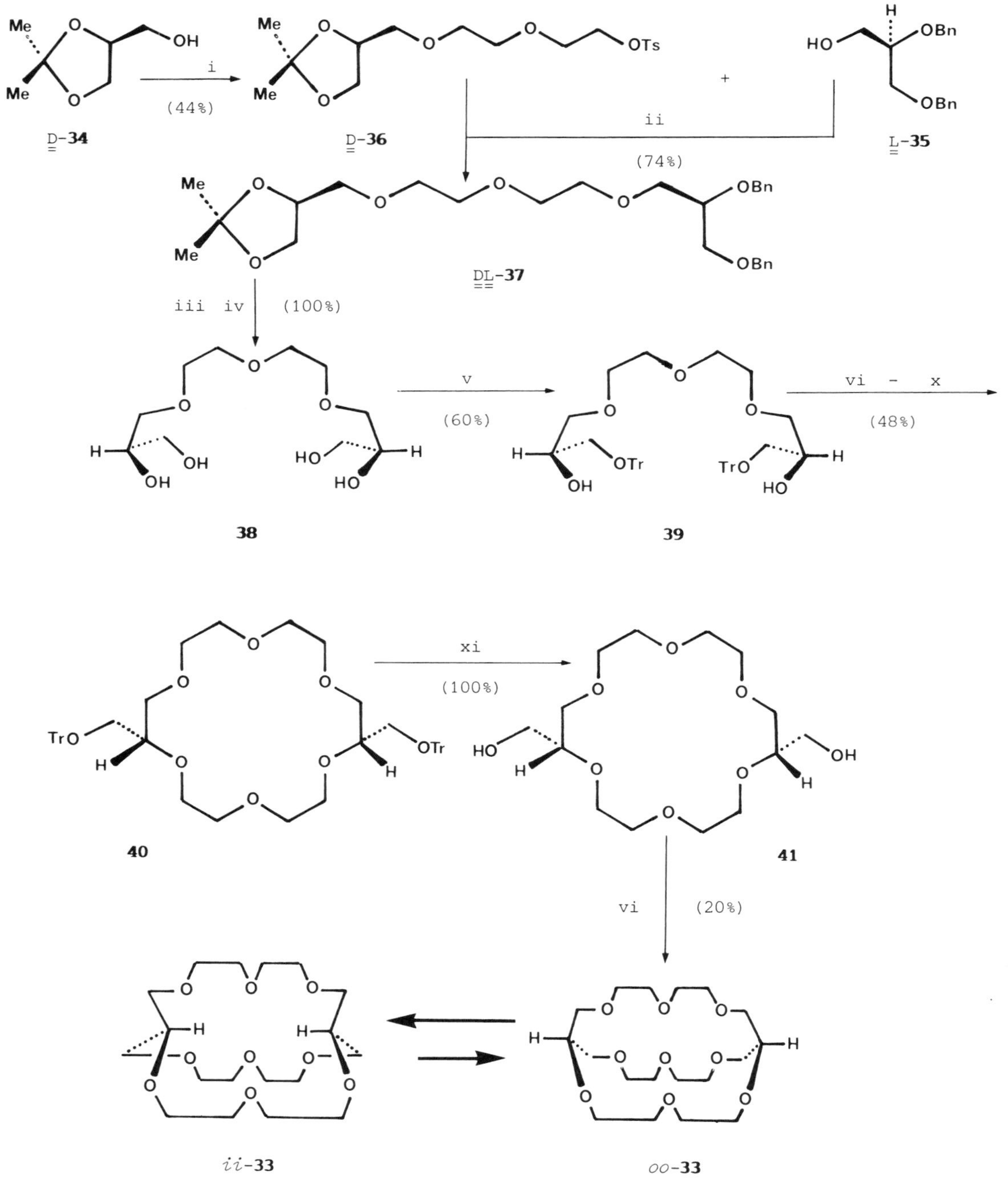

Scheme 6.   Reagents: i TsOCH$_2$CH$_2$OCH$_2$CH$_2$OTs, NaH, THF; ii NaH, THF; iii MeOH, Resin-H$^+$; iv H$_2$, Pd/C; v TrCl, Et$_3$N, DMAP, CH$_2$Cl$_2$; vi TsOCH$_2$CH$_2$OCH$_2$CH$_2$OTs, NaH, Me$_2$SO; vii BH$_3$NH$_3$, MeOH, n-C$_5$H$_{12}$; viii Me$_2$CO, NaOH; ix MeCN; x Δ (50°C); xi MeOH, CH$_2$Cl$_2$, HCl.

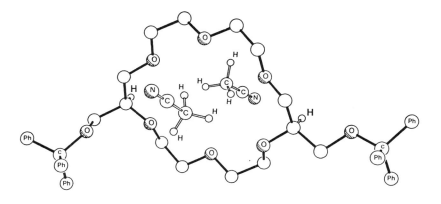

Figure 5. The supramolecular structure of **40**.2MeCN determined by *X*-ray crystallography.

Thus, addition of pentane to a solution of crude **40** and ammonia borane in methanol-chloroform yielded a precipitate [30], which was isolated and refluxed in basic aqueous acetone to regenerate **40** considerably purified but still containing impurities. Addition of acetonitrile to this partially purified sample afforded a crystalline **40**.2MeCN complex with m.p. 48-50°C. Single crystals of this complex, which were suitable for *X*-ray crystallography, were isolated from acetonitrile solution after layering with pentane. The crystal structure (Figure 5) of **40**.2MeCN confirmed the *cis* stereochemistry of the 18-crown-6 derivative **40**. Deprotection of **40** afforded the diol **41**, which was reacted with diethyleneglycol bistosylate [26] under basic conditions to afford the [3.3.3]cryptand **33** as an oil [31]. The *out-out* isomer *oo*-**33** was characterised as its barium thiocyanate complex by stirring a solution of **33** in dichloromethane with excess of the salt and then layering pentane on top of the extract. This procedure afforded single crystals of **33**.Ba(SCN)$_2$.CH$_2$Cl$_2$ (m.p. 90°C) suitable for *X*-ray crystallography. The crystal structure (Figure 6) demonstrates the *out-out* stereochemistry of the hydrogen atoms at the bridgehead positions of this [3.3.3]cryptate. Thin layer chromatography of **33** on alumina with ethyl acetate as the developing solvent indicated the presence of a major fast-moving (Fraction 1) and minor slow-moving (Fraction 2) component. ^{1}H And ^{13}C{^{1}H} nuclear magnetic resonance spectroscopy reveals that these components are slowly interconverting ($\Delta G^{\#}_{ii \rightarrow oo} \cong 24$ kcal mol^{-1}) conformational diastereoisomers [32]. Fractions 1 and 2 were separated by column chromatography on alumina at -5°C using ethyl acetate as eluant. They were characterised [33] as the *in-in ii*-**33** (Fraction 1) and *out-out oo*-**33** (Fraction 2) isomers on the basis of their relative chromatographic mobilities and ^{1}H and ^{13}C{^{1}H} nuclear magnetic resonance spectroscopic data.

Now that we have solved the stereochemical problem posed by the stereospecific synthesis of the [3.3.3]cryptand **33**, we are in a position to tackle the preparation of molecular receptor **4** in like-manner, *i.e.* stereospecifically.

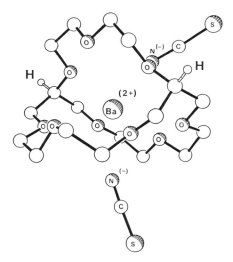

Figure 6. The supramolecular structure of **33**.Ba(SCN)$_2$ determined by *X*-ray crystallography.

CONCLUSIONS

In the synthesis of macrobicyclic and macropolycyclic polyethers such as **1 - 4** and **33**, the need for stereospecific routes arises for an important practical reason that is additional to those normally encountered in the synthesis of most other classes of organic compound where stereochemistry can or has to be controlled. The chromatographic separation of crown ethers and their derivatives is usually plagued by poor resolutions and low returns of products. The avoidance of isomer separations by employing stereospecific synthetic approaches is therefore not only attractive for aesthetic reasons but also desirable for practical purposes.

Finally, it should be noted that the stereospecific synthesis of the *achiral* [3.3.3]cryptand **33** involves the use of two suitably-substituted *chiral* glycerol derivatives, 2,3-*O*-isopropyl-idene-D-glycerol D-**34** and 2,3-di-*O*-benzyl-L-glycerol L-**35**, available in the enantiomerically-related series of such derivatives. By contrast, in the stereospecific synthesis (Scheme 5) of the *achiral cis-syn-cis*-DB36C12 derivative **3** from D-glucose, it is proposed to employ only *one* (D) enantiomer. These examples illustrate *two* different ways in which *chiral* (optically active) starting materials can be used to synthesise *achiral* products in a stereospecifically controlled manner.

ACKNOWLEDGEMENTS

We are grateful to Imperial Chemical Industries PLC and to the Science and Engineering, and Agricultural and Food, Research Councils in the United Kingdom for financial support.

REFERENCES AND NOTES

1. H.M. Colquhoun, J.F. Stoddart, D.J. Williams, J.B. Wolstenholme and R. Zarzycki, *Angew. Chem.*, **93**, 1093-1095 (1981); *Angew. Chem. Int. Ed. Engl.*, **20**, 1051-1053 (1981).

2. H.M. Colquhoun, S.M. Doughty, J.M. Maud, J.F. Stoddart, D.J. Williams and J.B. Wolstenholme, *Israel J. Chem.*, in press.

3. H.M. Colquhoun, E.P. Goodings, J.M. Maud, J.F. Stoddart, D.J. Williams and J.B. Wolstenholme, *J. Chem. Soc., Chem. Commun.*, 1140-1142 (1983).

4. H.M. Colquhoun, E.P. Goodings, J.M. Maud, J.F. Stoddart, J.B. Wolstenholme and D.J. Williams, *J. Chem. Soc., Perkin Trans.* 2, in press.

5. H.M. Colquhoun, S.M. Doughty, J.F. Stoddart and D.J. Williams, *Angew. Chem.*, **96**, 232-234 (1984); *Angew. Chem. Int Ed. Engl.*, **23**, 235-236 (1984).

6. C.J. Pedersen, *J. Am. Chem. Soc.*, **89**, 7017-7036 (1967). In some instances, the π-electron-rich benzo rings of the dibenzo-3*n*-crown-*n* ethers enter into charge transfer interactions with suitable π-electron-deficient guest species. See references 1 - 4 and R. Malini and V. Krishnan, *J. Phys. Chem.*, **84**, 551-555 (1980); *J. Chim. Phys.*, **78**, 503-509 (1981); *Spectrochim. Acta*, **40A**, 323-328 (1984).

7. R.C. Helgeson, G.R. Weisman, J.L. Toner, T.L. Tarnowski, Y. Chao, J.M. Mayer and D.J. Cram, *J. Am. Chem. Soc.*, **101**, 4928-4941 (1979).

8. For a discussion of this nomenclature system as it relates to macrobicyclic systems containing carbon bridgeheads, see C.H. Park and H.E. Simmons, *J. Am. Chem. Soc.*, **94**, 7184-7186 (1972).

9. M. Carmack and C.J. Kelley, *J. Org. Chem.*, **33**, 2171-2173 (1968); P.W. Feit, *J. Med. Chem.*, **7**, 14-17 (1964).

10. L.C. Hodgkinson and I.O. Sutherland, *J. Chem. Soc., Perkin Trans. 1*, 1908-1914 (1979); *cf.* D. Landini, F. Montanari and F. Rolla, *Synthesis*, 223-225 (1978).

11. R. Chénevert, N. Voyer and R. Plante, *Synthesis*, 782-785 (1982).

12. *Cf.* F.L. Cook, T.C. Caruso, M.P. Byrne, C.W. Bowers, D.H. Speck and C.L. Liotta, *Tetrahedron Lett.*, 4029- 4032 (1974); B.R. Bowsher and A.J. Rest, *J. Chem. Soc., Dalton Trans.*, 1157-1161 (1981); B.R. Bowsher, A.J. Rest and B.G. Main, *J. Chem. Soc., Dalton Trans.*, 1421-1425 (1984).

13. Recent *X*-ray crystallographic investigations have revealed [G. Shoham, W.N. Lipscomb and U. Olsher, *J. Am. Chem. Soc.*, **105**, 1247-1252 (1983); *J. Chem. Soc., Chem. Commun.*, 208-209 (1983)] that benzo-13-crown-4 and dibenzo-14-crown-4 derivatives form complexes with lithium thiocyanate.

14. E.P. Kyba, R.C. Helgeson, K. Madan, G.W. Gokel, T.L. Tarnowski, S.S. Moore and D.J. Cram, *J. Am. Chem. Soc.*, **99**, 2564-2571 (1977). An improved yield in the tosylation step was obtained by employing *N,N*-dimethyl-4-aminopyridine as a catalyst [J.D. White, M.A. Avery and J.P. Carter, *J. Am. Chem. Soc.*, **104**, 5486-5489 (1982)].

15. R.C. Hayward, C.H. Overton and G.H. Whitham, *J. Chem. Soc.*, *Perkin Trans. 1*, 2413-2418 (1976); R.B. Pettman and J.F. Stoddart, *Tetrahedron Lett.*, 457-460 and 461-464 (1979); D.A. Laidler, J.F. Stoddart and J.B. Wolstenholme, *Tetrahedron Lett.*, 465-468 (1979); N. Ando, S. Ohi, Y. Yamamoto, J. Oda and Y. Inouye, *Bull. Inst. Chem. Res.*, *Kyoto Univ.*, **58**, 293-307 (1980); D.R. Alston, J.F. Stoddart, J.B. Wolstenholme, B.L. Allwood and D.J. Williams, *Tetrahedron*, submitted.

16. Recently, the synthesis of binuclear chiral crown ethers from L-threitol has been reported: see A.V. Bogatsky, N.G. Lukyanenko, A.V. Lobach, N.Y. Nazarova and L.P. Karpenko, *Synthesis*, 139-140 (1984).

17. In principle, this problem could be solved if the di-*O*-isopropylidene derivative **17** could be partially hydrolysed to the mono-*O*-isopropylidene derivative. So far, all attempts to effect this partial hydrolysis have been unsuccessful.

18. R. Barker and D.L. MacDonald, *J. Am. Chem. Soc.*, **82**, 2301-2303 (1960).

19. S.A. Barker, A.B. Foster, A.H. Haines, J. Lehmann, J.M. Webber and G. Zweifel, *J. Chem. Soc.*, 4161-4167 (1963).

20. R.C. Hockett, D.V. Collins and A. Scattergood, *J. Am. Chem. Soc.*, **73**, 599-601 (1951).

21. D.G. Parsons, *J. Chem. Soc.*, *Perkin Trans. 1*, 451-455 (1978)

22. D.G. Parsons, *J. Chem. Soc.*, *Perkin Trans. 1*, 1193-1198 (1984).

23. For some selected approaches to the preparation of 2,3-*O*-isopropylidene-D-glycerol D-**34**, see J.J. Baldwin, A.W. Raab, K. Mensler, B.A. Arison and D.E. McClure, *J. Org. Chem.*, **43**, 4876-4878 (1978); H. Eibl, *Chem. Phys. Lipids*, **28**, 1-5 (1981); T. Makaiyama, Y. Tanabe and M. Shimizu, *Chemistry Lett.*, 401-404 (1984). See also *Nachr. Chem. Tech. Lab.*, **32**, 146-149 and 530 (1984). Recently, D-**34** and L-**34** have become commercially available from Aldrich.

24. 2,3-Di-*O*-benzyl-L-glycerol L-**35** was synthesised by a modification of a literature route [B. Wickberg, *Acta Chem. Scand.*, **12**, 1187-1201 (1958)].

25. The stereospecific synthesis of some *out-in* macrobicyclic polyethers with carbon bridge-heads from 2,3-*O*-isopropylidene-D-glycerol D-**34** has been reported [B.J. Gregory, A.H. Haines and P. Karntiang, *J. Chem. Soc.*, *Chem. Commun.*, 918-919 (1977) and A.H. Haines and P. Karntiang, *J. Chem. Soc.*, *Perkin Trans. 1*, 2577-2587 1979)].

26. J. Dale and P.O. Kristiansen, *Acta Chem. Scand.*, **26**, 1471-1478 (1972).

27. S.K. Chaudhary and O. Hernandez, *Tetrahedron Lett.*, 95-98 (1979); O. Hernandez, S.K. Chaudhary, R.H. Cox and J. Porter, *Tetrahedron Lett.*, **22**, 1491-1494 (1981).

28. Ammonia borane very readily forms crystalline complexes with crown ethers in general and 18-crown-6 and its derivatives in particular: see H.M. Colquhoun, G. Jones, J.M. Maud, J.F. Stoddart and D.J. Williams, *J. Chem. Soc.*, *Dalton Trans.*, 63-66 (1984); D.R. Alston, J.F. Stoddart, J.B. Wolstenholme, B.L. Allwood, and D J. Williams, *Tetrahedron*, submitted; B.L. Allwood, H. Shahriari-Zavareh, J.F. Stoddart, D.J. Williams and M.K. Williams, *J. Incl. Phenom.*, submitted; B.L. Allwood, H. Shahriari-Zavareh, J.F. Stoddart and D.J. Williams, *J. Chem. Soc.*, *Chem. Commun.*, in press.

29. *Cf*. G.W. Gokel, D.J. Cram, C.L. Liotta, H.P. Harris and F.L. Cook, *J. Org. Chem.*, **39**, 2445-2446 (1974) and H.M. Colquhoun, J.F. Stoddart and D.J. Williams, *J. Am. Chem. Soc.*, **104**, 1426-1428 (1982).

30. Elemental analysis indicates that this material is a 1:2 complex, *i.e.* **40**.$2NH_3BH_3$.

31. The [3.3.3]cryptand **33** may be regarded as the carbon bridgehead analogue of the [2.2.2]-cryptand with nitrogen bridgeheads [B. Dietrich, J.M. Lehn and J.P. Sauvage, *Tetrahedron*, **29**, 1647-1658 (1973)].

32. We believe that conformational interconversion between *ii*-**33** and *oo*-**33** involves passage of the shortest chain (seven atoms) through the largest ring (20-membered).

33. B.L. Allwood, S.E. Fuller, P.C.Y.K. Ning, A.M.Z. Slawin, J.F. Stoddart and D.J. Williams, *J. Chem. Soc.*, *Chem. Commun.*, in press.

New approaches to C-glycosides

Pierre Sinaÿ, Jean-Marie Beau, and Jean-Marc Lancelin

Laboratoire de Biochimie Structurale, E.R.A. 739, U.E.R. de Sciences
Fondamentales et Appliquées, 45046 Orléans Cédex, France

Abstract- Several novel stereoselective approaches to C-glycosides are described.
C-(alkyn-1-yl)- β -D-glucopyranosides are easily prepared and transformed
either in E or Z-vinyl- β-D-glucopyranosides. Reductive lithiation of an
anomeric chloride by lithium naphtalenide generates a reactive glycosyl-lithium
derivative which couples at -78°C so that electrophiles are introduced in
the axial position. Finally, the synthesis of C-glycosides from anomeric phenyl
sulphones is remarkable in that electrophiles may be selectively introduced
either in the axial or equatorial position.

INTRODUCTION

A variety of natural products may be regarded as "sugar-like molecules", inasmuch as functionnally
substituted C-glycosides occur as subunits. One of the best example is probably palytoxin (Ref.1),
where eight such C-glycosides are present.As a consequence, frantic efforts are currently devoted
to the direct stereoselective formation of α and β C-C bonds at the anomer centre of carbohydrates.
This lecture will describe recent approaches to C-glycopyranosyl derivatives developed in our
laboratory.

I.-SYNTHESIS AND CONVERSION OF C-(ALKYN-1-YL)- β-D-GLUCOPYRANOSIDES

Ambruticin (W-7783) (Ref.2) is an antifungal antibiotic with the following unique structure:

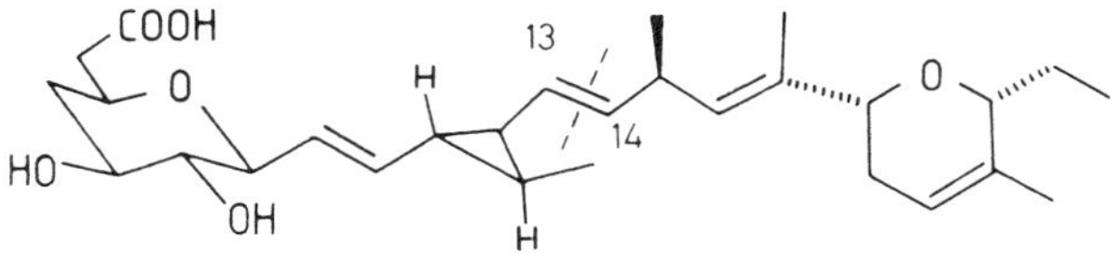

This compound shows in vitro and in vivo activity against a variety of pathogenic fungi and
represents a completely novel type of antibiotic. The structure was deduced from chemical and
spectral evidence, including single-crystal x-ray analysis. One tetrahydropyranoïd and one
dihydropyranoïd ring bearing alkyl substituents at positions 2 and 6, which may be regarded as
C-glycopyranosyl derivatives, are present in this structure.In connection with our efforts (Ref.3)
to synthesize a precursor of the west part of ambruticin, we developed the following stereospecific
transformation:

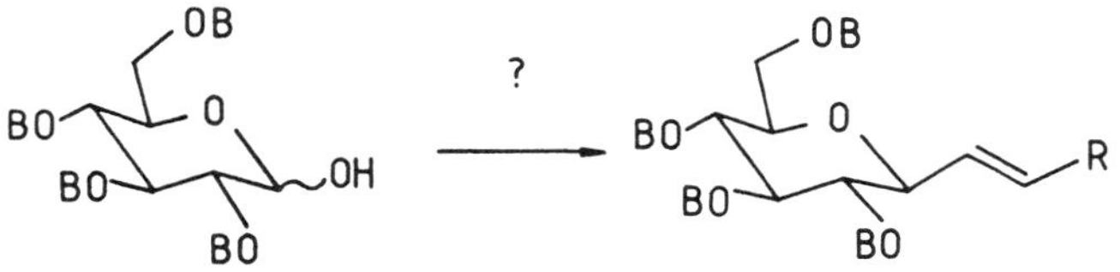

We restrict ourselves in this work to <u>direct</u> anchoring of the integrate appropriate unsaturated appendage at the anomeric centre of a carbohydrate. Various <u>indirect</u> formations of a substituted C-vinyl glycoside have been performed.

Ethynyl compounds have been shown to react with sugar lactones to give acetylenic lactols (Ref.4). On the other hand, triethylsilane with a Lewis acid is known to reduce acetals (Ref.5), a property which has elegantly been used by Kishi et al.(Ref.6) for the highly stereoselective synthesis of C-alkyl-β-D-glucopyranosides. As a model reaction combining these two results, treatment of lactone **1** with the anion of 1-benzyloxy-3-butyne (THF, -78° → -40°C, 1.5 h) gave a quantitative yield of the hemiketal **2**, which was stereospecifically reduced (Et$_3$SiH, BF$_3$.Et$_2$O in MeCN-CH$_2$Cl$_2$, -40°C, 1 h) into the β-D-C-glucoside **3** (72%). The stereochemical control is achieved by axially oriented addition of a hydride anion on the oxonium ion. The C-(alkyn-1-yl)-β-D-glucopyranoside **3** may be transformed either into the Z C-vinyl-β-D-glucoside **5** (75%) (H$_2$, Lindlar catalyst) or the E C-vinyl-β-D-glucoside **4** (57%) (LiAlH$_4$, DME, 90°C, 12 h).

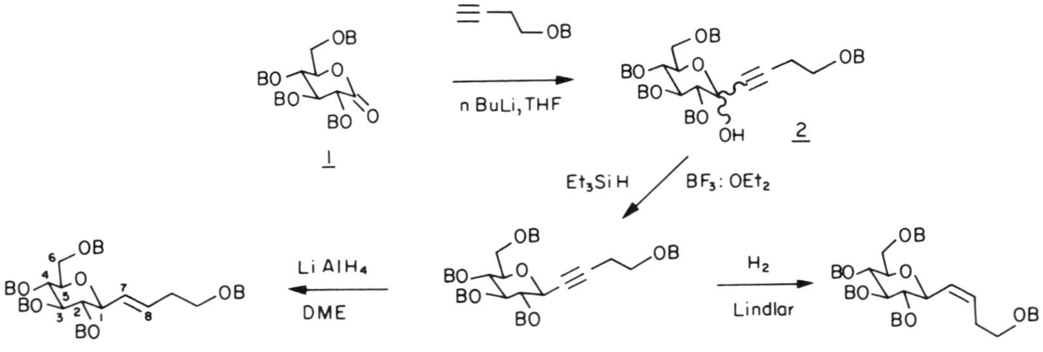

This methodology has first been applied to the synthesis of the E C-vinyl-β-D-glucoside **7**. The racemic dibromoolefin **6**, easily prepared from ethyl chrysanthemate, was transformed <u>in situ</u> into the corresponding lithio-alkynyl derivative (n-Buli, THF, -85°C for 1h, then 20°C for 1 h) and the three-steps procedure provided compound **7**, a model fragment of the west part of ambruticin.

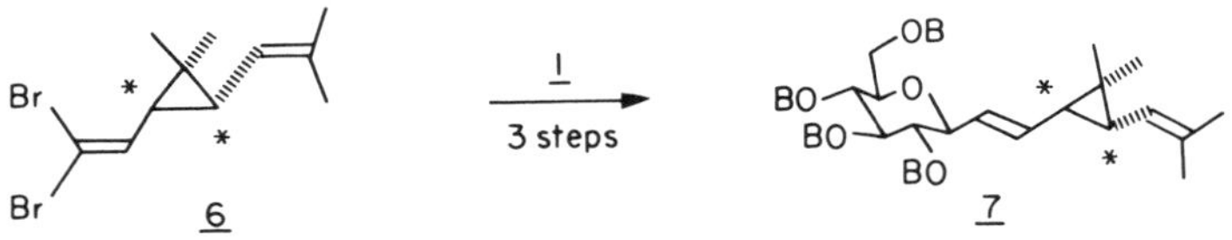

In order to synthesize a chiral direct precursor of the west segment of ambruticin (carbon atoms 1 to 13), the lactone **13** was prepared on one hand from methyl 2,3-di-O-benzyl-α-D-glucopyranoside **8**, according to the following route :

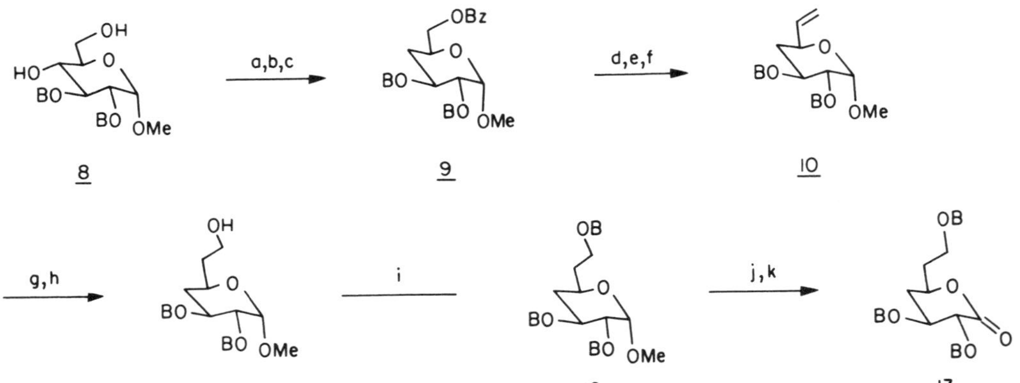

Reagents : (a) PhCOCN, CH$_2$Cl$_2$-pyridine, room temperature, 93%; (b) SO$_2$Cl$_2$, CH$_2$Cl$_2$-pyridine,

-15° C → room temperature, 7h, 67%; (c) Bu$_3$SnH, toluene, 90°C, 2h, 95%; (d) MeOH-MeONa, room temperature, 7h, 85%; (e) (COCl)$_2$-DMSO, CH$_2$Cl$_2$, Et$_3$N, -60°C, 30 min, 89%; (f) Ph$_3$PCH$_3$Br$^-$, NaNH$_2$, toluene, room temperature, 1h, 80%; (g) 9-BBN, THF, 75°C, 2h; (h) NaOH-H$_2$O$_2$, 0°C, 2h, 72%; (i) PhCH$_2$Br, NaH, DMF, room temperature, 1h; (j) H$_2$SO$_4$-AcOH, 80°C, 3h, 45%; (k) (COCl)$_2$-DMSO, CH$_2$Cl$_2$, Et$_3$N, -78°C, 10 min, 80%.

On the other hand, the chiral trisubstituted cyclopropane derivative **14** was easily prepared from D-glucose (Ref.7). It was selected by us as a starting material for the synthesis of the cyclopropane

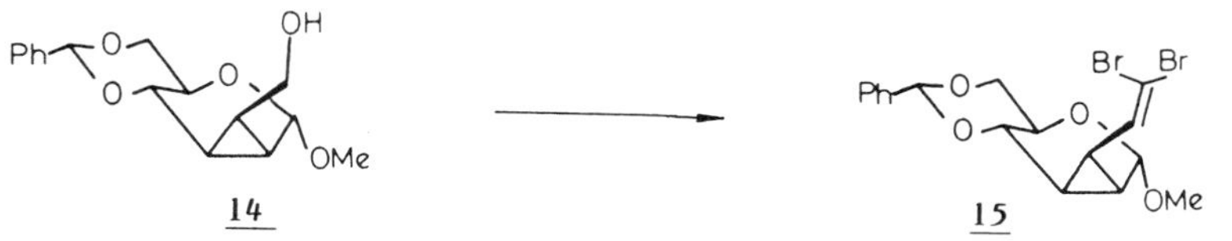

subunit of ambruticin. Oxidation of the primary alcohol into the corresponding aldehyde (PCC, CH$_2$Cl$_2$, 83%) was followed by conversion into the dibromoolefin **15** (Ph$_3$P, CBr$_4$, CH$_2$Cl$_2$, sym-collidine, 83%). Compound **15** was then transformed *in situ* into the corresponding lithio-alkynyl derivative and condensed with the lactone **1** to give the diastereoisomeric mixture **16** in excellent yield. During the reduction of the anomeric centre (Et$_3$SiH, BF$_3$.Et$_2$O, MeCN, molecular sieves 4 A, -40°C, 1 h), a sugar ring rearrangement took place, so that compound **17** was obtained as a diastereoisomeric mixture (two benzylidene stereoisomers). Acid hydrolysis (0.1 M HCl : dioxane, 1:1, reflux, 15 min), followed by reduction of the hemiacetal (NaBH$_4$, MeOH) and diol cleavage (NaIO$_4$, MeOH) gave compound **18** that we consider at this stage as a possible precursor of the west fragment of ambruticin. Synthesis of the east fragment is currently under investigation.

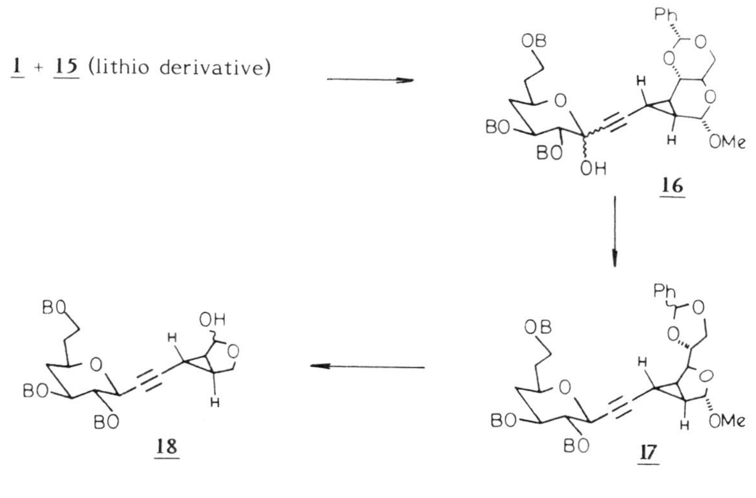

1 + **15** (lithio derivative) ⟶

The easy formation of C-(alkyn-1-yl)-β-D-glucopyranosides is of potential general utility and a few synthetic applications are now reported. After a period of fruitless attempts, three dimensional crystals of membrane proteins are now available (Ref.8). The use of "mild" detergents as β-D-octyl glucopyranoside during crystallization proved to be critical. A series of β-D-C-glucopyranosylalcanes have thus been prepared in order to test their detergent behavior. In a typical example heptanol was transformed into the corresponding dibromoolefin and condensed with the lactone **1**. Reduction (Et$_3$SiH) and catalytic hydrogenolysis provided β-D-C-glucopyranosyloctane **19** in excellent yield.

The alkyne may be replaced for this specific purpose by a Grignard reagent; for exemple, butylmagnesium bromide led to the β -D-C-glucopyranosylbutane derivative **20**.

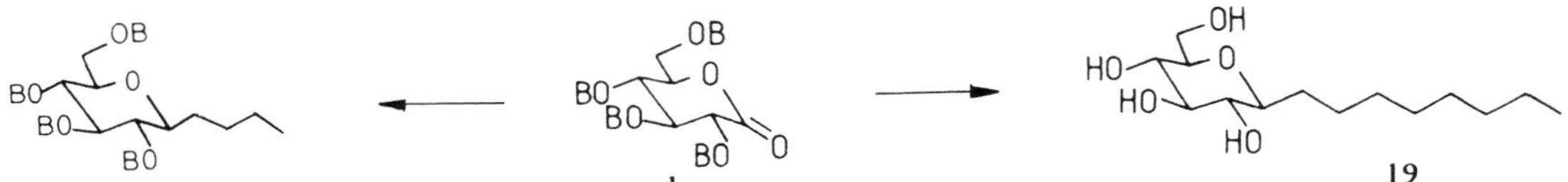

Compound **20** was easily transformed into the exo-methylene **21** then, on ozonolysis, into the lactone **22**, which was again submitted to the two-steps stereospecific introduction of an equatorially oriented alkyl chain to provide derivative **23**. This strategy thus provides a probably general entry to tetrahydropyranoïd rings bearing equatorial alkyl substituents at positions 2 and 6.

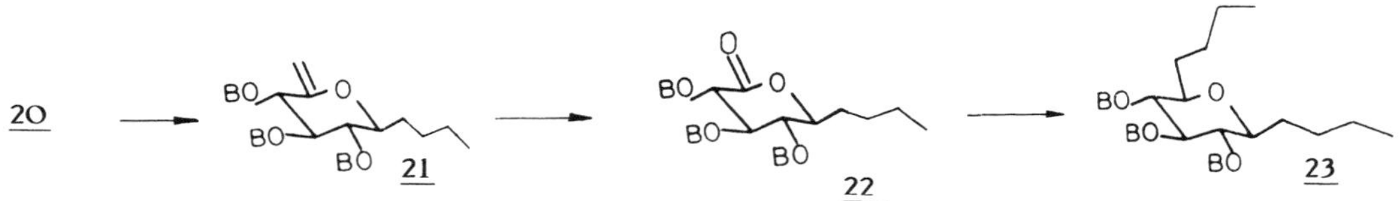

The C-glycoside **20** has also been transformed into the bromide **24**, then (Zn, propanol-water) into the open-chain derivative **25**. Such a transformation is of potential utility for the synthesis of chiral synthons.

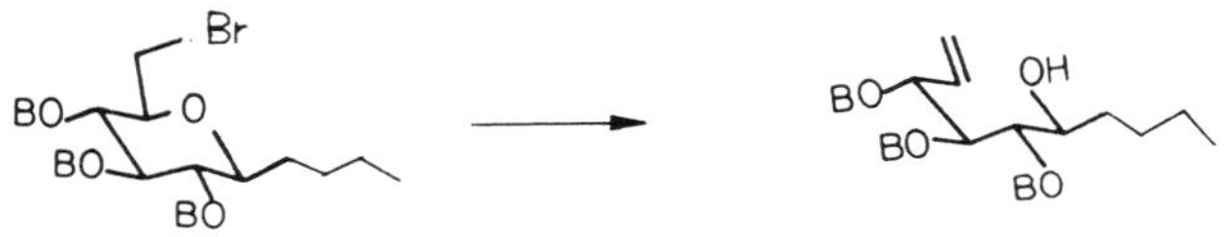

Although there is currently intense synthetic activity in the field of C-glycopyranosides, 'C-disaccharides', that is disaccharides in which the inter-unit oxygen atom is replaced by a methylene group, are still elusive compounds. Such compounds are highly in demand for studies of sugar metabolism and as enzyme inhibitors. Condensation of the lactone **1** with in situ generated acetylenic anion (n-Buli, THF, -50°C) derived from **26** gave, after reduction (Et_3SiH, $BF_3.Et_2O$ in $MeCN-CH_2Cl_2$, 0°C, 15 min) and catalytic hydrogenolysis (H_2, 10% Pd/C, AcOH, 24h), the crystalline 'C-disaccharide' **27.** This novel sequence, which provides the first entry to a 'C-disaccharide', should be of general utility for the preparation of C-analogues of β (1 → 6') di- or oligosaccharides.

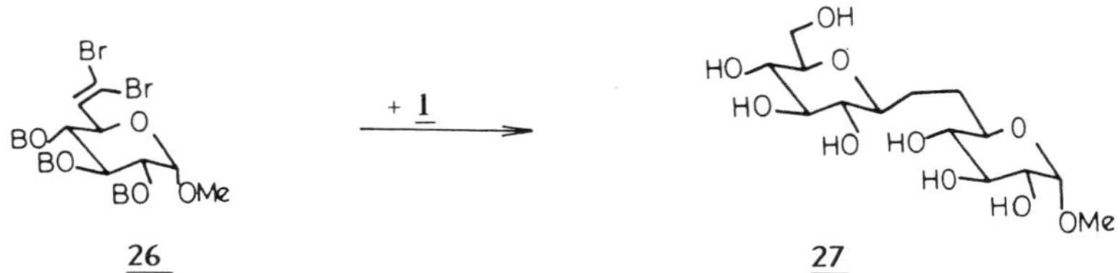

II.- SIMPLE GENERATION OF A REACTIVE GLYCOSYL-LITHIUM DERIVATIVE

The β-D-C-glycoside synthesis previously described rely on the electrophilic character of the anomeric centre of a carbohydrate. Less conventional recent approaches derive either from the generation and trapping of a glucosyl radical (Ref.9) or from the use of 1-deoxy-1-nitro sugars where the nitro group acts as an anion stabilizing substituent for the generation of a reactive anomeric anion under mild conditions (Ref.10). We would like to describe the generation and reactivity of 3,4,6-tri-O-benzyl-2-deoxy-α-D-arabino-hexopyranosyl lithium **34** (Ref.11), the first example to-date of a non-stabilized C-glycopyranosyl-lithium reagent.

Treatment of the benzylated thiophenylglucoside **28** with 2 equiv. of lithium naphtalenide (LN) (THF, -78° C, 15 min) resulted in quantitative formation of glucal **29** (Ref.12). This result indicates that highly selective reductive lithiation (Ref.13) occured at the anomeric centre of **28**, followed by a fast β-elimination (Ref.14). A two-steps hydrolithiation of **29**, as shown in the following scheme, followed by quenching with D$_2$O at -78°C provided selectively the deuteriated derivative **30** (80% from **29**).

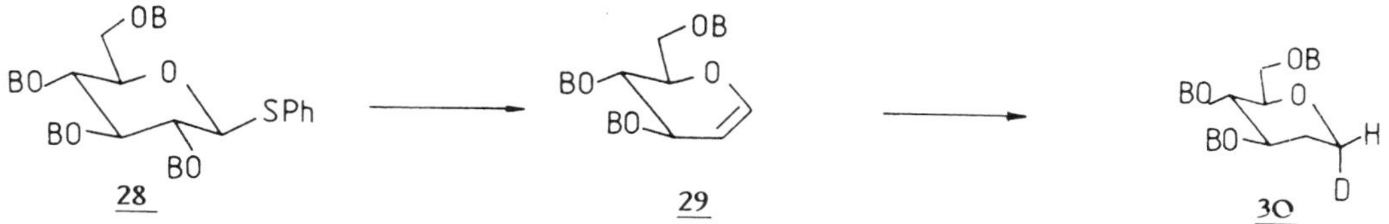

Similar results were obtained when the thiophenyl glycosides **31** or **32** were reductively lithiated (Ref.15). The initial step of the reaction was the generation of the <u>axial</u> C-1-glucosyl radical **33** (Ref.16), which was transformed into the reactive α-D-glucopyranosyl lithium **34**. Quenching of **34**

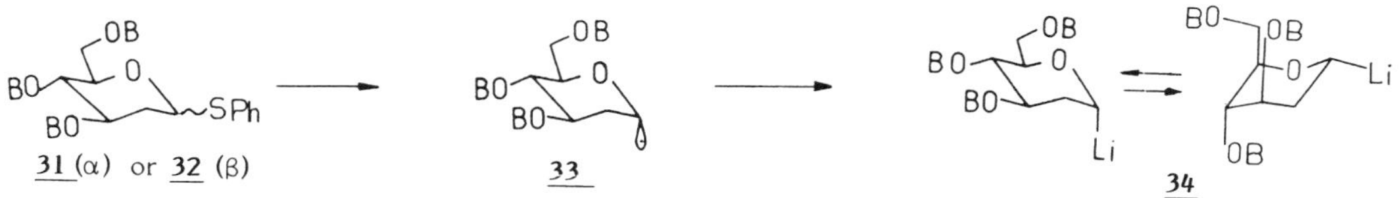

with electrophiles selectively provided α-D-C-glycosides. For example, treatment of **34** with p-anisaldehyde, followed by oxidation, gave a single crystalline product **35** in about 50% yield. Reaction of **34** with chlorotrimethylsilane gave the crystalline α-D-Si-glycoside **36** (36%), the first example of a Si-glycoside, whose chemistry is unknown.

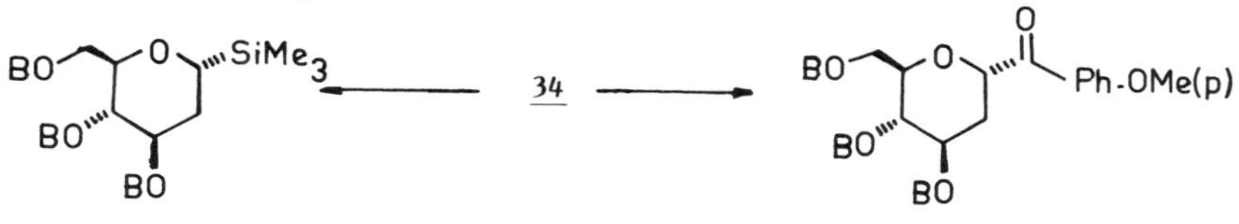

This methodology, combined with the previous one, provides an easy route to a new class of compounds (Ref.17), i.e. disaccharides wherein the two aldose rings are directly joined. An example is provided in the following scheme:

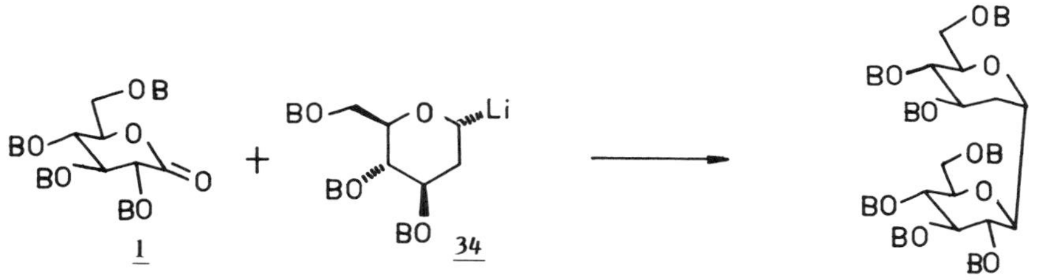

This approach nicely extends the chirality of a carbohydrate template.

In conclusion, the two-steps hydrolithiation of a glucal reverses the characteristic electrophilicity of the anomeric centre, opening a virgin field which is currently explored. The few examples reported in this lecture augur well.

III.- D-GLUCOPYRANOSYL PHENYL SULPHONES : VERSATILES INTERMEDIATES FOR THE SYNTHESIS OF C-GLYCOSIDES

Carbohydrate anomeric sulphones have already been prepared (Ref.18), but their chemistry is by-large unexplored. We would like now to emphasize the remarkable utility of this class of products for the selective synthesis of C-glycosides.

Anomeric sulphones of 2-deoxy-D-glucose are conveniently prepared by the new route shown in the following scheme:

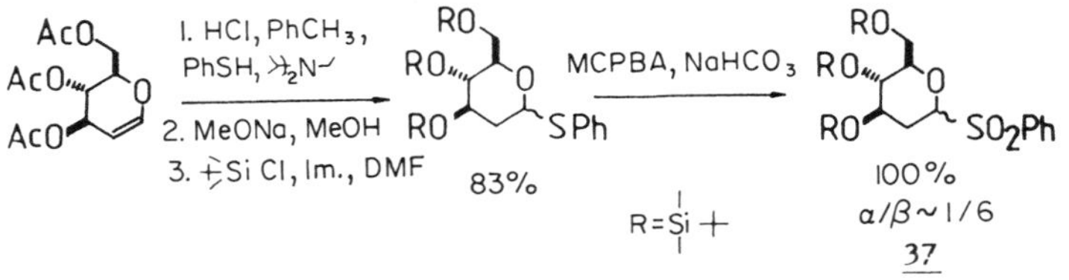

Although final protection of the hydroxyl groups by t-butyldimethylsilyl is depicted, this procedure is general and protections with methyl and benzyl ethers have also been used. The sulphone was obtained as an anomeric mixture, which is perfectly suitable for the development of the chemistry we would like to present; separation of both anomers by chromatography or crystallization (β-sulphones are highly crystalline) is possible, but offers no advantage.

When the anomeric sulphone **37** was treated with n-Buli or LDA at -78°C, deprotonation at the anomeric centre occured, as demonstrated after quenching with D_2O at -78°C, when an anomeric mixture **38** of deuterated sulphones was fished out. It is noticeable that the same anomeric composition (4:1) was obtained, regardless of the anomeric composition of the starting sulphone. The major isomer is the α-sulphone **38**   , which calls for glucopyranosyl lithium **39** as the main reactive species.

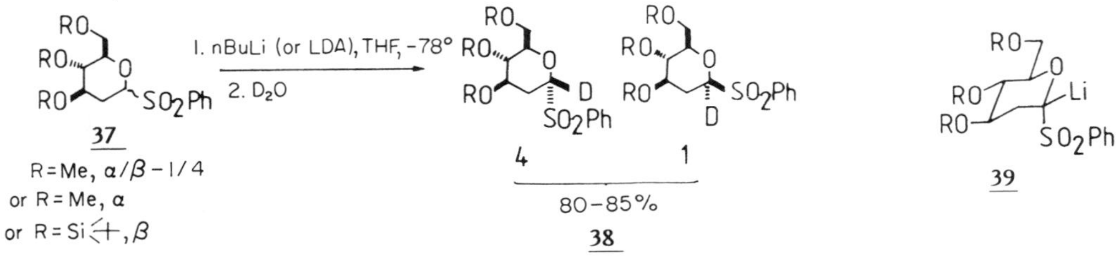

Therefore, when axial deprotonation was achieved on 3,4,6-tri-O-t-butyldimethylsilyl-2-deoxy- β-D-arabino-hexopyranosyl sulphone **37,** the anion stabilizing sulphone substituent triggers isomerization to the more stable species **39**. The strong anti-anomeric effect of the lithium anomeric substituent- or of the corresponding negative charge-overcomes the weaker anti-anomeric effect of the sulphone group.

A major result in connection with selective C-glycoside synthesis was obtained when Jean-Marie Beau realized that reductive lithiation of the anomeric mixture **38** with LN was operating, subsequent quenching with water providing **40** as the single expected product. The axial stereo-selective introduction of a proton is similar to the one which has previously been observed.

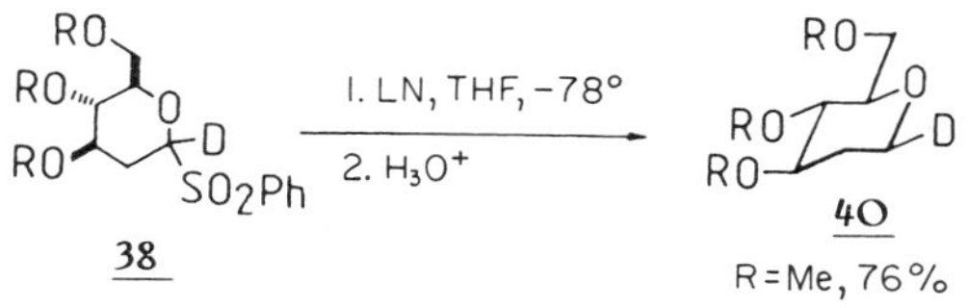

38 → 40 R = Me, 76%

A first straightforward consequence of this observation is that anomeric sulphones **37** can also be used for the <u>stereoselective synthesis of axially oriented C-glycosides</u>, using a reductive lithiation (LN)-alkylation sequence similar to the one previously developed with anomeric chloride. As anomeric chloride is a more readily available starting material than **37**, there was no real need for us to push the sulphone chemistry in that particular, although interesting direction. The selected following reaction demonstrates its viability.

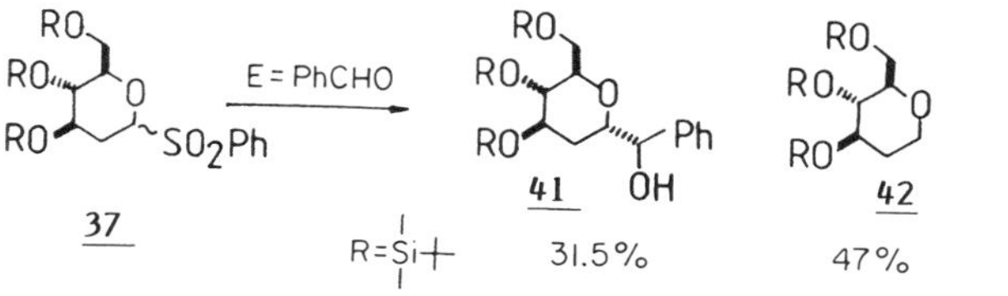

37   E = PhCHO   41   31.5%   42   47%

R = Si+

A practical serious restriction is the copious formation of **42**, a similar side-product being also formed during our previous study with anomeric chloride.

A practically much more useful and major consequence of this discovery is that a one-pot combination of sulphone promoted deprotonation-electrophile quenching and reductive lithiation-methanol quenching should provide a <u>novel stereoselective synthesis of equatorially oriented C-glycosides</u>, as shown on the following scheme:

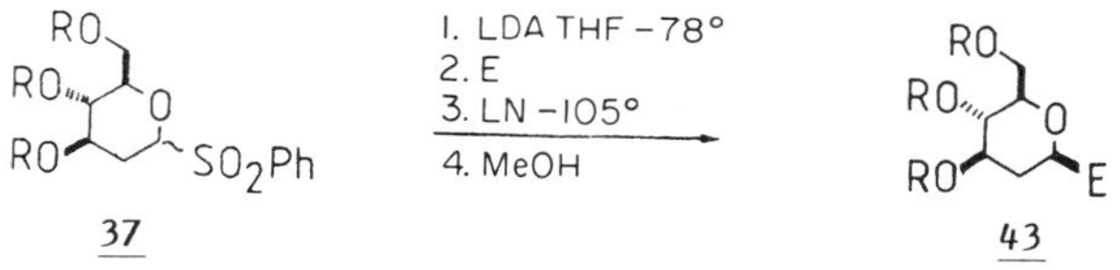

37   1. LDA THF −78° 2. E 3. LN −105° 4. MeOH   →   43

indeed, the use of D_2O as exploratory electrophile auspiciously gave **43** (E=D) as the only detectable isomer in about 80% yield and in a for-steps one-pot reaction from the anomeric sulphone **37** (R=t-butyldimethylsilyl). Table I illustrates the generality of this novel methodology (yields have not been optimized). In the case of entry 2, HMPT has been added to THF to improve the yield of compound **43** (E=Me). Entry 6 is of interest in connection with the stereoselective synthesis of polyhydroxylated natural products. It is noticable that a rather high diastereofacial selectivity has been observed (10:1); the absolute configuration of the exocyclic generated asymetric centre of the major isomer has not yet been established.

A further development of this reaction was realized when activated carboxylic acid derivatives were used as electrophiles. It was a nice surprise to us to discover that quenching with dimethyl-carbonate stereoselectively provided compound **44** in 72% yield. The structure of this C-glycoside was ascertained after comparison with the β-D-C analog **45**, which was prepared from the previously synthesized alcohol **42** (E=CH_2OH, entry 3 of table I).

Table I. Synthesis of Equatorially Oriented C-Glycosides

Entry	E	Product	Yield %
1	D₂O	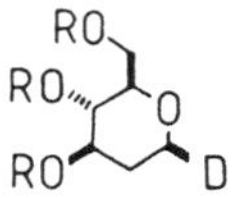	80
2	MeI	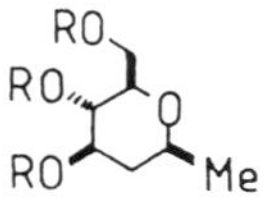	43
3	HCHO	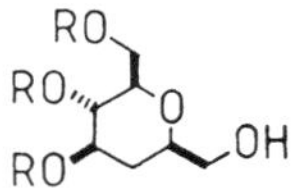	57
4	PhCHO	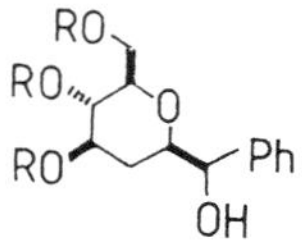	74
5		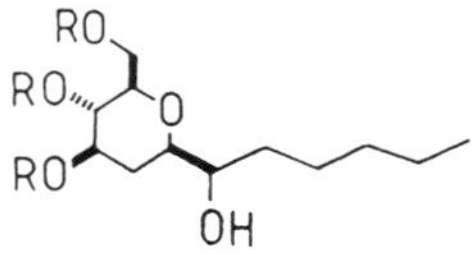	63
6	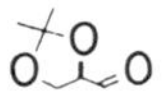	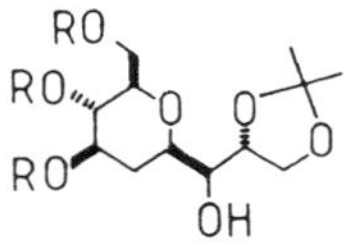	66

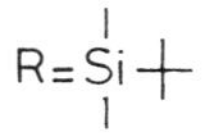

$R = Si\!+\!$

When phenylbenzoate was used as an electrophile, the α-D-C-glucoside **46** was obtained in 72% yield and found identical with the oxidation product of previously synthesized **41**. Oxidation of **43** (E=PhCHO, entry 4) gave a ketone which was clearly different from **46**.

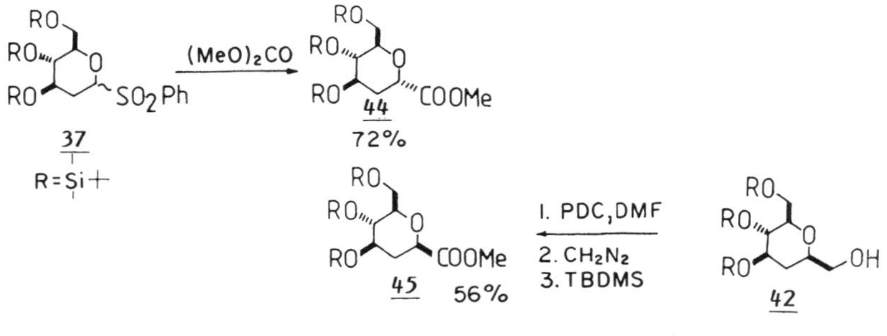

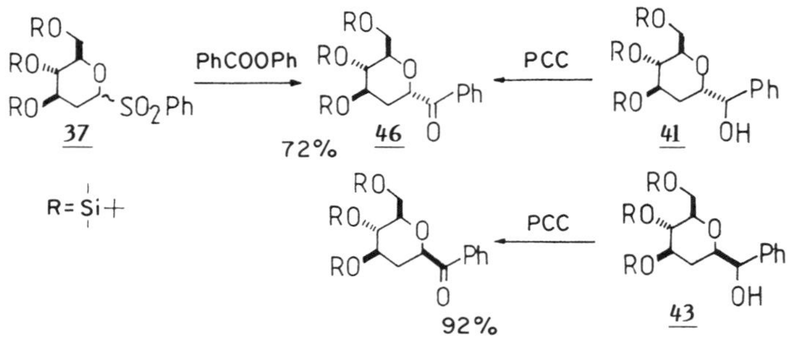

A significant example is shown in the following scheme, where the yield of compound **47** is at this stage moderated by the formation of the side-product **48**.

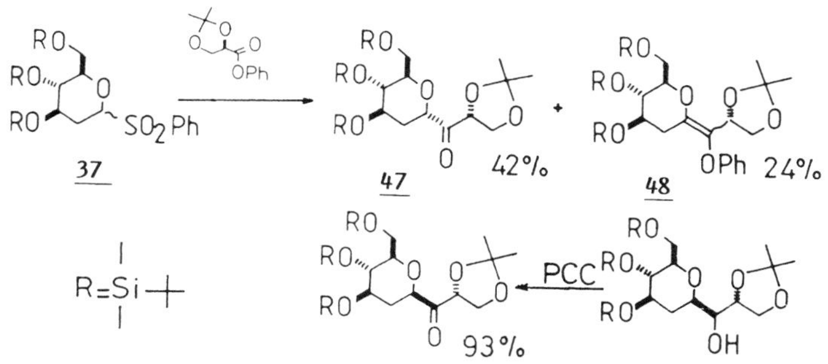

The formation of α-D-C-glycosides is tentatively explained in the following scheme:

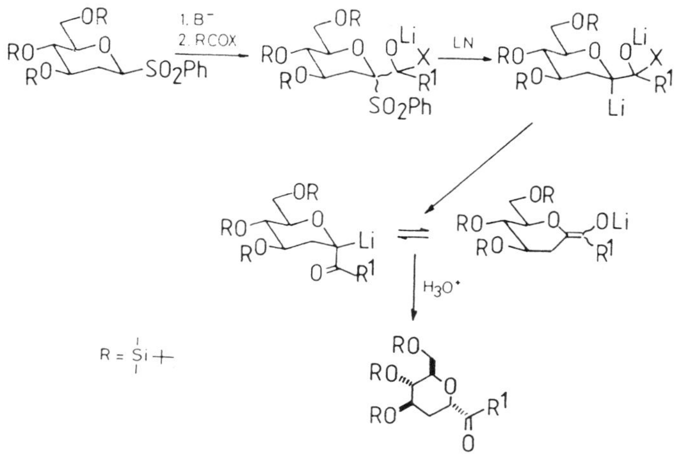

In conclusion easily released anomeric sulphones generate a rewarding chemistry which is currently under development in our laboratory.

Acknowledgments.- We thank Dr. Paul Henri Amvam Zollo and Mr. Dominique Rouzaud for their contributions to this programme, Mr. J. Esnault and Miss A.-M. Noirot for their technical helps, and the Centre National de la Recherche Scientifique for financial support.

REFERENCES

1. J.K.Cha, W.J. Christ, J.M. Finan, H. Jujioka, Y. Kishi, L.L. Klein, S.S. Ko, J. Leder,W.W. McWhorter, Jr., K.-P. Pfaff, and M. Yonaga, D. Uemura, Y. Hirata,J.Am.Chem.Soc., 104, 7369 (1982).

2. D.T. Connor, R.C. Greenough, and M. von Strandtmann, J.Org.Chem., 42, 3664 (1977);ibid, 43, 5027 (1978).

3. J.-M. Lancelin, P.H.Amvam Zollo, and P. Sinaÿ, Tetrahedron Lett., 4833 (1983).

4. H. Ogura, H. Takahashi, and T.Itoh, J.Org.Chem., 37, 72 (1972).

5. E. Frainnet and C. Esclamadon, C.R.Hebd.Scéances Acad.Sci., 254, 1814 (1962).

6. M.D. Lewis, J.K.Cha, and Y. Kishi, J.Am.Chem.Soc., 104, 4976 (1982).

7. B.J. Fitzsimmons and B. Fraser-Reid, J.Am.Chem.Soc., 101, 6123 (1972) and reference cited therein.

8. H. Michel, TIBS 8, 56 (1983).

9. B. Giese and J. Dupuis, Angew.Chem., Int.Ed.Engl., 22, 622 (1983).

10. B. Aebischer, J.H. Bieri, R. Prewo, and A. Vasella, Helv.Chim.Acta, 65, 2251 (1982).

11. J.-M. Lancelin, L. Morin-Allory, and P. Sinaÿ, J.Chem.Soc., Chem.Commun., 355 (1984).

12. I.D. Blackburne, P.M. Fredericks, and R.D. Guthrie, AustJ.Chem., 29, 381 (1976).

13. C.G. Screttas and M. Micha-Screttas, J.Org.Chem., 43, 1064 (1978).

14. S.J. Eitelman and A. Jordaan, J.Chem.Soc., Chem.Commun., 552 (1977);R.E. Ireland, C.S.
    Wilcox, S. Thaisrivongs,and N.R. Vanier, Can.J.Chem., 57, 1743 (1979).

15. T. Cohen and J.R. Matz, J.Am.Chem.Soc., 102, 6900 (1980).

16. R.M. Adlington, J.E. Baldwin, A. Basak, and R.P. Kozyrod,J.Chem.Soc., Chem.Commun.,
    944 (1983).

17. S.J. Danishefsky, C.J. Moring, M.R. Barbachyn, and B.E. Segmuller, J.Am.Chem.Soc., in press.

18. P.M. Collins and B.R. Whitton, Carbohydr.Res., 36, 293 (1974) and references cited therein.

Strategies in oligosaccharide synthesis

Hans Paulsen

Institute of Organic Chemistry, University of Hamburg
D-2000 Hamburg 13 (FRG)

Abstract - The selective synthesis of complex oligosaccharide
segments which are part of glycoconjugates has become of great
importance because usually the oligosaccharide section represents
the determinant being involved in the biological function of
glycoconjugates. The following three fundamental reactions have
proved to be most suitable for building up larger molecules with
the desired anomeric linkage:
1. The neighbouring group assisted procedure by which β-glycosidic linkages in the D-gluco and D-galacto serie as well as
   α-glycosidic linkages in the D-manno serie can be synthesized.
2. The in situ anomerisation procedure which permits the synthesis
   of an α-glycosidic linkage in the D-gluco, D-galacto and the
   D-manno serie.
3. The heterogeneous catalyst procedure by which β-glycosidic
   linkages in the D-manno serie as well as reactions with inversion can be accomplished without neighbouring group
   assistance.

For every glycosidation step, three parameters of the reactivity
must be carefully adjusted, namely the reactivity of the glycosyl
donor, of the catalyst and of the glycosyl acceptor. The application of the glycosidation methods is demonstrated by a serie of
syntheses of complex oligosaccharide chains of glycolipids and
glycoproteins.

INTRODUCTION

Within the last ten years the chemistry of glycoconjugates (Ref. 1) has gone
through a considerable progress. Modern analytic methods, as high field
n.m.r. spectroscopy and FAB mass spectroscopy allowed to find out the structure of still more compounds of this special type. Furthermore, we have got
new information about their most important biological functions. In glycolipids (Ref. 2) and glycoproteins (Ref. 3) the oligosaccharide structure
often represents the determinant which is responsible for the biological
functions of these substances. In this case, interactions between the carbohydrate residues and other proteins are possible, as for example the antigen-
antibody reaction or other receptor reactions playing an important role for
the cell communication (Ref. 4). Here the molecular structure of the carbohydrate sequence is of great importance because it determines the selectivity
of the reaction with a protein or an agglutinine. Available data indicate
that in this interaction predominantly hydrophobic attractive forces play a
greater part than do hydrogen bonds (Ref. 5).

Consequently the selective synthesis of such complex oligosaccharide segments
is of special interest. The corresponding compounds can be used for structural analyses and for inhibition tests. By modification of the determinants,
conclusions can be drawn regarding the kind of interaction of oligosaccharides with the protein as well as the form of the combining sides. Via a
spacer, oligosaccharides can be bound to large molecules or to solid phases
which leads to synthetic antigens or to suitable immuno absorbers. Thus there
will be a great number of biologically interesting fields of application for
selectively synthesized, complex oligosaccharides.

METHODS OF GLYCOSIDIC LINKAGES FOR THE SYNTHESIS OF OLIGO-
SACCHARIDES

In the recent decade great progress has also been made concerning the methods
of the selective synthesis of complex oligosaccharides (Ref. 6). For the
synthesis of larger oligosaccharides usually the glycosyl halides have proved
to be the best starting material, but also other leaving groups can be taken
into consideration. There are three different basic methods from which one
must be chosen according to the type of glycosidic linkage which shall be
prepared (Ref. 6):
1. The neighbouring group assisted procedure for the synthesis of β-glycosidic linkages in the D-gluco and D-galacto series as well as the α-
   glycosidic linkage in the D-manno serie.
2. The in situ anomerisation procedure by which α-glycosidic linkages in the
   D-gluco, D-galacto and D-manno series can be synthesized.
3. The heterogeneous catalyst procedure for the synthesis of β-glycosidic
   linkages in the D-manno serie and for linkages under inversion without
   neighbouring group assistance.

For the neighbouring group assisted procedure (Fig. 1) generally the stable
α-D-glycosyl halides are applied containing a neighbouring group active substituent at C-2 which are, for example, acyl groups as acetates or benzoates.

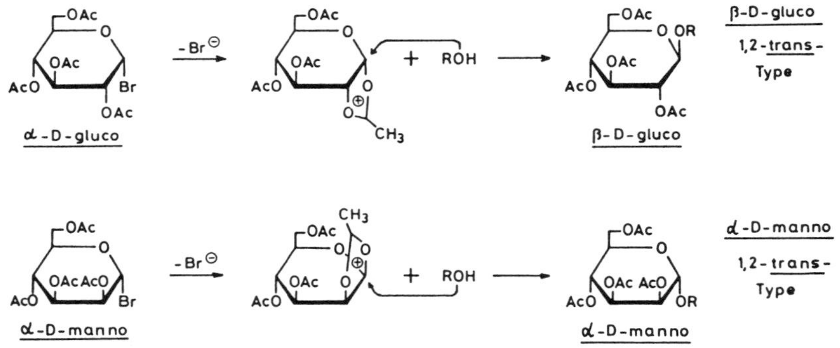

Catalyst : $Hg(CN)_2$, $HgCl_2$, $AgClO_4$, Ag triflate

Fig. 1. Neighbouring group assisted procedure: 1,2-trans-type
reaction for the synthesis of β-D-gluco-, β-D-galacto- and α-
D-mannopyranosides and saccharides with related stereochemistry

The elimination of a bromine anion or another leaving group caused by a catalyst with receptor activities yields a more stable cyclic dioxocarbenium ion
via a primary formed carboxonium ion. This new formed dioxocarbenium ion determines the stereoselective control of the reaction. As shown in Fig. 1, in
the D-gluco serie the cation at the anomeric centre can only be opened by a
nucleophile from the upper side which leads to the stereoselective formation
of a β-D-glucopyranoside (1.2-trans-type).

The stereochemistry of the 2-OH group in the D-manno serie is just the opposite so that, at a corresponding reaction, an inverted cyclic dioxocarbenium
ion is formed as intermediate (Fig. 1). In this case, the nucleophile can
open the ring at the anomeric centre only from below which leads to an α-D-
mannopyranoside (1.2-trans-type). Thus the reaction sequence in the D-gluco
and D-manno serie gives different anomeric linkages as it is determined by
the position of the neighbouring group active substituent. As a polar intermediate will be passed through, more polar solvents, such as nitromethane,
probably as a mixture with toluene, must be applied. It can be disposed of a
lot of catalysts which must only be well coordinated with each substrate and
with the reactivities of the components.

The synthesis of an α-D-glucosidic linkage is much more difficult. It is important to chose a starting compound which has a non-neighbouring group
active substituent at C-2 in order to avoid a reaction similar to that one
shown in Fig. 1. The β-D-halides, as for example (2), could be applied and
will react under inversion, but on the other hand, these halides are very
instable and their preparation will raise some difficulties. Furthermore,
they easily anomerise to the more stable α-D-halides so that this reaction

would result in a mixture of the two anomeric glycosides. An example for a
successful reaction is the conversion of (2) with the L-serine derivative (3)
yielding the α-D-glycoside (4) as a crystalline, relatively stable substance
(Ref. 7). In this case, (2) is obtainable by a kinetically controlled anomerisation of (1).

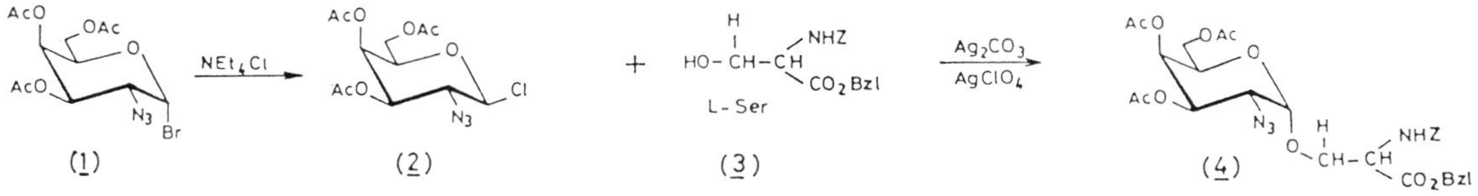

For the in situ anomerisation procedure, the more stable α-D-pyranosyl halide
with a non-neighbouring group active substituent at C-2 should be used. The
presence of a catalyst like tetraalkylammonium halides or metal salt catalysts causes an equilibration between the α-D-glycopyranosyl and the β-D-
glycopyranosyl halide (Ref. 8). This anomerisation shall proceed via different ion pairs (Fig. 2). However, the β-D-glucopyranosyl halide is destabil-
ized by the anomeric effect so that within the equilibrium a much higher
amount of α-D-glucopyranosyl bromide is present.

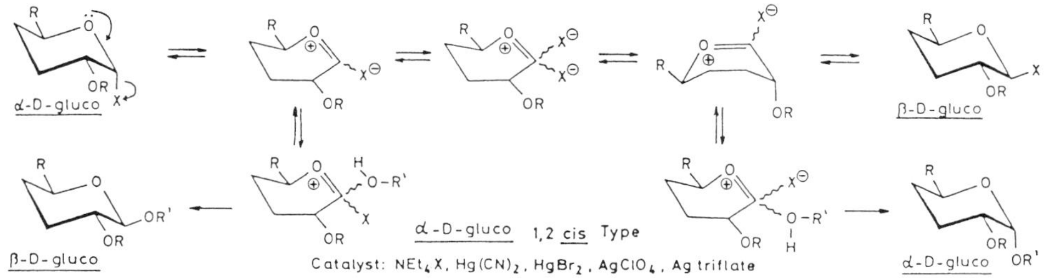

Fig. 2. In situ anomerisation procedure (Without neighbouring
group participation): 1,2-cis-type reaction for the synthesis
of α-D-gluco-, α-D-galacto- and α-D-mannopyranosides and saccharides with related stereochemistry.

This is, however, only the thermodynamic view of the equilibrium between the
halides. Regarding the kinetics of the reactions, it can be observed that the
reaction of the more instable β-D-halide to the α-D-glucopyranoside (right
reaction in Fig. 2) is a faster reaction compared with the one of the α-D-
halide which occurs under inversion to the β-D-glucopyranoside (left reaction
in Fig. 2). Utilizing this difference in the reaction rates, the reaction can
be accomplished almost selectively only to the α-D-glucopyranoside (1.2-cis-
type).

This indicates that the reaction must proceed under kinetic control and that
a sufficient difference between both reaction rates must be maintained. Even
small concentrations of β-D-glucopyranosyl halides make it possible to transfer the substance completely to the α-D-glucopyranoside. In this case, an
active catalyst must care for a continuous reforming of new β-D-glucopyranosyl halide from the α-D-glucopyranosyl halide via the equilibrium. It should
be mentioned that for this in situ anomerisation procedure not only tetra-
alkyl halides but also the more effective mercury cyanide, mercury bromide,
silver perchlorate, and silver triflate can be applied. Probably intermediates will be produced, as pyranosyl halide mercury complexes, pyranosyl
perchlorates or pyranosyl triflates. The reaction sequence of Fig. 2 is
generally applicable for these intermediates in a completely analogous way
and leads to the same results, concerning the selectivity. Following the so-
called azido method (Ref. 9), the in situ anomerisation procedure can also
be applied for the preparation of α-D-glycopyranosides of 2-amino-2-deoxy
sugars. As non-participating substituent at C-2 a 2-azido group is used
which, lateron, can easily be transferred to an amino group.

As the anomeric effect at the D-mannose is effective in the same way and the
substituent at C-2 is not involved in the in situ anomerisation procedure,
this method being applied at the D-mannose will also yield α-D-mannopyranosides which, however, were already prepared according to the neighbouring
group assisted procedure. Consequently, the two before mentioned procedures
are not suitable for the preparation of the β-D-mannopyranosides. This de-

monstrates that for the preparation of the β-D-mannosidic linkage the most difficult problems will arise.

For this reaction, the <u>in situ</u> anomerisation procedure, which occurs in the presence of Lewis-catalysts, must be avoided at any rate and a solid state catalyst must be applied for the reaction in heterogeneous phase. Besides silver carbonate, a silver silicate catalyst has proved to be an especially active catalyst which is to be dispersed on silica gel (Ref. 10). Herewith the α-D-mannopyranosyl bromide can be transferred under inversion to the β-D-mannopyranoside , following the heterogeneous catalyst procedure (Fig. 3).

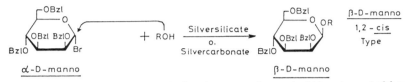

Fig. 3. Heterogeneous catalyst procedure (without neighbouring group participation): 1,2-<u>cis</u>-type reaction for the synthesis of β-D-mannopyranosides and other types which react under inversion at the anomeric centre.

In an oligosaccharide synthesis, a glycosyl donor is reacted with a glycosyl acceptor which has only one free hydroxyl group. For this reason, the blocking system for both components must carefully be tested. The reactivity of both compounds depends to a high degree on the blocking system and the further substituents. Therefore it is not possible to indicate standard conditions for an oligosaccharide synthesis, but each linkage reaction must be optimized according to the especially existing problem. Guide lines have, however, been developed for certain strategies by which the reactivities of the corresponding components can be estimated.

Table 1 shows the three most important parameters determining the selectivity and the yield of a glycosidation reaction. These are the reactivity of the halide, of the catalyst, and of the hydroxy component (Ref. 6). The reactivity of the halide can be varied to a considerable extent, by the choice of the blocking system of the substituents. Acyl substituents like acetyl groups increase the stability and reduce the reactivity whereas ether groups, like benzyl ether, reduce the stability and increase the reactivity of the halide. In this case, the number of ether groups is more important than their position in the molecule. So a wide range of possibilities of varying the reactivity is given for the halide.

TABLE 1. Parameters of reactivity, determining selectivity and yield of the oligosaccharide synthesis

Halogen	Catalyst	Alcohol
CH$_2$OR ... RO — O — OR — X — N$_3$	Et$_4$NBr / molecular sieve 4 Å Hg(CN)$_2$ Hg(CN)$_2$ / HgBr$_2$ HgBr$_2$ / molecular sieve 4 Å AgClO$_4$, Ag$_2$CO$_3$ Ag triflate , Ag$_2$CO$_3$	HO-CH$_3$ >> HO-CH$_2$R > 6-OH CH$_2$OH ... O ... OH ... HO — OR ... OH
R ~ Bzl > Bz ≳ Ac X ~ I > Br > Cl		6-OH >> 3-OH > 2-OH > 4-OH

As shown in Table 1, there is a serie of catalysts of different reactivities at our disposal which can be utilized for the neighbouring group participation procedure as well as for the <u>in situ</u> anomerisation procedure. As already mentioned before, the stated catalysts are not suitable for the heterogeneous catalyst procedure.

The most difficult problem is to vary the reactivity of the hydroxyl group because generally the glycosyl acceptor has already been predetermined. Usually primary hydroxyl groups are more reactive than secondary hydroxyl groups where the reactivity considerably depends on the further substituents in the molecule so that the reactivity can only be estimated with difficulties. For the <u>in situ</u> anomerisation procedure secondary hydroxyl groups with a medium reactivity are especially suitable and lead in the desired selectivity to α-D-glycosides. With reactive hydroxyl groups, like primary hydroxyl groups, difficulties can ·arise applying the in situ anomerisation reaction because the difference between the two reaction rates to both glyco- sides will not be sufficient and thus anomeric mixtures of the glycosides will be obtained. Furthermore, the solvent is an important parameter. As al- ready stated before, the neighbouring group assisted procedure, however, re- quires a more polar solvent like nitromethane whereas usually for the in situ anomerisation procedure and the heterogeneous catalyst procedure a more un- polar solvent, e.g. dichloromethane, would be advisable.

Besides glycosyl halides, also glycosyl compounds with other leaving groups can be used for the oligosaccharide synthesis, according to the existing problem. The O-acetyl group represents a leaving group with rather low re-

Fundamental sequence of O-glycoproteins : β-D-Gal(1→3)-α-D-GalNAc(1→O)-L-Ser(L-Thr)

(11)

activity which can, however, be eliminated by Lewis acid catalysts under neighbouring group participation so that this method is also suitable for the neighbouring group assisted procedure (Ref. 11). The reaction of (5) with the less reactive hydroxyl group of the glycoside (6) can be considered as an example (Ref. 12). In the presence of trimethylsilyl triflate the re- action of (5) with (6) yields the disaccharide (7). Generally it must be pointed out that this method requires acid-stable protecting groups and a corresponding choice of the blocking system is necessary. In this case, acyl substituents are especially suited as protecting groups. By acetolysis and bromination of (7) the compound (8) can be obtained which is an excellent synthetic block and, following the in situ anomerisation procedure, the re- action of (8) with (9) affords the α-D-glycosidically bound glycopeptide (10). Deblocking of (10) yields the sequence (11), representing the fundamental sequence of the O-glycoproteins (Ref. 13).

The trichloroaceto imidate method, developed by R.R. Schmidt (Ref. 14), can be considered as a further promising procedure for the oligosaccharide syn- thesis. Applying the trichloroaceto imidate group as a leaving group, it is a bit less reactive than a halide, but much more reactive than an O-acetyl group. Therefore, this group should be used for a neighbouring group assisted procedure in order to obtain good selectivities to the β-D-glycoside. An

R.R. Schmidt 1980

example for this is the reaction of (12) with (13) which in the presence of bortrifluoride etherate yields the disaccharide (14) under coupling with the less reactive hydroxyl group in (13) (Ref. 15). The phthalimido group possesses excellent neighbouring group assisted properties and directs the reaction always to the β-glycoside.

The trichloroaceto imidates can also be used for the reaction without neighbouring group assistance. Thus under inversion the α-trichloroaceto imidate (15) with the rather reactive hydroxyl group in (16) delivers in good yields the corresponding β-D-glycosidically bound compound (17) (Ref. 16). The inverted reaction of a β-trichloroaceto imidate to an α-D-glycoside raises some problems. Here, the stability between α- and β-compounds can be compared with that of the pyranosyl halides, because the β-trichloroaceto imidate is much more instable than the α-trichloroaceto imidate so that rearrangements from the β- to the α-form become possible (Ref. 17). It should be pointed out that the trichloroaceto imidates can only be prepared from a compound being unsubstituted at 1-OH. The reaction of the D-manno compound (18) with (19) demonstrates the limited field of application of this method. Though the free hydroxyl group in (19) has a relatively high reactivity, only the product of retention (20) can be obtained and not the desired β-D-mannopyranoside (Ref. 18). This reaction sequence is rather similar to the one of the in situ anomerisation procedure of a corresponding D-mannopyranosyl halide.

Another interesting procedure has recently been developed by Lönn (Ref. 19). With methyl triflate thio glycosides like (21) can be methylated to the instable intermediate sulfonium compounds (22) representing an easily separable leaving group. In the presence of glycosyl acceptors like (23), the oligosaccharide (24) will be achieved. Using thio glycosides with a neighbouring active group at C-2, this method yields oligosaccharides with good selectivity. Then the disaccharide synthesis selectively leads to the β-D-glycosidically linked compounds, as demonstrated by the reaction of (21) with (23) to (24). The corresponding reaction of thio glycosides like (25), containing a non-neighbouring active group at C-2, with (26) delivers a mixture of α- and β-glycosides, the proportion of which depends to a large extent on the solvent. The application of this method in its neighbouring group assisted form, however, enables the linkage of larger oligosaccharide units to further glycosyl acceptors. The tetrasaccharide (29) can be obtained by the reaction of the thio glycoside of the trisaccharide (27) with the D-manno derivative (28), forming the β-D-glycosidically linked product. During this reaction, some elimination compounds will be produced as by-products.

APPLICATION OF THE DISCUSSED METHODS FOR THE SYNTHESIS OF OLIGO-
SACCHARIDE CHAINS OF GLYCOLIPIDS

At first, the application of the discussed methods for building up complex oligosaccharides shall be backed up by an example of the synthesis of the pentasaccharide chain of the Forssman antigen (Ref. 20). The Forssman antigen (30) represents a glycolipid containing a pentasaccharide chain as a determinant. The lactose is directly bound at the ceramide residue, followed by

a further D-galactose residue and two D-galactosamine residues. For synthesizing this type of pentasaccharide, a block synthesis is preferably

$$\alpha\text{-D-GalNAc}(1\rightarrow3)\text{-}\beta\text{-D-GalNAc}(1\rightarrow3)\text{-}\alpha\text{-D-Gal}(1\rightarrow4)\text{-}\beta\text{-D-Gal}(1\rightarrow4)\text{-}\beta\text{-D-Glc-Ceramide}$$

(<u>30</u>)

Forssman - Glycolipid

indicated insted of a step by step synthesis; there are, however, different
block linkages possible (2+3 or 3+2). Especially favourable seems to be a
block linkage producing the $\alpha(1\rightarrow4)$-glycosidic linkage between the two D-
galactose residues. For this purpose, the <u>in situ</u> anomerisation procedure
will be well applicable which runs via a high reactive β-pyranosyl halide
resembling intermediate step (Fig. 2). This very effective linkage reaction
selectively leads to the α-glycoside and, compared with this reaction, the
preparation of a β-glycosidic linkage via a cyclic dioxocarbenium ion represents a less effective method. Consequently, a block linkage of a trisaccharide block with the lactose residue (3+2) will be advisable for synthesizing the pentasaccharide chain (Ref. 21).

The synthesis starts with an α-glycosidic linkage between (<u>31</u>) and (<u>32</u>),
following the <u>in situ</u> anomerisation procedure. Thus the α-glycosidically
bound disaccharide (<u>33</u>) is available which, after acetolysis and subsequent
functionalisation with titanium tetrabromide, can be transferred to the
halide (<u>34</u>). Now the compound (<u>34</u>) with a 2-phthalimido group enables a
linkage with neighbouring group assistance by which the desired new β-D-
glycosidic bond with the building unit (<u>35</u>) can be achieved. After this step,

the 1,6-anhydro compound of the reducing unit in the just obtained trisaccharide (<u>36</u>) must be opened by acetolysis in order to yield the acetate (<u>37</u>).
Conversion of (<u>37</u>) with titanium tetrabromide gives the trisaccharide block
(<u>38</u>) being functionalized as a bromide.

Following the in situ anomerisation procedure, the trisaccharide (38) can successfully be coupled with the 4'-OH group of the lactose derivative (39) to the desired pentasaccharide (40). This is a remarkable reaction regarding the low reactivity of the 4'-OH group as well as the fact that only the required α-glycosidically bound product becomes obtainable. The coupling of (38) with (39) delivers (40) in more than 30 % yield which can be considered as a really good result with respect to the low reactivity. By a serie of deblocking reactions, the pentasaccharide (41) can be achieved from compound (40).

There are different series of oligosaccharide segments containing lactose as a basic building unit. A choice of these structures have been summarized in Fig. 4. In the structures of the glycolipids the lactose is bound glycosidically to the corresponding ceramide residue. Concerning the sequences,

β-D-Galp-(1→3)-β-D-GlcpNAc-(1→3)-β-D-Galp-(1→4)-D-Glc   lacto-serie

β-D-Galp-(1→4)-β-D-GlcpNAc-(1→3)-β-D-Galp-(1→4)-D-Glc   neo-lacto-serie

β-D-GalpNAc-(1→3)-α-D-Galp-(1→4)-β-D-Galp-(1→4)-D-Glc   globo-serie

β-D-GalpNAc-(1→3)-α-D-Galp-(1→3)-β-D-Galp-(1→4)-D-Glc   iso-globo-serie

β-D-Galp-(1→3)-β-D-GalpNAc-(1→4)-β-D-Galp-(1→4)-D-Glc   ganglio-serie

Lactose

Fig. 4. Examples of lactose containing oligosaccharide chains

the stated globo serie corresponds to the Forssman antigen, with the difference of one D-galactosamine unit which fails to appear at the non reducing end of the Forssman antigen. Thus the synthesis of the globoside unit has been accomplished in a similar way as the Forssman antigen. In a final linkage step, the α-glycosidic linkage between both galactose residues can be prepared from the two disaccharide blocks using the in situ anomerisation procedure (Ref. 22).

As shown in Fig. 4, there are still different sequences with β-glycosidically bound residues to lactose at the 3'-OH and 4'-OH group. As already mentioned before, the β-glycoside synthesis represents a less effective method. The forming of a β-glycosidic linkage in the selectively blocked lactose derivative (39) will raise some difficulties because the 4'-OH group has proved to be of a low reactivity. Therefore, the lactose derivative (45) with two free hydroxyl groups has been applied in which the 3'-OH as well as the 4'-OH group is much more reactive. The reaction of (45) causes, however, the problem of the regioselectivity, for mixtures of compounds having a linkage at 3'-OH and 4'-OH can be formed. By comparison tests we have learnt that the 3'-OH group is more reactive than the 4'-OH group.

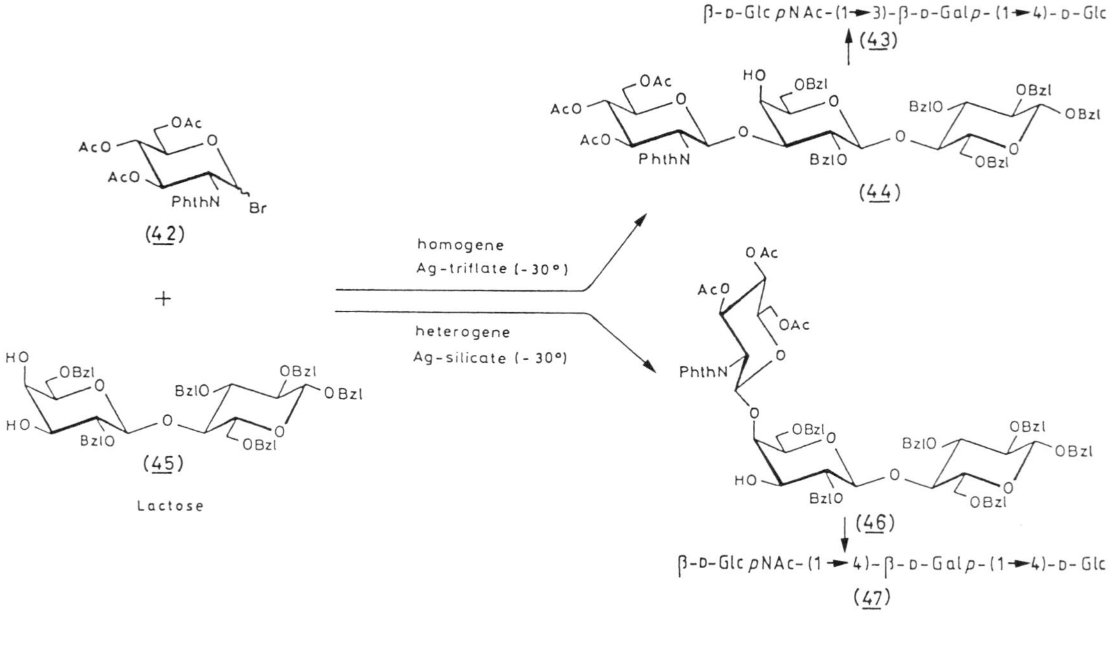

β-D-GlcpNAc-(1→3)-β-D-Galp-(1→4)-D-Glc

(43)

(44)

(42)

+

(45)

Lactose

homogene Ag-triflate (-30°)

heterogene Ag-silicate (-30°)

(46)

β-D-GlcpNAc-(1→4)-β-D-Galp-(1→4)-D-Glc

(47)

For a β-glycoside synthesis, an interesting regioselectivity can now be observed which can only be controlled by the choice of the catalyst system (Ref. 23). In the presence of silver triflate and under conditions of the neighbouring group assisted procedure, (45) will react with the glycosyl donor (42) in homogeneous phase at low temperatures regioselectively to the β(1→3)-glycosidically bound trisaccharide (44). In this way, the product (43) of the lacto serie becomes available (Ref. 24).

Under the same conditions, in heterogeneous phase with silver silicate as catalyst, the reaction of (45) with (42) yields regioselectively the β(1→4)-glycosidically linked product (46) which can also be deblocked to (47) (Ref. 24). This is a really unexpected regioselective reaction of the less reactive 4'-OH group which is difficult to explain. Probably the heterogeneous catalyst primary reacts with the more reactive 3'-OH group and, as a secondary reaction, the glycosylation at the less reactive 4'-OH group occurs within this complex. By this way, it can be disposed for the first time of various catalyst systems affording the different linked oligosaccharides from the same starting material. Of course, this new method represents an important progress because it reduces the necessary blocking steps in the glycosyl acceptor lactose.

$$\beta\text{-}D\text{-}Gal p\text{-}(1\rightarrow4)\text{-}\beta\text{-}D\text{-}Glc p NAc\text{-}(1\rightarrow3)\text{-}\beta\text{-}D\text{-}Gal p\text{-}(1\rightarrow4)\text{-}D\text{-}Glc$$

(48)

(49)

homogene Ag-triflate

heterogene Ag-silicate

(50) Lactosamin

(45) Lactose

(51)

$$\beta\text{-}D\text{-}Gal p\text{-}(1\rightarrow4)\text{-}\beta\text{-}D\text{-}Glc p NAc\text{-}(1\rightarrow4)\text{-}\beta\text{-}D\text{-}Gal p\text{-}(1\rightarrow4)\text{-}D\text{-}Glc$$

(52)

The regioselective glycosidation of (45) can also be applied on larger glycosyl donors. In homogeneous phase and in the presence of silver triflate, (50) reacts with (45) in an analogous way to the β(1→3)-glycosidically linked tetrasaccharide (49) whereas the same reaction in heterogeneous phase and with silver silicate as catalyst yields the regioselective formation of the β(1→4)-glycosidically bound tetrasaccharide (51) (Ref. 23). Also in this case, the different tetrasaccharides (48) and (52) are available from the same starting material, depending on the choice of the catalyst system.

Furthermore, it is interesting that also the 2-azido compound (53) allows a regioselective reaction. In heterogeneous phase and in the presence of silver silicate (53) reacts regioselectively with (45) to the β(1→4)-glycosidically linked trisaccharide (54) (Ref. 24). It is remarkable that under inversion the heterogeneous catalyst procedure also occurs without neighbouring group assistance. However, in homogeneous phase and under the now present conditions of the in situ anomerisation procedure, the reaction of (53) with (45) delivers only α-glycosidically bound products, as already expected before. But by this method, much more reactive intermediates are involved so that the regioselectivity can no longer be observed in this reaction and both α-glycosidically linked products (55) and (56) can be achieved in an equal proportion.

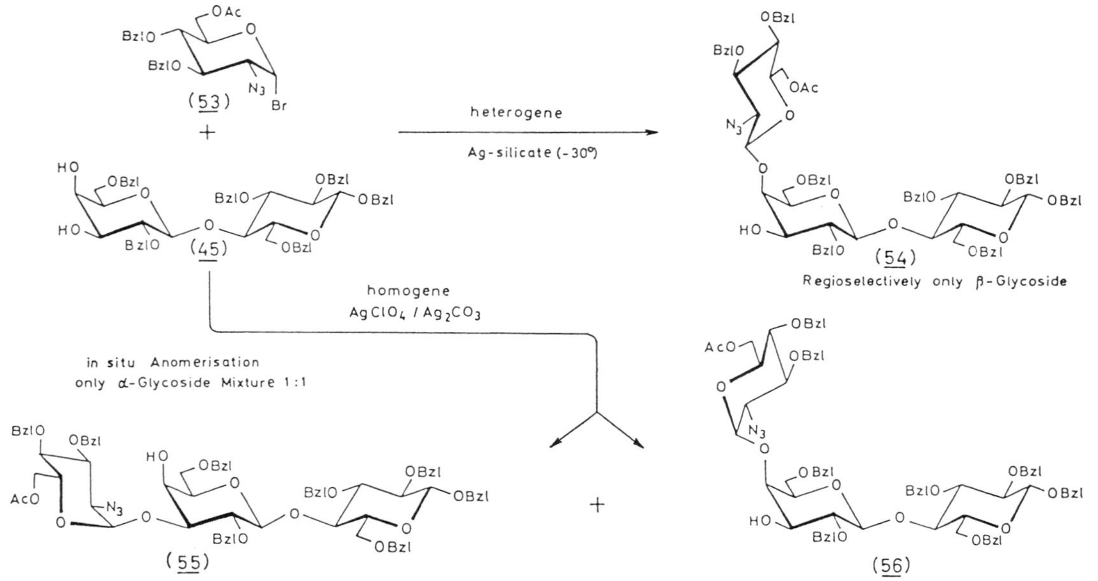

Some difficulties will, however, arise applying this synthesis method on galacto glycosyl donors like (57). In homogeneous phase (57) and (45) will

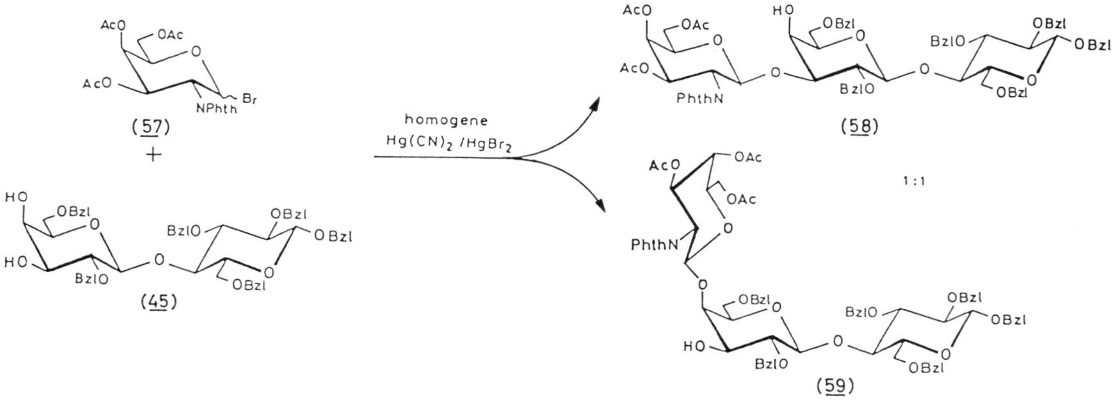

not react regioselectively, for a mixture of (58) and (59) will be obtained in a proportion of 1:1. The reason for this can be found in the generally higher reactivity of galacto compounds, compared with the one of gluco compounds so that a regioselective control of the reaction of (57) with (45) cannot be realized anymore. In order to solve this problem, the reactivity of the galacto compound must be reduced, as it has been accomplished, for example, with the β-acetate (60). Under neighbouring group assisted conditions and using trimethylsilyl triflate as catalyst, the reaction of (60)

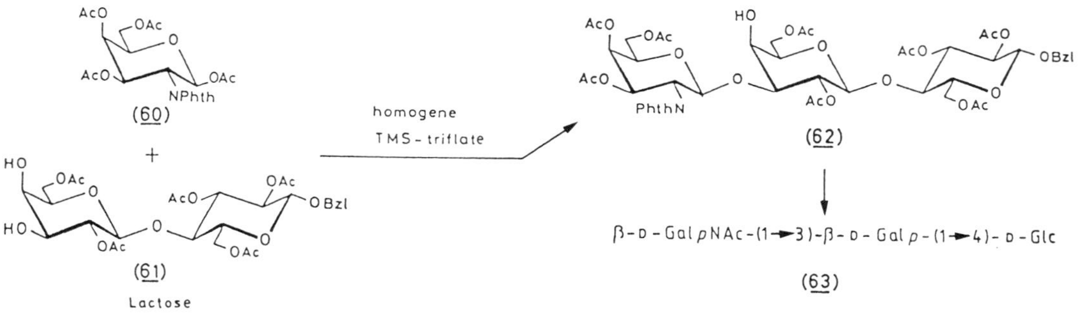

β-D-GalpNAc-(1→3)-β-D-Galp-(1→4)-D-Glc

(63)

with (61) in homogeneous phase now proceeds regioselectively to the β(1→3)-glycosidically bound trisaccharide (62). Here the less reactive O-actyl group has been applied as a leaving group. From (62) the deblocked compound (63) will be obtained (Ref. 25).

For a reaction in heterogeneous phase the phthalimido compound (57) is too reactive as well. A reduction of the reactivity can be achieved, replacing the phthalimido group by an azido group which results in the compound (64).

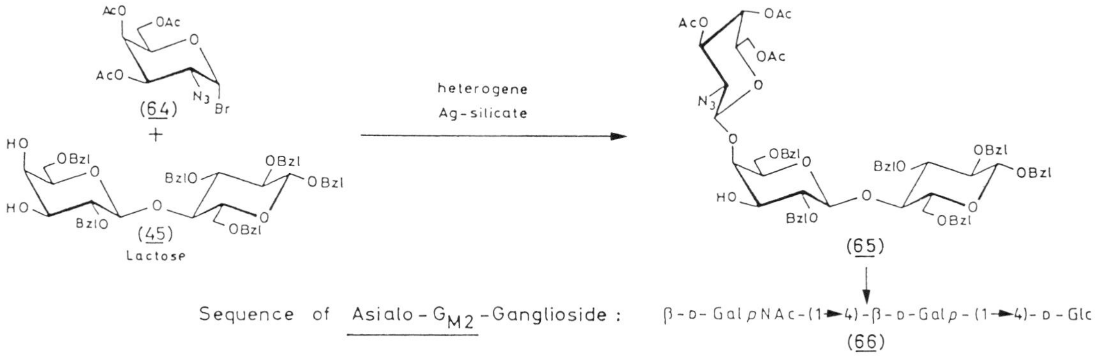

Sequence of Asialo-G_{M2}-Ganglioside : β-D-GalpNAc-(1→4)-β-D-Galp-(1→4)-D-Glc

(66)

In heterogeneous phase and with silver silicate as catalyst, (45) can also be caused to react regioselectively with (64) to the β(1→4)-glycosidically linked trisaccharide (65), leading to the important sequence of the asialo-GM2-ganglioside (66) (Ref. 23 a. 25).

The regioselectivity of the glycosidation reaction of (45) can also be controlled using disaccharide blocks of the galacto serie as glycosyl donors.

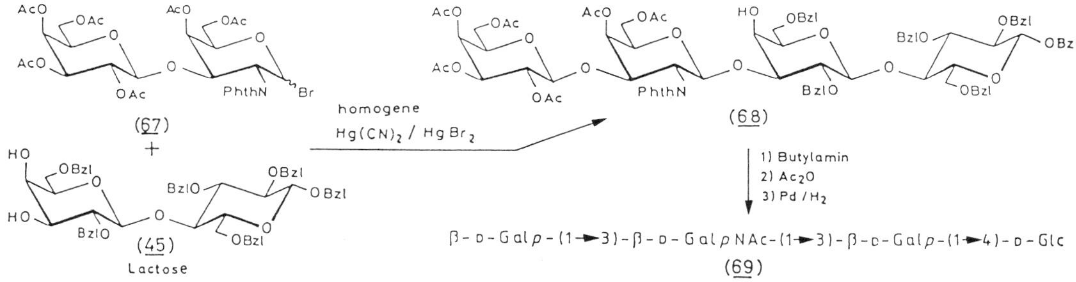

β-D-Galp-(1→3)-β-D-GalpNAc-(1→3)-β-D-Galp-(1→4)-D-Glc

(69)

Compared with the monosaccharide (57), the disaccharide (67) has a lower reactivity. In the presence of mercury salts, conditions of the homogeneous phase can now be found which enable a reaction of (67) with (45), yielding regioselectively the β(1→3)-glycosidically bound tetrasaccharide (68). The deblocking of (68) delivers the free tetrasaccharide (69). In heterogeneous phase this reaction can easily be accomplished with the azido compound (70).

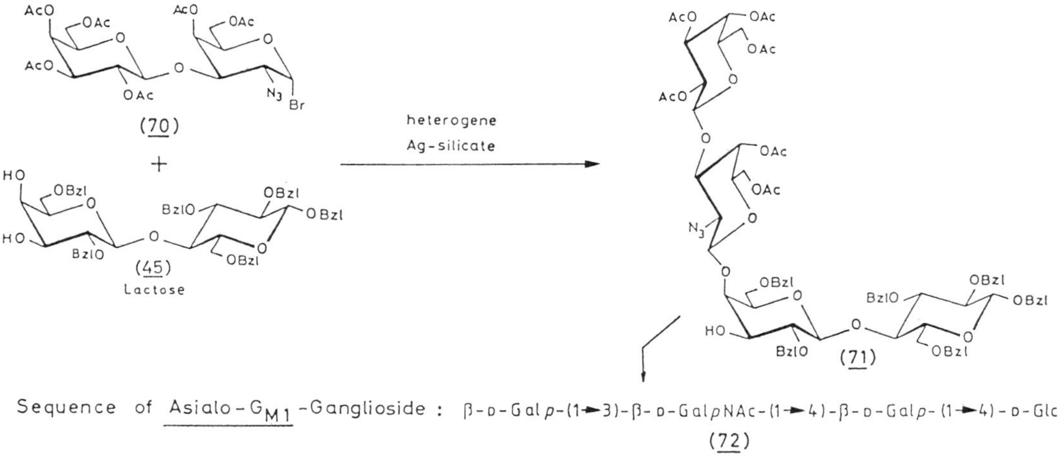

Sequence of Asialo-G_{M1}-Ganglioside : β-D-Galp-(1→3)-β-D-GalpNAc-(1→4)-β-D-Galp-(1→4)-D-Glc

(72)

In this case, the reaction of the disaccharide (70) with (45) affords regioselectively the β(1→4)-glycosidically linked tetrasaccharide (71) which can
be deblocked to the important sequence of the asialo-GM1-ganglioside (72)
(Ref. 23 a. 25). This new procedure is of great importance for the synthesis
of oligosaccharide chains as they are especially prevailing in the glycolipids.

APPLICATION OF THE DISCUSSED METHODS FOR THE SYNTHESIS OF OLIGO-SACCHARIDE SEQUENCES OF GLYCOPROTEINS

In the co-called N-glycoproteins of the lactosamine type, which is the most
important one, the carbohydrate chain bound to the L-asparagine contains the
structure shown in (73), representing a basic sequence which can be found in
most of the N-glycoproteins. The difference of the other N-glycoproteins can
be found in further branchings of this basic sequence with side chains. A
chitobiose residue is bound directly to the L-asparagine of the peptide
chain, followed by a branched trisaccharide unit of three D-mannoses. On
this   trisaccharide unit lactosamine antenna are bound, containing neuraminic acid as a terminal group. The most difficult problem for the synthesis of such a sequence represents the preparation of the β-D-mannosidic
linkage between the D-mannose and D-glucosamine as there is no direct way for
the synthesis of this sequence until now. Therefore, great efforts have been
made for preparing the key disaccharide unit β-D-Man(1→4)-D-GlcNAc [unit
3→2 in formula (73)].

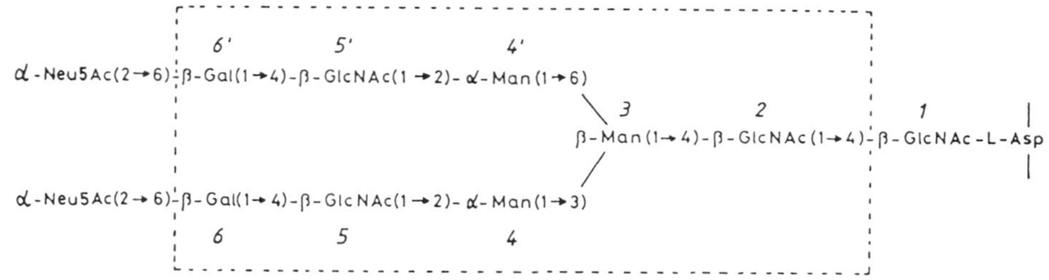

α-Neu5Ac(2→6)-β-Gal(1→4)-β-GlcNAc(1→2)-α-Man(1→6)

β-Man(1→4)-β-GlcNAc(1→4)-β-GlcNAc-L-Asp

α-Neu5Ac(2→6)-β-Gal(1→4)-β-GlcNAc(1→2)-α-Man(1→3)

(73)

According to the heterogeneous catalyst procedure, the β-D-mannosidic linkage
should be realizable for which the especially active silver silicate catalyst
has been developed (Ref. 10). The reaction of the highly reactive halide (74)
with the glucosamine derivative (75), however, yields almost completely the
undesired α-glycosidically linked disaccharide (76). The expected low reactivity of the 4-OH group in (75) causes a rather slow reaction. Probably it may

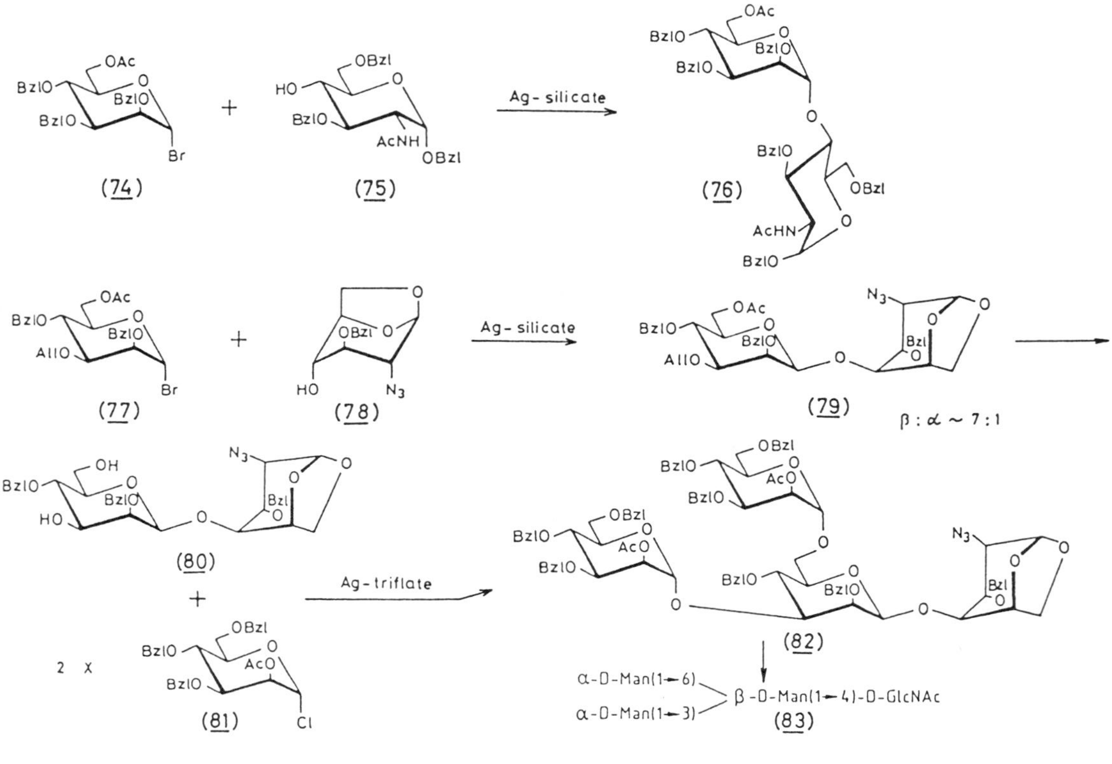

(74)   (75)   Ag-silicate   (76)

(77)   (78)   Ag-silicate   (79)   β:α ~ 7:1

(80)
+
2 X   Ag-triflate   (82)
(81)

α-D-Man(1→6)
                    β-D-Man(1→4)-D-GlcNAc
α-D-Man(1→3)   (83)

be the <u>in situ</u> anomerisation way which is responsible for the formation of the undesired compound (<u>76</u>). As a consequence, the reactivity of the 4-OH group of the glucosamine unit must be increased, for example by using the inverse conformation (<u>78</u>) for this reaction. Now the 4-OH group changes into an axial position and has a sufficient reactivity. In the presence of silver silicate, the coupling of (<u>77</u>) with (<u>78</u>) delivers the desired β-glycosidical-ly linked disaccharide (<u>79</u>) in high yields with a good selectivity (Ref. 26). Thus the key disaccharide (<u>79</u>) becomes available which has been used for all further building up syntheses.

The selective deblocking of (<u>79</u>) yields the disaccharide (<u>80</u>) with two free hydroxyl groups. Following the neighbouring group assisted procedure, the reaction with two moles of the <u>manno</u> glycosyl donor (<u>81</u>) allows the linkage of two D-mannose residues in one step, leading to the tetrasaccharide (<u>82</u>) from which the important branched key building unit (<u>83</u>) can be obtained (Ref. 26).

By selective deacetylation, the two acetyl residues, which formerly have been used for the neighbouring group assisted procedure, can be eliminated from (<u>82</u>) and the diol (<u>84</u>) will be achieved, enabling a block synthesis with 2 moles of the very effective lactosamine donor (<u>50</u>). Containing a 2-phthali-mido group, the compound (<u>50</u>) is well suited for a neighbouring group assisted procedure and usually affords a β-glycosidic linkage (Ref. 27). In the presence of silver triflate, the tetrasaccharide (<u>84</u>) can be caused to react with 2 moles of (<u>50</u>) to the octasaccharide (<u>85</u>) with two lactosamine antenna. The condensation step between (<u>84</u>) and (<u>50</u>) is extremely effective

1) NaOCH$_3$
2) NH$_2$NH$_2$
3) Ac$_2$O/Pyr
4) CF$_3$CO$_2$H/Ac$_2$O
5) Pd/H$_2$
6) Ac$_2$O/Pyr
7) K$_2$CO$_3$/CH$_3$OH

β-Gal(1→4)-β-GlcNAc(1→2)-α-Man(1→6)
β-Gal(1→4)-β-GlcNAc(1→2)-α-Man(1→3) β-Man(1→4)-GlcNAc
(<u>86</u>)

and yields (<u>85</u>) in more than 70 %, whereas the deblocking sequence raises some problems. All steps must carefully be optimized in order to avoid a splitting of the just formed glycosidic linkage in the sensitive molecule. A successful deblocking leads to the free octasaccharide (<u>86</u>) representing the most essential sequence of the complex oligosaccharide chain of the N-glyco-proteins (Ref. 28).

Furthermore, it has been tested whether a chain elongation of (<u>86</u>) would be possible as shown in the basic structure (<u>73</u>). The following steps demon-strate the possibilities of an elongation to the L-asparagine end by en-closure of a chitobiose unit. At first, the basic disaccharide (<u>79</u>) must be converted to (<u>87</u>). Then by acetolysis and halogenation, the synthesis block (<u>88</u>) will be available. Under neighbouring group assistance, in the presence of silver triflate/collidine, the reaction of (<u>88</u>) with (<u>75</u>) becomes possible yielding the trisaccharide (90). By selective deallylation of (90), the diol (89) can be obtained which enables a condensation with 2 moles of the before mentioned manno glycosyl donor (81), resulting in the pentasaccharide (91). From (91) the deblocked pentasaccharide (92) can be obtained which now contains the desired chitobiose unit and which represents the invariable core structure existing in all N-glycoproteins (Ref. 29).

The building unit (91) can be converted to a compound in which both 2-OH groups of the terminal mannose unit will be set free. Then a reaction with the lactosamine donor will result in the corresponding nonasaccharide. For this condensation step, however, it seems to be more interesting to develop a synthesis block containing a neuraminic acid residue. The desired neuraminic acid containing synthesis block can be obtained in the following way.

The glycoside synthesis with the halide of the neuraminic acid (93) usually raises considerable problems because to a large extent an elimination to the unsaturated neuraminic acid occurs which means an undesired competitive reaction. This elimination reaction can be limited by using mercury cyanide/mercury bromide as catalyst. In this way, more than 50 % of the neuraminic acid containing oligosaccharides can be achieved which are available as a separable mixture of the two possible anomers. This procedure could be applied for the synthesis of the trisaccharide unit α-D-Neu5Ac(2→6)-β-D-Gal(1→4)-β-D-GlcNAc (Ref. 30).

Under corresponding conditions, also the reaction of the modified derivative of the lactosamine (94) can be accomplished in which only the reactive hydroxyl group 6'-OH can be caused to react. The reaction of (93) with (94) leads to a mixture of the α-glycoside (95) together with the corresponding product in which the neuraminic acid is β-D-glycosidically linked in a proportion of 1:1. A chromatographic separation of the α-product (95) becomes possible and (95) can be isolated in about 25 % yield (Ref. 31).

For further reactions, a variation of the protecting groups will be necessary in order to come to the more stable, peracetylated compound (96). The considerably high stability of (96) allows a glycosylation according to the trimethylsilyl triflate procedure. In this case, the O-acetyl group serves as a leaving group which can be eliminated by a neighbouring group assisted reaction. As glycosyl acceptor, the D-manno compound (97) will be applied, containing sufficiently acid-stable protecting groups. In the presence of trimethylsilyl triflate, a coupling of (96) with (97) can be realized, yielding the tetrasaccharide (98). After a serie of deblocking steps, also the free compound (99) will be obtained (Ref. 31). Thus, (96) can be disposed as a neuraminic acid containing synthesis block which can be linked to corresponding glycosyl acceptors. Using the trimethylsilyl triflate

α-D-Neu$p$5Ac-(2→6)-β-D-Gal$p$-(1→4)-β-D-Glc$p$NAc-(1→2)-D-Man

(99)

procedure, each system of the protecting groups must exactly be chosen in order to avoid protecting groups which can be attacked under acid conditions of this coupling reaction. Compound (99) is also synthesized in a different way (Ref. 32).

These examples are intended to demonstrate the great progress in the selective synthesis of complex oligosaccharides. This progress has only become possible by consequent application of improved preparative techniques and, in this connection, also the new possibilities of chromatographic separation as well as the modern methods of instrumental analysis must be mentioned. The purity of all intermediate products of the oligosaccharide synthesis must be tested by high field n.m.r. spectroscopy. For larger complex molecules, the application of the new 2D-n.m.r. spectroscopy is indispensible in order to analyze the spectra as far as possible.

As in many cases the oligosaccharide chain of the glycoconjugates represents the determinant being responsible for the specificity of the biological reaction, its molecular shape, and that means its conformation, seems to be of great interest. There are two methods for determining the conformation of the oligosaccharide chain: a theoretical calculation in which the mutual steric interactions of the saccharide units are taken into consideration and the application of the modern methods of high field n.m.r. spectroscopy which gives important information about the conformation in solution.

In the calculations with HSEA (hard-sphere exo anomeric Effect) or GESA (geometric saccharide) programme, the pyranose ring is considered as a rigid chair-form and in this case, the corresponding data for the chair conformation can be gathered from X-ray structure analyses. Furthermore, the atomic radii and the exo anomeric effect are taken into account. Rotations giving different conformations will be carried out only about the different anomeric linkages. Here, glycosyl linkages with secondary hydroxyl groups enable a variation of two angles (ϕ and ψ) and by this variation the energy minimun will be determined. For glycosyl linkages with secondary hydroxyl groups a relatively steep minimum can be found so that one conformation is especially preferred in this type of linkage. Linkages with 6-OH groups enable an additional rotation around the C-5/C-6 axis (ω), causing several minima in this type of linkage so that variable parts of conformations will be present. The gg (gauche-gauche) and gt (gauche-trans) conformations represent the most important minima(Ref. 33 a. 34).

For the conformation analysis according to the n.m.r. method, the [1]H-n.m.r.
spectrum of the substance being analyzed must be assigned completely. For deblocked oligosaccharides this assignment will be rather difficult because of
the extensive overlapping of the signals of the ring protons of the various
oligosaccharide units. The assignment becomes, however, possible by a consequent application of the new 2D-n.m.r. spectroscopy. If the spectrum has
been solved, a difference n.O.e. (nuclear Overhauser effect) spectrum can be
set up and subsequently, it must be looked for interglycosidic n.O.e. effects
between protons of different saccharide units. Discovering the interglycosidic n.O.e. effects, statements can be made about the distance between observed protons of the various saccharide units thus leading to experimental
informations regarding the conformation of the oligosaccharide chain in
solution. The results of the n.m.r. experiments can be compared with the
theoretical calculations (Ref. 33 a. 34).

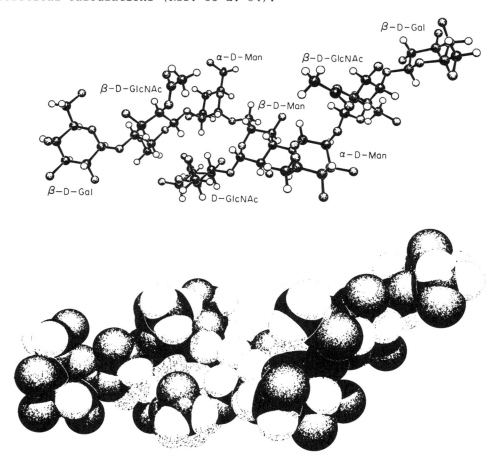

Fig. 5. Preferred conformation of the octasaccharide (86)
according to the GESA calculation, confirmed by n.m.r. experiments in solution. Above the molecular model, below the
space-filling model.

Both methods have been applied to the octasaccharide (86). By a GESA calculation it could be ascertained that one conformation is very much preferred
within the pentasaccharide part of (86) in which the lactosamine residue is
bound at O-3' on the branched β-D-mannose [units 2→ 3→ 4→ 5→ 6 in formula
(73)]. With the other pentasaccharide part of (86) in which the lactosamine
group is bound as a branch at the O-6' of the β-D-mannose [units 2→ 3→ 4'→
5'→ 6' in formula (73)], however, a mixture of gt and gg conformations can be
observed. This means the 4'(1→6)3 linked side arm [see formula (73)] is
much more flexible because of the one additional rotation possibility. The
energy calculation of the minima shows that in this case the gt conformation
should be the predominating one. In Fig. 5, this preferred gt conformation
can be seen as a molecular model and as a space-filling model. The n.m.r.
experiments indicate that this conformation exists also preferred in solution.

These experiments especially point out that the unit β-D-GlcNAc *2* as well as the unit α-D-Man *4'* in formula (73) are together in a rather close contact (Ref. 35).

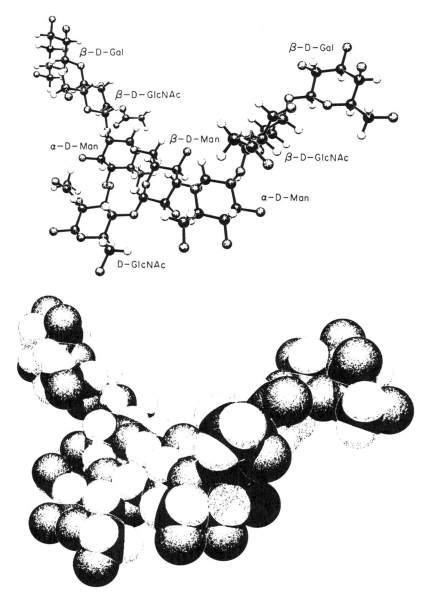

Fig. 6. Conformation of the octasaccharide (86) in crystal, according to the X-ray structure analysis of the immunoglobolin IgG1. The octasaccharide prevails in the N-glycosidically bound side chain in the Fc part. Above the molecular model, below the space-filling model. The protein chain is situated at the left, directing the saccharide chain bound at O-6' to an upper position.

Subsequently, also an X-ray structure analysis of the octasaccharide (86) becomes available. Of course, the octasaccharide itself cannot be crystallized, but Deisenhofer (Ref. 36) and Huber (Ref. 37) succeeded in crystallizing the immunoglobolin IgG1 which has been isolated from the human serum so that a complete X-ray structure analysis was possible. In the Fc part, the IgG1 contains a N-glycosidic carbohydrate chain as well as the sequence of the octasaccharide (86), for (86) represents the fundamental structure. By X-ray structure analysis, all amino acids and carbohydrate chains can be localized. A still more refined structure of the octasaccharide (86), found in the crystal of the immunoglobolin is shown in Fig. 6, as molecular model and space-filling model (Ref. 38).

A comparison of both conformations seems to be of great interest. In Table 2 all calculated torsion angles at the anomeric linkages of all saccharide units have been compared for the conformation in solution and for the conformation in crystal. Within the variation possibilities, the pentasaccharide arm with the lactosamine residue being bound at O-3' shows a rather similar conformation in solution and in crystal. The only important difference has been observed at the $4'(1 \rightarrow 6)3$ -glycosidic linkage [see formula (73)] of the

TABLE 2. Torsion angles for all glycosidic linkages of the octasaccharide (86) for the preferred conformation in solution and the conformation in crystal

e'(1→4)d'		d'(1→2)c'		c'(1→6)b		C5-C6	e(1→4)d		d(1→2)c		c(1→3)b		b(1→4)a		
φ	ψ	φ	ψ	φ	ψ	ω	φ	ψ	φ	ψ	φ	ψ	φ	ψ	
55°	2°	47°	0°	-54°	79°	38°	55°	2°	52°	23°	-47°	-14°	57°	-5°	Solution
75°	-4°	51°	27°	-55°	171°	68°	a		72°	28°	-49°	-4°	29°	-32°	X-Ray

a: not fixed because of insufficient electron density

lactosamine arm at O-6'. Here the ψ - angle is considerably larger in the crystal (171° in crystal, 79° in solution). In the crystal, another gt conformation is present, where the lactosamine arm, linked at O-6', has been turned to the upper side, as demonstrated by the picture in Fig. 6. A comparison of the conformation in Fig. 5 and Fig. 6 will obviously point to this difference which, however, can easily be explained. The turning of the arm linked at O-6' is caused by the protein chains which drive the arm into the corresponding position. Fig. 6 shows the position of the protein inducing the turning of the saccharide chain, linked at O-6'. This example demonstrates that by conformation analyses of the synthesized products important informations about the interactions between carbohydrate chains and protein chains can be achieved.

REFERENCES

1. M.I. Horowitz and W. Pigman, The Glycoconjugates, Vol. I-IV, Academic Press, New York 1977-1982.
2. S. Hakomori, Glycolipids of Animal Cell Membranes, Int. Rev. Sci. Org. Chem., Ser. Two 7, 223-249 (1976).
3. J. Montreuil, Adv. Carbohydr. Chem. Biochem., 37, 157-223 (1980)
4. E.J. Goldstein (Ed.), Carbohydrate Protein Interaction, ACS Symposium Series 88, Washington DC (1979).
5. R.U. Lemieux in IUPAC Frontiers of Chemistry, K. Laidler (Ed.), Pergamon Press, Oxford and New York (1982), p 3-34.
6. H. Paulsen, Angew. Chem., 94, 184-201 (1982); Angew. Chem., Int. Ed. Engl. 21, 155-173 (1982)
7. H. Paulsen and J.-P. Hölck, Carbohydr. Res. 109, 89-107 (1982).
8. R.U. Lemieux, K.B. Hendriks, R.v. Stick and K. James, J. Am. Chem. Soc. 97, 4056-4062 (1975).
9. H. Paulsen and W. Stenzel, Chem. Ber. 111, 2334-2347 (1978).
10. H. Paulsen and O. Lockhoff, Chem. Ber. 114, 3102-3114 (1981).
11. H. Paulsen and M. Paal, Carbohydr. Res. in press
12. H. Paulsen, M. Paal and M. Schultz, Tetrahedron Lett. 24, 1759-1762 (1983).
13. H. Paulsen, Chem. Soc. Rev. 13, 15-45 (1984).
14. R.R. Schmidt and J. Michel, Angew. Chem. 92, 763-764; Angew. Chem. Int. Ed. Engl. 19, 731-732 (1980).
15. R.R. Schmidt and G. Grundler, Angew. Chem. 95, 805-806 (1983); Angew. Chem. Int. Ed. Engl. 22, 776-777 (1983).
16. R.R. Schmidt and J. Michel, Angew. Chem. 94, 77-78 (1982); Angew. Chem. Int. Ed. Engl. 21, 72-73 (1982).
17. R.R. Schmidt, J. Michel and M. Roos, Liebigs Ann. Chem. 1343-1357 (1984).
18. P. Fügedi, A. Lipták, P. Nánási and A. Neszmélyi, Carbohydr. Res. 107, C5-C8 (1982)

19. H. Lönn, Dissertation University of Stockholm 1984; Abstr. XIIth Int. Carbohydr. Symp. 88, 116 (1984).
20. H. Paulsen and A. Bünsch, Liebigs Ann. Chem. 2204-2215 (1981).
21. H. Paulsen and A. Bünsch, Carbohydr. Res. 100, 143-167 (1982).
22. H. Paulsen and A. Bünsch, Carbohydr. Res. 101, 21-30 (1982).
23. H. Paulsen, M. Paal, D. Hadamczyk and K.-M. Steiger, Carbohydr. Res. 131, C1-C5 (1984).
24. H. Paulsen, D. Hadamczyk, W. Kutschker and A. Bünsch, Liebigs Ann. Chem. in press.
25. H. Paulsen and M. Paal, Carbohydr. Res., in press.
26. H. Paulsen and R. Lebuhn, Liebigs Ann. Chem., 1047-1072 (1983).
27. J. Arnap and J. Lönngren, J. Chem. Soc., Perkin Trans. I, 1841-1844 (1982).
28. H. Paulsen and R. Lebuhn, Carbohydr. Res. 125, 21-45 (1984).
29. H. Paulsen and R. Lebuhn, Carbohydr. Res. 130, 85-101 (1984).
30. H. Paulsen and H. Tietz, Carbohydr. Res. 125, 47-64 (1984).
31. H. Paulsen, H. Tietz and M. Armbrust, Carbohydr. Res., in manuscript.
32. T. Kitajima, M. Sugimoto, T. Nukada and T. Ogawa, Carbohydr. Res. 127, C1-C4 (1984).
33. H. Thøgersen, R.U. Lemieux, K. Bock and B. Meyer, Can. J. Chem. 60. 44-50 (1982).
34. H. Paulsen, T. Peters, V. Sinnwell, R. Lebuhn and B. Meyer, Liebigs Ann. Chem., 951-976 (1984).
35. H. Paulsen, T. Peters, V. Sinnwell, R. Lebuhn and B. Meyer, Liebigs Ann. Chem., in press.
36. J. Deisenhofer, Biochemistry 20, 2361-2370 (1981).
37. R. Huber, Klin. Wochenschr. 58, 1217-1231 (1980).
38. T. Peters, B. Meyer, V. Sinnwell, J. Deisenhofer, R. Huber and H. Paulsen Abstr. XIIth Int. Carbohydr. Symp., 453-454 (1984).